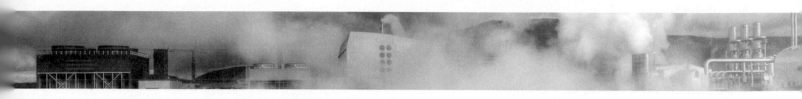

Renewable Energy Systems

David Buchla

Thomas Kissell

Thomas Floyd

Boston   Columbus   Indianapolis   New York   San Francisco   Upper Saddle River
Amsterdam   Cape Town   Dubai   London   Madrid   Milan   Munich   Paris   Montreal   Toronto
Delhi   Mexico City   São Paulo   Sydney   Hong Kong   Seoul   Singapore   Taipei   Tokyo

Editorial Director: Vernon R. Anthony
Acquisitions Editor: Lindsey Prudhomme Gill
Editorial Assistant: Nancy Kesterson
Director of Marketing: David Gesell
Senior Marketing Coordinator: Alicia Wozniak
Senior Marketing Assistant: Les Roberts
Program Manager: Maren Beckman
Project Manager: Rex Davidson
Full Service Project Manager: Penny Walker/iEnergizer Aptara®, Inc.
Procurement Specialist: Deidra Skahill
Lead Media Project Manager: Leslie Brado
Media Project Manager: April Cleland
Creative Director: Andrea Nix
Art Director: Jayne Conte
Cover Designer: Karen Noferi
Cover Image: Alterra Power Corp's Geothermal Power plant, Svartsengi, Iceland
Composition: Aptara® Inc.
Printer/Binder: Courier-Kendallville
Cover Printer: Lehigh/Phoenix Color Hagerstown
Text Font: Times Roman

Many of the designations by manufacturers and sellers to distinguish their products are claimed as trademarks. Where those designations appear in this book, and the publisher was aware of a trademark claim, the designations have been printed in initial caps or all caps.

Library of Congress Cataloging-in-Publication Data is Available from the Publisher Upon Request.

10 9 8 7 6 5 4 3 2 1

ISBN 10:   0-13-262251-3
ISBN 13: 978-0-13-262251-6

BRIEF CONTENTS

CONTENTS

Renewable Energy Systems is an introductory text with broad coverage of all the major renewable energy systems, resources, and related topics. The major types of systems are solar, wind, geothermal, biomass, and hydropower (including water in rivers and oceans). Related topics include electrical fundamentals, charge controllers, inverters, generators, fuel cells, the electrical power grid and the smart grid.

In the past few years, interest in renewable energy has grown in both the public and private sectors, with significant government support for research and for constructing large systems, such as solar installations, wind farms, and other projects. As a result of this research, exciting new technology is being developed that attest to a future with a variety of new opportunities. Innovations in the field include more efficient multijunction photovoltaic (PV) cells and cells with built-in charge storage using supercapacitors, high-flying wind turbines, systems for capturing tidal and wave energy in the ocean, and enhanced algae-growing methods, to name a few. Many people are optimistic about the future potential of renewable energy for supplying a larger portion of the worldwide energy needs.

Energy is a complex topic involving competing interests, but in general there is broad agreement by the public that the world needs to move toward more renewables for supplying energy needs. A single technology cannot solve all of the energy needs of the future or is best in all situations because resources vary significantly by location. In addition, different applications often compete for finite resources. For example agricultural land can be used for food production or for growing corn for ethanol or for a utility PV system. It is clear, however, that more people must understand renewable energy and its promise and limitations including all viable renewable energy technologies, sources, energy conversion systems and the environmental impacts (both positive and negative).

Renewable energy systems convert basic resources into usable forms of energy, including electricity, heat, and mechanical motion. In some cases, the renewable source is used to create a fuel, as in the case of ethanol or hydrogen fuel for use in fuel cells. Fuel cells in themselves are not a renewable source: Fuel cells may convert solar or other renewable source to the fuel used by the fuel cell but most fuel cells convert natural gas (a fossil fuel) to electricity. Fuel cells are becoming more important in providing electrical energy, particularly in remote areas or for backup power.

Many changes have occurred in recent years to renewable energy systems. For example, the capture of solar energy is shifting from smaller, customer-owned systems that are designed to supplement utility power, to large arrays that produce significant power (more than 1 megawatt [MW]) and are owned by utilities, the military, or industrial complexes. About half of solar power generation in the United States is in large utility power installations. Wind turbines have become massive with power outputs as high as 5 MW to 10 MW. Innovations in major systems like variable-speed generation and new power electronic and supervisory control systems are improving the efficiency of wind farms. New, enhanced geothermal production can capture more of the available energy with techniques such as fracking in hot dry rocks and injecting fluids in a closed-loop arrangement, which helps conserve water in desert environments. The organic Rankine cycle (ORC) is more prevalent in large systems because it can be used to obtain energy from lower-temperature sites. Future geothermal power plants may also be able to sequester carbon dioxide, actually helping the environment. Tidal energy has been captured in several parts of the world; a new crossflow turbine operates in the mouth of the Bay of Fundy, almost directly northeast of the coast of Maine, and a new wave energy system is being installed off the coast of Oregon. We have attempted to highlight these new systems and innovations throughout this text. In particular, the sidebars in many of the chapters feature innovative ideas; not all will eventually pan out, but they show the nature of this growing and dynamic field.

Chapter 1 presents an overview of the principal energy resources: fossil fuels, nuclear energy (including the future potential of fusion energy), and renewable energy, and where these resources are located. Most renewable energy is converted to electricity; for this reason Chapter 2 provides an introduction to electricity and magnetism, and discusses basic circuit laws, components, circuits, electrical measurements, and safety. Chapter 3 introduces the photovoltaic (PV) cell and its application to solar power systems. Solar modules and manufacturer's specification sheets are also covered. Chapters 4 and 5 continue the topic of solar energy with coverage of various types of solar energy systems and their operation, including concentrating systems and tracking methods for panels and heliostats. Chapter 6 discusses components that are important in many renewable energy systems, such as certain solar, wind, and fuel cell systems, with particular attention being paid to battery chargers, charge controllers, and inverters. Chapter 7 is an introduction to wind energy, including power in the wind, Betz's law, and an overview of wind turbines. Chapter 8 delves further into the operation of wind turbines with details of turbine control, measurements, and braking systems. Chapter 9 covers the major types of biomass and the systems used for converting biomass into oil or electrical power, and heat energy. Biomass has been used by humans for heat since our early ancestors first used wood fires for cooking and heat; today, biomass is used for power generation, heat, and biofuels (primarily ethanol). Chapter 10 gives an overview of geothermal energy and energy conversion systems for both electrical systems and heat pumps. Geothermal heat pumps are proving their worth in many areas, and geothermal energy is a significant resource that has barely been tapped. Chapter 11 covers various systems for extracting energy from moving water, including water from hydroelectric dams, streams, tides, and waves, and the methods for converting this energy to electricity. Chapter 12 discusses fuel cells and their applications, which have great potential to bring reliable power to remote locations of the world. Chapter 13 expands the discussion of magnetic theory, which was introduced in Chapter 2, and applies this to various types of electrical generators and how they are used in renewable energy systems. Finally, Chapter 14 gives an overview of the electrical grid, including the smart grid, and methods of power transmission, and it introduces the topic of three-phase ac and transformers.

Features

- The book is in full-color.
- Chapter openers have a chapter outline, objectives, key terms list, and introduction.
- Section openers in each chapter give a brief overview of what each section within a chapter covers.
- Section checkups contain questions related to each section within a chapter. Answers to section checkups are provided at the end of the chapter.
- Key terms are shown in bold and color in the running text. Definitions for key terms are provided at the end of the chapter and in the end-of-book glossary. Bold terms in black are defined in the end-of-book glossary only.
- Margin features are given throughout the book at appropriate places to highlight interesting innovations or historical information related to the topic being covered.
- Worked-out examples are provided throughout the text.
- Each chapter comes with abundant illustrations. Many are original and previously unpublished.
- Important formulas are numbered throughout each chapter, and they are listed at the end of each chapter for quick reference.
- A summary of the chapter discussion is provided at the end of each chapter.
- A true/false quiz, a multiple-choice quiz, and a set of chapter questions and problems appear at the end of each chapter.
- Answers to the chapter true/false multiple-choice quizzes are given at the end of the chapter.

- A suggested class discussion item appears at the end of each chapter under the heading For Discussion.
- A list of variables and their meanings is provided at the end of the book.

Instructor Supplements

To access supplementary materials online, instructors need to request an instructor access code. Go to www.pearsonhighered.com/irc to register for an instructor access code. Within 48 hours of registering, you will receive a confirming e-mail including an instructor access code. Once you have received your code, locate your text in the online catalog and click on the Instructor Resources button on the left side of the catalog product page. Select a supplement, and a login page will appear. Once you have logged in, you can access instructor material for all Pearson textbooks. If you have any difficulties accessing the site or downloading a supplement, please contact Customer Service at http://247pearsoned.custhelp.com/.

PowerPoint Slides

PowerPoint slides that support the text with illustrated chapter summaries and discussions, new examples, selected key terms, and a true/false quiz are available for instructors.

Online Instructor's Manual

Answers and solutions to all questions and problems are in the Instructor's Manual.

Illustration of Features

- **Chapter opener** Each chapter begins with a full-page opener.

FIGURE P-1 Typical Chapter Opener

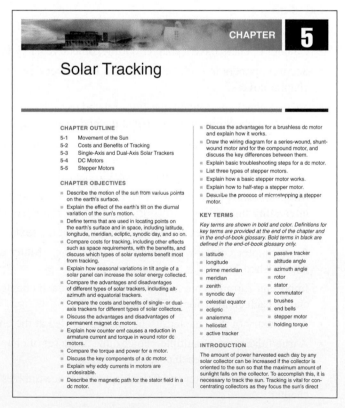

- *Section opener* Each section within a chapter begins with a brief introduction and overview.

FIGURE P-2 Typical Section Opener

- *Section Checkup* Each section within a chapter ends with a series of questions related to the section.

FIGURE P-3 Typical Section Checkup

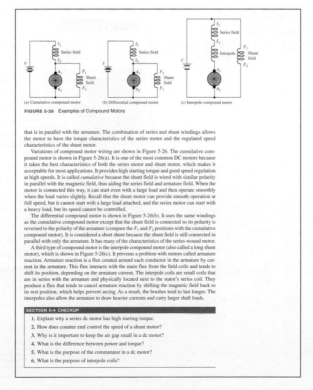

- **Examples**   Worked-out examples illustrate and clarify basic concepts and procedures.

FIGURE P-4   Typical
Examples

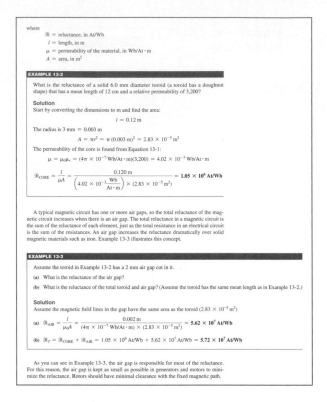

- **Margin features**   Interesting sidebars are presented throughout the book. Their topics range from items with historical interest to new technology.

FIGURE P-5   Typical
Margin Feature, Key
Term, and Glossary Term

- *Key terms and other glossary terms* Key terms are highlighted in bold and blue; these terms are defined at the end of each chapter *and* in the end-of-book glossary. Other important terms are shown in bold and black throughout the book; these terms are defined in the end-of-book glossary only. See the examples in Figure P-5.

Acknowledgments

The authors wish to thank Vern Anthony, Lindsey Gill, Rex Davidson, and Maren Beckman at Pearson Education; Penny Walker at Aptara; and Marianne L'Abbate at Double Daggers Editing Services for their help in bringing this book to reality. We relied heavily on the expertise of numerous reviewers who contributed to the content and coverage of the topics. These reviewers include James A. Buck, Wayne Technical & Career Center; Lazaro Hong, Pima Community College; and Dana Veron, University of Delaware.

Finally, we gratefully acknowledge the support of our wives, Lorraine, Kathleen, and Sheila. We dedicate this book to them.

David Buchla
Tom Kissell
Tom Floyd

Energy Sources and Environmental Effects

CHAPTER OUTLINE

CHAPTER OBJECTIVES

- Describe fossil fuel sources and major sectors for which it is used.
- Explain how a pressurized water reactor generates electricity.
- Discuss the advantages of fusion reactors as a future option for energy production.
- Explain the locations that you would expect to find the best resources for solar, wind, and geothermal energy.
- Compare the environmental impacts of various energy sources, including various renewable resources.
- List the type of information you would need to plan for a solar or wind power station.
- Explain what is meant by *biomass*. What are various biomass sources?
- Discuss the distribution of biomass resources.

KEY TERMS

Key terms are shown in bold and color. Definitions for key terms are provided at the end of the chapter and in the end-of-book glossary. Bold terms in black are defined in the end-of-book glossary only.

- fossil fuels
- hydrocarbon
- diagenesis
- kerogen
- bitumen
- catagenesis
- breeder reactor
- pressurized water reactor (PWR)
- Tokamak
- inverse square law
- solar constant
- global horizontal irradiance (GHI)
- direct normal irradiance (DNI)
- diffuse horizontal irradiance (DHI)
- biomass
- ethanol

INTRODUCTION

The Industrial Revolution ushered in an era of prosperity unlike any other in the long history of humanity. With it, came a surge in energy usage throughout the world, which has continued to increase to this day. The rate of energy consumption increased dramatically over a period of just a few generations near the end of the nineteenth century. The principal source of energy in the United States at the start of the nineteenth century was wood, with other sources (water and wind) playing a minor role. At the close of the nineteenth century, coal became the dominant energy resource because of its high energy content, ready availability, and demand from the railroads. During the twentieth century, coal continued to be an important resource, but petroleum moved up dramatically as motor vehicles became widespread. In the last part of the twentieth century, nuclear energy and natural gas became important parts of the energy mix.

It became obvious that the world was using fossil fuels at an increasing rate and that there is a finite limit to fossil fuels, so people have been looking for alternative energy sources. Renewable energy use has grown, but it still accounts for only a small fraction of the world's total energy consumed. Abundant renewable resources are available, but the issue is to find ways for them to be competitive economically with fossil fuels. The specific renewable energy that is best for a given location varies, so one best resource is not available for all places. This chapter focuses on the supply side of the major renewable resources and effects on the environment. Nearly all of the renewable resources have much less environmental impact than fossil fuels do, but many other issues must be considered (such as land use, costs, and resource availability) before developing a specific resource. This chapter focuses on the resources and explores these other issues. The bottom line is that no renewable source is entirely free of negative effects on the environment, and many issues are complicated by competing factors.

One important factor is *energy payback time*. It is important to keep in mind that all energy sources should be required to produce more energy than was expended in developing the sources. All sources require a certain amount of material for development. It requires energy to make the steel to build the drill rig to develop a natural gas well, and it requires more steel to build the pipeline to transport it. A certain amount of the natural gas that is produced must offset this energy cost. The time required to produce enough energy from the source is the payback time. Payback times vary widely with the particular source, but they need to be considered to evaluate any source of energy fairly.

1-1 Fossil Fuels: Oil, Coal, and Natural Gas

Fossil fuels are fuels formed over millions of years from decaying plants and marine diatoms. They are considered to be nonrenewable because they required millions of years to form and they are consumed at a rate that far exceeds their rate of formation. Fossil fuels account for a significant fraction of worldwide energy usage. Nuclear and renewable resources represent the remainder. Although this section is primarily devoted to discussing fossil fuels, a short overview of renewable sources is included.

Fossil Fuels

Fossil fuels are fuels formed from decaying plant and animal matter that was primarily formed over millions of years. Fossil fuels include coal, oil (petroleum), and natural gas. These fuels have been the primary energy source for over 200 years, but they will eventually have to be replaced by other sources because the world supply of fossil fuels is finite. The amount of each type of fossil fuel that is left is debated, but it is clear that the continued rate of use for any fossil fuel cannot be sustained. Today, fossil fuels supply approximately 80% of the energy consumed in the United States and about 87% worldwide according to the BP Statistical Review of World Energy.

An interesting tool for visualizing the sources of energy and ultimate outcome is the so-called spaghetti diagram, which was developed and used by Lawrence Livermore National Laboratory. Figure 1-1 shows the diagram with line width proportional to the percentage of each source; this one is for a recent year in the United States. It clearly shows the dependency on fossil fuels. The left side shows the various sources and the right side shows the demand sectors. Electricity is an intermediate step and not a demand in itself. Notice that the largest source of energy comes from petroleum. Approximately 72% of the petroleum is used in the transportation sector. In the case of electricity, 48% of electrical energy comes from coal-fired plants and, as you can see, the process creates a lot of rejected heat. Rejected heat is due to the inefficiencies in generating and distributing electrical power. Notice that electricity production is the primary user of coal. The best electrical generation plants are less than 50% efficient, and 5 to 10% of the energy is lost in transmission lines. It is important to be aware of the inefficiencies in the generating and distribution process because it affects other demand areas.

Renewable energy sources—biomass, hydroelectric, geothermal, wind, and solar—constitute only a small percentage of the overall mix at this time. All are covered in more detail throughout this text, but a quick overview of the current mix of renewable sources in the United States is shown in Figure 1-2. The mix has been changing rapidly in recent years, with significantly more wind energy. Biomass, the largest source, includes a variety of sources, including wood, waste, garbage, and even plants that are grown for fuel. Hydroelectric is the second largest source today because of the huge infrastructure of dams and power plants. Other sources (wind, geothermal, and solar) account for 14% of renewable use in the United States.

Worldwide consumption is continuing to rise for all fossil fuels. Fossil fuels include coal, petroleum and natural gas.

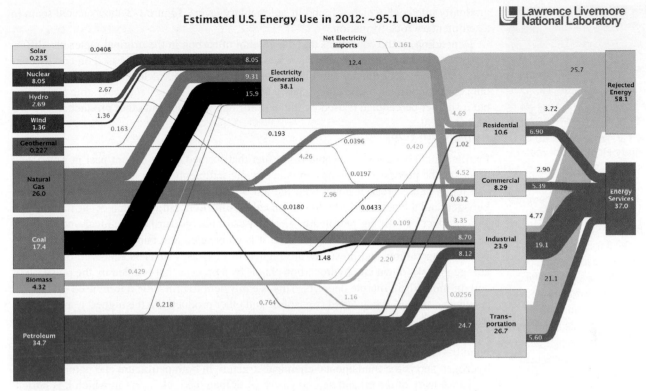

FIGURE 1-1 Energy Flow for the United States   (*Source:* Lawrence Livermore National Laboratory. Reproduced by permission.)

Coal

Coal has been used as a fuel for more than 2,000 years, and historical records document its usefulness to the early Greeks, Romans, and Chinese as well as other cultures. Large-scale mining of coal was brought on by the Industrial Revolution with the development of the steam engine and improvements in steelmaking. Coal is a combustible rock that is composed mostly of carbon and hydrocarbons. A **hydrocarbon** is a molecule containing only hydrogen and carbon (but not all hydrocarbons are fuels). Coal is derived from ancient plant life, mainly trees. It is believed that these ancient terrestrial forests were flooded rapidly and eventually sank, where layer upon layer of dying plants was covered by sediments. Mild heat and pressure condensed the organic material into peat in a process called **diagenesis**. If enough heat and pressure is supplied, the organic material will undergo physical and chemical change to form coal. This process takes several million years and turns the peat

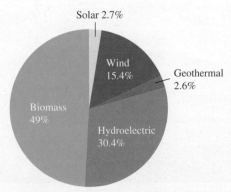

FIGURE 1-2 Renewable Energy Sources in the United States   (*Source:* US Energy Information Administration.)

FIGURE 1-3 A Coal Seam in Sedimentary Rock (*Source: David Buchla.*)

La Brea Tar Pits

The La Brea Tar Pits in Los Angeles are part of an ancient reservoir that was not trapped by cap rock but migrated to the surface. The oil that seeped up became a tarlike substance as lighter fractions degraded and evaporated. Over thousands of years, animals that had come to drink surface waters were entrapped in the gooey tars. Predators were also trapped when they investigated the cries of trapped animals. The tar pits preserved the bones of these animals and today, the tar pits are the home of the world's only active urban Ice Age excavation site.

gradually into coal, which is found in sedimentary layers. Figure 1-3 shows a coal seam in a sedimentary rock formation.

Depending on the conditions and the amount of carbon in the original materials, different types of coal formed. Coal is classified into four main categories based on energy content: anthracite, bituminous, sub-bituminous, and lignite (anthracite has the highest energy, but is not as common as the other types.). Most coal today is used for generating electricity, but a smaller percentage is used by industry for making steel and other products.

Petroleum (Oil)

Petroleum is also a nonrenewable fossil fuel that formed in the distant past in a two-step process. The process starts when aquatic organic sediments are compacted and when heat and temperature break it down with the aid of microbes into a waxy material known as **kerogen** and a black tarlike hydrocarbon called **bitumen**. Bitumen can occur naturally or as a product of refining petroleum. Kerogen can undergo further chemical and physical change in a process called **catagenesis** if it is compacted and buried deeper underground where temperatures and pressures are higher. In this case, water is squeezed out and the kerogen breaks down into hydrocarbon chains by a process that is aided by the presence of certain minerals in marine deposits. This is equivalent to *cracking,* a term used by refineries when crude oil is converted to gasoline and other products. At the highest temperatures, natural gas forms. If the temperature is lower, oil forms. If the temperature is lower still, the kerogen remains unaltered. Carbon, with four electrons in its outer shell, has the ability to bond to other carbon atoms and form long chains and complex atomic arrangements with hydrogen and is the fundamental chemical structure in both petroleum and natural gas.

The density of the oil and natural gas is lower than the rock layers in which it is buried, so these substances would normally migrate to the surface. Instead, they are trapped by a layer of impervious rock called cap rock, which is typically shale. The cap rock traps the gas and oil in porous sedimentary rock formations. Large volumes of natural gas and viscous liquid oil are trapped in these underground regions called **reservoirs.** The natural gas is under pressure and escapes when the formation is drilled. Figure 1-4 illustrates how oil and gas and sometimes water are trapped underground by the cap rock. There are various types of oil and natural gas traps, but the common feature is that oil moves through the porous rock layer and is trapped by an impervious layer in the underground reservoir. Reservoirs that contain very hot water under pressure are useful as a geothermal heat source; geothermal resources are discussed in Section 1-5.

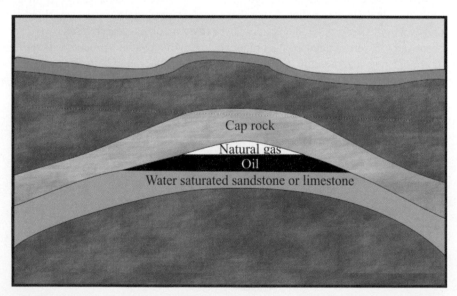

FIGURE 1-4 Underground Reservoir. The particular reservoir shown is an anticline. Other types have been identified by geologists.

The principal use for petroleum is in the transportation sector. In the United States, approximately 72% of all petroleum is used to make gasoline, diesel, and other products for vehicles. One important use for petroleum that is often overlooked is in chemical feedstock. It is also used in the manufacture of many products including lubricants, waxes, solvents, asphalt, hydraulic fluid, and vinyl, to name a few.

Natural Gas

The primary constituent of natural gas is methane, the simplest hydrocarbon. Although natural gas is nonrenewable, methane, a principal component of natural gas, can be produced by various processes such as the decomposition of waste in landfills and by anaerobic (without oxygen) decay of organic matter such as manure or biomass. The chemical formula for methane is CH_4. This formula indicates a single carbon atom is bound to four hydrogen atoms. When methane burns, it reacts with oxygen in the air to release energy and form carbon diode and water as products. This basic chemical reaction for burning methane is:

$$CH_4 + 2O_2 \rightarrow CO_2 + 2H_2O + energy$$

The reactants (methane and oxygen) are written on the left side and the products of the reaction (carbon dioxide and water) are written on the right side. This equation not only shows the reactants and products, it also shows the ratios of molecules involved. For each molecule of methane, two molecules of oxygen combine to release one molecule of carbon dioxide and two molecules of water plus energy. As shown by the chemical equation, a byproduct of this reaction is carbon dioxide, which typically escapes into the atmosphere as a by-product of combustion. In addition to methane, natural gas also contains other, more complex hydrocarbons as well as some other undesirable materials, including sulfur. Carbon dioxide and methane are each considered to be a **greenhouse gas.** Greenhouse gases contribute to the greenhouse effect by absorbing short wavelength infrared energy and reradiating it at longer wavelengths. Other greenhouse gases include nitrous oxide (NO_2), fluorinated gases (CFC, etc.), and water vapor (H_2O).

From the chemical formula for the reaction of burning methane, the ratio of masses of the reactants and products can easily be determined by determining their molecular weights. The atomic weights are given on the periodic table of the elements and other sources. In a chemical reaction, mass is always conserved; that is, the mass of reactants is equal to the mass of the products. To determine the relative weight of the reactants and products, look up the atomic weights of each atom, then determine molecular weights by multiplying the number of atoms of each type by its atomic weight. The following example illustrates the idea.

Carbon Dioxide and Methane

Carbon dioxide emissions result from burning of fossil fuels and other fuels, as well as certain other chemical reactions such as those during production of cement. An example of a naturally occurring emission is volcanic eruption.

Methane enters the atmosphere primarily from the production of fossil fuels (coal, oil, and natural gas), livestock, and the decay of organic waste. The Environmental Protection Agency (EPA) estimates that one-half of all methane enters the atmosphere from human activities.

EXAMPLE 1-1

What is the mass of carbon dioxide (CO_2) that is released to the atmosphere if 1,000 kg of methane (CH_4) is burned? (One thousand kg is approximately 2,200 pounds, or 1.1 tons.)

Solution

$$CH_4 + 2O_2 \rightarrow CO_2 + 2H_2O + energy$$

The equation shows that each molecule of CH_4 reacts with 2 molecules of O_2 to produce 1 molecule of CO_2 and 2 molecules of H_2O. Start by determining the molecular weight of each reactant and product. By expressing the molecular weight in grams, you will obtain the *relative* masses of the reactants and products. The atomic weight of carbon is 12.0 g, hydrogen is 1.0 g, and oxygen is 16.0 g.

CH_4 = 1 carbon and 4 hydrogen = $1(12.0\,g) + 4(1.0\,g) = 16.0\,g$

$2O_2$ = 2 molecules of oxygen, each with 2 atoms = $4(16.0\,g) = 64\,g$

CO_2 = 1 carbon and 2 oxygen = $1(12.0\,g) + 2(16.0\,g) = 44.0\,g$

$2H_2O$ = 2 water molecules each with 2 hydrogen and 1 oxygen = $4(1.0\,g)$
            $+ 2(16.0\,g) = 36\,g$

Thus, 16 g of CH_4 reacts with 64.0 g of O_2 to produce 44 g of CO_2 and 36 g of H_2O. You can check that the mass on each side is the same, as it must always be in chemical reactions. From this you can set up a proportion:

$$\frac{(16 \text{ g CH}_4)}{(1{,}000 \text{ kg CH}_4)} = \frac{(44 \text{ g CO}_2)}{X \text{ kg CO}_2}$$

Solving for X (the unknown quantity of CO_2), gives $X = 2{,}750$ kg CO_2, which is approximately 3 tons. This illustrates that burning methane creates a weight of carbon dioxide that is 2.75 times the weight of the original methane gas.

Applications for natural gas include heating for homes and businesses. Most of the heat from the reaction is available for heating; a small fraction escapes with flue gases. Natural gas is also used as an alternative to gasoline for automobiles and trucks and is widely used in electrical power generation. As an electrical power energy source, it emits the lowest amount of carbon dioxide of any fossil fuel per unit of energy produced. Natural gas turbines are used to supplement renewable sources when they are not available because natural gas turbines have quick start-up times and thus can be brought on line rapidly. Another natural gas application is as a fuel source for some fuel cells; it may find more widespread use when fuel cell vehicles become more common in the future.

Environmental Concerns

Kingston Coal Ash Spill

(*Source:* Tennessee Valley Authority.)

The Kingston Coal Ash Spill was a disastrous slurry spill in Kingston, Tennessee. It was caused by a combination of high rainfall and cold temperatures that weakened an earthen dam that had contained the ash. The ash had liquefied in layers of water-saturated materials. A massive cleanup effort, which will continue for years, has removed most of the ash, but the memories and horrific images remain in people's minds.

The environmental effects of using fossil fuels can have a serious impact on the quality of life. The process of burning fossil fuels combines carbon in the fuel with oxygen in the air to form carbon dioxide (CO_2) that is released to the atmosphere. This is true for all fossil fuels and more so with coal and oil. The United States produces over 6 billion tons of CO_2 each year that is released into the atmosphere; 35% of this is from coal-fired power stations. Today, carbon dioxide levels are at historic high levels, the evidence of which can be traced back 420,000 years from analysis of ice cores taken at Vostok, Antarctica. Carbon dioxide is a natural constituent of the atmosphere, but it is also a greenhouse gas that contributes to global warming. It is not a simple matter to sequester carbon dioxide and prevent it from getting in the atmosphere, so levels will continue to build for the foreseeable future until economic and reliable carbon capture can be implemented. Carbon dioxide capture is an ongoing research area, and commercial scale CO_2 capture has been tested in a 25 megawatt (MW) coal-fired plant in Alabama, with over 100,000 tons of CO_2 captured to date.

Carbon dioxide gets the most attention for burning fuels, but other pollutants are also released. When gasoline is burned in a motor vehicle, it produces carbon monoxide, a poisonous gas, and releases nitrous oxides, ozone, and benzene. Coal is not pure carbon or hydrocarbons; it also contains small amounts of sulfur and other toxic materials that are released when it is used as a fuel. Sulfur dioxide is another atmospheric pollutant that contributes to forming acid rain and the acidification of oceans and lakes. Coal burning also results in fly ash in the form of a fine residue, which gets in the atmosphere, and bottom ash, which is the residue in the burners. Coal ash also contains traces of poisonous heavy metals such as arsenic, lead, and mercury; these pollutants are hazardous to human health, as well as fish and other wildlife.

Atmospheric CO_2 gets most of the attention, but another aspect to increases in CO_2 is acidification of the oceans. When atmospheric CO_2 is absorbed by water, it creates acid. There is clear evidence of an increase in the acidity of the world's oceans, which will ultimately have very serious consequences to marine life, particularly shellfish, coral reefs, and certain types of plankton at the bottom of the food chain. The shells of shellfish can dissolve in a more acidic environment, and corals stop building reefs. The loss of corals will have profound effects on the marine food chain, ultimately affecting fisheries.

In addition to pollution caused by burning fossil fuels, the process of obtaining these fuels has led to some major environmental disasters, such as the infamous *Exxon Valdez* spill and the British Petroleum (BP) Gulf of Mexico oil spill. Coal has been particularly hazardous to obtain. One serious disaster was the Partizansk, Russia, coal ash dam break in 2004 that released 160,000 cubic meters of toxic slurry coal ash into a drainage canal. Another dam break occurred in Kingston, Tennessee, on December 22, 2008, when a failure of a power plant ash storage dam caused the largest industrial waste spill in US history. It damaged forty-two homes and did serious environmental damage to the Clinch and Emory rivers. Coal mining has always been dangerous, and many lives have been lost over the years as a result of explosions in coal mines. In addition, the processing, transportation, and use of coal have added to the environmental impact and safety of workers and the public.

Even natural gas, which is the cleanest burning of the fossil fuels, contributes to environmental damage and health problems. Not all methane in natural gas burns, so some methane is released to the atmosphere. The overall impact of greenhouse gases is determined by several factors, including the residence time in the atmosphere—a longer residence time has a greater impact. Ton for ton, methane has a greater impact as a greenhouse gas than carbon dioxide; however, it has a shorter residence time and there is much less methane released than carbon dioxide. Carbon dioxide has an average residence time of over a century before it is removed, whereas methane's residence time is only ten years. The net impact on global climate change is higher for carbon dioxide than for methane. The burning of fossil fuels adds carbon dioxide and methane (as well as other pollutants) to the atmosphere.

In appliances such as stoves, natural gas contributes to significant indoor pollution because stoves are generally not vented to the outside, thus trapping the gas inside. This can create a health hazard, particularly for people with respiratory problems.

Obtaining natural gas via a technique called *hydraulic fracking* is controversial. Fracking has its origins in a natural process where fluid seeps into rocks, and the resulting pressure breaks the rocks and enables natural gas and oil to find an easier path to seep out. Human fracking consists of drilling a well over a mile below the earth's surface to reach shale deposits. The drilling rig then turns the drill horizontally and drills another 4,000 feet through the shale deposit. The final step in this process is to pump millions of gallons of water, sand, and chemicals into the well to shatter the shale, which allows trapped natural gas, oil, propane, and butane to the surface where it is captured and sent through existing transportation pipelines. Sometimes material injected into the shale is a mixture that includes sand, ceramics, or sintered bauxite as a fracturing fluid, or it may be in the form of a foam, gel, or gas.

The procedure is considered to be controversial because it is not possible to know all of the environmental effects associated with fracking, but there is evidence of natural gas contamination of groundwater in Pennsylvania and there may be other potential environmental impacts, such as effects on the water table. There is also evidence that fracking has been responsible for two rather small earthquakes in England in the spring of 2011.

Water usage and disposing of warm water with all fossil-fueled power plants, and with thermal cycle power plants in general, are also issues. In any heat engine, the temperature difference between the heat input and the heat rejected is related to the work done. A greater efficiency is obtained when the temperature difference is higher. Water is the most widely used coolant for power plants because it has very high heat capacity and can remove heat efficiently. Water is an excellent coolant, but steam-operated power plants require massive amounts of cooling water. One method of cooling, called once-through cooling, takes water from a source (usually a river) and puts warmed water back into the source. The warmed water can have a deadly effect on aquatic life.

A widely used alternative to river water for cooling is evaporative cooling, which uses huge quantities of water. Waste heat is removed by evaporating the water in large cooling towers rather than returning it to its source. Water is a precious resource, so it is an issue that must be considered for power plants. In drought conditions, such as those that occurred in the summer of 2012, the lack of water at some power stations required several power plants to cut back production. In hot dry summers, this problem becomes especially acute, with power plants, farmers, ranchers, and fisheries all competing for this valuable resource.

SECTION 1-1 CHECKUP

1. How were fossil fuels formed?

2. What is methane and what is its chemical formula?

3. What are the four categories for coal?

4. What are applications for oil as a chemical feedstock?

5. What is a greenhouse gas?

6. What are environmental concerns with burning fossil fuels?

7. What are hazards for obtaining fossil fuels?

1-2   Nuclear Energy

Nuclear energy derives its power from the enormous amount of stored binding energy in the nucleus of the atom, which is the energy that holds the nucleus of the atom together. By tapping this energy, a substance such as water is heated and converted to steam to drive a turbine and produce electricity.

Binding Energy per Nucleon

A **nucleon** is an atomic particle—either a proton or a neutron—found in the nucleus of an atom. Heavy atoms like lead (Pb) or uranium (U) have many nucleons, whereas light atoms, like hydrogen (H) and helium (He), have very few nucleons. The **binding energy** is the energy that holds the nucleus together. It is different for each atom and depends on the number of nucleons. Figure 1-5 is a plot of the binding energy in million electron volts (MeV) per nucleon. Notice that the highest binding energy per nucleon occurs with element 26, which is iron, with 56 nucleons. Above or below iron, the binding energy per nucleon drops.

Albert Einstein showed that mass and energy are directly related in his famous equation $E = mc^2$. The graph in Figure 1-5 shows that the most stable nucleus is iron because it has the highest energy (least mass) per nucleon. If a heavier nucleus breaks apart, the

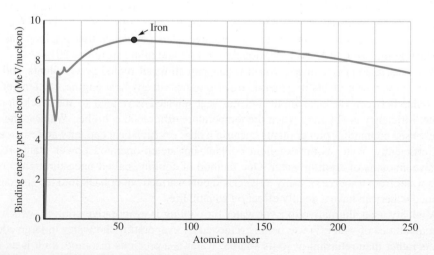

FIGURE 1-5   Binding Energy per Nucleon

"extra" mass is converted to energy, and this is a huge amount for a single atomic event, several million times greater than the energy released in a chemical reaction like burning methane. The process of breaking a heavy nucleus into smaller fragments and releasing energy is called **fission.** All commercial reactors that are used for producing electricity are fission reactors and rely on breaking apart heavy atomic nuclei into fragments. The process of putting together (or *fusing*) light nuclei to produce energy is called **fusion.** Fusion is the ongoing process in the core of the sun that accounts for the prodigious amount of energy it produces.

Nuclear Fission

As you can see from the binding energy curve, nuclear fission can release large amounts of energy. The problem is that only a very few materials can be broken down and release more energy than we put in, although most heavy elements can be broken apart if the energy of the incoming projectile is high enough. In 1939, it was discovered that a specific isotope of uranium (^{235}U) could be broken apart with low-energy neutrons (called *thermal neutrons*), and would release energy and immediately release additional neutrons. Natural uranium contains only 0.7% ^{235}U; to be useful for reactors, the uranium has to be enriched. **Enrichment** is a complicated process that involves separating ^{235}U from natural uranium, which is mostly ^{238}U. The amount of enrichment for a reactor varies, depending on the type of reactor, but is usually between 1% and 2%.

Neutrons emitted during the fission process are called **prompt neutrons** because they are emitted as the heavy nucleus breaks up, but they tend to have high energies, which are not efficient for continuing the fissioning process. These neutrons are reduced in energy by a **moderator,** which is a material that absorbs energy from fast-moving neutrons and slows them down. On average, 2.46 neutrons are emitted by fissioning of ^{235}U. If each of these neutrons triggered another fission event, the process would grow exponentially and lead to an explosion. This type of reaction is called a **chain reaction,** which is a self-sustaining reaction. The reaction continues if at least one neutron triggers another fission event. Figure 1-6 illustrates a nuclear fission chain reaction. Each fission reaction leads to additional neutrons that can continue the reaction and to additional fission fragments, which are the lighter nuclei left over from the fissioning. Most of the fragments are radioactive and produce radiation over varying half-lives (a **half-life** is the time required for one-half of the substance to decay).

Plutonium is another fissionable material used in certain types of reactors. Although plutonium occurs only in small amounts naturally, it can be made in reactors in a two-stage process from ^{238}U, during which it captures a fast neutron and emits an electron to become ^{239}Pu, which can support a chain reaction. In particular, a type of reactor called a **breeder reactor** is designed to produce plutonium, which could extend uranium supplies considerably because it can convert the otherwise unusable ^{238}U into a fissionable fuel. Breeder reactors have not been used for generating power in the United States. One issue is that

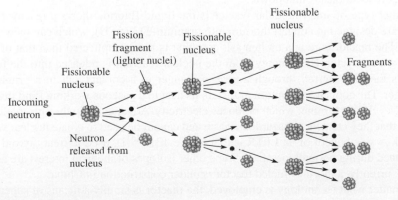

FIGURE 1-6 A Fission Chain Reaction

plutonium is particularly useful in nuclear weapons, so the widespread use of breeder reactors would greatly complicate issues of controlling nuclear weapons.

Nuclear Reactors

All commercial reactors rely on nuclear fission, and they must be able to control the rate at which fissioning takes place. The chain reaction described previously would be almost impossible to control except for one important fact. Not all of the neutrons that are the result of fissioning are prompt (instantaneous) neutrons. A small fraction of the neutrons are emitted many seconds later by certain fission fragments. These **delayed neutrons** are critically important for controlling a reactor. By carefully controlling the number of neutrons that are present in the core of a reactor, and hence the neutron multiplication factor, the power level can be adjusted up or down. Prompt neutrons produced from the fission reaction have a high probability of being lost and not contributing to the continuation of the reaction. If this were true for all the neutrons, the reaction would die out. A way around this problem is to slow down the prompt neutrons, producing thermal neutrons. This is done by mixing the fuel in reactors with a moderator, which is typically water or graphite. To be an effective moderator, a substance needs to have a lightweight nucleus, as does hydrogen. Water is an effective moderator because each molecule contains two hydrogen nuclei (H_2O).

There are various types of reactors. A **pressurized water reactor (PWR)** is the most common type. In a PWR, the reactor fuel is loaded in rods called fuel rods, which are long thin tubes loaded with the fissionable material [typically enriched uranium oxide (UO_2)]. The fuel rods are inserted in a pressure vessel, which is filled with water that is under high pressure to prevent it from boiling. Control rods are also in the vessel and contain a good absorber of neutrons, like cadmium (Cd) or boron (B). These rods can be moved in or out to control the reaction. The water in the vessel serves as both a moderator and a means to move hot water to a heat exchanger and eventually to a steam-driven turbine and an alternator.

A much more efficient reactor is the fast breed er reactor, which can use the much more abundant ^{238}U as a fuel. This uranium fuel creates ^{239}Pu and extends uranium fuel supplies for thousands of years. Many issues surround the use of the fast breeder reactor, including the use of the plutonium in nuclear weapons. Also, liquid water cannot be used as a coolant to keep high-energy neutrons in the system, so the more dangerous liquid sodium must be used instead. So far, fast breeder reactors have been used only in limited numbers, and some countries have shut down their breeder programs.

A new type of reactor that is being developed is the small modular reactor (SMR), defined as a reactor that is less than 350 MW. SMRs will have a standardized design and will include passive safety procedures that allow shutdown without electrical power. Most of the reactor can be constructed in a shop rather than onsite. One design has the reactor's core located completely underground, beyond the reach of potential terrorists and able to absorb both earthquakes and tsunamis without the threat of overheating. Instead of large reactor cooling pumps, gravity induced flow will be used. The reactor will not need power from offsite locations to shut down..

Another type of small modular reactor is the liquid-fluoride thorium reactor (LFTR). LFTRs are designed to convert thorium into uranium-233 (^{233}U), which can then undergo fission. The reactor core is a molten salt core that is less complicated than that of a pressurized water reactor. Heat energy from the nuclear reaction is released into the fuel salt, which is then transferred through a heat exchanger to a coolant salt in a primary heat exchanger. The coolant salt, in turn, transfers its heat to a gaseous working fluid that drives a closed-cycle gas turbine, which produces electricity. Advantages to thorium-fueled reactors are that they create less hazardous waste and cannot be used to create nuclear weapons material. A side benefit of an LFTR is that iodine-131 (^{131}I), used to treat thyroid cancer, is produced during normal operation. Some other isotopes of medical interest are also produced. Currently, a thorium-fueled reactor is under construction in China.

No matter what technology is employed, the reactor is simply a means of superheating water to well above the boiling point and converting it to steam. The process from this

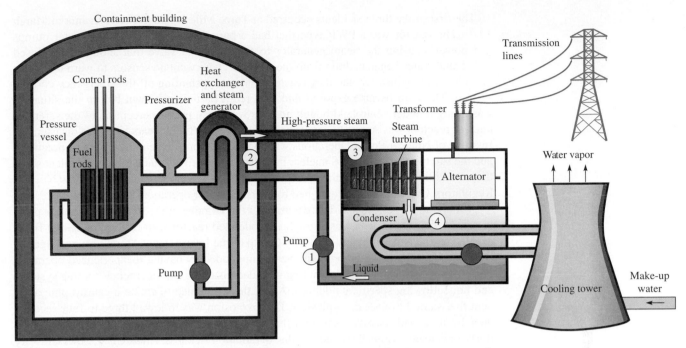

FIGURE 1-7 Diagram of a Pressurized Water Reactor. The numbered points correspond to the steps in the Rankine cycle.

point is similar to almost all thermal power stations, namely, converting the steam energy to mechanical motion and eventually electricity. The **turbine** is a machine that converts the kinetic energy in a fluid into rotary motion. There are various types of turbines, including steam, wind, and water; they will be discussed throughout this text. The turbine is mechanically linked to an alternator, which produces electricity. An **alternator** is simply an alternating current generator, and you will see both terms used interchangeably. Figure 1-7 shows a pictorial diagram of a typical pressurized water reactor for generating electricity.

The portion of the PWR that holds the steam once the steam has been produced uses a standard Rankine cycle. The **Rankine cycle** is a four-step, thermodynamic, closed-loop process that is used to convert energy in fluid to mechanical motion. Briefly, the steps in the Rankine cycle are (1) pressurize the working fluid with a pump; (2) heat the working fluid to boiling or beyond boiling, thus converting it to a dry saturated vapor; (3) expand the dry vapor in a turbine, which extracts a large portion of the energy from the vapor and cools it, causing it to become a saturated vapor at reduced pressure; and (4) condense the wet vapor back to a low-pressure liquid by cooling it in a condenser at a constant temperature.

Environmental Concerns

The environmental concerns associated with nuclear reactors are very different than with fossil fuel–burning power plants. On the positive side, reactors do not emit greenhouse gases or other atmospheric pollutants. They also use significantly less fuel in terms of weight than coal-fired plants. Reactors for electricity production have had an excellent safety record for the most part, and they have been used in many countries generally without incident. Three major accidents mar that record; they occurred at Three Mile Island, Pennsylvania, in 1979; Chernobyl, Ukraine, in 1986; and Fukushima, Japan, in 2011. In addition, several experimental reactors have experienced partial meltdowns and other accidents in the early years of reactor development. As a result of these accidents, approval for new plants, particularly in the United States, has been slowed and the public perception of nuclear power tends to be negative. In spite of these setbacks, new reactors are now being constructed and a number of reactors in the United States have been upgraded in order to increase power output, and more upgrades are planned for the future.

The first of the three incidents occurred at Three Mile Island in Pennsylvania in March 1979. The reactor was a PWR type that had a partial meltdown when feedwater pumps lost power, causing the steam generator to shut down. A valve that should have closed failed, and water began to drain from the reactor. There were no sensors to warn operators of an impending disaster, and they reacted incorrectly, shutting off the emergency cooling system. The core became exposed and suffered a partial meltdown before the situation was brought under control. Although no lives were lost, public perception of the safety of nuclear reactors was seriously compromised as a result. In the years since the disaster, the Institute of Nuclear Power Operations was formed to monitor the industry's best practices, and the safety record of the US nuclear industry was improved and has been excellent since then. In addition, federal regulation of nuclear plants was strengthened and a new Nuclear Regulatory Commission was assigned to oversee reactor operations.

A much more serious accident occurred in the Ukraine at the Chernobyl nuclear plant in April 1986. The reactor was a graphite-moderated reactor with two major design flaws as cited by the Nuclear Safety Advisory Group of the International Atomic Energy Agency (IAEA). The flawed design problem was compounded by having inexperienced operators perform a questionable experiment that included disabling the emergency cooling systems and not following specific guidelines. All of this combined to create a catastrophic accident that started as a steam explosion. This explosion was followed three to four seconds later by a second massive explosion that killed two engineers immediately and caused thirty-one deaths within three months due to radiation sickness. Hundreds of others, mostly emergency responders, were diagnosed with radiation sickness, and an entire nearby town was permanently abandoned due to high levels of radiation. Other environmental effects included radiation in groundwater that affected fish, and radiation sickness and cancers in cattle, horses, and wildlife. The effects continue, with latent cancers and sickness still showing up. At Chernobyl, the melted fuel rods have been entombed in tons of concrete and a huge confinement sarcophagus was begun in 2010. Other reactors on the site are still operating but an exclusion zone that is 31 km (19 mi) in radius around the site is still largely uninhabited.

Another major disaster involving nuclear reactors took place in Japan in March 2011. At this time, an earthquake—the largest on record—was centered off the coast of Japan and triggered a devastating tsunami that swept over cities and farmland in northern Japan. It struck the Fukushima Daiichi Nuclear Power Station, a complex of six reactors. The effects of the tsunami were devastating, but they were compounded by the nuclear reactors losing cooling water as a result of backup power failures; explosions and leaks of radioactive gas occurred in three of the Fukushima reactors, and a partial meltdown was recorded in at least one of the plants. Large quantities of radioactive material were released into the atmosphere, and radioactive water was released into the ocean. Nearby farmland was contaminated with radiation, resulting in restrictions on the distribution and consumption of foods from the area. Some 80,000 people in nearby towns have been evacuated, and many will never be allowed back in the area. A few weeks after the tsunami, the disaster was placed at the same level of severity (level 7) as the Chernobyl meltdown. Public confidence in nuclear energy was severely shaken as a result, and the accident has forced numerous countries to rethink their nuclear ambitions. At least twenty-five reactor projects have been shut down or cancelled in Europe in the aftermath, although China and other countries have continued to pursue nuclear power with a number of new projects. The industry has implemented safety enhancements and upgrades from lessons learned from Fukushima. In the United States, a complete review of safety and emergency preparedness was implemented for various possible disasters, including earthquakes, fires, explosions, and terrorist acts. Interest in nuclear power continues worldwide, and the United States is building new reactors in South Carolina and Georgia.

Disposal of radioactive waste from reactors is a huge issue that has not been resolved. Radioactive waste includes spent fuel rods and all other waste, and it is classified by the level of radioactivity. Products with short half-lives tend to be more radioactive because they disintegrate rapidly over time. High-level waste includes spent fuel rods that are highly radioactive and thermally hot. Spent fuel rods are stored in water for approximately ten years to cool them to safe temperatures for handling and to reduce the radiation. After about ten years, radiation levels subside enough for the fuel rods to be reprocessed, extracting leftover fuel and plutonium for reuse, but the remaining high-level waste is dangerous for thousands of years (plutonium, for example, has a half-life of 24,000 years.) Because of concerns with proliferation of nuclear weapons, reprocessing of ^{238}U is banned in many countries, including the United States.

Much of the high-level waste from reactors has been allowed to accumulate onsite and in reprocessing plants, postponing the eventual requirement to have a disposal plan in place. The problem is that it must be contained safely for thousands of years. Geological burial of waste is the only viable solution, but a site needs to be identified that is geologically stable, and safe transportation of the waste needs to be addressed. There is no experience base for predicting geological events for tens of thousands of years. In the United States, Yucca Mountain in Nevada (100 miles northwest of Las Vegas) was the leading candidate for years, but there is considerable opposition to this site in part because of groundwater, transportation, and potential ground movement (earthquakes). Groundwater could be contaminated if it mixes with waste and can contribute to corrosion problems for the containers. After more than $10 billion was spent on this project, there is still no determination on what the United States should do for long-term storage of waste. The Yucca Mountain project has an uncertain future because of litigation; meanwhile, the amount of radioactive waste continues to accumulate at scattered locations and the problem gets worse.

Nuclear power plants use steam-driven turbines and massive amounts of cooling water. The water issue is the same as that discussed for fossil fuel–burning plants. The temperature of outgoing water from one plant (Alabama's Brown's Ferry plant) has been too warm in several recent summers, requiring the plant to cut power output (in one case, for five consecutive weeks). Incoming water can also be too hot to provide the required cooling, which can force cutbacks.

Nuclear Fusion

Fusion reactors offer the promise of nearly limitless clean energy. Fusion, as already defined, is the process of putting together light nuclei to form heavier nuclei and release energy. While fusion reactions are common in particle accelerators, the goal for energy production is to produce more energy than is consumed in a controlled reaction, an elusive goal that has not been achieved. Fusion reactors require two charged particles that repel each other to come together and react, a very difficult process requiring extreme temperatures like those in the interior of the sun.

The fuel for fusion reactors is almost endless. One fuel is a mixture of deuterium (written as ^2H or D) and tritium (written as ^3H or T). **Deuterium** and **tritium** are isotopes of hydrogen (one proton), with one neutron for deuterium and two neutrons for tritium. Deuterium is readily available in seawater and tritium can be produced from lithium, which is available in quantities that would last literally millions of years. The basic reaction between deuterium and tritium is written as follows:

$$D + T \rightarrow He + n$$

where

n is a high-energy (14.1 MeV) neutron

This reaction, called the D-T reaction, produces a high-energy neutron, but it is not a chain reaction (the neutron does not induce further fusion events).

There are two approaches to making a fusion reactor. One is a pellet method, in which small pellets of D-T fuel are compressed by high-powered lasers, creating helium and releasing energy; this method is referred to as an inertial fusion project. The major effort for this approach is at the National Ignition Facility (NIF), which is part of the Lawrence Livermore National Laboratory. The NIF brings 192 high-power lasers to bear on a tiny target of D-T located at the center of a reaction chamber, replicating conditions at the center of stars.

Fusion-Triggered Fission

Fusion-triggered fission is being investigated at the National Ignition Facility (NIF). The idea is to use lasers to trigger fusion reactions in tiny pellets of D-T that react in a chamber. The new twist is to line the walls of the chamber with a blanket of uranium or other fuel. The fast neutrons from the fusion would cause fission to occur in the blanket, thus increasing the power over fusion energy alone. One of the significant benefits from this method is that it can help eliminate reactor waste.

A major milestone was reached in the summer of 2012, when a laser pulse with a peak power of 522.6 terawatts (TW) was fired; however, the goal of achieving ignition continues to be elusive and very expensive. The NIF continues to be a major program element in the United States for developing fusion energy by advancing the science of fusion.

The second approach to making a fusion reactor is to confine super-hot plasma (10^8 K) of D-T fuel long enough for an appreciable fraction of the D-T to react and produce energy. There are various efforts at confinement. One project is called Alcator C Mod, which is a **Tokamak** design used by researchers at the Massachusetts Institute of Technology (MIT) to investigate the stability, heating, and transport properties of plasmas. The Tokamak has a doughnut-shaped reaction area; it was originally a Russian idea for confining the plasma. The most ambitious Tokamak to date is a huge cooperative project called the international thermonuclear experimental reactor (ITER), which is under construction in Cadarache, France, and is funded by various countries. The goal of ITER is to produce more power than it consumes and to show that fusion energy is an achievable goal. The ITER Tokamak heats the plasma using neutral beam injection and two sources of high-frequency electromagnetic waves. Energy is extracted by slowing down high-energy neutrons from the fusion reaction in a blanket that surrounds the reaction and using the resulting heat to convert water to steam. Figure 1-8 shows a cutaway view of the planned reactor. The original scheduled date for deuterium-tritium operation is March 2027, with the goal of ultimately taking ITER to 500 MW of fusion power.

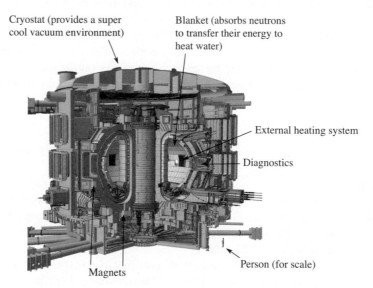

FIGURE 1-8   The ITER Tokamak Fusion Reactor. The reaction takes place in the hollow, doughnut-shaped area in the center.   (*Source:* Courtesy of ITER Organization.)

Fusion-based reactors have many advantages, including no greenhouse gas emission. There are no radioactive fission fragments or fuel rods to recycle, but some radioactive waste is produced due to the activation of materials by high-energy neutrons. The biggest advantage is safety because there is no chain reaction; the fuel is consumed much like the gas in a kitchen stove. A failure in a fusion reactor will simply shut it down, so there is no possibility of a Chernobyl-like accident.

SECTION 1-2 CHECKUP

1. What is a chain reaction?

2. Why is it necessary to enrich uranium to be used as a reactor fuel?

3. What are the basic steps to making electricity in a pressurized water reactor?

4. What is a breeder reactor?

5. What are the positive and negative environmental effects of generating electricity with nuclear fission reactors compared to coal-fired power plants?

6. What are two approaches to creating a viable fusion reactor?

7. What advantage does a fusion reactor have over a fission reactor?

1-3   The Solar Resource

The solar energy directed at earth is considerable. The sun has far more energy associated with it than we need, but its energy is spread out over the surface of the earth. The solar constant of 1,368 W/m² is the amount of energy received on a surface that is outside the atmosphere and oriented perpendicular to the sun's rays at the mean distance of the earth from the sun. When averaged over the earth's entire surface for a 24-hour period (day and night), the solar constant is one-fourth of the solar constant or 342 W/m² at the outer edge of the atmosphere and much less at the earth's surface. Several factors affect this number, including clouds that reflect sunlight back into space, the scattering of sunlight in the earth's atmosphere, and absorption of energy in the earth's atmosphere.

The Solar Spectrum

The earth receives energy from the sun in the form of radiation. A portion of this energy is in the visible range, which is in the wavelengths from 400 nm (blue) to 700 nm (red). Much of the energy we receive is in the infrared region (wavelengths longer than 700 nm) and some is in the ultraviolet region (wavelengths shorter than 400 nm). The spectral distribution of energy from the sun at the top of the atmosphere approximates a black body at a temperature of 5,770 K. (K stands for kelvin, which is the absolute temperature scale. K = °C + 273.15.) In physics, a **black body** is one that absorbs all radiation falling on it and then reemits it in a wavelength that is related to its absolute temperature. The actual temperature of the sun's photosphere varies from about 4,400 K at the lowest level to 7,500 K at the highest level. The basic shape of the spectrum is described by an equation called Planck's energy distribution, which gives a good approximation of the measured spectrum. Figure 1-9 shows the measured spectrum at the top of the atmosphere and at sea level. The sea level plot varies depending on location and amount of water, clouds, and other materials in the atmosphere; a significant fraction of the incident radiation is scattered and

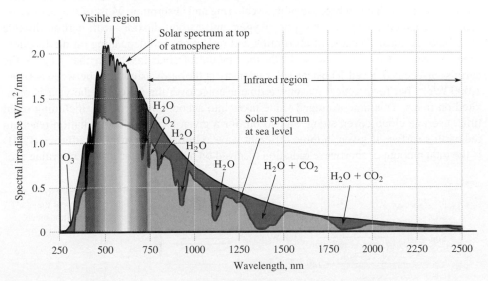

FIGURE 1-9   Solar Spectrum at the Top of the Atmosphere and at Sea Level

absorbed by the atmosphere. At shorter wavelengths (ultraviolet), the incident radiation is significantly attenuated by ozone (O_3) in the stratosphere, which absorbs the ultraviolet portion of the solar spectrum. This absorption is critical to life on earth, so maintaining this shield is crucial.

In the ultraviolet (UV) region (wavelengths less than 400 nm), ozone is responsible for much of the absorption, a fact that helps protect us against harmful UV rays. Water vapor and carbon dioxide are the primary gases that contribute to global warming by absorbing shorter wavelength radiation from the sun and reemitting it at longer infrared wavelengths. Thus, the earth is warmed by these gases and they contribute to the average temperature on earth. This is why an increase in atmospheric carbon dioxide, which has risen steadily for the past 200 years, has many scientists concerned about its effect on climate.

The sun emits energy in all directions, so the incident radiation at the top of the atmosphere follows a familiar law in physics known as the **inverse square law**, which states that the flux from an isotropic point source is reduced by the square of the distance from the source to the receiver (the sun is at such a great distance, the inverse square law is quite accurate). Recall that the **solar constant** is defined as the amount of energy received on a surface that is outside the atmosphere and oriented perpendicular to the sun's rays at the mean distance of the earth from the sun. From the power output of the sun (approximately 3.847×10^{26} W) and the inverse square law, it is simple to calculate the intensity of sunlight at the top of the atmosphere (solar constant, S):

$$S = \frac{P_{sun}}{4\pi r^2} = \frac{3.847 \times 10^{26} \text{ W}}{4\pi(149.6 \times 10^9 \text{ m})^2} = 1,368 \text{ W/m}^2$$

where

$$S = \text{the solar constant (W/m}^2)$$

$$P_{sun} = \text{the power emitted by the sun (W)}$$

$$r = \text{the radius of a sphere equal to the sun–earth mean distance } (149.6 \times 10^9 \text{ m}).$$

The calculation is illustrated in Figure 1-10. Notice that the unit for the solar constant or *insolation*, is power (W) divided by area (m^2). Insolation (from *incident solar radiation*) can also be measured in units of kWh/m²/day, which is easier to relate to electrical units (1 W/m² = 0.024 kWh/m²/day). The value of S is 32.8 kWh/m²/day at the outer edge of the atmosphere. Because of the earth's rotation, the intensity of sunlight received on its surface is averaged over a twenty-four-hour period. Clouds and the atmosphere, which vary significantly by location, have a major effect on the amount received on the surface.

As the solar energy passes through the atmosphere, photons are scattered and absorbed by molecules and atoms in the atmosphere. By the time it reaches the surface of the earth, the spectrum has changed because of this scattering and absorption. Most of the absorption can be accounted for by water vapor, oxygen, and carbon dioxide, with carbon dioxide absorbing the most. Because of absorption and reflection, and depending on the location, time of year, cloud cover, and so forth, the amount of solar insolation that reaches the earth's surface is reduced. The peak solar intensity at the earth's surface is between 800 and 1,000 W/m², but for practical reasons, a time-average over the course of the year is often cited on maps. This time-averaged value takes into account all variables, including night time, average cloud cover, smog, and so on, for a given location. The radiation reaching the earth's surface is given in three different ways. **Global horizontal irradiance (GHI)** is the total amount of shortwave radiation received on a horizontal surface. This value is of

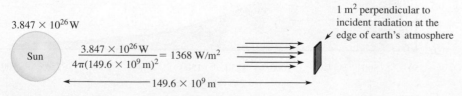

FIGURE 1-10   Calculation of the Solar Constant, S

particular interest for photovoltaic installations. The portion of GHI that comes in a straight line from the sun is called **direct normal irradiance (DNI).** Some of the scattered radiation arrives at the earth's surface from indirect paths and is called **diffuse horizontal irradiance (DHI).** The GHI value includes both the direct and diffuse irradiance. Both direct irradiance and diffuse irradiance are useful for solar applications like flat-plate collectors, but only direct radiation can be focused using concentrating collectors. Maps are available that show the direct radiation suitable for determining the solar input for concentrators, or the GHI that is useful for determining the solar input for flat panels.

EXAMPLE 1-2

Assume a flat solar collector is 150 cm wide by 180 cm high and is oriented so that it is perpendicular to the sun. Its active area is 90% of the panel size. What is the peak power delivered to the active area of the collector if it is in a location that receives solar insolation of 1,000 W/m^2 peak?

Solution

$$\text{Active area} = 90\% \times 1.5 \text{ m} \times 1.8 \text{ m} = 2.43 \text{ m}^2$$

$$\text{Power} = 1,000 \text{ W/m}^2 \times 2.43 \text{ m}^2 = 2.43 \text{ kW}$$

The solar resource depends on weather and location on earth as well as other factors. Figure 1-11 shows the solar resource, averaged over a full year, for the United States. The area shown in red is the best area for locating concentrator collectors because it has the most direct radiation.

While the map gives a good general indication of locations that can benefit from solar energy, more detailed information is needed for planning a specific project. For the United

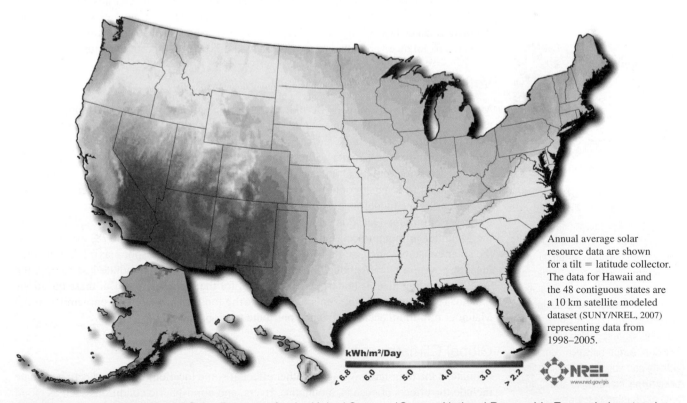

Annual average solar resource data are shown for a tilt = latitude collector. The data for Hawaii and the 48 contiguous states are a 10 km satellite modeled dataset (SUNY/NREL, 2007) representing data from 1998–2005.

kWh/m²/Day

< 6.8   6.0   5.0   4.0   3.0   > 2.2

FIGURE 1-11   Average Annual Solar Resource for the United States   (*Source:* National Renewable Energy Laboratory.)

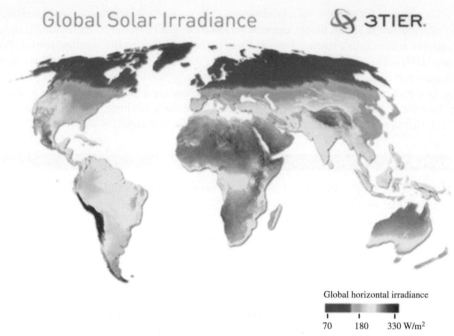

Global Solar Irradiance   &3TIER.

Global horizontal irradiance

70   180   330 W/m²

FIGURE 1-12   Average Annual Global Solar Irradiance   (*Source:* Courtesy of 3TIER.)
Copyright © 2010 3TIER Inc. All Rights Reserved.

States, the **National Solar Radiation Database (NSRDB)** is a document available from the National Renewable Energy Laboratory (NREL) that describes the amount of solar energy available at any location. It includes data for each component of solar radiation.

World data for annual average solar irradiance is shown in the map in Figure 1-12. As in the case of the NREL data, the data in this map is derived from satellites but has higher resolution (approximately 3 km) and covers a time period of at least eleven years. The detailed data for any given site can be downloaded from 3TIER to determine a range of expected irradiance values and to provide an assessment of any particular site in the world.

Variations in the Solar Constant

Much progress has been made in understanding the sun and its magnetic field, which is responsible for sunspots. Sunspots are places on the surface of the sun where the magnetic fields break through the surface. Sunspot frequency has a normal eleven-year cycle, going from minimum activity to maximum. A period occurred between 1645 and 1715 in which sunspot activity almost ceased. This period, known as the Maunder minimum, corresponded to a much cooler global climate, including much colder winters. A NASA computer climate model suggests that there was a reduction in the solar constant during the Maunder minimum. Questions still linger about the variability of the solar output and why the Maunder minimum occurred. From satellite measurements since 1978, we know that the solar constant varies, and it is about 0.15% higher during sunspot maximum than during sunspot minimum. Although the sunspots are cooler areas that radiate less energy, the area surrounding them is hotter, which accounts for the higher output. But these questions remain: How constant is the sun's output over the long term? Can an event similar to the Maunder minimum occur again and contribute to global cooling?

Global Climate

Climate is the long-term average weather effect at some location. Factors affecting climate include the effects of oceans, mountains, rivers, and atmospheric conditions such as cloud cover and wind. Very little effect on climate can be attributed to solar variations over the

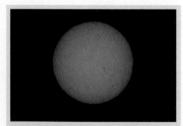

(*Source:* David Buchla.)

This photo, taken in 2012, shows the sun and its typical sunspots. During the Maunder minimum, from 1645 to 1715, there were almost no sunspots. At this time, temperatures were so cold that the Thames River in London, England, froze over regularly, and the ice was thick enough that winter festivals and skating parties were held. The connection between sunspot activity and solar output has not been firmly established and is an area of ongoing research.

more than 50-year timeframe that satellites have observed the sun. Global warming is well documented, but factors influencing global warming include greenhouse gases in the atmosphere rather than variations in the solar output.

Environmental Effects of Solar Energy

All sources of energy, including renewable energy, have some environmental effects. The process of converting sunlight to usable energy generates no pollution, but the chemicals used to create certain solar materials (panels, inverters, etc.) include hazardous materials. Chemical spills can cause pollution; in September 2011, fluoride from solar cell waste storage contaminated a river. Other dangerous chemicals are used in the production of solar cells and can create an environmental hazard if they leech into a landfill. A consideration in the full life cycle of solar production is the disposal of waste in the production process and waste disposal when the product is at the end of its usable life.

Water use is another environmental issue that needs to be addressed prior to construction of solar installations. Steam generators are used in utility power generation, and they can be a problem for water tables in arid regions, which tend to be best for solar insolation. Solar systems vary widely in the amount of water required; photovoltaic (PV) systems require water only for cleaning, whereas most power tower systems require water for driving turbines. Some systems require additional cooling water.

Another environmental effect is land use. Most small solar PV systems are located on roofs and over parking areas, so they have little or no impact on land usage. Some large arrays use solar collectors for utility power generation and thus require many acres of land. This amount of land use can have a large impact on certain species of wildlife. Typically, five acres of land are required for each megawatt of capacity from the solar collectors. Concentrating PV systems and power towers are the most efficient solar systems when considering the use of land. A commercial solar power plant has a much larger footprint than a comparable coal-fired plant, but if the entire area required for mining coal and storing waste is factored in, the solar footprint is actually less.

Another land-use issue involves locating solar arrays on valuable agricultural areas. This is important in areas like California's inland valleys, where prime growing areas exist. For example, Fresno, Imperial, and Kern counties in California have some of the most productive agricultural land in the United States, partially because of the abundance of sunshine. Planners are proposing restrictions on solar projects that affect food production in these areas. One way to have agriculture and solar electricity co-exist is being tested in Japan, where special open structures for solar collectors have been constructed that allow enough sunlight to reach the ground for plants to grow underneath.

Another important issue for both solar and wind power is the fact that the resource is not available at all times. Grid operators cannot be assured that solar and wind power is available when they need it, which means backup generation capability must be available. Solar power is reasonably predictable, but it is affected by clouds or storms in the daytime and cannot serve as baseline (always ready) power that utilities need unless significant storage capability is part of the project.

SECTION 1-3 CHECKUP

1. What is meant by the solar constant? What are two common measurement units for it?

2. What does the term *insolation* mean?

3. Explain how atmospheric carbon dioxide affects climate.

4. What is global horizontal irradiance (GHI)?

5. What is the difference between direct normal irradiance (DNI) and diffuse horizontal irradiance (DHI)? Which one is not useful for solar concentrators?

6. What is the Maunder minimum? Is there any evidence that it affected the solar constant?

1-4   The Wind Resource

Sailing ships have been using the wind for thousands of years. The wind was thought to be first used for pumping water by the ancient Persians. They adopted the idea of using the sail to capture the wind in a vertical-style windmill. The Persians and later the Romans also used the wind to power mills for grinding grain, hence the term windmill. Today, the wind offers a clean renewable energy source that can turn the shaft of a generator to create electricity or be harnessed directly to drive a pump.

Global Patterns of Wind

The earth's prevailing winds are the result of two major factors: differential heating and the earth's rotation. A three-cell model of global circulation is a starting point for understanding the prevailing winds, but keep in mind that seasonal changes and landmasses have a major effect on these patterns.

Heating warms the moist air near the equator, which rises because of lower density. This moist air cools and dries as it moves away from the equator. As the warm air moves up in the atmosphere, cooler air moves in to replace it. In the northern hemisphere, this cooler air moves from north to south at low altitudes; in the southern hemisphere, it moves from south to north. This process forms an atmospheric circulation cell that is called the **tropical cell** (also called the Hadley Cell). The process forms a high-pressure region in the vicinity of 30° latitude, where the air sinks. In addition to the tropical cells on either side of the equator, each hemisphere has a **mid-latitude cell** (also called the Farell cell) and a **polar cell.** The predominate air movement from these cells is shown in Figure 1-13. As air in the mid-latitude cells move toward their respective poles, colder air from the polar region is encountered, so the warmer air moves up and over the air from the poles, creating a lower-pressure region. In the polar regions, the cold dense air tends to circulate away from the poles at lower elevations and forms the polar cells. Warmer air rises above the cold polar air and tends to move from a region of high pressure to low pressure; this pressure gradient difference creates a force that moves air in the cells.

When the earth's rotation is taken into account, the winds are deflected by the **Coriolis force,** which adds an east–west component to the movement of the air in the cells. The Coriolis force is a fictitious force that explains how the rotation of the earth creates an accelerating reference frame for an observer on earth; as a result, the observer concludes that the resulting motion is due to a force. To understand the Coriolis force, imagine sitting on one side of a spinning platform and throwing a ball across the center of the platform to someone sitting on the opposite side. You and the person on the other side are said to be on an accelerating reference frame. When the ball leaves your hand, it does not experience any force other than gravity (ignoring air friction). The person on the other side would say the ball had curved away because that person was rotating as well. He or she could explain

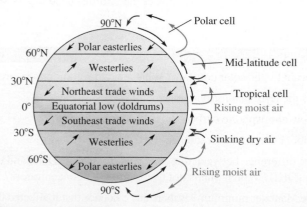

FIGURE 1-13   Global Wind Patterns

the ball's curving away by assuming he or she was stationary and the ball had experienced a force, that is, the Coriolis force.

In the regions north and south of the equator, the trade winds blow primarily from the east to the west due to the Coriolis force combined with the pressure gradient difference in the circulating tropical cell. The trade winds were so named because they enabled trade between Europe and the New World in sailing days. When wind direction is specified, it is always named as the direction from which it blows. The trade winds are easterlies and are primarily responsible for the movement of hurricanes that are spawned in the equatorial regions. At the equator, there is no Coriolis force, so the winds diverge toward the equator and create a low-pressure region with very calm wind, which was given the name *doldrums*. Early sailors obviously wanted to avoid the doldrums.

In the mid-latitude cells, the circulation is opposite that of the tropical cells, and with the addition of the Coriolis force, the surface winds blow primarily from west to east. Thus, the mid-latitude winds are referred to as westerlies. On the west coast of the United States, the westerlies tend to move the weather from the Pacific to the coast, and they produce wind that tends to channel into mountain passes. The westerlies are also responsible for bringing wet weather from the Atlantic across western Europe. The mid-latitude winds tend to vary more because of many factors, including high- and low-pressure regions that tend to form and last several days. The types of air masses that form in these regions vary depending on whether they are formed over a continent or out at sea and whether they have a tropical or polar origin. The boundaries between air masses are called **fronts,** and these fronts mark the most active regions of weather.

In the polar regions, the polar easterlies are cold, dry winds that flow from high-pressure regions at the poles toward low-pressure areas in the mid-latitude regions. These winds tend to be weaker irregular winds.

Superimposed on the global patterns are effects due to landmasses and large bodies of water. Air temperature varies because the earth's surface heats up at different rates. Landmasses tend to heat and cool more quickly than water does, and mountains tend to be cooler than valleys. Latitude, season, cloud cover, and other factors also cause temperature and pressure variations and create variable conditions for wind and weather. Wind always tends to flow from a high-pressure area toward a low-pressure area, and these changes depend on many factors, as you have seen.

In addition to surface winds, the advent of high-flying aircraft led to the discovery of a high, fast-moving river of air moving from west to east known as the **jet stream,** which is created by temperature gradients in the atmosphere. The jet stream has a significant impact on weather and the location of high- and low-pressure regions in the atmosphere. Each hemisphere has a jet stream that marks the boundary between cold polar air and warmer mid-latitude air. It constantly changes its pattern both daily and seasonally, so it is monitored constantly by weather forecasters. In the winter, the jet stream tends to move closer to the equator; in the summer, it moves away from the equator. Because the speed of the winds is extremely high, there is some interest in developing the ability to harvest its energy, but plans for this are still in the concept stage and it will be years before any actual device is ready for the jet stream. Currently, prototype models of tethered wind turbines have been flown by several innovators to show the feasibility of flying turbines in the high-altitude winds but these are far below the jet stream. At 500 m above the ground, wind speeds tend to double and have roughly 8 times the energy of ground level winds. Even more energy is available higher up, so the potential for significant energy harvesting is enticing. Chapter 7 discusses this in more detail.

As in the case of the solar resource, maps are available for wind resources based on historical patterns. The United States has a large wind resource that, if fully exploited, could easily meet all of the electricity demand. So far, however, wind has not proven suitable for baseline power. According to a report from the National Renewable Energy Laboratory (NREL), there are potentially 4,150 gigawatts (GW) of potential wind power available within the United States and within 50 nautical miles of the coast. This is based on a wind speed of at least 7 mi/s (16 mph) and within a height of 90 m (295 feet). The major areas for

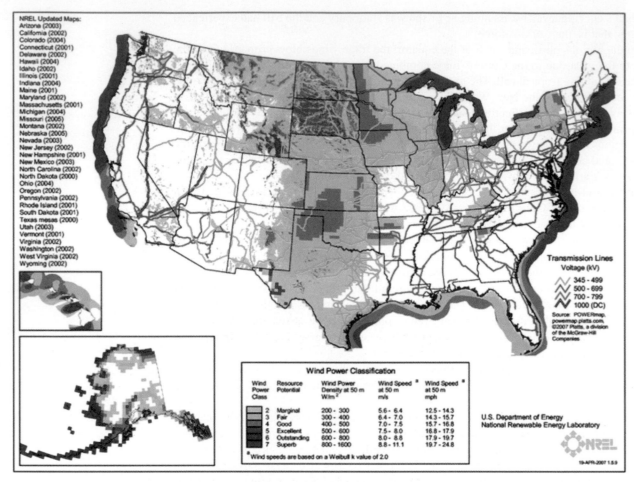

FIGURE 1-14   Wind Resource and Transmission Lines for the United States   (*Source:* National Renewable Energy Laboratory.) The remaining states use data from the 1987 "Wind Energy Atlas of the United States."

wind resources tend to be on both coasts, the Great Lakes, and in the Midwest to the east of the Rockies. Figure 1-14 shows the US wind resource based on the NREL data along with major transmission lines.

Maps such as the one in Figure 1-14 are useful to gain an overview of the resource, but the map lacks the details necessary for specific site selection and evaluation. For these kinds of assessments, maps with much finer detail are available for almost any location in the world that can pinpoint wind (or solar) conditions for a given site based on observational data and computer modeling. For the wind resource, hourly average wind speeds for each month and at various tower heights are only one of the many options to consider for locating a wind farm. Table 1-1 shows an example of hourly wind data for each month at a given site that enables managers to determine optimum operation.

World data for wind between 60°S and 70°N is shown in the map in Figure 1-15 shown on page 24 with a resolution of approximately 5 km. Some of the data is based on various models of the atmosphere and small-scale terrain features. Data over the ocean (not shown on the maps) is available based on satellite observations of waves using an instrument called a scatterometer, which uses radar scattering to measure wave height and surface winds. The detailed data, including hourly wind speed, wind direction, air temperature, and air pressure for three different heights above the ground, for any given site can be downloaded from 3TIER to evaluate any particular site for wind energy. The key areas where wind energy is being developed include the Midwest and some inner coastal areas of the United States and offshore in Europe. In terms of offshore wind, the United Kingdom and Denmark are far ahead of other countries. As already mentioned in regard to solar power,

TABLE 1-1

Wind Speed as a Function of the Time of Day and Month   (*Source:* Courtesy of 3TIER).

Hour	Jan	Feb	Mar	Apr	May	Jun	Jul	Aug	Sep	Oct	Nov	Dec	Avg
00	7.3	7.3	7.6	8.1	7.4	6.9	6.3	5.8	5.9	6.8	7.1	7.3	7.0
01	7.6	7.6	7.8	8.2	7.6	7.2	6.9	6.5	6.7	7.5	7.6	7.6	7.4
02	7.9	7.9	8.2	8.7	8.4	8.0	7.8	7.4	7.4	7.9	7.9	7.8	7.9
03	8.1	8.1	8.4	9.2	9.1	8.8	8.5	8.0	7.9	8.3	8.1	8.0	8.4
04	8.3	8.4	8.7	9.4	9.6	9.4	9.1	8.4	8.2	8.6	8.2	8.1	8.7
05	8.4	8.4	8.8	9.5	9.9	9.9	9.5	8.6	8.4	8.7	8.4	8.2	8.9
06	8.5	8.4	8.9	9.6	10.1	10.1	9.6	8.7	8.4	8.8	8.4	8.2	9.0
07	8.5	8.4	8.9	9.6	10.1	10.1	9.6	8.7	8.4	8.8	8.5	8.3	9.0
08	8.5	8.3	9.0	9.7	10.1	10.0	9.5	8.6	8.4	8.7	8.5	8.3	9.0
09	8.5	8.2	9.0	9.6	10.0	9.8	9.3	8.4	8.2	8.6	8.4	8.3	8.9
10	8.5	8.2	9.0	9.5	9.9	9.6	9.0	8.1	8.1	8.6	8.3	8.3	8.8
11	8.5	8.1	8.9	9.4	9.8	9.4	8.8	8.0	7.9	8.5	8.1	8.3	8.6
12	8.4	8.1	8.8	9.3	9.6	9.2	8.6	7.8	7.8	8.3	8.0	8.3	8.5
13	8.5	8.2	8.7	9.4	9.3	8.7	8.4	7.7	7.8	8.6	8.2	8.5	8.5
14	8.5	8.1	8.6	8.7	8.7	8.3	7.8	6.7	7.0	8.3	8.1	8.4	8.1
15	8.2	7.8	8.2	8.6	8.5	7.9	7.4	6.4	6.6	7.7	7.8	8.2	7.8
16	7.8	7.5	8.2	8.7	8.3	7.3	6.6	5.8	6.3	7.5	7.3	7.7	7.4
17	7.5	7.5	8.2	8.5	7.8	6.7	5.7	5.1	5.8	7.3	7.1	7.4	7.1
18	7.5	7.5	8.2	8.4	7.5	6.3	5.2	4.7	5.4	7.1	7.2	7.4	6.9
19	7.6	7.5	8.1	8.2	7.2	6.1	5.1	4.6	5.2	6.9	7.2	7.4	6.8
20	7.6	7.5	8.0	8.2	7.1	6.0	5.1	4.7	5.2	6.8	7.1	7.4	6.7
21	7.5	7.5	7.9	8.2	7.1	6.2	5.2	4.8	5.3	6.7	7.0	7.4	6.7
22	7.3	7.4	7.9	8.4	7.2	6.4	5.5	5.1	5.4	6.6	7.0	7.3	6.8
23	7.1	7.3	7.8	8.4	7.4	6.6	5.8	5.4	5.6	6.7	6.9	7.1	6.8
Avg	8.0	7.9	8.4	8.9	8.7	8.1	7.5	6.8	7.0	7.8	7.8	7.9	**7.9**

wind power cannot generally serve as baseline power without significant (and expensive) storage capability; however, some offshore locations may be able to offer very reliable power. (Samso Island, a small island in the North Sea off the coast of Denmark, is trying to go 100% wind power for its 4,000 residents).

Over 240 GW of installed power from wind turbines worldwide can provide over 500 terawatt-hours of energy on an annual basis. This is roughly 3% of the electricity consumption worldwide. The World Wind Energy Association estimates that installed capacity will be over 1 terawatt by 2020. Currently, the five largest markets are China, the United States, Germany, Spain, and India, with China accounting for 44% of the world wind market.

Environmental Effects of Wind Energy

Wind energy has many attributes that make it environmentally friendly, including no emission of carbon dioxide like fossil fuels produce, but it does have some drawbacks. The land required for wind turbines is substantial, and some people criticize the appearance of wind farms as a visual eyesore. Wind turbines are typically the tallest structures in an area, so they are visible for miles. These structures may also create a hazard for small aircraft. Newer wind turbines tend to be even taller, compounding the problem. In addition to the

5km Global Wind   3TIER.

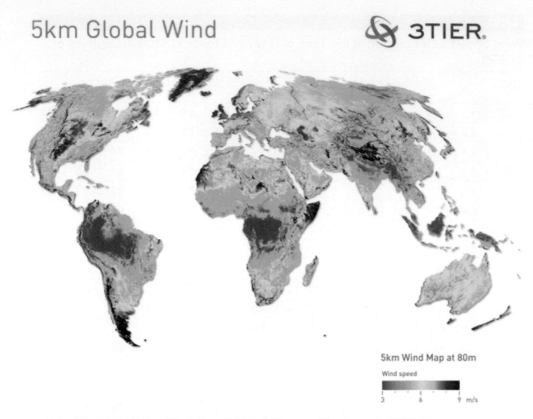

5km Wind Map at 80m

Wind speed

3   6   9 m/s

FIGURE 1-15   Global Wind Speed Map   (*Source:* Courtesy of 3TIER.)
Copyright © 2010 3TIER Inc. All Rights Reserved.

height and area requirements for wind turbines, they can have negative impacts on historical, recreational, and cultural sites.

Another issue is noise. Wind turbines produce noise that can be annoying to nearby residents (generally those within 2 miles). The source of noise is twofold: (1) mechanical parts that vibrate and (2) aerodynamic noise from the blades. Wind machine manufactures have made excellent progress in reducing mechanical noise by damping vibrations, adding insulation and special gear boxes, and engineering every part of the wind machines for low-noise operation. The aerodynamic swishing sound of the turbines can cause sleep disturbance in some cases, and some people complain of other health-related issues. Aerodynamic noise is highly dependent on rotor speed; newer turbines rotate more slowly and create much less noise than early models. Helical turbines (such as the Quietrevolution models in England) are designed for quiet running and work well in environments where people are nearby.

Another environmental effect is that of bird strikes by the wind machines. The numbers of birds killed by wind machines is in dispute because the numbers from different sources vary widely. Data from the American Wind Energy Association indicates that wind power causes far fewer losses of birds (approximately 108,000 a year) than other human-caused deaths: buildings (100 million to 1 billion), power lines (130 million), cars (60 to 80 million), pesticides (67 million), domestic cats (at least 10 million), and radio and cell towers (4.5 million). Although the number of bird kills is low compared to other sources, expansion of transmission lines could increase the number. The number of total kills from wind machines masks the fact that certain species appear to be more vulnerable than others. In one study, the wind farm at Altamont Pass in California is responsible for the death of 80 golden eagles per year (a protected species) and thousands of other birds.

But the goal is to reduce the number of bird strikes, and this effect has been reduced significantly in recent years because of slower-moving turbine blades that birds can see and avoid. In addition to slower blades, wind turbine manufacturers have tested different paints and reflective devices on the blades. The sound of the turbine itself can help birds avoid the blades. There is also concern about locating wind machines in pathways used by migratory birds because this could alter the routes used by the birds. One way to help cut down bird strikes is to use monopole towers rather than the lattice towers used on older machines, which helps to cut down nesting sites on and around the wind turbine itself.

Many of the best locations for wind farms are miles away from the end users of the electricity. As a result, transmission lines need to be constructed; these have their own environmental impact, including land requirements, complaints about them acting as visual eyesores, and bird electrocution. One option to alleviate this condition is underground transmission lines, but this is expensive.

Another impact is the appearance of wind turbines. The visual impact of wind turbines is hard to alleviate; however, larger and more efficient models may help. Locating wind farms farther from population centers and highways, and limiting the number and height of wind turbines in residential or commercial areas are other ways to alleviate eyesores. For offshore wind turbines, locating the wind farm further from shore also has less visual impact but much higher capital costs for both the towers and the transmission lines to bring power to shore. New installations have run into public acceptance issues, particularly with visual impact and the effect on recreational boating. On the positive side, offshore wind turbines can be located closer to population centers, and they alleviate the need for long transmission lines and the power losses associated with them.

SECTION 1-4 CHECKUP

1. What is the primary direction of wind in the mid-latitude cells?

2. What are the doldrums? Why did early sailors avoid them?

3. Why does air from the equator eventually sink, forming the tropical cells?

4. Which areas of the United States have the best potential for harvesting wind energy?

5. What environmental issues are associated with wind power?

1-5 Geothermal Resources

Geothermal energy, available around the clock, is the energy within the earth that can be used for creating reliable electrical power. The energy originates from the enormous temperature at the center of the earth (thought to be 5,000°C). This heat dates back to the formation of the earth and from continuous radioactive decay. Heat radiates out from deep within the earth and can reach the surface along fractures, through the thinner regions of the earth's crust, and between continental plates. The total geothermal resource is enormous, but heat flow near the surface varies significantly by region.

Geothermal Energy

Geothermal energy is a clean source of renewable energy with many desirable attributes. It has a small carbon footprint and low emissions, and it can provide continuous baseline electrical power (unlike solar and wind, which are intermittent sources). Geothermal energy does not burn fuel and is available anywhere in the world, but the highest underground temperatures that are within easy reach of current drilling technology are along geological plate boundaries or so-called hot spots, such as the Hawaiian Islands. In these locations, the earth's crust is thin enough to allow heat from hot magma deep within the earth to move closer to the surface and heat rocks and underground water. Geothermal hot spots tend to be more prevalent around the Pacific Rim, a region around the Pacific Ocean that has many volcanoes and is seismically active.

When referring to the geothermal resource, the reference is usually to those concentrations of heat that can be tapped for uses such as electrical power generation and heating.

Geothermal energy is a bountiful resource that could easily power all of our electrical needs if enough can be tapped. Current technology makes sources deeper than 4 km of questionable economic value, but new drilling methods may extend that depth.

Plate Tectonics

Plate tectonics is a geological theory that explains how the earth's surface moves and changes over time. The earth's surface is divided into large areas called *plates* that interact with each other. This outer surface is approximately 60 miles (100 km) thick; it consists of the crust and upper mantle, and it is called the **lithosphere.** The lithosphere is made of a number of plates that move or float over the mantle. Plate movement can be defined by three types of activity. Plates that are colliding are called **convergent plates;** plates that are moving apart are called **divergent plates.** When plates slide laterally past each other, they are called **transform plates.** The plates interact along **boundaries,** so the three types of boundaries are called **convergent boundaries, divergent boundaries,** and **transform boundaries.** The boundaries are where plates interact to produce earthquakes, volcanoes, and mountains. These boundaries are places where the earth's internal heat can move closer to the surface because of cracking that occurs along the boundaries. This is why the nations that border tectonic plates have the best geothermal resources.

Earth's Interior

Figure 1-16 shows a diagram of the inner regions of the earth. The earth looks like a giant onion, with five regions or layers. At the center of the earth is the **inner core;** it is approximately 6,378 km (3,963 mi) from the surface. The inner core is extremely hot (about the same temperature as the surface of the sun) and dense; it is composed mostly of solid iron (85%) and solid nickel (5%) that descended to the center during earth's formation. It is a solid because of the intense pressure. Surrounding the inner solid core is a hot molten **outer core** that is thought to be the same compositionally as the inner core (mostly iron and nickel). It starts approximately 5,100 km (3,200 mi) from the surface. Surrounding this is

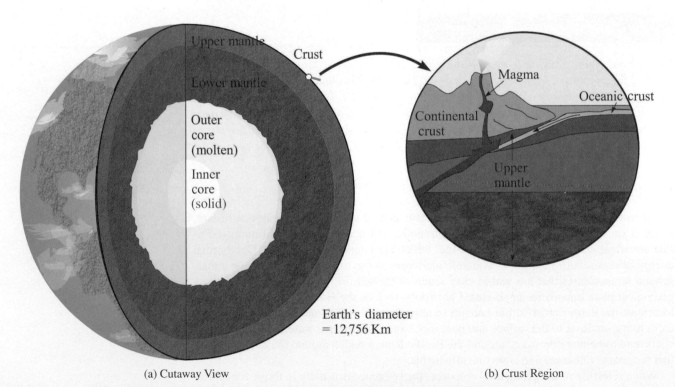

(a) Cutaway View   (b) Crust Region

FIGURE 1-16   Earth's Interior. The crust is the thin outer layer; it is shown in more detail in part (b).

the **lower mantle,** which starts approximately 650 km (404 mi) from the surface. The next region is the **upper mantle,** which runs approximately from 400 km to 650 km deep and is partially molten. The **continental crust** (or lithosphere) is a region that varies in thickness: It is about 30 to 50 km thick under the continents and much thinner under the oceans. The deepest hole ever made by humans is a hole drilled in Russia that is 12.3 km (7.5 mi) from the surface, which is only a fraction of the thickness of the crust. The mantle rock remains out of reach. Drillers had to stop on the deep hole before the goal of 15 km was reached because the temperatures of 180° C (356° F) and pressures at that depth squeezed surrounding rock, closing off the hole. Because drills have never reached the mantle, almost all that has been learned about the regions below the crust have come from earthquakes and underground nuclear weapons tests.

Source of Geothermal Energy

Heat from deep in the center of the earth is primarily caused by the decay of radioactive isotopes of uranium, potassium, and thorium, which occur naturally. The heat is intense enough to melt rocks, which can flow as magma. Some of the magma moves to the surface through volcanic activity, and earthquakes and other separation of the plates allow surface water from rains to circulate down to approximately 2 km (6,000 ft) to the hottest material. When the water comes into contact with the molten material, it can create underground hot water and steam reservoirs.

The best hydrothermal resources have several details in common: high average thermal gradients, high rock permeability and porosity, sufficient fluids in place, and an adequate reservoir of recharge fluids to replenish the water and steam removed. High-quality geothermal sources of steam and hot water are available only in a small area of the world and, to be cost effective, the steam and water must be within a few thousand feet of the surface. Drilling technology is continually improving so that deeper wells are being drilled. Deeper wells allow more geothermal energy to be tapped in locations that were once thought unusable because of their depth. In areas where hot water and steam are not available, other geothermal energy is available for use with heat pumps and other heating applications. The temperatures just a few meters below the surface remain constant, at approximately 10° C (50° F), which is much warmer than the temperatures aboveground during much of the winter.

Locations with lower-quality heat can be used for space heating in buildings, agricultural applications such as crop drying, and various industrial applications. A steady supply of lower-temperature heat exists at depths ranging from a few meters to hundreds of meters throughout the world. Relative to conventional systems, geothermal heat pumps (GHPs) are an efficient way to exploit this resource to heat homes and buildings. They work on the basic refrigeration cycle, like any heat pump, and can move heat into or from a building. Standard heat pumps are not efficient at very cold outside temperatures. The main difference between a normal heat pump and a GHP is that the GHP has the earth as a heat reservoir, so when outside temperatures are very cold, the GHP is still efficient. Because temperatures underground do not fluctuate nearly as much as surface temperatures, GHPs can be run in the summer to pump heat into the ground and provide cooling. They use electricity to *move* heat, not to produce it. Chapter 10 covers resources including GHP in more detail.

World Geothermal Resources

The world geothermal resources are concentrated at the boundaries of some of the tectonic plates that form the earth's crust. The most notable region is the so-called Pacific ring of fire, which follows the Pacific Ocean and encompasses the Pacific Plate, and the Nazca Plate along the west side of South America. The plate boundaries are the most likely spots for volcanoes and earthquakes and have the best locations for geothermal energy. Figure 1-17 shows a world map of volcanoes, earthquakes, impact craters, and plate boundaries.

Iceland, which is well known for its hot springs and volcanic activity, is located in the rift zone between the North American Plate and the Eurasian Plate and produces 30% of its

Hot Springs

(*Source:* David Buchla.)

Hot springs are associated with both active volcanic areas as well as nonvolcanic areas. In volcanic areas, water is heated by coming in contact with magma or hot rocks. The hot spring shown in the photo is bubbling near an ancient volcanic caldera that forms the harbor at Rabaul in Papua New Guinea. Other hot springs can be heated from hot rocks located deep within the earth and may not be near active volcanic areas.

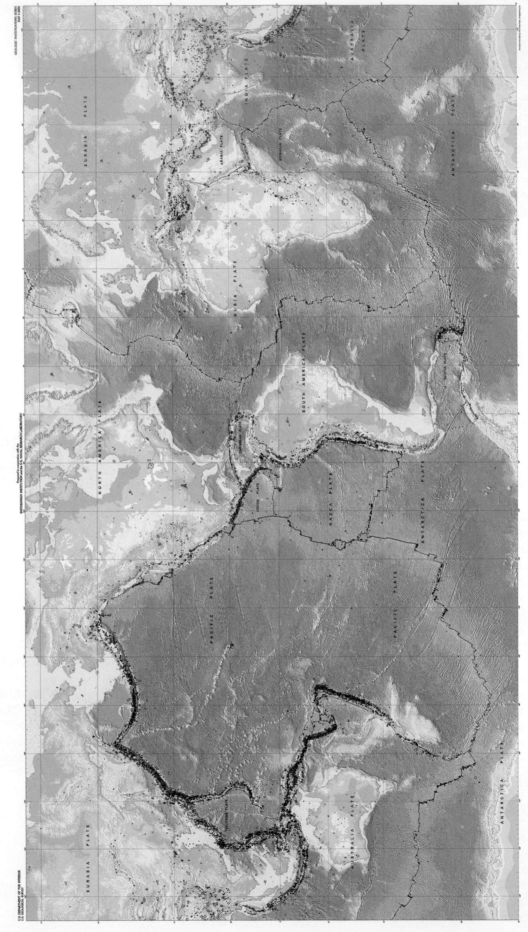

FIGURE 1-17 World Map of Volcanoes, Earthquakes, Impact Craters, and Plate Boundaries. The map shows notable volcanoes (red triangles), earthquakes (round black dots), tectonic plates, and impact craters. Elevations are keyed to colors. (*Source:* Courtesy of USGS.)

electrical energy from geothermal sources. Icelanders also use geothermal sources to heat their homes and businesses with hot water. Countries that produce 25% or more of their electricity from geothermal sources are the Philippines, Iceland, and El Salvador, all of which are on or next to a plate boundary. The United States has the largest installed electrical generating capacity using geothermal energy. The next largest producers, in order from largest to smallest, are the Philippines, Indonesia, Mexico, Italy, New Zealand, and Iceland. Currently over twenty-four countries use geothermal energy to produce electrical power.

US Geothermal Resources

In the United States, high-grade geothermal resources are concentrated in the West. California currently has approximately 80% of the country's geothermal electrical production. Figure 1-18 shows a map of the geothermal resources in the United States. Obtaining geothermal energy from these sites requires drilling; most current geothermal fields are less than 3 km (10,000 ft) deep. In general, drilling is more expensive for geothermal exploration than for petroleum exploration because of the types of rock and the temperatures encountered, the larger diameter holes required, and a higher degree of uncertainty in underground formations. As the depth is extended, costs go up significantly, so the depth of the sources is an important economic consideration when looking for economically viable sites. In the United States, the largest production by far is at the geysers in Northern California, with over 80% of US production. The geysers are situated over a very large, naturally occurring steam reservoir. They have a total generating capacity of 725 megawatts (MW), enough to power roughly 725,000 homes. Most of the remaining plants are in Nevada, with a small amount of production in other states.

Environmental Effects of Geothermal Plants

The small carbon footprint, very low gaseous emissions, and the ability to provide continuous (baseline) power make geothermal energy one of the more attractive renewable options. Although geothermal plants are one of the best options for clean renewable power,

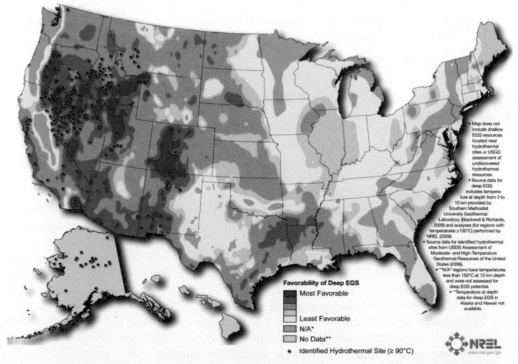

FIGURE 1-18 Geothermal Resources of the United States (*Source:* National Renewable Energy Laboratory.) This map was produced by the National Renewable Energy Laboratory for the US Department of Energy. October 13, 2009 Author: Billy J. Roberts.

they do have some negative environmental impacts. Most geothermal plants inject water underground to create steam. Closed-loop systems usually inject water condensed from steam back into the ground, so they do not require significant input of fresh water. Some closed-loop systems use a two-stage process where heat from the ground is transferred to a working fluid that is used to drive the turbines. Open-loop systems require a water supply because the water is injected continuously into the ground. The water requirement can be met with treated water from waste treatment plants. These systems can generate solid waste and release gases contained in the steam into the atmosphere. The geysers in California include a small amount of hydrogen sulfide, which is released with the steam and accounts for an occasional rotten egg smell in the area, but it is a relatively small amount and no residents live nearby to be affected by it. Some carbon dioxide is released, but only about 5% of the amount that would be generated from a comparable coal-fired plant. In addition, some sludge is produced that contains hazardous materials, from the process of condensing steam. This material needs to be disposed of safely.

Another impact is the occurrence of microquakes, which are small earthquakes. Generally these are too small to be felt, but occasionally one will reach magnitude 4.0 on the Richter scale. Earthquakes tend to be found along the earth's plate boundaries, which is the same region that geothermal energy plants are most efficient. The microquakes are related to the amount of injection water, so plant operators monitor seismic activity carefully to minimize these quakes.

One proposal—with a positive benefit to the environment—is the possibility of sequestering carbon dioxide (CO_2) underground. If CO_2 can be absorbed and used as a heat-transfer fluid, there could be a means to sequester carbon in stable formations, resulting in negative carbon dioxide emissions, which would help reduce greenhouse gases.

SECTION 1-5 CHECKUP

1. What are common attributes of a high-grade geothermal site?

2. What is the most significant advantage of geothermal energy over solar or wind energy for generating electricity?

3. Why is the western United States a better region for geothermal energy than the eastern part of the country?

1-6   Hydroelectric Resources

Hydroelectric resources are used almost exclusively for electricity production. Hydroelectricity is primarily made by converting the power from moving or falling water into electrical power. Moving water is derived from storing water behind dams and using the energy it produces as it is released and falls. Of course, moving water can also be found in rivers, streams, ocean tides, and waves. In addition to moving water, some future possibilities for hydropower exploit temperature differences in the ocean. This section discusses hydroresources, including the environmental effects.

Hydroelectricity is generated much like other sources of electrical energy in that a turbine spins a generator, which produces electricity for distribution. The major difference is that the turbine is spun by water power rather than steam, as in coal-fired, nuclear, and geothermal plants, so water is considered the **prime mover** (initial agent or source of energy). Most hydroelectric power is generated at sites with dams that can back up a reservoir of water to be released as needed. Some hydroelectricity is generated in rivers and streams. Dams have the advantage of a large hydrostatic head compared to most rivers and streams. The amount of power that can be generated is directly proportional to the head and the flow rate.

The global hydroelectric resource is a complicated issue if oceans are included. Excluding oceans, the best sites for hydropower are generally mountainous regions with above

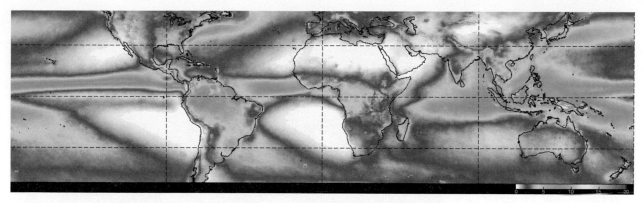

FIGURE 1-19 World Rainfall Patterns Averages from 1998 to 2011. The highest rainfall is shown in green and yellow. (*Source:* Courtesy of NASA.)

average rainfall. Figure 1-19 shows the world rainfall averages. In general, the tropical regions have the most rainfall. The World Energy Council, which considers all natural water flows for potential development, has estimated the world hydropower potential at more than 16,400 terawatt-hours (TWh), of which approximately 19% has been developed. The resource is unevenly distributed, with the top ten countries accounting for two-thirds of the world's current generation capability.

Power is generated by tides and waves; however, this resource has generally been expensive to develop, so development lags far behind dams and rivers for generating power. The realistic resource that can be recovered from tides and waves is smaller than the freshwater resource in dams, run-of-the-river (ROR) systems, and streams. Currently, the largest system for generating energy from tides is a barrage tidal power system in South Korea; it generates 260 MW. This system uses a barrier called a **barrage,** which is essentially a relatively low dam that traps the inflow water of the tides and releases it through turbines to generate power. Regions where large systems like this are economically viable are limited.

Other types of ocean projects can take advantage of the energy in waves. Coastal areas with waves of sufficient height to generate energy economically are along the coasts of Scotland, northern Canada, southern Africa, and Australia; the northwestern coast of the United States; and a few other isolated locations like Hawaii.

Today, hydroelectric power provides approximately 17% of the world's electricity. The top five producing countries (in order from largest to smallest) are China, Canada, Brazil, the United States, and Russia. The world's largest hydroelectric power plant is at Three Gorges on China's Yangtze River. This dam is 2.3 km (1.4 mi) wide and 185 m (607 ft) high and produces 22,500 MW of power, which is enough to supply 11% to 14% of China's need (see Figure 1-20). The largest hydroelectric facility in the United States is the Grand Coulee Dam on the Columbia River in northern Washington. The Grand Coulee Dam is 1.59 km (1 mile) wide and 167 m (550 ft) high. It produces 6,800 megawatts of power that is sent to eleven western states and Canada. The power plant has over three times the capacity of Hoover Dam on the Colorado River.

The capital investment for dams is very high because of the cost of the dam and acquiring land for the reservoir. After the initial investment, the cost of generating electricity is low because the major cost is for maintenance and salaries for operations. The water is renewed continually by rainfall and snow melt. In addition to generating electricity, many hydroelectric dams also control flooding on rivers and provide recreational opportunities.

Figure 1-21 shows the locations of existing hydroelectric power plants in the United States. As you can see, the principal areas for generating hydroelectric power are mountainous regions in the west and in the Appalachians. The largest facilities are in the northwestern United States, along the Colorado River, and in the Tennessee Valley.

Ocean Thermal Energy

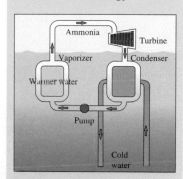

A tremendous amount of low-quality heat is stored in the oceans. Ocean thermal energy conversion (OTEC) systems have been proposed for years, but interest in them has decreased recently. A dual-thermal cycle runs between two different ocean temperatures. Ammonia is used in a closed-loop vaporization-condensation system to drive a turbine. Warm surface temperatures vaporize the ammonia; cooler water from deeper depths condenses it. The vapor drives a turbine. In addition to electrical production, OTEC can support related applications, including desalinization, high-density culturing of marine food species, and mineral extraction.

FIGURE 1-20   Three Gorges Dam in China   (*Source:* David Buchla.)

Environmental Effects of Hydroelectric Plants

Hydroelectric systems have both positive and negative impacts on the environment. The most important impacts are those from hydroelectric dams and their effect on the ecosystem. Dams and the reservoirs they produce generally require large land areas and often involve dislocations of people, and destruction of forest wildlife habitat and natural rivers. In the case of the Three Gorges Dam in China, huge areas of farmland were taken out of production, so other valuable uses for the land are excluded. In addition to the land lost to

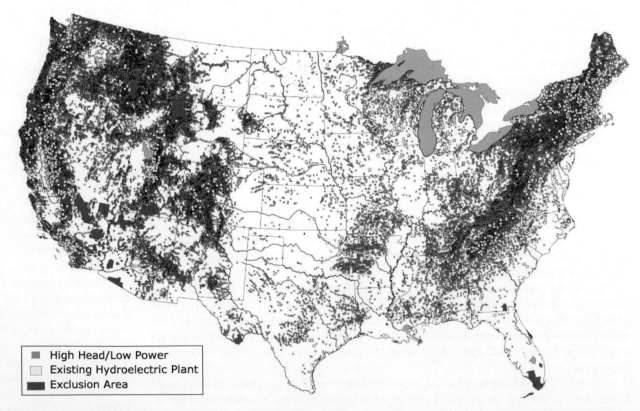

■ High Head/Low Power
☐ Existing Hydroelectric Plant
■ Exclusion Area

FIGURE 1-21   Hydroelectric Plants in the United States   (*Source:* http://www.nationalatlas.gov/articles/people/ a_energy.html.)

the reservoir, the loss of silting downstream because of flood control removes essential soil nutrients for agricultural needs, reducing the land's natural fertility.

A problem occurs upstream as moving river water slows when it reaches a reservoir: It deposits silt in the lake rather than in the sandbars along the river or on farmland. Silting, which once made the Nile valleys into rich farmland, has ceased since the construction of the Aswan Dam. Instead of annual floods, the silt builds in the reservoirs. If silt builds sufficiently, the reservoir volume is reduced; eventually, the dam could become worthless. Rivers such as the Yangtze in China carry huge amounts of silt. The precise rate of silting behind the Three Gorges Dam is uncertain, but it is known to be rapid. The Three Gorges project has special silt-scouring gates built in, but these gates have never been tested on a dam so large or on this section of the Yangtze, so their effectiveness has yet to be determined. The reservoir has also experienced many landslides due to weakened riverbanks, which have contributed to the problem of silting.

Water quality and flow are also affected by dams and the resulting reservoirs. Reservoirs tend to have low levels of dissolved oxygen compared to rivers, which is not good for fish. Slow-moving water in reservoirs also has stratified temperature layers, with the lowest levels of dissolved oxygen in the cold bottom layers. Organic material (plants and trees) tend to rot in the bottom, which leads to carbon dioxide and methane production; both are greenhouse gases. The result is stagnant water that has lower quality and offers an ideal breeding ground for mosquitoes and certain snails, which carry disease.

The loss of flow in the lower part of the river reduces the sandbars and fish habitat in that region. Some dam managers have made high-flow releases to simulate these natural conditions and help restore the sandbars. Fish in particular are affected when a river is blocked by a dam. Wild fish such as salmon are blocked from moving upriver to their spawning grounds, which can significantly reduce fish populations. Many fish are killed or injured trying to pass turbine blades. Strides have been made in designing turbines that are fish-friendly by studying the physical effects fish experience (using special mechanical fish that measure the forces) during passage through hydroelectric turbines, high-discharge outfalls, pumps, and other dam passage routes. Some of these effects can be alleviated using screens to prevent young fish from entering turbines in the first place and by using so-called fish ladders to allow returning fish to migrate past dams. Fish hatcheries can help mitigate the effects of diminishing numbers with species such as salmon. In addition to the effect that dams have on fish populations, many other factors are important considerations, such as overfishing, food sources for fish, climate and habitat changes, and pesticides. On the positive side, some additional habitat is created for certain freshwater lake fish such as bass and catfish. Dams can also provide a benefit for flood control, provide irrigation water, and create recreational opportunities.

Sometimes the impact of river and stream diversion for hydropower is less than that of a huge dam, such as in the case of the Tazimina project in Alaska. In this project, only a small portion of a river is diverted to generate power, and the water is returned to the river (see Figure 1-22). In this case, no dam was required. In other cases, a large fraction of the river is diverted for a project; this affects river flows in the area and may have an adverse impact on fish, including fish migration. Some ROR installations create a long path between the diversion and the power plant, which can create additional problems. Infrastructure, including substations, power lines, roads, diversion tunnels, and canals, can also have an impact on the environment.

The hydropower with the least environmental impact is probably tidal power. The largest projects to date involve barrage dams, which can generate a large amount of power from the tides. Another option for capturing tidal power is putting underwater windmills in swiftly moving tidal currents. The largest of these systems is the SeaGen project. After several years of operation in Northern Ireland, a final environmental impact report has been issued. The report found no significant detrimental impact on harbor seals, underwater biological communities, or seabirds. Some decrease in porpoise population was observed but could not be attributed to the presence of the turbines. Seals and porpoises transit regularly past the operating turbines, and it appears they can easily avoid the relatively slow-moving propellers.

FIGURE 1-22   Tazimina Project in Alaska   (*Source:* US Department of Energy.)

Wave power was mentioned previously and has had limited development to date, but several installations are producing commercial power. The potential for wave energy is of great interest, particularly in remote coastal areas such as the islands of Scotland. The biggest obstacle to new development is capital and maintenance costs, which increase in deeper water and thus limit applications primarily to selected coastal areas. The environmental impact varies with projects because of substantial different construction and area requirements. Each project and type of system must be evaluated for cost, effect on marine life, visual and noise effects, and effects on shipping and recreational boating. Chapter 11 covers some of these projects and other forms of hydropower in more detail.

SECTION 1-6 CHECKUP

1. What is a barrage?

2. What are three basic methods of using moving water to generate electricity?

3. What are four negative environmental impacts from dams?

4. What steps can help maintain migratory fish populations?

5. What type of system is SeaGen?

1-7   Biomass and Biofuel Resources

Biomass resources include a wide variety of organic materials that can be used as fuels. In many cases, the fuels are used just for heat; in other cases, the heat is used to turn a turbine and generate electricity. Biomass is the oldest source of fuel: Wood has been used to cook food and to heat spaces for centuries. Today, even the fumes from landfills create a biofuel: The fumes are primarily composed of methane, a constituent of natural gas.

Biomass is organic material that is commonly used for heating fuel, power generation, or making liquid fuels useful in transportation. Many industries require process heat for making certain products (for example, bricks). In many cases, the fuel for providing needed heat can be plant material or animal waste. In some cases, heat can be used for more than one purpose, such as generating electricity and operating a dryer. This process of using heat for electrical generation and the lower temperature waste heat for another purpose is known as **co-generation.** The process of burning waste products to produce electrical energy and heat is a side benefit to waste management.

Biomass fuels comprise the largest segment of the renewable energy sector (refer again to Figure 1-2). They include various waste products such as wood chips and waste from the paper industry, agricultural waste such as almond and rice hulls, organic and food waste,

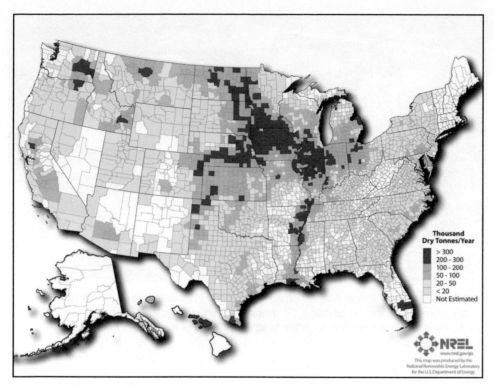

FIGURE 1-23 Crop Residue Resources for the United States   (*Source:* National Renewable Energy Laboratory.) Author: Billy Roberts–September 23, 2009.

as well as various energy crops specifically planted for their value as fuels. Many of these fuels are combined with traditional fossil fuels in plants such as electrical power stations to reduce the amount of fossil fuels needed by the power station.

The biomass resource and its location are different for each product. NREL has maps for the United States for residues from crops, forests, and mills, as well as wood waste and various methane resources. Figure 1-23 shows the overall crop residue resource—a composite of several different biomass fuels—for the United States.

One use for biofuels is to fire the biofuel itself with coal in electrical power stations. This enables coal-fired power electrical plants to alleviate some of their CO_2 emissions because biomass is considered almost neutral with respect to CO_2 (it absorbs as much during the growing process as it releases during burning, less some for production). Co-fired electrical plants burn a variety of fuels, including the cellular walls of plants, which can reduce the amount of fuel burned per energy unit output as well as reducing emissions.

Liquid biofuels are derived from starchy biomass such as corn and grain crops that are essentially plant sugar. Cellular products like corn stalks are difficult to break down, so they have not been used commercially for producing ethanol. **Ethanol** is the primary biofuel used as a gasoline additive and is a type of alcohol; methanol and biodiesel are other fuels. **Biodiesel** is another biofuel, but it is made from vegetable oil, animal fat, or cooking grease that is combined with alcohol and other ingredients to form the final product. Ethanol and methanol used as a gasoline additive help reduce certain volatile vehicle emissions that contribute to ozone, a pollutant at ground level, and replace a certain fraction of the fuel.

The raw material used in biomass energy conversion is called **feedstock.** To make ethanol from feedstock, the feedstock is basically fermented and distilled. The feedstock is first milled to a fine powder and processed with water and enzymes to break it down. The resulting starches and sugars are fermented for fifty hours with the addition of yeast. After extracting the alcohol by distillation, water is removed and the solid waste by-products are dried to produce cattle feed or fertilizer. Figure 1-24 shows a typical ethanol plant for producing ethanol from corn.

FIGURE 1-24   An Ethanol Plant. The principal steps in ethanol production are (1) offloading the corn, (2) grinding, (3) treatment with sulfuric acid to decompose the feedstock into sugars, (4) fermentation, and (5) storage of fuel.   (*Source:* Courtesy of Poet Bio Refining.)

A thermochemical process has been developed in recent years for producing methanol and ethanol from other sources; however, it has not been widely implemented. This process converts nonfood biomass (mostly cellulose) into gas using heat, pressure, and steam to form a synthetic gas. The gas is converted to biofuels by passing it over a catalyst, and the biofuels are separated into a variety of low-carbon biofuels. This process can use various products like paper pulp, rice straw, and other cellulose organic materials, which are widely available. Cellulose is fibrous and more difficult to break down than feedstocks (grains), but this new technology could help increase ethanol production if it goes into effect. In addition to materials containing cellulose, other waste (such as animal manure and sewage sludge) can be used to produce ethanol. Research is also being done on finding enzymes that can speed up ethanol production. Enzymes, which are vital for life, are specialized proteins that are catalysts for certain chemical reactions.

Environmental Effects of Energy Crops

Ideally, biofuel should be sustainable, renewable, and in excess supply. The production of corn and other feedstocks such as sorghum, wheat, barley, and sugar cane have competing uses. In a recent year, the US production of corn was allocated as follows: livestock feed (36%); ethanol (31%), exports (14%), human food, seed, and industrial use (9%); and held over as surplus (11%). Certain biofuels use cellular plant materials that are not competitive with food sources and other uses, but they have other negative impacts, such as pesticide use in production.

Traditional high-yield methods adopted for energy crops can have several negative impacts. More land under cultivation means reduced biodiversity because of habitat destruction for wildlife. Large-scale single-crop farming often requires increased pesticide use and can deplete soils of certain nutrients, requiring much more fertilizer to be applied. Although plants grown for ethanol production are not always food crops, they may use valuable farmland that could be put to other uses. Diverting grains such as corn for ethanol production has an impact on supply and price for other applications, including food.

The process of converting the raw product to ethanol involves a lot of heat. Some plants obtain this heat from natural gas, cutting into the overall savings of lessening dependence on fossil fuels. In addition, farmers require a certain amount of fuel for tractors and harvesters to bring the raw product to market.

Air pollution is another issue with biomass. Air pollution from the combustion of biomass depends on the specific source of the biomass. In general, most plants that burn biomass have less pollution than comparable fossil fuel–burning plants. Plants that burn municipal waste often have nonorganic (nonbiomass) materials mixed in with other wastes and thus may have plastics, toxic metals, and other materials that pollute the atmosphere when incinerated. With strictly plant biomass combustion, the carbon dioxide emitted from burning is equivalent to the carbon dioxide that was absorbed by the plant when it was growing, so the net result is that biofuels are carbon neutral (excluding their production) in the long term. Carbon monoxide emission from automobiles running on ethanol is reduced compared to that of automobiles burning gasoline.

Ethanol plants require large quantities of water for growing the crop and processing it, and for the required cooling systems. The amount varies widely, depending on irrigation needs, but as ethanol production expands, the land used for production tends to need more irrigation water. In cases where a large amount of irrigation water is required, the amount of water may exceed the ethanol produced by a factor of 1,000 or more. This consideration is serious for areas hit by drought.

Biomass energy crops can have beneficial effects, depending on the specific crops, how they are managed, and how much impact they have on land taken out of production for food and on water requirements. Certain energy crops can lead to productive and sustainable farming that includes regular crop rotation. Some nitrogen-fixing crops can be rotated into production to provide a natural fertilizer and crop diversity that can reduce pesticide requirements. With certain crops such as switchgrass, the need for disruptive tilling is reduced, which helps to preserve soil. Trees are also an excellent energy crop that stabilizes soil for erosion control with their root systems and leaf litter. Some fast-growing species, such as poplars, can be incinerated for heat content. Wood biomass has a positive side in helping foresters manage forests because diseased and damaged trees can be removed for biofuel, leaving behind the healthiest specimens to propagate. This can also have the positive impact of making a forest less prone to devastating forest fires.

SECTION 1-7 CHECKUP

1. What is biodiesel made from?
2. Where is most ethanol produced in the United States?
3. What are important environmental issues about using corn for producing ethanol?
4. What are the negative impacts of using municipal waste for generating electricity?
5. What are the positive impacts of using wood biofuels?

CHAPTER SUMMARY

- Fossil fuels account for approximately 80% of the energy used in the United States.
- Biomass and hydroelectric resources account for most of the renewable energy used in the United States.
- A by-product of combustion is carbon dioxide, which is a greenhouse gas.
- Fission reactors account for all nuclear power in the world; they use enriched uranium or other fissionable material as a fuel and are designed to control a chain reaction.
- Research continues to develop a viable fusion reactor. Fusion reactors will have major advantages over fission reactors. There are two approaches to fusion reactors: inertial fusion and magnetic confinement.
- The solar spectrum is considerably changed due to absorption and scattering as it passes through the earth's atmosphere. Water vapor and carbon dioxide absorb energy at specific places in the spectrum, especially in the infrared region.
- The solar constant is the power per square meter from the sun as measured at the outer atmosphere. The mean value is 1,368 W/m^2.
- The solar resource on earth depends on location. The irradiance is highest in locations that are cloud-free and closer to the equator.
- Three major wind circulation cells—the tropical cell, the mid-latitude cell, and the polar cell—in each hemisphere of the world generally determine prevailing wind direction.

- Wind is caused by pressure differences in the atmosphere and is also affected by specific conditions such as landmasses, temperature differences, cloud cover, and terrain.

- Wind resources can be shown on maps and calculated in detail for a specific site using computer programs that are checked against actual measurements.

- In the United States, major wind resources are concentrated on the coasts and the Midwest.

- Geothermal energy is available anywhere in the world; however, the best sites are near tectonic plate boundaries.

- In the United States, the western states have the best geothermal sites.

- Moving water has kinetic energy. The potential energy of reservoirs is changed to kinetic energy as the water falls out of reservoirs. Rivers and tides also have kinetic energy that can be exploited for power generation.

- Hydroelectric power is a renewable resource that provides clean energy but has negative impacts on fish and wildlife habitat, and it requires large areas of land for reservoirs. Many dams have positive impacts such as providing flood protection, irrigation water, and recreational opportunities.

- Biomass is currently the largest renewable resource in terms of energy obtained. As an energy source, biomass refers to any organic material used for a fuel.

- Ethanol is a renewable resource made primarily from corn; it has the ability to replace a portion of the gasoline required.

- Biofuel production has negative environmental impacts on land use, food and feedstock supplies, and water requirements.

KEY TERMS

biomass Organic material that is commonly used for fuels for heating, power generation, or making liquid fuels useful in transportation.

bitumen A black tarlike hydrocarbon classified as pitch; it can occur naturally or after refining petroleum.

breeder reactor A nuclear reactor designed to produce plutonium. It could extend uranium supplies considerably because it can convert the otherwise unusable ^{238}U into a fissionable fuel.

catagenesis The cracking process that results in the conversion of kerogens into hydrocarbons, including natural gas and oil.

diagenesis The process of converting constituents to a different product through application of heat and pressure.

diffuse horizontal irradiance (DHI) The portion of global horizontal irradiance that comes in indirectly (scattered radiation) from the sun.

direct normal irradiance (DNI) The portion of global horizontal irradiance that comes in a straight line from the sun.

ethanol The primary biofuel used as a gasoline additive; it is a type of alcohol.

fossil fuels Fuels that formed from decaying plant and animal matter and were primarily formed over millions of years. Fossil fuels include coal, oil (petroleum), and natural gas.

global horizontal irradiance (GHI) The total amount of short-wave radiation received on a horizontal surface.

hydrocarbon A molecule containing only hydrogen and carbon.

inverse square law A physics law that states that the flux from an isotropic point source is reduced by the square of the distance from the source to the receiver.

kerogen A mixture of organic chemicals that are part of the organic matter in sedimentary rocks.

pressurized water reactor (PWR) A nuclear reactor that includes fuel rods, control rods, and a moderator in a vessel that is filled with high-pressure water to prevent it from boiling. The water in the vessel serves as both a moderator and a means to move hot water to a heat exchanger and eventually a steam-driven turbine.

solar constant The power emitted by the sun that falls on 1 m^2. It is generally cited as 1,368 W/m^2.

Tokamak A fusion reactor used by researchers to investigate properties of plasmas; the goal is to create a fusion energy reactor that can be used for electrical power generation.

CHAPTER TRUE/FALSE QUIZ

Determine whether each statement is true or false. Answers are at the end of the chapter.

1. The most common use for coal is home heating.

2. A hydrocarbon molecule contains only hydrogen and carbon atoms.

3. When natural gas burns, carbon dioxide is released.

4. All reactors used for generating power are breeder reactors.

5. Delayed neutrons are not important for controlling reactors.

6. A Tokamak is a type of fusion reactor that attempts to confine the plasma.

7. The ozone in the upper atmosphere is particularly important because it absorbs infrared radiation.

8. The portion of global horizontal irradiance that comes in a straight line the sun is called direct normal irradiance.

9. The solar constant is not affected by sunspots.

10. The prevailing wind direction in mid-latitudes is from east to west.

11. The wind resource in the United States is mainly in the eastern states.

12. *Transform plates* is the name given when two tectonic plates slide laterally past each other.

13. Low-quality heat from geothermal sources is not useful.

14. The largest installed electrical generating capacity for geothermal energy is in Iceland.

15. Iceland has a large geothermal resource because of its location at a boundary between tectonic plates.

16. Geothermal energy has no adverse environmental impacts.

17. Hydroelectric power is generated primarily by turbines turning a generator.

18. By definition, biomass consists of only nonorganic materials.

19. The primary crop used to produce ethanol is corn.

20. An advantage of burning biomass is that it does not produce air pollution.

CHAPTER MULTIPLE-CHOICE QUIZ

Complete each statement by selecting the one correct answer. Answers are at the end of the chapter.

1. All fossil fuels are composed of
   a. methane
   b. carbon dioxide
   c. sulfur
   d. hydrocarbons

2. Burning methane produces
   a. carbon dioxide and sulfur dioxide
   b. sulfur dioxide and water vapor
   c. carbon dioxide and water vapor
   d. none of these

3. A fossil fuel that releases carbon dioxide into the atmosphere when it is burned is
   a. coal
   b. petroleum
   c. natural gas
   d. all of these

4. A by-product of the fission process is
   a. hydrogen
   b. helium
   c. various radioactive fragments
   d. carbon dioxide

5. A serious environmental issue with nuclear reactors is
   a. radioactive waste
   b. carbon dioxide emission
   c. air pollution
   d. all of these

6. A Tokamak is a type of
   a. fusion reactor
   b. breeder reactor
   c. pressurized water reactor
   d. laser

7. Ozone in the stratosphere
   a. is considered to be a greenhouse gas
   b. absorbs greenhouse gases
   c. absorbs ultraviolet radiation
   d. contributes to global warming

8. Global horizontal irradiance is composed of
   a. direct normal irradiance and diffuse horizontal irradiance
   b. all irradiance received on a vertical surface
   c. ultraviolet irradiance and infrared irradiance
   d. visible light and infrared irradiance

9. The solar constant is known to vary slightly in response to
   a. wind conditions on earth
   b. sunspot activity
   c. cloud cover
   d. changes in stratospheric ozone

10. A common unit for expressing the solar constant is the
    a. W/m^2
    b. kWh/day
    c. W
    d. nm

11. Of the following forms of radiation, the one with the shortest wavelength is
    a. blue light
    b. red light
    c. infrared radiation
    d. ultraviolet radiation

12. The sun's output approximates a black body. This means that the sun
    a. has many absorption lines in its spectrum
    b. is nearly a perfect emitter of radiation at some temperature
    c. is hotter than most stars
    d. is approximately a point source of radiation

13. Two gases that absorb a large fraction of the infrared radiation from the sun are
    a. sulfur dioxide and water vapor
    b. ozone and carbon dioxide
    c. carbon dioxide and water vapor
    d. ozone and water vapor

14. For the past 200 years, carbon dioxide levels in the atmosphere have
    a. fallen
    b. stayed about the same
    c. risen

15. The winds in the mid-latitude cell are called
    a. easterlies
    b. westerlies
    c. northeast trade winds
    d. southeast trade winds

16. The strongest winds are, in general,
    a. at the poles
    b. at the equator
    c. in the desert
    d. in the jet stream

17. The doldrums exist
    a. near the poles
    b. at the equator
    c. around the Pacific rim
    d. in the jet stream

18. In the United States, high-grade geothermal resources are concentrated in the
  a. Pacific Northwest
  b. western states
  c. upper Midwest
  d. Southeast

19. The most important environmental effect of hydroelectric energy is
  a. the visual eyesore of dams

  b. the effect on fish and wildlife habitat
  c. noise
  d. air pollution

20. Ethanol is primarily used as a
  a. gasoline additive
  b. cattle feed
  c. fertilizer
  d. pesticide

CHAPTER QUESTIONS AND PROBLEMS

1. What are two products of burning fossil fuels?

2. What are diagenesis and catagenesis?

3. What is true about the mass of reactants and the mass of products in all chemical reactions such as combustion?

4. Why is water vapor considered to be a greenhouse gas?

5. What substance has the highest binding energy per nucleon?

6. What is deuterium and where is it found?

7. Why is a potential fusion reactor inherently safer than a fission
reactor?

8. What is ITER and where will it be constructed?

9. What is the inertial fusion project?

10. How is the solar spectrum changed as it passes through the
atmosphere?

11. Assume a 6-foot by 3-foot solar panel is oriented so that it has
a maximum solar insolation of 1,000 W/m^2. What is the peak
power delivered to the panel (3.048 feet = 1 meter)?

12. What is the inverse square law in physics?

13. Show that the solar constant of 1,368 W/m^2 is equal to
32.8 kWh/m^2/day.

14. Compare the positive and negative environmental impacts of
solar and wind energy.

15. What are geothermal heat pumps?

16. What is the purpose of a barrage dam?

17. What are two ways to obtain energy from tides?

18. Summarize the important benefits and negative impacts of
large dams.

19. What are the positive and negative environmental impacts of
using ethanol as a gasoline supplement?

20. In addition to ethanol, what are other uses for biofuels?

FOR DISCUSSION

What renewable resources are the best fit for the area you live in?

ANSWERS TO CHECKUPS

Section 1-1 Checkup

1. Fossil fuels formed from decaying plant and animal matter that
was primarily formed over millions of years.

2. Methane is a primary constituent of natural gas; its chemical
formula is CH_4.

3. The categories for coal are based on energy content and are
anthracite, bituminous, sub-bituminous, and lignite.

4. Oil is used in the manufacture of lubricants, waxes, solvents,
asphalt, hydraulic fluid, and vinyl.

5. Greenhouse gases are gases that contribute to global warming
by absorbing short wavelength infrared energy and reradiating
it at longer wavelengths.

6. Burning fossil fuels puts carbon dioxide, a greenhouse gas,
in the air. Depending on the specific fuel, pollutants include

sulfur, nitrous oxides, benzene, and others, which contribute
to acid rain. Coal also produces a large amount of fly ash that
creates a disposal problem. Other concerns include obtaining
and transporting fossil fuels and safety issues.

7. Safety hazards exist for workers, particularly in the coal industry, and for the public in the form of air pollution. Transporting fuels can create hazards in the form of spills, such as the
infamous oil spill by the *Exxon Valdez*.

Section 1-2 Checkup

1. A chain reaction is a self-sustaining reaction used in fission
reactors to continue the process.

2. Normal uranium will not have a chain reaction. Enrichment
separates useable ^{235}U, which can be used as a fuel from regular uranium.

3. Fuel rods are inserted along with control rods (good absorbers of neutrons). The control rods are moved out to increase the reaction or moved in to slow it. Hot pressurized water is converted to steam in a heat exchanger and drives a steam turbine and generator to make electricity.

4. A breeder reactor is a type of reactor designed to convert ^{238}U into ^{239}Pu, which is then used as a fuel.

5. Nuclear power produces almost no greenhouse gas emission or air pollution, and it is a more concentrated form of energy, so significantly less fuel is required than is the case with coal-fired plants. Its disadvantages include radioactive waste disposal and hot water releases to rivers and waterways. A reactor accident can have serious radioactive consequences and thus poses a safety issue.

6. One approach is the inertial fusion project, in which pellets of D-T fuel are dropped into a chamber and hit with high-power lasers, triggering the fusion reaction. The second approach is to keep a super-hot plasma of D-T fuel confined long enough to cause a reasonable fraction of the fuel to fuse.

7. The advantages of fusion reactors over fission reactors are that they produce significantly less radioactive waste and that no safety issues can result from losing control of the reactor.

Section 1-3 Checkup

1. The solar constant is the energy received on an area that is outside the atmosphere and oriented perpendicular to the sun's rays at the mean distance of the earth from the sun. Common units are W/m^2 and $kWh/m^2/day$.

2. The term *insolation* means incident solar radiation.

3. Carbon dioxide absorbs solar radiation and reemits it at longer wavelengths that do not escape the atmosphere as easily. As a result, more of the sun's energy is absorbed by the atmosphere.

4. Global horizontal irradiance is the total amount of shortwave radiation received on a horizontal surface.

5. Direct normal irradiance is the radiation that comes in a straight-line path from the sun; diffuse horizontal irradiance is scattered radiation and comes from all directions. DHI is not useful for solar concentrators.

6. The Maunder minimum is a period between 1645 and 1715 in which sunspot activity almost ceased, which led to much colder winters. Evidence from satellite measurements indicates that the solar constant is lower during periods of low solar activity.

Section 1-4 Checkup

1. The mid-latitudes are affected primarily by westerlies.

2. The doldrums are very calm winds near the equator. Ships required wind power to move, so sailors wanted to avoid these areas.

3. The air rises to the upper atmosphere, where it cools and becomes denser; hence it sinks.

4. The major wind resource areas in the United States are in the Midwest.

5. (1) Land use; (2) height, which creates aircraft hazards; (3) noise; (4) bird strikes; and (5) transmission line requirements crossing many miles of land.

Section 1-5 Checkup

1. High-grade geothermal sites have high underground temperatures within easy reach of drilling (less than 3 km) and have a large amount of stored geothermal energy. High-grade sites tend to be concentrated along geological plate boundaries and hot spots.

2. Geothermal power is good baseline power because it is a constant reliable source.

3. The western United States is closer to a tectonic plate, which provides a path for subterranean heat to escape.

4. Open-loop systems require a water source so that water can be injected into the ground. Closed-loop systems condense the steam and recycle it in the ground.

Section 1-6 Checkup

1. A barrage is a barrier such as a low dam that traps the inflow water of the tides.

2. Three methods for generating power from water are (1) storing runoff behind a dam and releasing it through a turbine, (2) using the moving water of a river (run of the river) to turn a turbine, (3) using tides by either trapping the inflow behind a barrier or turning a turbine in tidal current.

3. Negative environmental impacts include (1) loss of land for other purposes besides the reservoir; (2) silt loss downstream, which affects soil fertility; (3) water quality; and (4) fish blocking, kills, and loss of fish habitat.

4. Fish hatcheries, fish ladders, and screens around turbines.

5. SeaGen is a tidal current generator.

Section 1-7 Checkup

1. Biodiesel is made from vegetable oil, animal fat, or cooking grease that is combined with alcohol.

2. The upper Midwest, especially the Dakotas, Iowa, Illinois, and Indiana.

3. Some important issues for producing ethanol from corn include (1) diversion of corn from producing a feed and food crop; (2) reduced biodiversity because of single-crop farming; (3) more pesticide use; (4) increased water use because of the need for irrigation, production, and cooling; and (5) removal of wildlife habitats. A benefit is reducing dependence on oil and fossil fuels.

4. Municipal waste often includes nonorganic materials such as plastics, toxic metals, and similar materials that pollute the atmosphere.

5. Wood biofuels can help foresters manage forests by removing diseased and damaged trees, and help remove excess fuels for forest fires.

ANSWERS TO TRUE/FALSE QUIZ

1. F 2. T 3. T 4. F 5. F 6. T 7. F 8. T 9. F   17. T 18. F 19. T 20. F
10. F 11. F 12. T 13. F 14. F 15. T 16. F

ANSWERS TO MULTIPLE-CHOICE QUESTIONS

1. d 2. c 3. d 4. c 5. a 6. a 7. c 8. a 9. b   17. b 18. b 19. b 20. a
10. a 11. d 12. b 13. c 14. c 15. b 16. d

Electrical Fundamentals

CHAPTER OUTLINE

CHAPTER OBJECTIVES

- Define energy, electrical charge, and voltage and their corresponding units.
- Explain how energy, charge, and voltage are related.
- Define current and its unit.
- Define resistance, and describe how voltage, current, and resistance are related.
- Define power and energy.
- Discuss how power, current, and voltage are related.
- Identify series and parallel circuits, and determine their characteristics.
- Discuss the basic idea of a magnetic field.
- Explain how a magnetic field is produced.
- Discuss the basic operation of relays, generators, and motors.
- Discuss the differences among conductors, insulators, and semiconductors.
- Describe capacitors and inductors, and discuss their functions.
- Explain how to measure voltage, current, and resistance in electrical circuits.

KEY TERMS

Key terms are shown in bold and color. Definitions for key terms are provided at the end of the chapter and in the end-of-book glossary. Bold terms in black are defined in the end-of-book glossary only.

- energy
- joule (J)
- charge
- voltage
- volt
- sinusoidal wave
- current
- ampere
- load
- resistance
- ohm
- Ohm's law
- power (P)
- watt (W)
- kilowatt-hour (kWh)
- Watt's law
- series circuit
- parallel circuit
- conductors
- insulators
- semiconductors
- magnetic flux
- magnetic flux density (B)
- capacitance (C)
- inductance
- Lenz's law
- digital multimeter (DMM)

INTRODUCTION

In this chapter, you will learn some basic electrical concepts that are essential to working with any system that utilizes a renewable energy source. For example, a solar panel converts light energy from the sun directly to electrical energy, and wind turbines use generators to convert wind energy into electricity. A hydroelectric system converts the energy in moving water to electrical energy via a generator. These systems can be tied into the electrical grid (called grid-tie systems) to simplify distribution of energy to a large number of users. Simple systems can use the electricity produced by a renewable resource to run an appliance or electrical device such as a water pump directly. These systems are common in remote applications, where tying to the electrical grid is not practical.

Anyone working with renewable energy must have some basic knowledge of electrical circuits and components and their operation. This knowledge provides a foundation with which to acquire the skills necessary to install, operate, and maintain these systems.

2-1   Energy, Charge, and Voltage

These three quantities—energy, charge, and voltage—are closely related. It is difficult to visualize or measure energy directly because it is an abstract quantity and represents the ability to do work. Electrical charge can be positive or negative, and it can do work when it moves from a point of higher potential to one of lower potential. Voltage is a measure of the energy per unit of charge and can be measured easily with common instruments. Voltage is one of the electrical quantities that you will work with in most renewable energy systems.

Energy

Work is done whenever an object is moved by applying a force over some distance. To do work, you must supply energy. **Energy** is the ability or capacity for doing work; it comes in three major forms; potential, kinetic, and rest. Stored energy is called **potential energy** and is the result of work having been done to put the object in that position or in a configuration such as a compressed gas. For example, the water stored behind a dam has stored (potential) energy because of its position in a gravitational field. **Kinetic energy** is the ability to do work because of motion. The moving matter can be a gas, a liquid, or a solid. For example, wind is gas in motion. Falling water is a liquid in motion; a moving turbine is a solid in motion. Each of these processes is a form of kinetic energy because of the motion. **Rest energy** is the equivalent energy of matter because it has mass. Einstein, in his famous equation $E = mc^2$, showed that mass and energy are equivalent.

Unit of Energy

Because energy is the ability to do work, energy and work are measured in the same units. In all scientific work, the **International System of Units (SI)** is used. SI stands for Système International, from French. These units are the basic units of the metric system. Energy, force, and many other units are *derived* units in the SI standard, which means they can expressed as a combination of the seven fundamental units. The most common derived units are those using three fundamental units, which are the meter, kilogram, and second (mks). This forms the basis of the mks system of units, which are the most common derived units in the SI system. Another derived set of units is based on the centimeter, gram, and second (cgs). These smaller units are referred to as the cgs system.

The SI unit for energy is the **joule (J)**, which is defined to be the work done when 1 newton of mechanical force is applied over a distance of 1 meter. A newton is a small unit of force, equivalent to only 0.225 pound. The symbol W is used for energy, and we will use W_{PE} or W_{KE} to specify potential energy and kinetic energy, respectively, to be consistent with W. (You may see E for energy in some cases, such as Einstein's $E = mc^2$, or PE and KE for potential energy and kinetic energy, respectively). The equation for gravitational potential energy is

$$W_{PE} = mgh \hspace{4cm} \text{Equation 2-1}$$

where
$$W_{PE} = \text{potential energy in J}$$
$$m = \text{mass in kg}$$
$$h = \text{height in m}$$

The equation for kinetic energy is

$$W_{KE} = \frac{1}{2}mv^2 \qquad \text{Equation 2-2}$$

where

W_{KE} = kinetic energy in J

m = mass in kg

v = velocity in m/s

Electrical Charge

Charles Augustus Coulomb (1736–1806) was the first to measure the electrical forces of attraction and repulsion of static electricity. Coulomb formulated the basic law that bears his name and states that the force between two point charges is proportional to the product of the charges and inversely proportional to the square of the distance between them. His name was also given to the unit of **charge**, the **coulomb (C)**.

Coulomb's law works for like charges or unlike charges. If the signs (+ or −) of both charges are the same, the force is repulsive; if the signs are different, the force is attractive. Long after Coulomb's work with static electricity, J. J. Thomson, an English physicist, discovered the electron and found that it carried a negative charge. The **electron** is the basic atomic particle that accounts for the flow of charge in solid conductors. The charge on the electron is very, very tiny, so literally many trillions of electrons are involved in practical electrical circuits. The charge on an electron was first measured by Robert Millikan, an American physicist, and found to be only 1.60×10^{-19} C. The power of ten, 10^{-19}, means that the decimal point is moved back 19 decimal places.

Voltage

Voltage (V) is defined as energy (W) per unit charge (Q). The **volt** is the unit of voltage symbolized by V. For example, a battery may produce twelve volts, expressed as 12 V. The basic formula for voltage is

$$V = W/Q \qquad \text{Equation 2-3}$$

One volt is the potential difference between two points when one joule of energy is required to move one coulomb of charge from one point to another.

Sources of Voltage

Various sources supply voltage, such as a photovoltaic (solar) cell, a battery, a generator, and a fuel cell, as shown in Figure 2-1. Huge arrays of solar modules can provide significant power for supplying electricity to the grid.

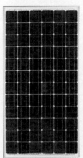

(a) Solar module   (b) Storage battery   (c) AC generator   (d) Solid-oxide fuel cell for a remote location

FIGURE 2-1 Sources of Voltage (*Source:* Part (a), National Renewable Energy Laboratory; part (c), Courtesy of Aerostar; part (d), National Renewable Energy Laboratory.)

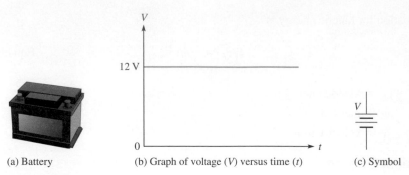

(a) Battery   (b) Graph of voltage (*V*) versus time (*t*)   (c) Symbol

FIGURE 2-2   DC Voltage Source

DC Voltage

Voltage is always measured between two points in an electrical circuit. Many types of voltage sources produce a steady voltage, called dc or direct current voltage, which has a fixed polarity and a constant value. One point always has positive polarity and the other always has negative polarity. For example, a battery produces a dc voltage between two terminals, with one terminal positive and the other negative, as shown in Figure 2-2(a). Figure 2-2(b) shows a graph of the ideal voltage over time. Figure 2-2(c) shows a battery symbol. In practice the battery voltage decreases some over time. Solar cells and fuel cells also produce dc voltage.

AC Voltage

Electric utility companies provide voltage that changes direction, or alternates back and forth between positive and negative polarities with a certain pattern. AC generators produce alternating voltage or alternating current (ac) voltage. In one cycle of the voltage pattern, the voltage goes from zero to a positive peak, back to zero, to a negative peak, and back to zero. One cycle consists of a positive and a negative **alternation** (half-cycle). The cyclic pattern of ac voltage is called a **sinusoidal wave** (or sine wave) because it has the same shape as the trigonometric sine function in mathematics. In North America, ac voltage alternates one complete cycle 60 times per second; in most other parts of the world, it is 50 times per second. The number of complete cycles that occur in one second is known as the **frequency** (*f*). Frequency is measured in units of **hertz (Hz),** named for Heinrich Hertz, a German physicist. Figure 2-3 illustrates the definition of frequency for the case of three cycles in one second, or 3 Hz.

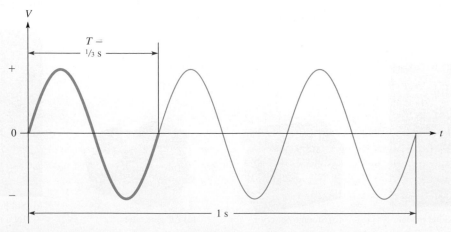

FIGURE 2-3   Example of an AC Sinusoidal Voltage. The frequency is 3 Hz, and the period (*T*) is ⅓ s.

The **period** (*T*) of a sine wave is the time required for 1 cycle. For example, if there are 3 cycles in one second, each cycle takes one-third second. This is illustrated in Figure 2-3, where one cycle is shown with a heavier curve. From this definition, you can see that there is a simple relationship between frequency and period, which is expressed by the following formulas:

$$f = 1/T \qquad\qquad \text{Equation 2-4}$$

$$T = 1/f \qquad\qquad \text{Equation 2-5}$$

EXAMPLE 2-1

(a) What is the voltage if the energy available for each coulomb of charge is 100 J and the total charge is 5 C?

Solution

$$V = W/Q = 100\,\text{J}/5\,\text{C} = \textbf{20 V}$$

(b) If the period of an ac voltage 0.01 s, determine the frequency.

Solution

$$f = 1/T = 1/0.01\,\text{s} = \textbf{100 Hz}$$

(c) If the frequency of an ac voltage is 60 Hz, determine the period.

Solution

$$T = 1/f = 1/60\,\text{Hz} = \textbf{0.0167 s} = \textbf{16.7 ms}$$

SECTION 2-1 CHECKUP

1. What is energy, and what is its unit?
2. What is the smallest particle of negative electrical charge?
3. What is the unit of electrical charge?
4. What is voltage, and what is its unit?
5. Name two types of voltage.
6. Define *frequency* and *period*.

2-2 Electrical Current

The original idea of current was based on Benjamin Franklin's belief that electricity was an unseen substance that moved from positive to negative. Conventional current is defined based on this original assumption of positive to negative. The original definition is widely used, although there is another definition of current called electron flow. Electrons move from the negative to the positive point through the wires and components in a circuit, just opposite to the defined direction of conventional current. Some people prefer to think in terms of conventional current, while others prefer the electron flow definition.

Definition of Current

Current is the flow of electrical charge past a specified point in a circuit. The word *current* implies a flow, so it is redundant to say "current flow." Current is symbolized

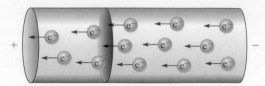

FIGURE 2-4 Current. Current is defined as the charge per time passing a point. If one coulomb of electrons passes through the cross-sectional area in 1 second, the current is 1 ampere.

by I and its unit is the **ampere (A)**. One ampere of current is defined as a rate of charge flow equal to one coulomb per second, as illustrated in Figure 2-4. This is equivalent to 6.25×10^{18} electrons passing a point in one second, an enormous number of electrons. The power of ten, 10^{18}, means that the decimal point in the number is moved 18 places to the right.

The basic formula for current is

$$I = Q/t$$ Equation 2-6

where

I = current, in amperes

Q = charge, in coulombs

t = time, in seconds

This equation is rarely applied to actual circuit work because the quantity of charge is generally not as important as the current itself. For practical reasons, the *rate* at which a charge moves past a point is defined, rather than the quantity of charge.

Direct Current

Current that is uniform in one direction is called **direct current (dc).** A dc voltage source such as a battery causes direct current in a circuit. To have current in a circuit, the circuit must be complete, which means it has a voltage source, a load, and a closed path between the voltage source terminals. The **load** is a component such as a lamp or a heater that uses the power provided by the source. A simple complete dc circuit is shown in Figure 2-5(a), with a battery as the voltage source and a lightbulb for the load. The current in this circuit is steady, as illustrated in Figure 2-5(b).

Because the conductors connecting the battery to the lamp are solid wires, electrons comprise the moving charge in the circuit itself, but ions (charged particles) move in the fluid of the battery. The flow of electrons is from the negative terminal of the battery, through the load, and back to the positive terminal of the battery, where they complete a chemical reaction by combining with positive ions.

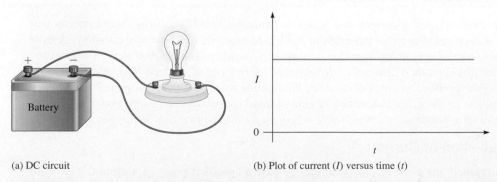

(a) DC circuit

(b) Plot of current (I) versus time (t)

FIGURE 2-5 Complete DC Circuit. The current is dc because the voltage source is a battery.

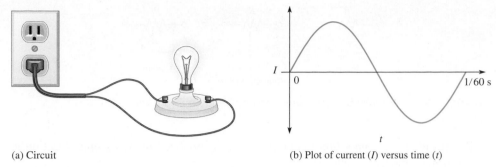

(a) Circuit   (b) Plot of current (I) versus time (t)

FIGURE 2-6   In a circuit with an ac voltage source, the current reverses direction during one cycle.

Alternating Current

As mentioned earlier, electric utility companies produce sinusoidal ac voltage. When the ac voltage from the utility company is connected as a source to complete an electric circuit, a sinusoidal current with the same frequency as the voltage results. Figure 2-6 shows a circuit where the lamp is connected to the utility voltage from a wall outlet. The **alternating current (ac)** through the wires and lamp reverses direction during each cycle of the ac voltage, as indicated.

Comparison of AC and DC

You might wonder how a steady current (dc) and a cyclic current (ac) can be compared. In ac, the charge reverses direction many times in a second, so current is constantly changing. The fact that it moves in one direction and then the other averages to zero, yet energy is still delivered to the load.

To make sense of ac and compare it to dc, engineers devised a measurement of current and voltage based on the equivalent electrical power that is delivered. If the ac is specified as **root-mean-square (rms) current** and **root-mean-square (rms) voltage,** the result is equivalent to a steady direct current and direct voltage of the same values. This is the way ac is normally specified (unless stated otherwise). For example, 1 A of ac (rms) is equivalent to 1 A of dc and 1 V ac (rms) is equivalent to 1 V dc. So ac and dc can be compared directly when the ac is stated as an rms value (the usual case).

SECTION 2-2 CHECKUP

1. What is current, and what is its unit?

2. If an ac voltage source with a frequency of 60 Hz is connected to a circuit, what is the frequency of the current?

3. Define *load*.

4. What does rms stand for?

5. Explain the difference between dc and ac.

2-3   Resistance and Ohm's Law

With the exception of a special class of materials called superconductors, all materials have resistance, which is the opposition to current. Conductors are generally metallic materials with low resistance, whereas insulators are materials with high resistance. When there is current through a resistive material, heat is produced by the collisions of electrons and atoms. Even wire, which has a very small resistance, can become warm when there is sufficient current through it.

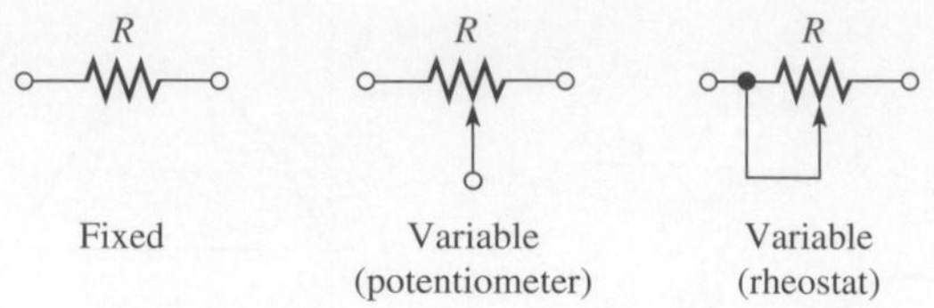

$$R \quad\quad\quad R \quad\quad\quad R$$

Fixed   Variable   Variable
                 (potentiometer)   (rheostat)

FIGURE 2-7 Schematic Symbols for Fixed and Variable Resistors

Resistance is the opposition to current and is symbolized by R. Its unit is the **ohm**, symbolized by the Greek letter Ω.

One ohm (1 Ω) of resistance exists when there is one ampere (1 A) of current in a material with one volt (1 V) applied across the material.

Resistors

Resistors are components that are designed to have a certain amount of resistance between their leads or terminals. The principal applications of resistors are to limit current; divide voltage; and, in certain cases, generate heat. Although there are many types of resistors and they come in many shapes and sizes, they can all be placed in one of two main categories: fixed or variable. The symbols for a fixed resistor and two variable resistors are shown in Figure 2-7. Variable resistors are subdivided into **potentiometers (pots)**—with three terminals—and **rheostats**—with two terminals. Notice in Figure 2-7, however, that a potentiometer can be connected as a rheostat by connecting the center terminal to one side. Potentiometers are used to control voltage; rheostats are used to control current.

Fixed Resistors

Fixed resistors have resistance values that cannot be altered. They are available with a large selection of values and power ratings that are set by the manufacturer. Fixed resistors are constructed using various methods and materials. A common type of fixed resistor is the carbon-composition type, which is made with a mixture of finely ground carbon, insulating filler, and a resin binder. Figure 2-8 shows the construction of a typical carbon-composition resistor.

Another type of fixed resistor for low-resistance precision applications is the wire-wound resistor. Wire-wound resistors are available either as small precision resistors or as large power resistors. They have excellent low-frequency characteristics, but they are not suitable for use at high frequencies. They are constructed by wrapping a resistive wire around an insulating rod and then sealed.

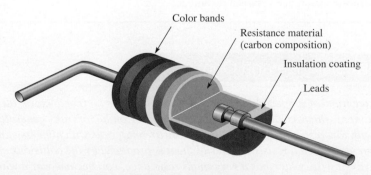

Color bands

Resistance material
(carbon composition)

Insulation coating

Leads

FIGURE 2-8 Cutaway View of a Carbon-Composition Resistor

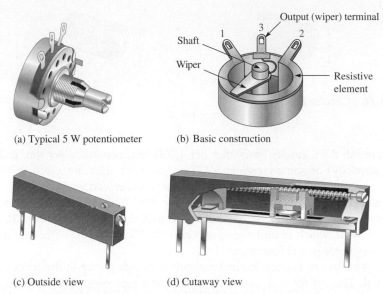

(a) Typical 5 W potentiometer   (b) Basic construction

(c) Outside view   (d) Cutaway view

FIGURE 2-9   Examples of Potentiometers with Construction Views

Variable Resistors

Variable resistors have resistance values that can be varied. They are also available in different types, including carbon-composition, film, and wire-wound. Potentiometers are constructed with a rotating shaft connected to a wiper, similar to the one shown in Figure 2-9(a). The center terminal is connected to the wiper arm, which is controlled by rotating the shaft. For applications requiring more precise control, potentiometers are available with multiple turns and in a multi-turn linear arrangement, as shown in Figure 2-9(c).

Wire Resistance

Although copper is an excellent conductor and is widely used in electronic circuits, it has some resistance. For many applications, the resistance of wire can be ignored; but in some cases, it can affect a circuit and needs to be considered. You may have seen low-voltage yard lights that grow dimmer the farther they are from the source. In this case, the wire resistance has affected the circuit. The problem is reduced if a larger-diameter wire (greater cross-sectional area) is used or if the lights are moved closer to the source (because a shorter wire has less resistance). A wire can overheat due to the resistance when there is too much current. Wire gauges are used throughout the world to specify the size of wire.

American Wire Gauge

The size and type of wire that is appropriate for a given application depends on a number of factors, including the length, the maximum amount of current, and the environment that the wire is in, among other factors. The diameter of wire is arranged according to standard gauge numbers called **American Wire Gauge (AWG)** sizes.

The larger the gauge number, the smaller the wire and the greater the resistance per unit length. For example, 12 gauge is common in household wiring, and 24 gauge is useful for telephone circuits. Wire as small as a human hair is number 40 gauge; even smaller sizes are used in certain applications. The AWG is related to the cross-sectional area of the wire; the common English unit for cross-sectional area of wires is the **circular mil (CM).** One circular mil is the area of a wire that is one thousandth of an inch (one mil) in diameter. A circular mil is illustrated in Figure 2-10.

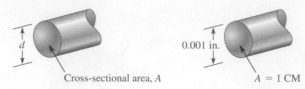

FIGURE 2-10   Circular Mil

For reference, the diameter, resistance per 1,000 feet, resistance per km, and current capacity (ampacity) of some common AWG copper wire sizes are given in Table 2-1. **Ampacity** is the maximum current a wire can carry under certain specified conditions. When the diameter of the wire becomes larger, the AWG number is smaller and the wire is rated for more current. After the AWG size gets to 1, the next size wire is 1/0, which is called a *one ought*, which in turn means one zero (0). The next three sizes are two ought (2/0), three ought (3/0), and four ought (4/0).

Because wire has resistance, current through it causes power in the form of heat to be dissipated. Thus, if there is too much current, the wire overheats. Ampacity ratings vary with temperature and the type of cable. Other ampacity ratings are also provided based on the National Electric Code (NEC). For example, AWG 12 is rated at 25 A in free air and 20 A as part of a three-conductor cable for a temperature of 30° C. Also, the NEC provides specifications that include the maximum number of amperes and fuse or breaker requirements for various wires (normally referred to as conductors in the code). In addition, there are specific requirements for conditions such as buried conductors, conductors exposed to sunlight, or conductors installed in damp locations such as those used in solar panel installations. The current capacity is derated for these conditions and more when there are a number of wires in a conduit or bundle. This is very important in renewable energy systems as bundled wires in hot locations are common. Always check the NEC for ratings in specific applications and the code requirement in your area before installing any wiring,

TABLE 2-1

Normal NEC Current Rating and Wire Resistance of Selected Sizes of Copper Wire

American Wire Gauge (AWG) Wire Size	Current Capacity (A) of Copper Wire	Resistance (Ω) per 1,000 Feet	Resistance (Ω) per Kilometer
16	15	4.016	13.176
14	20	2.525	8.284
12	25	1.588	5.210
10	30	0.9989	3.277
8	40	0.6282	2.061
6	55	0.3951	1.296
4	70	0.2485	0.815
3	85	0.187	0.614
2	95	0.1563	0.513
1	110	0.1239	0.406
1/0	125	0.0983	0.323
2/0	145	0.0779	0.256
3/0	165	0.0618	0.203
4/0	195	0.049	0.161

EXAMPLE 2-2

Find the resistance of 2,500 feet of AWG 12 copper wire.

Solution

From Table 2-1, the resistance of AWG 12 copper wire is 1.588 Ω/1,000 ft. This means that 2,500 ft. will have a resistance that is

$$R = (2,500 \text{ ft})(1.588 \ \Omega/1,000 \text{ ft}) = \textbf{3.97} \ \boldsymbol{\Omega}$$

Ohm's Law

Georg Simon Ohm formulated **Ohm's law**, which is named in his honor. Ohm's law expresses the relationship among voltage (V), current (I), and resistance (R). Ohm's law can be expressed in three equivalent forms: one for voltage, one for current, and one for resistance. It allows you to determine any one of the quantities (V, I, or R) if you know the other two.

Formula for Voltage

You can use Ohm's law to find the voltage when you know the current and resistance. Voltage equals current times resistance.

$$V = IR \hspace{4cm} \text{Equation 2-7}$$

Formula for Current

You can use Ohm's law to find the current when you know the voltage and resistance. Current equals voltage divided by resistance.

$$I = V/R \hspace{4cm} \text{Equation 2-8}$$

Formula for Resistance

You can use Ohm's law to find the resistance when you know the current and voltage. Resistance equals voltage divided by current.

$$R = V/I \hspace{4cm} \text{Equation 2-9}$$

The three formulas ($V = IR$, $I = V/R$, and $R = V/I$) are, of course, equivalent. Figure 2-11 can help you remember the relationship among voltage, current, and resistance.

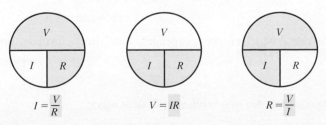

FIGURE 2-11   Memory Aid for Ohm's Law

EXAMPLE 2-3

(a) Determine the voltage when the current is 2 A and the resistance is 10 Ω.

Solution

$$V = IR = (2\text{ A})(10\ \Omega) = \textbf{20 V}$$

(b) Determine the current when the voltage is 48 V and the resistance 20 Ω

Solution

$$I = V/R = 48\text{ V}/20\ \Omega = \textbf{2.4 A}$$

(c) Determine the resistance when the voltage is 120 V and the current is 8 A.

Solution

$$R = V/I = 120\text{ V}/8\text{ A} = \textbf{15 }\Omega$$

SECTION 2-3 CHECKUP

1. Define resistance and name its unit.

2. What are the two categories of resistor?

3. Name two types of variable resistor and draw the schematic symbol for each.

4. What is a circular mil?

5. What does the AWG number of a wire indicate?

6. What is meant by the term *ampacity*?

7. Write the Ohm's law formula for voltage in terms of current and resistance.

8. Write the Ohm's law formula for current in terms of voltage and resistance.

9. Write the Ohm's law formula for resistance in terms of voltage and current.

10. If the voltage in a circuit is fixed and the resistance is increased, what happens to the current?

2-4   Power and Watt's Law

When there is current through a resistance, electrical energy is converted to heat or other form of energy such as light. A common example of this is the incandescent lightbulb that becomes too hot to touch when it has been on for a short time. The current through the filament of the bulb produces light energy, but heat energy is also produced as a by-product because the filament has resistance. Electrical components must be able to dissipate a certain amount of heat energy in a given period of time.

Power

Power (*P*) is the rate at which energy is expended. Rate always involves time (*t*), so power is expressed as

$$P = W/t \qquad\qquad \text{Equation 2-10}$$

This formula can be stated in two other equivalent ways:

$$W = Pt \qquad\qquad \text{Equation 2-11}$$

$$t = W/P \qquad\qquad \text{Equation 2-12}$$

The unit of power is the **watt (W)**. Notice that the unit of power is a roman W where energy is represented by an italic W. Recall that the unit of energy is the joule (J) and the unit of time is seconds (s).

One watt is the amount of power when one joule of energy is used in one second.

The Kilowatt-Hour Unit of Energy

If you multiply power in watts and time in seconds, the result is energy in joules. Another way to express energy, called the **kilowatt-hour (kWh)** is commonly used by utility companies because they deal in huge amounts of energy. The kilowatt-hour is more practical than the joule in applications involving large amounts of power. The power company bills you for kilowatt-hours, not joules. One kilowatt-hour of energy is dissipated when you use 1,000 watts of power for one hour.

EXAMPLE 2-4

(a) Determine the power in watts if 72 kJ of energy are used in 1 hour.

Solution

$$1 \text{ hr} = 3,600 \text{ s}$$

$$P = W/t = 72,000 \text{ J}/3,600 \text{ s} = \textbf{20 W}$$

(b) Determine the amount energy in kilowatt-hours if a 100 W load is left on for 24 hours.

Solution

$$W = Pt = (100 \text{ W})(24 \text{ h}) = 2,400 \text{ Wh} = \textbf{2.4 kWh}$$

(c) Determine the power in watts if a certain electrical heater uses 12 kWh of energy in 12 hours.

Solution

$$P = W/t = 12 \text{ kWh}/12 \text{ h} = 1 \text{ kW} = \textbf{1,000 W}$$

(d) Find the time it takes for a 50 W lightbulb to use 1 kWh of energy.

Solution

$$t = W/P = 1 \text{ kWh}/50 \text{ W} = \textbf{20 h}$$

Watt's Law

James Watt formulated the law, called **Watt's law** in his honor, that expresses the relationship among power (P), voltage (V), current (I), and resistance (R). Watt's law can be expressed in three equivalent forms. It allows you to determine any one of the quantities, P, V, I, and R. You can use Watt's law to find the power when you know the voltage and current. Power equals voltage times current.

$$P = VI \qquad\qquad \text{Equation 2-13}$$

You can use Watt's law to find the power when you know the current and resistance. Power equals current squared times resistance.

$$P = I^2R \qquad\qquad \text{Equation 2-14}$$

You can use Watt's law to find the power when you know the voltage and resistance. Power equals voltage squared divided by resistance.

$$P = V^2/R \qquad \text{Equation 2-15}$$

In certain ac cases, these equations are modified to reflect a phase shift between ac voltage and current.

EXAMPLE 2-5

(a)   Determine the power when 120 V produces 5 A.

Solution

$$P = VI = (120 \text{ V})(5 \text{ A}) = \mathbf{600 \ W}$$

(b)   Determine the amount of power when the current through a 1,000 Ω resistor is 2 A.

Solution

$$P = I^2 R = (2 \text{ A})^2(1,000 \ \Omega) = (4 \text{ A}^2)(1,000 \ \Omega) = \mathbf{4,000 \ W}$$

(c)   Determine the power if a 120 V source is connected across a 10 Ω heating element.

Solution

$$P = V^2/R = (120 \text{ V})^2/10 \ \Omega = \mathbf{1,440 \ W}$$

SECTION 2-4 CHECKUP

1. What is power and what is its unit?

2. Define one watt.

3. What is a kilowatt-hour and what is the abbreviation?

4. Express Watt's law in three ways.

5. Compare a 100 W load that is on for one hour with a 75 W load that is on for two hours. Which one used the most power? Which one used the most energy?

2-5   Series and Parallel Circuits

Series and parallel circuits are two basic arrangements of the source and loads. Series and parallel circuits are often combined to form complex circuits that can be understood on the basis of simpler series and parallel arrangements. In this section you will learn the differences between these two important circuit configurations.

Series Circuits

A series connection is one with a single path. A **series circuit** has a single complete path (forming a string) from the voltage source through the load (or loads) and back. A series circuit can always be identified by the fact that it has only one current path. Figure 2-12 shows three resistive loads ($R_1, R_2,$ and R_3) in series with a dc voltage source. All three circuits are electrically the same, but they have been drawn differently. Of course, there are other ways of drawing these three load resistances in series. The important point to remember with a series connection is that only *one* current path exists from the source through the loads and back to the source.

In a series circuit, the current is the same throughout the circuit.

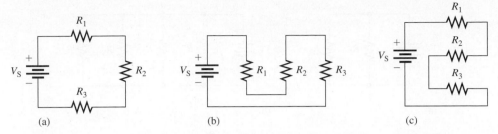

FIGURE 2-12 Series Resistive Circuit Drawn Three Ways. All three resistive loads are equivalent electrically. The standard schematic symbol for a dc source is used in each circuit.

A good analogy can be made with plumbing, where water running through a single pipe with several valves is the same through each of the valves. If one or more valves are closed, there is no water through any of them. If all the valves are completely or partially open, the same amount of water flows through all of them.

When two or more resistive loads are connected together in series, the total load resistance is the sum of the individual load resistances. The total resistance (R_T) of a series circuit is the sum of the individual resistances.

$$R_T = R_1 + R_2 + R_3 + \ldots + R_n \qquad \text{Equation 2-16}$$

where R_n is the resistance of the last resistor.

Voltage Sources in Series

When several voltage sources such as batteries or solar cells are connected in series, the *total output voltage is the sum of the individual voltages.*

$$V_{OUT} = V_1 + V_2 + V_3 + \ldots + V_n \qquad \text{Equation 2-17}$$

For example, four 12 V batteries connected in series produces 48 V. Figure 2-13 illustrates a series connection of *batteries* and a series connection of *photovoltaic cells* (solar cells).

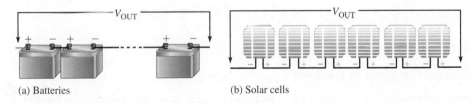

(a) Batteries   (b) Solar cells

FIGURE 2-13 Examples of Series Voltage Sources

Parallel Circuits

A parallel connection is one in which two or more components are connected across the same two points (called nodes). A **parallel circuit** has two or more loads connected across a common voltage source. Each load provides a separate path for current. When two or more loads are connected in parallel, the total load resistance is less than the value of the smallest-value load resistor. With a parallel circuit, you can always trace a complete path from the voltage source to any load and back to the voltage source without going through another load. That is, the voltage source has two or more paths for current. Figure 2-14 shows examples of parallel circuits. Figure 2-14(a) shows two resistive loads connected across a dc source. The current from the source divides between R_1 and R_2. Figure 2-14(b) shows four resistive loads across a dc source. The current from the source divides among R_1, R_2, R_3, and R_4. Figure 2-14(c) shows two resistive loads across a dc source, and Figure 2-14(d) is two resistive loads across an ac source. Ideally, there is no limit to the number of

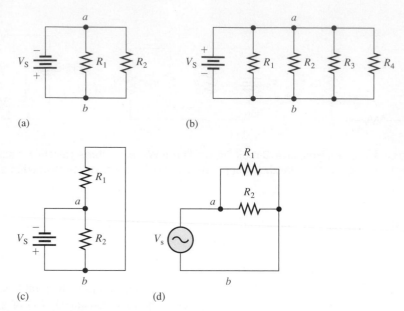

FIGURE 2-14   Four Examples of Parallel Circuits

resistive loads that can be in a parallel circuit, but in practice, there is a limit to how much current a source can provide. Notice that in each circuit, the voltage source is connected directly across two or more components. The current divides among the parallel paths in inverse proportion to the resistor value. The lowest-value resistor has the most current and the highest-value resistor has the least. If all the resistors are the same, there is an equal amount of current in each.

The total resistance (R_T) of a parallel circuit is reciprocal of the sum of the reciprocals of the individual resistances and is always less than the smallest-value resistor. The *reciprocal of R is 1/R*.

$$R_T = \cfrac{1}{\cfrac{1}{R_1} + \cfrac{1}{R_2} + \cfrac{1}{R_3} + \ldots + \cfrac{1}{R_n}} \qquad \text{Equation 2-18}$$

where R_n is the resistance of the last resistor.

Voltage Sources in Parallel

When several voltage sources, such as batteries or solar cells, are connected in parallel, the *total voltage is the same as a single battery or solar cell*. However, the capacity for producing current to a load increases. For example, when four batteries are in parallel, they can produce four times as much current as a single battery to the same load. Figure 2-15 illustrates a parallel connection of batteries and a parallel connection of solar (photovoltaic) cells.

When a load is connected across the output of a circuit with parallel voltage sources, the total current through the load is still determined by Ohm's law; however, each battery needs to supply only part of the total current. As a result, battery life can be extended because each battery has less current drain.

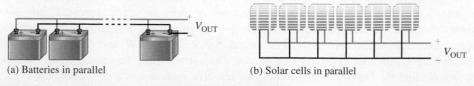

(a) Batteries in parallel   (b) Solar cells in parallel

FIGURE 2-15   Examples of Parallel Voltage Sources

EXAMPLE 2-6

(a) What is the total resistance of the following series resistors: 10 Ω, 15 Ω, 33 Ω, and 100 Ω?

Solution

$$R_T = 10 \ \Omega + 15 \ \Omega + 33 \ \Omega + 100 \ \Omega = \mathbf{158 \ \Omega}$$

(b) Determine the total voltage produced by 100 solar cells connected in series if each solar cell generates 0.5 V.

Solution

Because all of the solar cells produce the same voltage, simply multiply as follows. This is equivalent to adding 0.5 V one hundred times.

$$V_T = 100(0.5 \ \text{V}) = \mathbf{50 \ V}$$

(c) Find the total resistance of the resistors in part (a) if they are all connected in parallel.

Solution

$$R_T = 1/(1/10 \ \Omega + 1/15 \ \Omega + 1/33 \ \Omega + 1/100 \ \Omega)$$
$$= 1/(0.100 + 0.067 + 0.030 + 0.010) = 1/0.207 = \mathbf{4.83 \ \Omega}$$

(d) Determine the currents in the smallest and the largest resistors in part (a) if they are connected to the solar cells in part (b).

Solution

The 10 Ω resistor has the most current, which is, by Ohm's law:

$$I = V/R = 50 \ \text{V}/10 \ \Omega = \mathbf{5 \ A}$$

The 100 Ω resistor has the least current, which is, by Ohm's law:

$$I = V/R = 50 \ \text{V}/100 \ \Omega = \mathbf{0.5 \ A}$$

SECTION 2-5 CHECKUP

1. What is a series circuit?

2. Three resistors, a 10 Ω, a 4.7 Ω, and a 8.2 Ω, are connected in series with a 12 V source that is producing 0.524 A of current. What is the current in each of the resistors?

3. What is a parallel circuit?

4. Four resistors having different values are connected in parallel with a voltage source. Which resistor has the most current?

5. How would you connect four 12 V batteries to get more voltage? More current capacity?

2-6 Conductors, Insulators, and Semiconductors

Electricity consists of moving charge. When we think of the path for electricity, we think of wires most of the time. In metallic solids, electrons make up the charge that moves; however, charge can also move through certain liquids and gases as well as space itself. In this section, you will learn the basic concepts of conductors, insulators, and semiconductors.

Conductors

Conductors are materials that allow the free movement of charge and can be composed of solids, liquids, or gases. Nearly all circuits use metallic solids such as copper or aluminum to provide a path for electricity. Wires and traces on pc boards are the most common form of solid conductor for electrical signals. Wires are either solid or stranded. Stranded wire is used for meter leads or in lamp cords where flexibility is important. No matter the size, stranded copper wire has the same current capacity as solid copper wire as long as the gauge is the same in both.

For electronic circuits, copper is the most widely used material because it is an excellent conductor, can easily be drawn into wires, is corrosion resistant, and is cost effective. Copper is such a good conductor because of its atomic structure. Solid copper has a regular arrangement of positive ions (atoms or groups of atoms that have lost one or more outer shell electrons). This arrangement forms a basic metallic crystal. These electrons are free to move about but tend to distribute themselves to maintain overall electrical neutrality, as shown in Figure 2-16.

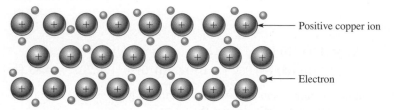

FIGURE 2-16   Metallic Bonding in Copper

The negatively charged electrons hold the positive ions of the metal together, forming the metallic bond. The "sea" of free electrons accounts for copper's excellent electrical properties and is responsible for its metallic luster and other properties. Other solid metals have a similar structure to copper. Most are excellent conductors and have other properties that make them suited to specific electrical and electronic applications. Gold, for example, can be drawn into extremely fine wires; it is noncorrosive, so it is used in switch contacts and as plating on plugs. Alloys (mixtures of two or more metals) also have many electrical and electronic applications. In the past, most solder used in electrical work was an alloy of lead and tin. After July 1, 2006, the *Restriction of Hazardous Substances Directive* (RoHS) became effective and most manufacturers moved to lead-free solders composed of alloys of other metals. For most electronics work, the lead-free solders contain tin alloyed with bismuth, copper, silver, or other metals. These solders are widely used for making solid connections on circuit boards, cables, and connectors.

In liquids, the moving charge is composed of positive and negative ions, never electrons. Materials known as **electrolytes** form ions in water solution and are good conductors. The three general categories of substances that form ions in water solution are acids, bases, and salts. An example of a common acid that is used in lead-acid batteries is sulfuric acid, an example of a common base is sodium hydroxide (lye), and an example of a salt is ordinary table salt. In addition to water solutions of salts, molten salts can conduct electricity also. Ordinary table salt is composed of a crystalline structure of sodium (Na^+) ions and chlorine ions (Cl^-). When table salt is added to water, the crystal breaks apart and separate ions enter the solution. In solution, the sodium and chlorine ions are the charge carriers. The solution can conduct electricity due to the movement of these ions.

Gases can also conduct electricity when they are broken down into ions and electrons. Unlike liquid solutions, the charge carriers in a gas include electrons. A good example of a gas conductor is the fluorescent lamp, shown in Figure 2-17. The tube contains low-pressure argon gas (not shown) and a small quantity of mercury. The instant the lamp is turned on, there are no free electrons or ions to allow current. An alternating voltage is applied to the electrodes, through the ballast, which controls the starting and regulates the current. The voltage on the electrodes heats the electrodes rapidly, causing electrons to boil off and start

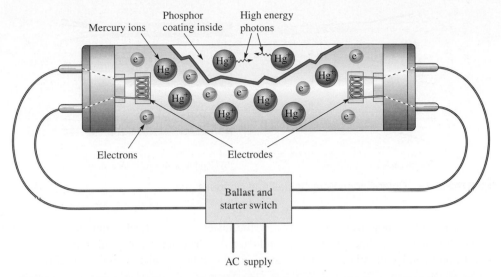

FIGURE 2-17   Fluorescent Lamp. In a fluorescent lamp, mercury ions emit high-energy photons, which are converted by the phosphor coating to visible light.

the process. These electrons move back and forth rapidly between the electrodes because of the voltage difference between them. The electrons ionize the mercury atoms in the tube, causing them to emit high-energy photons. Finally, the high-energy photons from the mercury interact with the phosphor coating on the inside of the bulb, emitting light composed of low-energy photons. All of this happens very rapidly in modern lamps, so rapidly that the light appears to come on almost instantaneously.

Insulators

Insulators are materials that prevent the free movement of charge; thus, they are considered nonconductors. While no material is a perfect insulator, many materials are such poor conductors that we classify them as nonconductors. The atoms in these materials tend to form bonds with other atoms that leave no "extra" electrons in the structure. The most common class of electronic insulating material is plastics. Other common solid materials used as insulators include ceramics, paper, and glass.

Plastics are widely used as the insulation layer on wiring. The insulating coating provides isolation from other conductors and is also important for protection against shock hazard. Sometimes a second coating is added to provide moisture or flame resistance to wiring. The type of insulation required depends on the application; for example wiring that is used in solar applications needs to be rated for moisture and heat resistance. The maximum rated voltage, type of wire, AWG size, and manufacturer are all printed on the wire.

Wire Protection

Two or more insulated wires are commonly bundled together within a plastic or armored sheathing to form a wire cable. See Figure 2-18(a). The individual insulated wires are color-coded for identification. Some local codes require armored cable or conduit (neither of which are necessarily made of insulating material) wherever wiring is exposed. **Conduit** is essentially a pipe through which the insulated wires are run. It offers the best protection against wire damage. Also, conduit makes it easy to change or install new wires by simply pulling the wires or cables through the conduit without having to run new cable. Examples of metal and plastic conduits with wires run through them are shown in Figure 2-18(b).

Semiconductors

A **semiconductor** is a crystalline material that has four electrons in its valence shell and has properties between those of conductors (metals) and insulators (nonmetals). Chemically,

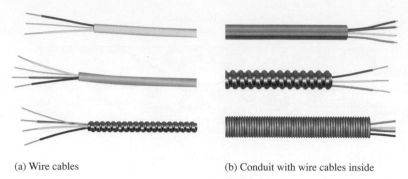

(a) Wire cables   (b) Conduit with wire cables inside

FIGURE 2-18   Color-Coded Wire Cables and Conduits

silicon acts like a nonmetal, but it has metallic luster and electrical behavior closer to a metal. It is the most widely used semiconductor for electronics. Although pure silicon is not a good conductor, silicon can become a good conductor when certain impurities are added to it.

Two different types of impurities can be added to the crystal structure of the semiconductor. One common impurity is phosphorous, which has five electrons in its outer shell. It is called an ***n*-type impurity** (*n* stands for "negative") because four of the outer shell electrons bond with adjacent silicon atoms in the crystal, leaving an extra one available for conduction. A different type of impurity is represented by boron, which has only three electrons in its outer shell. This is called a ***p*-type impurity** (*p* stands for "positive"). When it is introduced into the silicon crystal, one bonding position for the silicon is missing an electron. This missing position is called a *hole*. (The silicon atom and impurities are discussed in more detail in Chapter 3.)

Diode

When the *p*-materials and *n*-materials are manufactured together as one solid, some electrons on the *n* side move to fill holes on the *p* side, creating a region in the middle called the *depletion region*. Because of this depletion region, the resulting device is a good conductor if the voltage is applied in one direction and a good insulator if the voltage is applied in the other direction. Thus, current can pass in only one direction. This type of device is called a **diode,** a very important electronic device. Because diodes allow current in only one direction, they are used in power supplies as a key component for converting ac to dc. The schematic symbol for a rectifier diode is shown in Figure 2-19(a) and (b). When voltage is applied with a polarity as shown in Figure 2-19(a), the diode conducts current (forward bias). When the polarity of the voltage is reversed as shown in Figure 2-19(b), the device blocks current (reverse bias). Another type of diode is the **light-emitting diode (LED),** so called because it emits light when current passes through it. The symbol is shown in Figure 2-19(c).

Photovoltaic Cell

The photovoltaic (PV) cell or **solar cell** is actually a type of diode designed specifically to allow sunlight to penetrate the semiconductor regions. Energy from sunlight releases electrons within the cell, which can drift across the depletion region, which, in turn, will

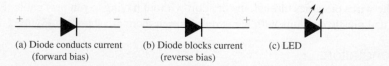

(a) Diode conducts current   (b) Diode blocks current   (c) LED
    (forward bias)   (reverse bias)

FIGURE 2-19   Symbols of a Semiconductor Rectifier Diode Showing Its Two Modes of Operation and a Light-Emitting Diode (LED)

cause a voltage to be generated. Chapter 3 provides thorough coverage of the solar cell. The schematic symbol is shown in Figure 2-20.

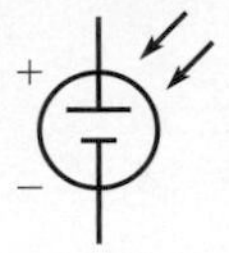

FIGURE 2-20 Schematic Symbol of a Solar (Photovoltaic) Cell

Transistor

Many other electronic devices, such as the **transistor,** are constructed from semiconductors. The first successful transistor, invented at Bell labs, was a semiconductor that used three layers of alternating p-type and n-type materials. This type of transistor, called the bipolar junction transistor (BJT) is said to be a current amplifier because a small input current controls a large output current. Bipolar transistors are used in amplifiers, which are devices that increase the signal strength of a small signal such as the signal from a microphone.

Another type of transistor is the field-effect transistor (FET), which has become the dominant type of transistor. This type of transistor uses a voltage input to control output current. It is particularly useful for switching applications, where the transistor acts like a voltage-controlled switch that can open or close on a voltage command. The most important type of FET is the metal-oxide semiconductor field-effect transistor (MOSFET). Because of its importance in logic and in switching applications, the MOSFET is widely used in renewable energy systems. It is the primary component in integrated circuits, which form complex devices such as microprocessors and controllers for renewable energy systems.

A newer type of transistor called the insulated gate bipolar transistor (IGBT) combines the best attributes of the BJT and the FET for high-power switching applications. The IGBT combines the high-current capability of BJTs with the voltage control of a FET. The IGBT has made possible high-power devices that are much more efficient than older analog switching devices. For this reason, you are likely to encounter IGBTs when you work on high-voltage, high-current switching devices, which are common in large renewable energy systems.

Thyristor

The **thyristor** is another type of semiconductor device that is widely used. It is typically constructed of four semiconductor layers. Examples of thyristors include the diac, triac, silicon-controlled rectifier (SCR), and silicon-controlled switch (SCS). Thyristors can be used to control the amount of ac power to a load and are used in lamp dimmers, motor-speed controls, ignition systems, and charging circuits. These and other semiconductor components are studied in detail in courses dealing with electronic devices.

SECTION 2-6 CHECKUP

1. What type of particle forms the structure of metallic solids?
2. Why are metals good conductors?
3. What is the name for liquids that are good conductors?
4. What are two purposes for the insulated coating on wiring?
5. What is a semiconductor?
6. Explain what a diode does.
7. What does PV stand for? What is a PV cell?
8. What is the main purpose of an IGBT?
9. What do the letters SCR stand for?

2-7 Magnetism and Electromagnetic Devices

In this section, you will learn about basic magnetism and electromagnetism. Three important electromagnetic devices are introduced: the relay, the generator, and the motor. All of these devices are important in renewable energy systems. Relays are commonly used in many applications to switch current on and off. Generators and motors are rotating electromagnetic machines that generate electrical power (in the case of the generator) and use electrical power to produce mechanical motion (in the case of the motor). Generators are used in wind and hydroelectric systems, and motors are used in solar tracking systems.

The Magnetic Field

A permanent magnet, such as the bar magnet shown in Figure 2-21, has a magnetic field surrounding it. All magnetic fields have their origin in moving charge, which in solid materials is caused by moving electrons. In certain materials, such as iron, atoms can be aligned so that the electron motion is reinforced, creating an observable field that extends in three dimensions.

To explain and illustrate magnetic fields, Michael Faraday drew lines of force, or flux lines, to represent the unseen field. Flux lines are widely used as a description of a magnetic field, showing the strength and direction of the field. The flux lines never cross. When lines are close together, the field is more intense; when they are farther apart, the field is weaker. The flux lines always extend from the north pole (N) to the south pole (S) of a magnet. Although the number of lines for any given magnet can be extremely large, only a few representative lines are usually shown in drawings.

When unlike poles of two permanent magnets are placed close together, their magnetic fields produce an attractive force, as indicated in Figure 2-22(a). When two like poles are brought close together, they repel each other, as shown in Figure 2-22(b).

Magnetic Flux

The group of force lines going from the north pole to the south pole of a magnet is called the **magnetic flux** and is symbolized by φ. Several factors determine the strength of a magnet, including the material and physical geometry as well as the distance from the magnet. Magnetic flux lines tend to be more concentrated at the poles.

The mks unit of magnetic flux is the weber (Wb), which is a very large unit. In most practical situations, the microweber (μW), which is equal to 100 flux lines, is more appropriate. The **magnetic flux density (B)** is the amount of flux, φ, per unit area (A) perpendicular to the magnetic field. Flux density is defined mathematically as follows;

$$B = \phi/A \qquad\qquad \text{Equation 2-19}$$

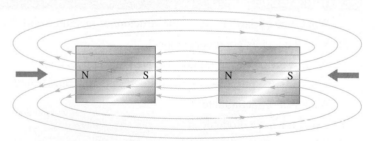

(a) Unlike poles attract

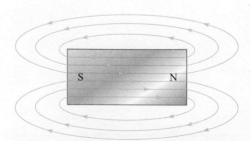

Blue lines represent only a few of the many magnetic lines of force in the magnetic field.

FIGURE 2-21 Magnetic Lines of Force Around a Bar Magnet

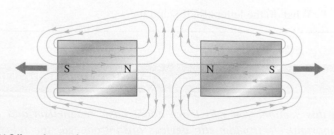

(b) Like poles repel

FIGURE 2-22 Magnetic Attraction and Repulsion

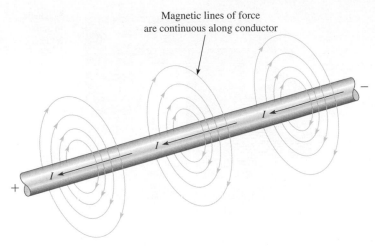

FIGURE 2-23   Magnetic Field Around a Current-Carrying Conductor. The red arrows show the direction of electron current.

The flux density is in Wb/m^2 when the magnetic flux is in webers and the area is in square meters. One Wb/m^2 defines the **Tesla (T),** which is the mks unit. The Tesla represents a large unit; the strongest permanent magnets are above 5 T. The **Gauss (G)** is the much smaller cgs unit for flux density (10^4G $= 1$T). The meter used to measure flux density is named the gaussmeter (rather than the teslameter).

Electromagnetism

Current produces a magnetic field around a conductor, as illustrated in Figure 2-23. The invisible lines of force of the magnetic field form a concentric circular pattern around the conductor and are continuous along its length. Although the magnetic field cannot be seen, it is capable of producing visible effects. For example, if a current-carrying wire is inserted through a sheet of paper in a perpendicular direction, iron filings placed on the surface of the paper arrange themselves along the magnetic lines of force in concentric rings, as illustrated in Figure 2-24(a). Figure 2-24(b) shows that the needle of a compass placed in the magnetic field points in the direction of the lines of force.

Electromagnet

Figure 2-25(a) illustrates a basic magnetic circuit with a coil of wire around a magnetic material. The current through the coil creates a magnetic field represented by flux lines along the magnetic path. An electromagnet works on the same principle except that an air gap exists in the magnetic material so that the magnetic field set up by the current in the coil of wire extends from the north pole to the south pole. When the current through the coil

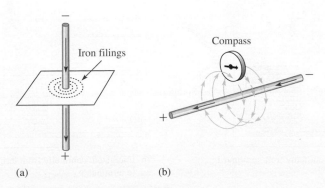

(a)   (b)

FIGURE 2-24   Visible Effects of a Magnetic Field

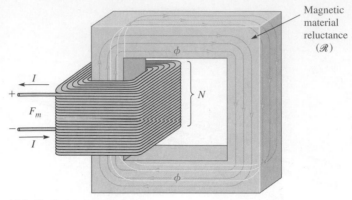

(a) Basic electromagnetic circuit

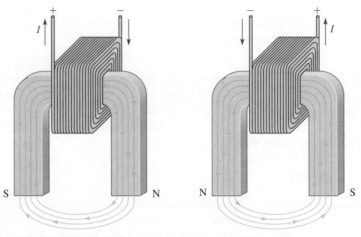

(b) Magnetic field reversing direction in an electromagnet

FIGURE 2-25   Basic Electromagnetic Circuit and Electromagnet

reverses direction, the magnetic field also reverses direction, as shown in Figure 2-25(b). An electromagnet can have various configurations, but a U-shape magnetic core is shown.

Relay

Figure 2-26 shows the basic operation of an armature-type relay with a normally open (NO) contact and one normally closed (NC) contact. When there is no current through

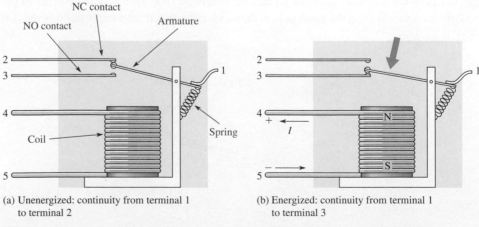

(a) Unenergized: continuity from terminal 1 to terminal 2

(b) Energized: continuity from terminal 1 to terminal 3

FIGURE 2-26   Basic Structure of a Single-Pole, Double-Throw Armature Relay

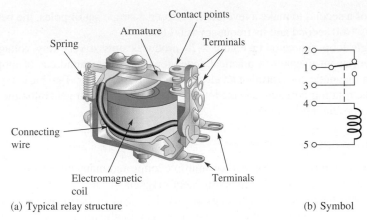

(a) Typical relay structure   (b) Symbol

FIGURE 2-27   Typical Armature Relay

the coil, the armature is held against the upper contact by the spring, thus providing continuity from terminal 1 to terminal 2, as shown in Figure 2-26(a). When energized with coil current, the armature is pulled down by the attractive force of the magnetic field and makes a connection with the lower contact to provide continuity from terminal 1 to terminal 3, as shown in Figure 2-26(b). A typical armature relay and its schematic symbol are shown in Figure 2-27.

Another widely used type of relay is the reed relay. Like the armature relay, the reed relay uses an electromagnetic coil. The contacts are thin reeds of magnetic material and are usually located inside the coil. Reed relays are faster, are more reliable, and produce less contact arcing than armature relays. However, they have less current-handling capability and are more susceptible to mechanical shock.

AC Generators

The ac generator is an electromagnetic machine that produces a sinusoidal voltage. The basic principle of an ac generator can be understood using a simplified single-loop model, as shown in Figure 2-28. The loop is mechanically driven by a rotating force from a motor shaft, wind turbine blades, or water-driven turbine blades. As the loop rotates through the magnetic field, a voltage is induced across the slip rings. When a load is connected via the brushes, a current is produced and power is delivered to the load.

Each revolution of the loop produces one cycle of a sine wave. The positive and negative peaks occur when the loop cuts through the maximum number of flux lines. The rate at which the loop spins determines the time for one complete cycle and the frequency. If it

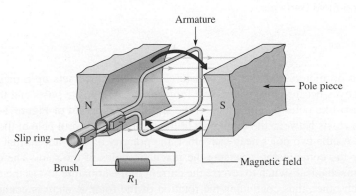

FIGURE 2-28   Simplified AC Generator

takes 1/60 of a second to make a revolution between a single set of poles, the period of the sine wave is 1/60 second and its frequency is 60 Hz.

The single-loop generator in Figure 2-28 produces only a very tiny voltage. Instead of using permanent magnets, a practical generator usually has hundreds of loops that are wound on a magnetic core forming an electromagnet for the rotor. Two basic types of generator are the rotating-armature and the rotating-field. In a motor or generator, the **armature** is the power producing component.

Rotating-Armature Generator

In a rotating-armature generator that has multiple loops and many pole-pairs, the magnetic field is stationary and is supplied by permanent magnets or electromagnets operated from dc. With electromagnets, field windings are used instead of permanent magnets. These windings provide a fixed magnetic field that interacts with the rotor coils. The output power is taken from the rotor through the slip rings and brushes.

Rotating-Field Generator

Figure 2-29 show how a rotating-field generator can produce three-phase sine waves with an electromagnetic rotor. A permanent-magnet rotor is shown for simplicity. AC is generated in each set of windings as the north pole and the south pole of the rotor alternately sweep by a stator winding. The stator winding are separated by 120°, causing the sine wave outputs also to have 120° between them. Three-phase (abbreviated 3φ) is generated by power companies, which is then converted to single-phase (1φ) for residential users. AC generators are discussed in more detail in Chapter 13, and the conversion process from three-phase to single-phase is discussed in Chapter 14.

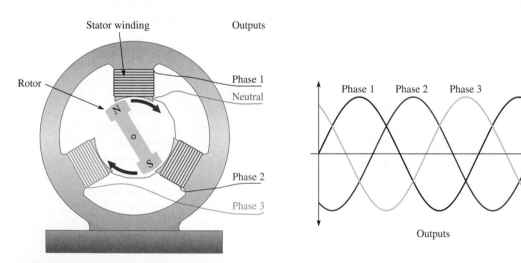

FIGURE 2-29   Rotating-Field Generator

Motors

The rotor in dc motors contains the armature winding, which sets up a magnetic field. The rotor moves because of the attractive force between opposite poles and the repulsive force between like poles, as illustrated in the simplified diagram of Figure 2-30. A force of attraction exists between the south pole of the rotor and the north pole of the stator (and vice versa). As the two poles near each other, the polarity of the rotor current is suddenly switched by the commutator, reversing the magnetic poles of the rotor. The commutator serves as a mechanical switch to reverse the current in the armature just as the unlike poles are near each other, thus continuing the rotation of the rotor. A shaft is connected to the rotor; as the rotor moves, the shaft turns to provide mechanical torque.

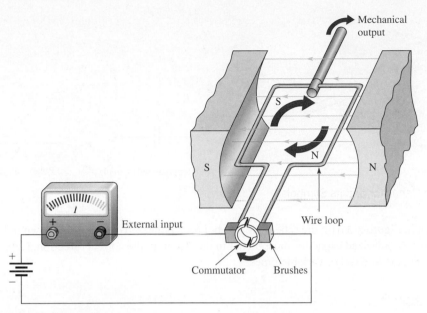

FIGURE 2-30   Simplified DC Motor

Some dc motors do not use a commutator to reverse the polarity of the current. Instead of supplying current to a moving rotor, the magnetic field is rotated in the stator windings using an electronic controller. Motors are covered in more detail in Chapter 5 in relation to solar tracking systems.

SECTION 2-7 CHECKUP

1. Discuss the repulsive and attractive forces in magnets.

2. How is an electromagnetic field produced?

3. Describe the difference between a generator and a motor.

4. Name two basic types of ac generator.

2-8   Capacitors, Inductors, and Transformers

A resistor presents a fixed value of resistance (opposition to current) that is independent of the type of voltage (dc or ac) applied to it. Capacitors and inductors are two types of electrical components classified as reactive, which means that their opposition to current depends on the type of voltage and the frequency of the applied ac voltage. The opposition to current that a capacitor or inductor presents in a circuit is called reactance. When reactance and resistance are both present in a circuit, the combination is called impedance.

Capacitors

A **capacitor** is an electrical device that stores energy in the form of an electric field established by electrical charge. I n its most basic form, the capacitor is constructed of two conductive plates placed physically in parallel and separated by an insulating material called the **dielectric.** Connecting leads are attached to the parallel plates. Basic capacitor structure is shown in Figure 2-31(a). The schematic symbol for a nonpolarized capacitor is

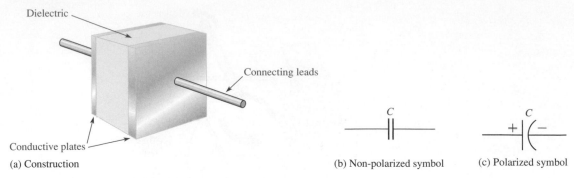

(a) Construction   (b) Non-polarized symbol   (c) Polarized symbol

FIGURE 2-31   Basic Capacitor and Its Symbols

shown in Figure 2-31(b); the schematic symbol for a polarized capacitor is shown in Figure 2-31(c). A polarized capacitor must be put in the circuit in the correct direction. Capacitors are available in fixed or variable types.

Capacitance

The amount of charge that a capacitor can store per unit of voltage across its plates is its **capacitance (C)**. The more charge per unit of voltage that a capacitor can store, the greater its capacitance, as expressed by the following formula:

$$C = Q/V$$ Equation 2-20

where

> C is capacitance in farads
>
> Q is charge in coulombs
>
> V is voltage in volts

Capacitance depends on the area of the plates and the distance between them. Some typical capacitors are shown in Figure 2-32.

Unit of Capacitance

The **farad (F)** is the unit of capacitance. One farad is the amount of capacitance when one coulomb (C) of charge is stored with one volt across the plates. Smaller values of capacitance are expressed in microfarads (μF) or in picofarads (pF). A microfarad is one-millionth of a farad; a picofarad is one-trillionth of a farad. Capacitors are available from very small to very

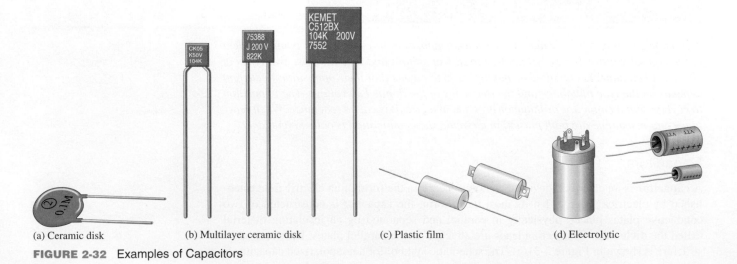

(a) Ceramic disk   (b) Multilayer ceramic disk   (c) Plastic film   (d) Electrolytic

FIGURE 2-32   Examples of Capacitors

large, both in capacitance and physical size. Generally, very large capacitors ($>1~\mu F$) are polarized.

How a Capacitor Stores Energy

Energy is stored in the form of an electric field that is established by the opposite charges stored on the two plates. The electric field is concentrated in the dielectric of the capacitor. When a capacitor is connected across a voltage source, the plates acquire a charge that is positive on one plate and negative on the other. This creates the electric field between the plates, which stores energy. The energy stored is proportional to the capacitance and the square of the voltage across the plates, as expressed by the following formula:

$$W = \frac{1}{2}CV^2$$

Equation 2-21

When capacitance is in farads and voltage is in volts, energy is in joules.

Applications

Capacitors are used in a wide variety of applications. They can be used to block direct current and pass or reject certain ac signals with certain frequencies. They are used in ac to dc converters that change an ac input to a dc output. Large capacitors are also used to store energy in solar power applications and other renewable resource applications to provide power when the resource is not available; however, cost remains a major factor in using capacitors for energy storage.

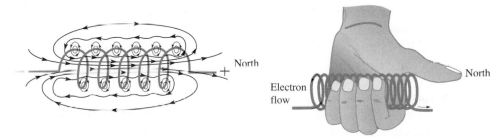

FIGURE 2-33   Coil of Insulated Wire Forms an Inductor. When there is current through the coil of wire, a three-dimensional magnetic field is created. With the fingers of the left hand wrapped toward electron flow, the thumb will point to the North pole.

Inductors

Current through a wire creates a magnetic field around the wire, as discussed in Section 2-7. An inductor is basically a length of insulated wire formed into a coil. Current through the coil creates a much stronger magnetic field than the magnetic field for a straight wire. Figure 2-33 illustrates an inductor with current through it; Figure 2-34 shows some typical inductors.

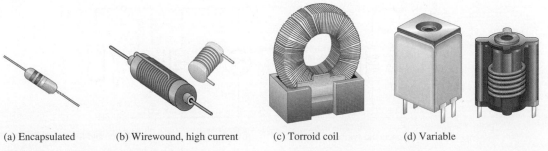

(a) Encapsulated   (b) Wirewound, high current   (c) Torroid coil   (d) Variable

FIGURE 2-34   Several Typical Inductors

Inductance

Inductance is the ability of a wire conductor to oppose a change in current. Inductance is a property of all conductors, but it has only a tiny effect in straight wire conductors. It is much more pronounced in coils because of the enhanced magnetic field surrounding a coil. In fact, coils are usually referred to as inductors because they are selected for their inductive properties. Motors, generators, and transformers are all devices that depend on coils to operate. These topics are covered later in this text.

Unit of Inductance

The unit of inductance is the **henry (H).**

> **The inductance of a coil is one henry when the current through the coil changing at one ampere per second induces one volt across the coil.**

Smaller values of inductance are expressed in millihenries (mH) or in microhenries (μH).

When current starts in an inductor because of an applied voltage, an electromagnetic field expands outward from the coil. This field stores energy. As the field expands, a voltage is induced across the coil and acts in opposition to the original voltage. As a result, it opposes a change in the original current. Once the current has reached a steady value, the electromagnetic field no longer expands, and no voltage is induced in the coil. This fundamental action is defined by **Lenz's law**, which states:

> **When the current through a coil changes, an induced voltage is created across the coil in a direction that always opposes the change in the current.**

You have seen that opposition to current can be from resistance, capacitance, or inductance. For capacitors and inductors, the opposition is called reactance. The combination of reactance and resistance is called impedance.

Transformers

A **transformer** is a device formed by two or more coils (commonly called windings) around a common core that are magnetically coupled to each other and provide for transfer of ac power electromagnetically from one winding to the other. Most power transformers are used to change ac from one ac voltage to another. A transformer will not transfer dc from one winding to the other. A schematic symbol is shown in Figure 2-35. One coil is called the **primary winding,** and the other coil is called the **secondary winding.** An ac source voltage is applied to the primary winding, and a load is connected to the secondary winding, as shown in Figure 2-35(b).

Several power transformers are illustrated in Figure 2-36; all are power types that use an iron core but are very different in size. The basic construction of a power transformer is shown in Figure 2-36(a). Figure 2-36(b) shows a small power transformer such as you might find in an electronic controller or other electronic device. Three utility pole transformers connected to a commercial building are shown in Figure 2-36(c), and very large transformers at a power plant are shown in Figure 2-36(d).

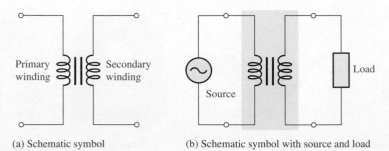

(a) Schematic symbol   (b) Schematic symbol with source and load

FIGURE 2-35 Transformer Schematic

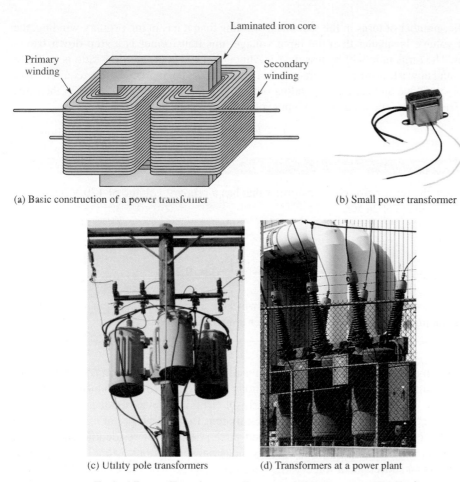

Primary winding

Laminated iron core

Secondary winding

(a) Basic construction of a power transformer

(b) Small power transformer

(c) Utility pole transformers

(d) Transformers at a power plant

FIGURE 2-36 Typical Power Transformers. Power transformers range in size from very small transformers to huge power plant transformers. (*Source:* Parts (b), (c), and (d) David Buchla.)

Transformer Action

Recall that a changing magnetic field can produce an induced voltage, which is the basic idea behind transformer action. A changing current on the primary side produces an electromagnetic field in the core, which follows the changing current. This changing electromagnetic field induces a voltage across the secondary coil. Keep in mind that a transformer can work *only* with an ac input because it is necessary for the magnetic field in the core to change in order to induce the voltage in the secondary winding.

The secondary voltage may be larger, smaller, or equal to the primary voltage; it depends on the ratio of the number of turns in the secondary winding to the number of turns in the primary winding. In some cases, there may be more than one secondary. The number of turns in a given secondary divided by the number of turns in the primary is called the **turns ratio,** given by the following equation:

$$n = N_{sec}/N_{pri} \hspace{3cm} \text{Equation 2-22}$$

where

n is turns ratio

N_{sec} is number of turns in the secondary winding

N_{pri} is number of turns in the primary winding

When the transformer has a larger number of turns in the secondary winding than it has in the primary winding, the output voltage is greater than the input voltage; this transformer is a **step-up transformer.** The turns ratio will be greater than 1. When the transformer has a smaller number of turns in the secondary winding than it has in the primary winding, the output voltage is smaller than the input voltage; this transformer is a **step-down transformer.** The turns ratio will be less than 1. Although the idea of the turns ratio is useful to understand how a transformer operates, most transformers are marked by the required input (primary) voltage and the corresponding output (secondary) voltage rather than the turns ratio. The ratio of output voltage to input voltage also equals the turns ratio:

$$n = V_{sec}/V_{pri}$$

Equation 2-23

EXAMPLE 2-7

What is the turns ratio of a transformer that has a primary voltage of 120 V and a secondary voltage of 24 V?

Solution

$$n = \frac{V_{sec}}{V_{pri}} = \frac{24 \text{ V}}{120 \text{ V}} = \textbf{0.2}$$

The turns ratio is less than 1, which is always the case when the output voltage is less than the input voltage for electronic power transformers.

Transformers range in size from the thumbnail-size coupling transformers used in electronics to huge units weighing hundreds of tons that are used to interconnect portions of power grids. Transformers are found in nearly all electronic devices using utility voltage and are essential for high-voltage power transmission. Two basic types of transformers are used in power applications: single-phase and three-phase. Most residential applications are single-phase; three-phase is used in many industrial applications. More information on transformers and coverage of three-phase transformers is provided in Chapter 14.

SECTION 2-8 CHECKUP

1. What is a capacitor? What is the unit of capacitance?
2. How does a capacitor store energy?
3. What is an inductor?
4. What happens when the current changes in an inductor?
5. What is the unit of inductance?
6. What is Lenz's law?
7. What is impedance?

2-9 Protective Devices

Fuses and circuit breakers are placed in the current path of an electric circuit. They are used to create an open (break in the circuit) when the current exceeds a specified number of amperes due to a malfunction or other abnormal condition. The basic difference between a fuse and circuit breaker is that when a fuse is "blown" by excessive current, it must be replaced, but when a circuit breaker trips (opens), it can be reset and reused repeatedly. Both of these devices protect against damage to a circuit due to excessive current and prevent a hazardous condition created by the overheating of wires and other components. Because fuses cut off excess current more quickly than circuit breakers, fuses are used whenever delicate electronic equipment needs protection.

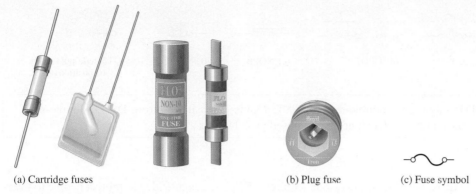

(a) Cartridge fuses   (b) Plug fuse   (c) Fuse symbol

FIGURE 2-37 Typical Fuses

Fuses

Two basic categories of fuses in terms of their physical configuration are cartridge and type and plug. Cartridge fuses have housings of various shapes, with leads or other types of contacts as shown in Figure 2-37(a). A typical plug fuse with a screw-in base is shown in Figure 2-37(b), and the schematic symbol for a fuse is shown in Figure 2-37(c). A **fuse** is a nonresettable protective device that opens the circuit when there is excessive current. Fuse operation is based on the melting temperature of a wire or other metal element. As current increases, the fuse element heats up, and when the rated current is exceeded, the element reaches its melting temperature and opens, thus removing power from the circuit.

Two common types of fuses are the fast-acting and the time-delay (slow-blow). Fast-acting fuses are type F; time-delay fuses are type T. In normal operation, fuses are often subjected to intermittent current surges that may exceed the rated current, such as when power to a circuit is first turned on. A slow-blow fuse can tolerate greater and longer-duration surges of current than a fast-acting fuse can.

Circuit Breakers

Typical circuit breakers are shown in Figure 2-38(a), and the schematic symbol in Figure 2-38(b). Generally, a **circuit breaker** is a device that detects excessive current either by the heating effect of the current or by the electromagnetic field that the current creates and opens the circuit. In a circuit breaker based on the heating effect, a bimetallic spring opens the contacts when the rated current is exceeded. Once opened, the contact is held open by mechanical means until it is manually reset. In a circuit breaker based on an electromagnetic field, the contacts are opened by a sufficient magnetic force created by excess current. These contacts must also be manually reset.

Ground Fault Circuit Interrupter

The ground fault circuit interrupter (GFCI) is a type of circuit breaker that is used to protect from severe or fatal electric shock. A GFCI detects a ground fault, which is an unintentional electric path between a source of current and a grounded surface, resulting in a leakage current. (If your body provides that path to ground for the leakage current, you could be electrocuted.)

(a) Typical small circuit breakers   (b) Schematic symbol

FIGURE 2-38 Typical Circuit Breakers and Schematic Symbol (*Source:* Part (a) image Tom Kissell.)

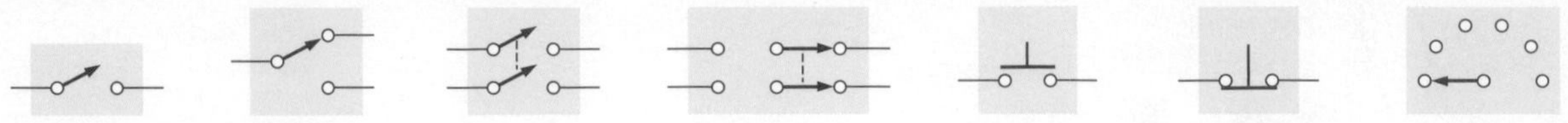

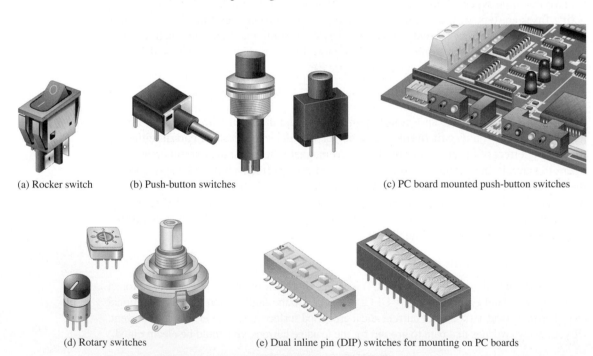

(a) SPST   (b) SPDT   (c) DPST   (d) DPDT   (e) NOPB   (f) NCPB   (g) Single-pole rotary (6-position)

SPST: single-pole, single throw, SPDT: single-pole, double-throw, DPST: double-pole, single-throw, DPDT: double-pole, double throw, NOPB: normally open push button, NCPB: normally closed push button.

FIGURE 2-39 Switch Symbols

The National Electrical Code requires GFCI devices intended to protect people to interrupt (trip) the circuit if the leakage current exceeds 4 to 6 mA of current within 25 ms. A GFCI device that protects equipment (equipment protective device) is allowed to trip at currents as high as 30 mA. GFCI receptacles can be found in kitchens, bathrooms, and other places that can be wet, such as outdoor locations.

Mechanical Switches

Mechanical switches are commonly used for controlling the opening or closing of circuits, such as turning a lamp on or off. When a switch is off, the circuit is open; when a switch is on, the circuit is closed. Switches are usually classified according to the number of positions and contacts. Figure 2-39 shows the types of switches. A switch pole is a contact, and a throw is the movable part. Figure 2-40 shows several varieties of switches.

(a) Rocker switch   (b) Push-button switches   (c) PC board mounted push-button switches

(d) Rotary switches   (e) Dual inline pin (DIP) switches for mounting on PC boards

FIGURE 2-40 Typical Mechanical Switches

SECTION 2-9 CHECKUP

1. What is the purpose of a fuse? Of a circuit breaker?
2. Name two types of fuses in terms of operation.
3. How does a circuit breaker differ from a fuse?
4. Name two ways that circuit breakers detect excessive current.

2-10   Basic Electrical Measurements

To test and troubleshoot circuits, it is necessary to understand how to use basic electronic measuring instruments. The digital multimeter (DMM) is the most widely used electronic measuring instrument. Also, the clamp-type ammeter is useful for measuring current, especially in power applications An older analog meter called the volt-ohm-milliammeter (VOM) is still used in some applications but has largely been replaced by the DMM.

Digital Multimeter

A **digital multimeter (DMM)** is an instrument that can measure several basic electrical quantities and shows the measurement with a number in a display. All multimeters measure ac and dc voltage (voltmeter), ac and dc current (ammeter), and resistance (ohmmeter). Some can also measure other electrical quantities such as frequency or even temperature. A typical DMM with probes is shown in Figure 2-41.

DC Voltage Measurements

To measure voltage, connect the meter leads across the component to be measured (this is a parallel connection). With digital meters, the meter normally sets the polarity automatically, and the sign is indicated on the display. It is convenient to use a red lead in the voltage jack and a black one on common. Then positive or negative readings are easy to interpret. Many DMMs are capable of **autoranging,** a feature in which the DMM selects the optimum range automatically for displaying the reading. The range is the maximum voltage that can be displayed with a particular setting. Lower-priced meters have **manual ranging,** which requires the user to select an appropriate range for the measurement. To make a voltage measurement, first move the selector switch to DC VOLTS; if the meter is a manual-ranging type, choose a range larger than the expected voltage (this is not necessary with an autoranging meter). Sometimes the

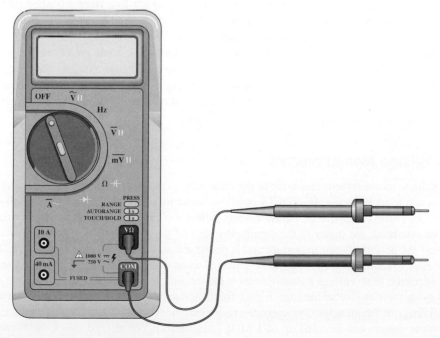

FIGURE 2-41   Typical DMM with probes

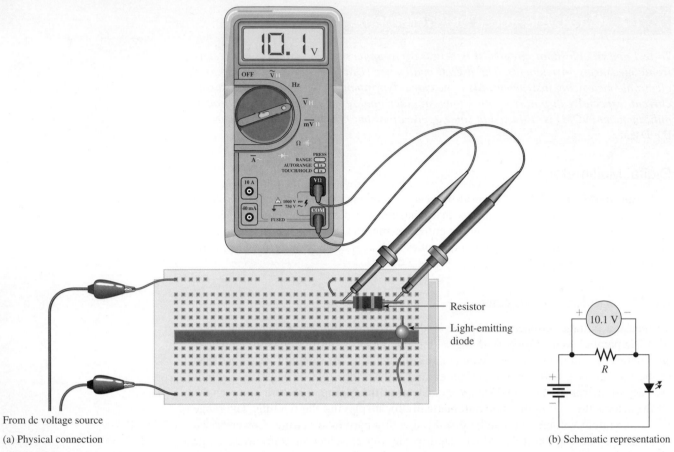

From dc voltage source

(a) Physical connection

(b) Schematic representation

FIGURE 2-42   DC Voltage Measurement. The meter is connected directly across the component to be measured; in this case, the reading is the voltage across the resistor.

dc volts position is indicated by the letter V and a straight line over a dashed line (to remind you of dc). Only after selecting the proper function (DC VOLTS) and range, should you connect the meter to the circuit.

Most meters indicate the voltage no matter which way the leads are connected to the circuit. If the positive meter lead is connected to the more positive voltage, then the voltage reading is displayed as a positive value; otherwise, it is indicated as a negative value. An example of measuring the voltage in a basic circuit on a protoboard is shown in Figure 2-42(a). The schematic equivalent is shown in Figure 2-42(b).

AC Voltage Measurements

AC voltage measurements are done in the same way as dc measurements except the selector switch must be in the AC VOLTS position. Some meters indicate ac with a small sine wave. As is the case with dc measurements, select the function and range (if necessary) before you hook the meter to the circuit. Because ac goes between positive and negative values, the polarity is not important and is displayed as a positive value in any case. It is important to realize that the reading is displayed as an *rms value* for ac. Another important consideration in ac voltage measurements is the frequency you are measuring. DMMs vary widely in their ability to measure higher frequencies; many are accurate only for a very small range of frequencies. Typically, the range may be from 45 Hz to 1 kHz (1 kilohertz), but some meters can be used up to 1 MHz (megahertz). You should check the accuracy specification of a meter.

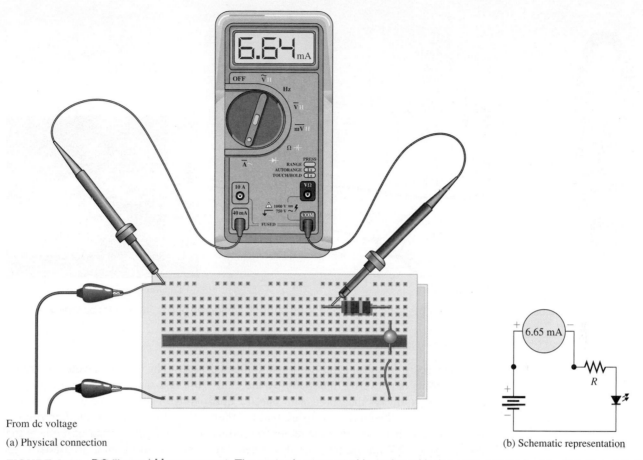

From dc voltage

(a) Physical connection

(b) Schematic representation

FIGURE 2-43   DC Current Measurement. The meter is connected in series with the components by breaking the circuit. In this case, the reading is the current through the resistor and LED.

Current Measurements

For DMMs, select either the AC or DC function (it is necessary to move the probes to separate current jacks on the meter). As in any current measurement, connect the meter in series (in line) with the circuit under test. It is necessary to break a connection in the circuit in order to insert the meter. In this case, the meter is inserted in series between the voltage source and the resistor. Note that, if you are measuring ac, it will be shown on the display as the rms value. A current measurement on a protoboard is shown in Figure 2-43(a); its schematic equivalent is illustrated in Figure 2-43(b).

Caution! If the meter is inadvertently connected across a voltage source, there will be very high current through the meter, causing either a fuse to blow or damage to the meter.

Clamp Meter

A useful meter for measuring alternating and direct current is the **clamp meter,** which does not require the circuit to be broken to insert the meter. The meter may be a stand-alone unit or part of a traditional DMM. Figure 2-44(a) shows a typical clamp meter, and Figure 2-44(b) shows a technician using a clamp meter (shown held in the technician's right hand). The sensing element is a set of jaws that can be opened and closed around a single conductor. The meter senses the magnetic field from the current and converts it directly to a current reading.

Resistance Measurements

DMMs measure resistance by using an internal current source to provide an accurate, fixed current through the unknown resistance. This develops a voltage across the

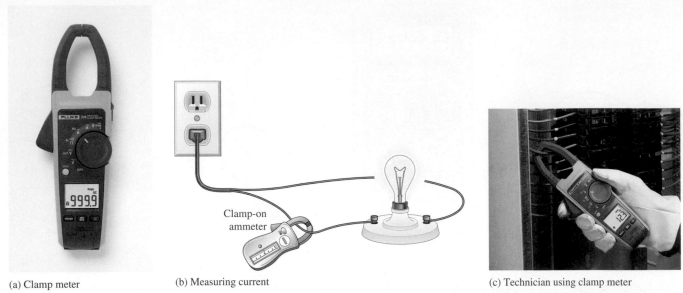

(a) Clamp meter   (b) Measuring current   (c) Technician using clamp meter

FIGURE 2-44   Clamp Meter. The clamp meter makes it easy to measure current.   (*Source:* Parts (a) and (c) courtesy of Fluke Corp.)

unknown resistance that is proportional to the resistance. The voltage developed across the unknown is converted internally by the meter to a resistance value and displayed. Resistance is always measured with one (or both) ends of the resistor under test disconnected from the circuit to ensure that another path in the circuit or another voltage is not present that can cause the meter to produce an erroneous reading or damage the meter. Figure 2-45(a) on page 81 illustrates a resistance reading on a protoboard. Its schematic equivalent is shown in Figure 2-45(b). Notice that one end of the resistor is disconnected from the source.

DMMs frequently have a continuity test as part of the ohmmeter function. A **continuity test** is a test of an electrical path to verify the path. An audible beep indicates a conducting path between the probes, making the test a quick check of wiring because you don't have to look at the display. The continuity tester is a handy feature for tracing wiring and for testing for open or shorted paths on circuit boards or other electrical systems. Some meters use two tones to differentiate between a reading that is less than 1% of the selected range and a reading that is less than 10% of the selected range.

Certain renewable energy applications require resistance measurements that are beyond the range of most DMMs. For example, if you need to measure the contact resistance of a switch or circuit breaker, a precision instrument that can measure resistance in the μohm region is required. In other cases, such as the determination of the insulation resistance of cables, extremely high resistance measurements may be needed. Again, specialized instruments are available for these types of measurements.

SECTION 2-10 CHECKUP

1. What measurements can be made with a basic DMM?

2. What are two procedures that must be done when connecting a DMM for a current measurement?

3. Why is it necessary to disconnect a voltage source when using a DMM to measure resistance?

4. Is it easier to measure current with a DMM or a clamp-on ammeter?

5. What is a continuity check?

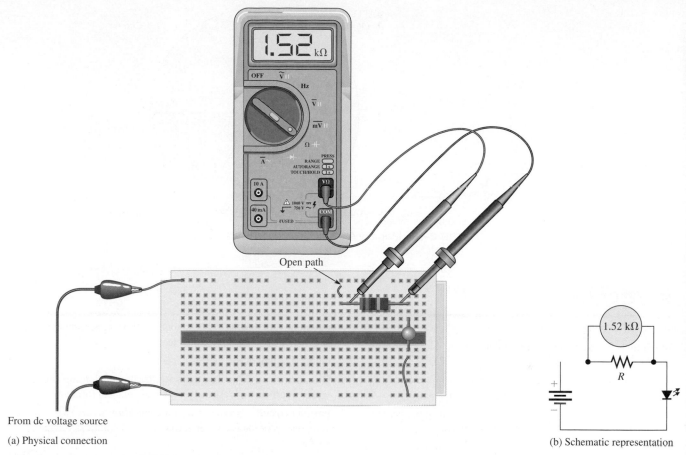

Open path

From dc voltage source

(a) Physical connection

(b) Schematic representation

1.52 kΩ

R

FIGURE 2-45 Resistance Measurement. The meter is connected across the resistor after first disconnecting one end of the resistor. In this case, the reading is the resistance of the resistor.

CHAPTER SUMMARY

- Energy is the ability to do work, and power is rate at which energy is used. The unit of energy is the joule or kilowatt-hour; the unit of power is the watt.

- Voltage is the energy used to move charge from one point to another.

- The electron is the basic element of negative charge.

- Current is the rate of flow of charge. Current is produced by voltage across a load.

- Resistance is the opposition to current. Ohm's law states the relationship among voltage, current, and resistance.

- Power can be expressed in terms of voltage, current, and resistance. Watt's law specifies this relationship.

- Total resistance is increased with a series connection. Total resistance is decreased with a parallel connection.

- Conductors are used to carry current. Insulators are used to prevent current. Semiconductors are materials with properties between insulators and conductors.

- Voltage, current, and resistance can be measured with a DMM. Voltage is measured across a component, current is measured in series with a component, and resistance is measured across a

- component with one end of the component disconnected from the circuit.

- Alternating current can be measured with a clamp meter, which goes around a conductor without electrical contact.

- Opposite poles of a magnet attract and like poles repel.

- A magnetic field is produced when there is current through a conductor.

- When a conductor moves through a magnetic field, a voltage is produced.

- A capacitor is an electrical component that stores energy in an electric field.

- An inductor is an electrical component that stores energy in a magnetic field. It is often called a coil.

- Reactance is the opposition to alternating current exhibited by a capacitor or an inductor.

- Impedance is the opposition to alternating current caused by a combination of resistance and reactance.

- Fuses and circuit breakers are protective devices that create on open circuit when the current exceeds a specified value.

- A memory aid for Ohm's and Watt's laws is provided in Figure 2-46.

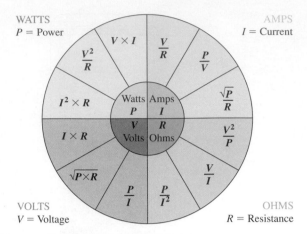

FIGURE 2-46   Memory Aid for Ohm's and Watt's Laws

KEY TERMS

ampere   The unit of current.

capacitance   The amount of charge that a capacitor can store per unit of voltage across its plates.

charge   A property of matter resulting from the presence of an extra electron or the absence of an electron in the atom.

conductors   Materials that allow the free movement of charge. They can be composed of solids, liquids, or gases.

current   The flow of electrical charge.

digital multimeter (DMM)   An instrument that can measure voltage, current, and resistance.

energy   The ability or capacity for doing work.

inductance   The property of a wire conductor to oppose a change in current.

insulators   Materials that prevent the free movement of charge. Also known as *nonconductors*.

joule   The SI unit of energy. The work done when 1 newton of mechanical force is applied over a distance of 1 meter.

kilowatt-hour   A unit of energy. The energy used when 1,000 watts of power are expended in one hour.

Lenz's law   The law that states that when the current through a coil changes, an induced voltage is created across the coil in a direction that always opposes the change in the current.

magnetic flux   The group of force lines going from the north pole to the south pole of a magnet.

magnetic flux density   The amount of flux, ϕ, per unit area (A) perpendicular to the magnetic field.

ohm   The unit of resistance.

Ohm's law   A circuit law that specifies the relationship among voltage, current, and resistance as a mathematical formula.

parallel circuit   A type of circuit connection where two or more components or loads are connected across a common voltage source.

power   The rate at which energy is expended.

resistance   The opposition to current.

semiconductors   A crystalline material that has four electrons in its valence shell and has properties between those of conductors (metals) and insulators (nonmetals).

series circuit   A type of circuit connection in which there is a single complete path (forming a string) from the voltage source through the load (or loads) and back.

sinusoidal wave   The cyclic pattern of ac voltage or current. Also known as *sine wave.*

volt   The unit of voltage.

voltage   Energy per unit charge.

watt   The unit of power.

Watt's law   A circuit law that expresses the relationship of voltage, current, resistance, and power as a formula.

FORMULAS

Equation 2-1	$W_{PE} = mgh$	Potential energy
Equation 2-2	$W_{KE} = \frac{1}{2}mv^2$	Kinetic energy
Equation 2-3	$V = W/Q$	Voltage
Equation 2-4	$f = 1/T$	Frequency
Equation 2-5	$T = 1/f$	Period
Equation 2-6	$I = Q/t$	Current
Equation 2-7	$V = IR$	Ohm's law for voltage
Equation 2-8	$I = V/R$	Ohm's law for current
Equation 2-9	$R = V/I$	Ohm's law for resistance

Equation 2-10	$P = W/t$	Power
Equation 2-11	$W = Pt$	Energy
Equation 2-12	$t = W/P$	Time
Equation 2-13	$P = VI$	Watt's law
Equation 2-14	$P = I^2R$	Watt's law
Equation 2-15	$P = V^2/R$	Watt's law
Equation 2-16	$R_T = R_1 + R_2 + R_3 + \ldots + R_n$	
	Resistances in series	
Equation 2-17	$V_T = V_1 + V_2 + V_3 + \ldots + V_n$	
	Voltages in series	

Equation 2-18
$$R_T = \cfrac{1}{\cfrac{1}{R_1} + \cfrac{1}{R_2} + \cfrac{1}{R_3} + \ldots + \cfrac{1}{R_n}}$$
Resistances in parallel

Equation 2-19	$B = \phi/A$	Magnetic flux density
Equation 2-20	$C = Q/V$	Capacitance
Equation 2-21	$W = \dfrac{1}{2}CV^2$	Energy stored in a capacitor
Equation 2-22	$n = N_{sec}/N_{pri}$	Transformer turns ratio
Equation 2-23	$n = V_{sec}/V_{pri}$	Transformer voltage ratio

CHAPTER TRUE/FALSE QUIZ

Determine whether each statement is true or false. Answers are at the end of the chapter.

1. Energy cannot be changed from one form to another.
2. The electron is the smallest particle of negative charge.
3. Voltage is independent of charge.
4. In a solid conductor, current is the flow of electrons.
5. The unit of current is the coulomb.
6. Current will decrease if the resistance increases with the same voltage.
7. If voltage is increased across a given resistance, the current increases.
8. Power and energy are the same.
9. The unit of power is the watt.
10. Electrical energy can be measured in either joules or kilowatt-hours.
11. When batteries are connected in series, the voltage decreases.
12. When solar cells are connected in parallel, more current is available.
13. Copper is one of the best conductors of electricity.
14. AWG is a standard for specifying wire sizes.
15. AWG 10 wire is smaller than AWG 12.
16. The diode, PV cell, transistor, and SCR are all examples of semiconductor devices.

CHAPTER MULTIPLE-CHOICE QUIZ

Complete each statement by selecting the one correct answer. Answers are at the end of the chapter.

1. The unit of energy is the
   a. watt
   b. joule
   c. coulomb
   d. kilowatt

2. A coulomb is the unit of
   a. current
   b. energy
   c. power
   d. charge

3. Voltage is defined as
   a. current per unit of charge
   b. energy per unit of charge
   c. charge per unit of energy

4. Current is defined as the flow of
   a. charge
   b. protons
   c. power

5. The unit of current is the
   a. ohm
   b. joule
   c. ampere
   d. watt

6. Resistance is the
   a. opposition to a change in current
   b. opposition to a change in voltage
   c. opposition to current
   d. ability to conduct current

7. When a fixed voltage is applied across a resistance and the resistance is increased, the current
   a. increases
   b. decreases
   c. stays the same

8. When the voltage across a fixed resistor is decreased, the current
   a. increases
   b. decreases
   c. stays the same

9. Power is the
   a. same as energy
   b. rate at which energy is used
   c. same as force

10. Power can be determined if you know the
    a. voltage
    b. current
    c. resistance
    d. any two of these

11. The total resistance of several resistors connected in series is
    a. equal to the largest resistance
    b. less than the smallest resistance
    c. the sum of all the resistances

12. When several solar cells are connected in parallel,
    a. the voltage is increased
    b. the available current is increased
    c. both voltage and current are increased

13. The most commonly used conductor for electrical applications is
    a. silver
    b. aluminum
    c. copper
    d. gold

14. The wire with the greatest diameter is
    a. AWG 10
    b. AWG 24
    c. AWG 18

15. A component that stores energy in an electric field is the
    a. inductor
    b. resistor
    c. capacitor
    d. thyristor

16. A transformer can
    a. step up ac voltage
    b. step down ac voltage
    c. operate with dc voltage
    d. both (a) and (b)

17. When a fuse blows, the current
    a. increases
    b. decreases
    c. ceases
    d. stays the same

CHAPTER QUESTIONS AND PROBLEMS

1. Determine the voltage when 500 J of energy is used to move 10 C of charge.

2. A certain voltage sine wave goes through 5,000 cycles in 2 seconds. Determine the frequency.

3. The frequency of the ac utility voltage is 60 Hz. What is the period?

4. Determine the current when 50 coulombs of charge move past a point in 0.5 seconds.

5. What is the equivalent dc voltage for 50 V rms?

6. Find the resistance of 50 feet of 14 gauge wire.

7. Calculate the voltage across a 10 Ω resistor if the current is 1.5 A.

8. Calculate the current through a 33 Ω resistor if there is 12 V across it.

9. Calculate the resistance if $V = 24$ V and $I = 1.2$ A.

10. How much energy is dissipated when a 100 W lightbulb is on for 24 hours?

11. Determine the power in each of the resistances in problems 7, 8, and 9.

12. Find the total resistance for the following resistors in series: 10 Ω, 47 Ω, 75 Ω, and 100 Ω.

13. If the resistors in Problem 12 are connected in parallel, what is the total resistance?

14. Assume a solar cell produces 0.5 V at peak sunlight. If a total of 12 V is needed, what is the minimum number of solar cells required and how must they be connected?

15. How would you triple the current capacity of the solar cell array in problem 14 if you keep the same voltage?

16. Determine the energy stored by a 10 μF capacitor with 24 V across it.

17. Determine the power dissipated if 25 V is measured across a 100 Ω resistance.

18. What is the ampacity of an AWG 6 copper wire?

19. Find the resistance for 500 ft of an AWG 18 copper wire.

20. Calculate the voltage drop across the wire in problem 21 if there are 2 A in it.

21. Determine the minimum AWG size of copper wire if the application requires 8 A.

22. Determine the minimum fuse rating required for an electric heater that operates on 120 V and the power in the heater is 1,200 W.

FOR DISCUSSION

What role, if any, will superconductors play in saving energy in the future?

ANSWERS TO CHECKUPS

Section 2-1 Checkup

1. Energy is the ability or capacity for doing work; it is measured in joules in the SI system.
2. The electron
3. The coulomb
4. Energy per charge; the unit is the volt, symbolized by V.
5. DC voltage and ac voltage
6. Frequency is the number of cycles per second measured in hertz. Period is the time for one cycle, measured in seconds.

Section 2-2 Checkup

1. Current is the flow of electrical charge past a specified point in a circuit; its unit is the ampere.
2. 60 Hz
3. Load is a component that uses the power produced by a circuit.
4. rms = root-mean-square
5. DC is continuous; ac is cyclic.

Section 2-3 Checkup

1. Resistance is opposition to current; the unit is the ohm.
2. Fixed resistors and variable resistors
3. Potentiometers and rheostats; the schematic symbols are shown in Figure 2-7.
4. A circular mil is the area of a wire that is one thousandth of an inch (one mil) in diameter.
5. AWG stands for "American Wire Gauge" and the system uses a number related to the cross-sectional area of the wire.
6. Ampacity is the maximum current a wire can carry under certain specified conditions.
7. $V = IR$
8. $I = \dfrac{V}{R}$
9. $R = \dfrac{V}{I}$
10. The current will decrease.

Section 2-4 Checkup

1. Power is the rate of doing work; the unit for power is the watt.
2. One joule per second
3. The kilowatt-hour is a unit of energy; the abbreviation is kWh.
4. $P = VI, P = I^2R, P = V^2/R$
5. The power is 100 W and 75 W, so the 100 W load dissipates higher power. The energy used is higher for the 75 W load because it is on for twice as long (100 Wh versus 150 Wh).

Section 2-5 Checkup

1. A series circuit is one with one path for current.
2. 0.524 A
3. A parallel circuit is one with two or more paths for current.

4. The smallest value resistor has the largest current.
5. To get more voltage, connect the batteries in series; to increase current capacity, connect them in parallel.

Section 2-6 Checkup

1. Positive ions
2. Electrons are free to move within the structure and are not bound to any given atom.
3. Ionic solutions
4. Insulation is used for isolation from other conductors and to avoid shock hazard.
5. A crystalline material that has properties between a conductor and an insulator
6. A diode is a device that allows current in only one direction.
7. PV stands for "photovoltaic." A PV cell (solar cell) is a type of diode that allows sunlight to penetrate the semiconductor regions, which causes atoms to release electrons. When electrons cross the *pn* junction, a voltage is generated.
8. It is a transistor used for high-voltage, high-current switching.
9. Silicon-controlled rectifier

Section 2-7 Checkup

1. All magnets have two poles: north and south. Like poles repel; unlike poles attract.
2. Moving charge (current) in a conductor creates a magnetic field.
3. A motor converts electrical energy to mechanical (rotational) energy. A generator does the opposite.
4. Rotating armature and rotating field.

Section 2-8 Checkup

1. A capacitor is a component with the ability to store charge. The unit of capacitance is the farad.
2. Capacitors store energy in the electric field between the plates.
3. An inductor is an insulated coil of wire that has a magnetic field when current is in it.
4. Changing current induces a voltage across the inductor.
5. The henry
6. A changing current in a coil induces a voltage that opposes the change in the current.
7. Impedance is total opposition to ac from the combination of resistance and reactance.

Section 2-9 Checkup

1. To protect a circuit by breaking the path when current exceeds a certain specified amount
2. Cartridge and plug types
3. A circuit breaker can be reset after the condition that tripped it is corrected; a fuse is a one-time device and must be replaced.
4. Circuit breakers can operate based on a heating effect or by detecting the electromagnetic field.

Section 2-10 Checkup

1. Voltage (ac and dc), current (ac and dc) and resistance

2. Probes are moved to a current jack and the circuit is broken to insert the meter.

3. If the voltage is connected, the current will cause an erroneous reading or damage to the meter.

4. Generally, a clamp on an ammeter is easier to use because the circuit does not need to be broken (but the current needs to be isolated).

5. A check of an electrical path

ANSWERS TO TRUE/FALSE QUIZ

1. F 2. T 3. F 4. T 5. F 6. T 7. T 8. F   14. T 15. F 16. T
9. T 10. T 11. F 12. T 13. T

ANSWERS TO MULTIPLE-CHOICE QUESTIONS

1. b 2. d 3. b 4. a 5. c 6. c 7. b   14. a 15. c 16. d 17. c
8. b 9. b 10. d 11. c 12. b 13. c

Solar Photovoltaics

CHAPTER OBJECTIVES

- Discuss the atom and describe the crystalline structure of silicon.
- Describe the doping process in a semiconductor.
- Describe the *pn* junction and how it functions.
- Describe photovoltaic (PV) cell structure and operation.
- Define and describe several characteristics of a PV cell.
- Calculate the conversion efficiency of a PV cell.
- Discuss how PV cells are used to create solar modules and arrays, and calculate output current, voltage, and power to a load.
- List and describe five or six important specifications on solar module data sheets.
- Discuss various semiconductor materials used in PV cells.
- Explain multijunction cells and laser scribing.
- Discuss concentrating photovoltaic (CPV) systems.

KEY TERMS

Key terms are shown in bold and color. Definitions for key terms are provided at the end of the chapter and in the end-of-book glossary. Bold terms in black are defined in the end-of-book glossary only.

- *pn* junction
- valence electrons
- doping
- free electron
- hole
- photovoltaic (PV) cell
- photovoltaic effect
- photon
- band gap
- crystallinity
- thin-film photovoltaic
- multijunction
- efficiency
- fill factor (*FF*)
- solar module
- solar array
- concentrating photovoltaic (CPV)
- high-concentration photovoltaic (HCPV)

INTRODUCTION

In this chapter, you will learn about the atomic structure of silicon and how silicon and other materials are used to make photovoltaic (PV) cells that convert energy from the sun to electrical energy. The terms *PV cell* and *solar cell* refer to the same device and are often used interchangeably. PV cells are the basic component of all solar power systems and they are available in a variety of materials, shapes, and sizes. When many cells are arranged and connected together, the result is a solar module. Modules are then connected together to form solar arrays, which can vary from a small residential unit to large grid-connected solar systems that may cover many acres.

Several basic types of solar renewable energy systems are in service throughout the world: flat-panel arrays that use wafer-type PV cells, arrays that use thin-film PV cells and can have a variety of surface shapes, and concentrating PV systems that use lenses and/or mirrors to focus light onto PV cells. Each system has advantages and disadvantages.

3-1   The *PN* Junction

The fact that certain solid materials are sensitive to light was a chance discovery during an investigation of silicon as a radar detector by Russell Ohl in 1940. A seam in the slab of silicon he was using separated it into a p-region and an n-region. Ohl discovered that exposure to bright light caused a considerable increase in current. Thus was born a basic photovoltaic device. In this section, the pn *junction, which forms the basis of photovoltaic devices, is introduced.*

Silicon (Si) is the most common element used in the construction of photovoltaic solar cells. Recall that the addition of impurities to pure silicon (or other semiconductors) can change the conduction properties and form *p* and *n* materials depending on the specific impurity. If part of a small block of silicon is doped with an *n*-type impurity and the other part with a *p*-type impurity, the boundary created between them is called a ***pn* junction**. The *pn* junction is essential to the operation of solar cells and many other semiconductor devices.

The Silicon Atom and Crystal Structure

Silicon is one of the most abundant elements on earth and is found in sand and quartz as well as other natural sources. Silicon can form a crystalline structure (designated c-Si), or it can form an amorphous type of silicon that has a random arrangement of atoms (designated a-Si). Both types are useful in solar cells. In this section, we will focus on c-Si and its use.

In c-Si, there is an orderly arrangement of covalently bonded silicon atoms. According to the Bohr model, an atom consists of a nucleus of protons and neutrons surrounded by orbiting electrons at various distances from the nucleus called shells. The neutral silicon atom has 14 protons, 14 to 16 neutrons, and 14 electrons. The electrons in the outer shell of an atom are known as **valence electrons**, as shown in the atomic model for silicon in Figure 3-1(a) on page 89. This outer shell of a semiconductor or insulator is known as the valence band and includes electrons that are bound to the atom. The valence electrons participate in bonding to other atoms. In a silicon crystal, each silicon atom shares an electron with its four neighboring atoms, creating a four covalent bond as shown in Figure 3-1(b). Each atom in the crystal is bonded to its neighbors. Figure 3-1(c) illustrates the structure of a silicon crystal that has multiple atoms held together by the covalent bonds.

Doping a Silicon Crystal

As mentioned, silicon in its pure (intrinsic) state is a poor electrical conductor. The process of **doping** is used to increase the conductivity in a precise and controlled way. The doping process adds impurity atoms in very low concentrations to the silicon to create an excess of free electrons (*n*-type impurity). The place where a deficiency occurs is called a hole. A *pentavalent* impurity atom has five valence electrons and, when added to a silicon crystal, four of its electrons form covalent bonds with four adjacent silicon atoms. One electron is left and becomes an unbonded or **free electron**, as illustrated in Figure 3-2(a). Examples of pentavalent atoms are phosphorous (P), arsenic (As), antimony (Sb), and bismuth (Bi).

A *trivalent* impurity atom has three valence electrons and, when added to a silicon crystal, three of its electrons form covalent bonds with four adjacent silicon atoms. An electron vacancy or **hole** is created in one of the bonds, as illustrated in Figure 3-2(b). Examples of trivalent atoms are boron (B), indium (In), and gallium (Ga).

Formation of the *PN* Junction

Figure 3-3(a) shows the formation of the *pn* junction between *n*-type and *p*-type regions of a piece of silicon that has been doped with phosphorous in the top part and with boron in the bottom part. The free electrons in the *n* region are randomly drifting in all directions.

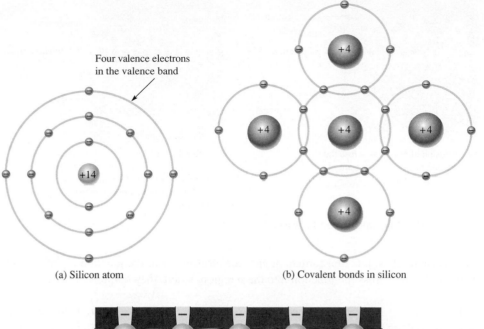

(a) Silicon atom

(b) Covalent bonds in silicon

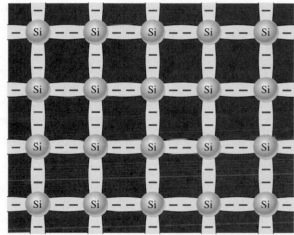

(c) Bonding diagram for a silicon crystal

FIGURE 3-1   Silicon Atom and Covalent Bonding Forms a Crystal Structure

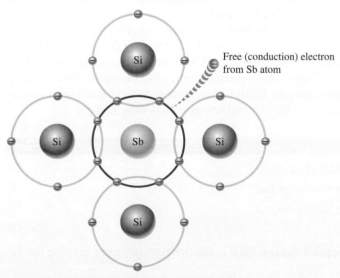

(a) Pentavalent impurity atom (antimony) in a silicon crystal. The extra electron becomes a free electron.

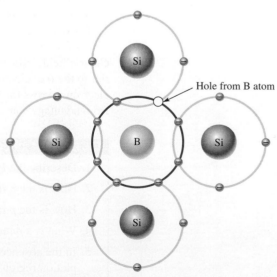

(b) Trivalent impurity atom (boron) in a silicon crystal. A hole (electron vacancy) is created.

FIGURE 3-2   Doping a Silicon Crystal

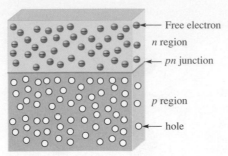

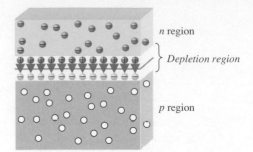

(a) Diffusion of electrons across the junction begins at the instant of junction formation.

(b) After the initial surge of diffusion, a depletion region forms near the junction. A state of equilibrium is established and there is no further diffusion.

FIGURE 3-3 *PN* Junction and Depletion Region

At the instant of *pn* junction formation, the free electrons near the junction in the *n* region begin to diffuse across the junction into the *p* region, where they combine with holes near the junction.

Before the *pn* junction is formed, there are as many electrons as protons in the *n*-type material, making the material neutral in terms of charge. The same is true for the *p*-type material. When the *pn* junction is formed, the *n* region loses free electrons as they diffuse across the junction. This creates a layer of positive charges (pentavalent ions) near the top side of the junction. As the electrons move across the junction, the *p* region loses holes as the free electrons and holes combine. This action creates a layer of negative charges (trivalent ions) near the bottom side of the junction. These two layers of positive and negative charges form the depletion region, as shown in Figure 3-3(b). The depletion region is depleted of charge carriers (electrons and holes).

Formation of the Depletion Region

The depletion region is very thin compared to the *p* and *n* regions. Its width is exaggerated in Figure 3-3(b) for purposes of illustration. After the initial surge of free electrons across the *pn* junction, the depletion region expands to a point where equilibrium is established and there is no further diffusion of electrons across the junction. This process occurs because, as electrons continue to diffuse across the junction, more and more positive and negative charges are created near the junction and the depletion region forms. A point is reached where the total negative charge in the depletion region repels any further diffusion of electrons into the *p*-region and the diffusion stops.

The depletion region has many positive charges and many negative charges on opposite sides of the *pn* junction. The force between the positive and the negative charges form an electric field, indicated by the blue arrows in Figure 3-3(b). The electric field acts as a barrier to the free electrons in the *n* region, and external energy must be applied to move these electrons across the depletion region. In diodes and transistors, the external energy comes from a voltage source; in solar cells, the external energy comes from sunlight.

SECTION 3-1 CHECKUP

1. Describe the Bohr model of an atom.

2. How is the *n* region in silicon created?

3. How is the *p* region in silicon created?

4. What is a *pn* junction?

5. In the absence of an applied voltage, what keeps the electrons from crossing the depletion region?

3-2   Photovoltaic Cell Structure and Operation

The key feature for conventional PV (solar) cells is the pn *junction that was discussed in the previous section. In the* pn *junction solar cell, sunlight provides sufficient energy to the free electrons in the* n *region to allow them to cross the depletion region and combine with holes in the* p *region. This energy creates a potential difference (voltage) across the cell. When an external load is connected, the electrons flow through the semiconductor material and provide current to the external load.*

PV Cell Structure

Although there are other types of solar cells and continuing research promises new developments in the future, the crystalline silicon PV cell is by far the most widely used. A silicon **photovoltaic (PV) cell** converts the energy of sunlight directly into electricity—a process called the photovoltaic effect—by using a thin layer or wafer of silicon that has been doped to create a *pn* junction. The depth and distribution of impurity atoms can be controlled very precisely during the doping process. As shown in Figure 3-4, the thin silicon circular wafers are first sliced from an ingot of ultra-pure silicon and then the surface of the wafer is textured for better light absorption. The circular wafer may be trimmed to an octagonal, hexagonal, or rectangular shape for maximum coverage when fitted in a module. One commonly used process for creating an ingot is called the Czochralski method. In this process, a seed crystal of silicon is dipped into melted silicon. As the seed crystal is withdrawn and rotated, a cylindrical ingot of silicon is formed.

FIGURE 3-4   Making a Wafer from a Silicon Ingot

The silicon wafer is doped to create the *pn* junction structure. The *n* region is much thinner than the *p* region to permit light penetration. As shown in Figure 3-5(a), a grid of very thin conductive contact strips is deposited on top of the wafer by methods such as photoresist or silkscreen. The contact grid must maximize the surface area of the silicon wafer that will be exposed to the sun in order to collect as much light energy as possible. The conductive grid is necessary so that the electrons have a shorter distance to travel through the silicon when an external load is connected. The more distance an electron travels through a material, the more the energy loss due to the inherent resistance of the material. A solid contact covering the bottom of the wafer is then added, as indicated in Figure 3-5(a).

After the contacts are incorporated, an anti-reflective coating is placed on top of the contact grid and *n* region. The purpose of the coating is to allow the PV cell to absorb as much of the sun's energy as possible by reducing the amount of light energy reflected away from the surface of the cell. The thickness of the PV cell compared to the surface area is greatly

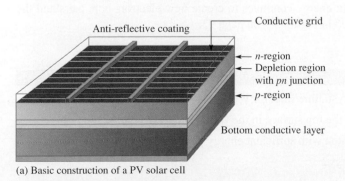

(a) Basic construction of a PV solar cell

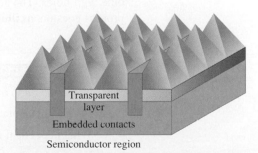

(b) Example of a type of transparent surface texturing (greatly enlarged)

FIGURE 3-5   Basic Construction of a PV Solar Cell and an Example of Transparent Surface Texturing

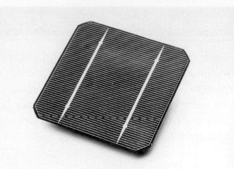

FIGURE 3-6   Complete PV Solar Cell   (*Source:* National Renewable Energy Laboratory.)

exaggerated for purposes of illustration. In some PV cells, the contact grid is embedded in a textured surface consisting of tiny pyramid shapes that result in improved light capture. A small segment of a cell surface is illustrated in Figure 3-5(b). A complete PV cell with a standard surface grid is shown in Figure 3-6.

Operation of a PV Cell

Sunlight is composed of **photons**, or packets of energy. The sun produces an astonishing amount of energy. The small fraction of the sun's total energy that reaches the earth is enough to meet all of our power needs many times over if it could be harnessed. Sufficient solar energy strikes the earth each hour to meet worldwide demands for an entire year.

The *n*-type layer of a PV cell is very thin to allow light penetration into the *p*-type region. The thickness of the entire cell is actually about the thickness of an eggshell. When a photon penetrates either the *n* region or the *p* region and strikes a silicon atom near the *pn* junction with sufficient energy to knock an electron out of the valence band, the electron becomes a free electron and leaves a hole in the valence band. This is called an electron–hole pair. The amount of energy required to free an electron from the valence band of a silicon atom is called the **band gap** and is 1.12 eV (electron volts). The electron volt is a unit of energy used for convenience at the atomic level and specifically for electrons. In the *p* region, the free electron is swept across the depletion region by the electric field into the *n* region. Electrons accumulate in the *n* region, creating a negative charge. In the *n* region, the hole is swept across the depletion region by the electric field into the *p* region. The electrons that accumulate in the *n* region create a negative charge, and holes that accumulate in the *p* region, create a positive charge. A voltage is developed between the *n* region and *p* region contacts.

When a load is connected to a PV cell, the free electrons flow out of the *n* region to the grid contacts on the top surface, out the negative contact, through the load, back into the positive contact on the bottom surface, and then into the *p* region, where they can recombine with holes. The sunlight energy continues to create new electron–hole pairs and the process goes on, as illustrated in Figure 3-7 on page 93.

SECTION 3-2 CHECKUP

1. Name the parts of a crystalline silicon PV cell.
2. Why is the contact on the top surface in the form of a grid?
3. What happens if a photon with sufficient energy penetrates into the *p* region?
4. What is band-gap energy?
5. What is an electron–hole pair?

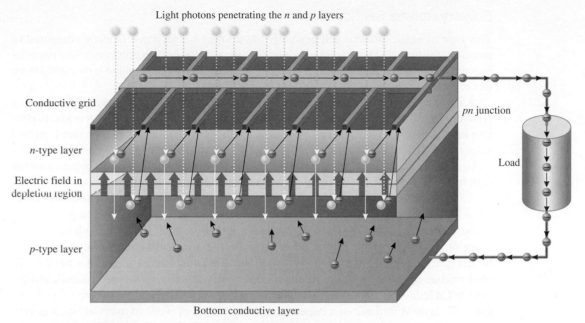

Light photons penetrating the *n* and *p* layers

Conductive grid

pn junction

n-type layer

Load

Electric field in
depletion region

p-type layer

Bottom conductive layer

FIGURE 3-7   Basic Operation of a PV Cell with Incident Sunlight

3-3   Types of Photovoltaic Technologies

The preceding sections focused on wafer-type crystalline silicon PV cells because they are the most widely used. The three main types of photovoltaic materials include two types of crystalline semiconductors and amorphous silicon thin film. These three types account for the most market share. Two other types of PV cells that do not rely on the pn *junction are dye-sensitized solar cells and organic photovoltaic cell. PV technology is a rapidly growing field and many improvements, especially in efficiency and cost, can be expected.*

Basic Types of Photovoltaic Materials

Photovoltaic cells are made from a variety of semiconductor materials that vary in performance and cost. Basically there are three main categories of conventional solar cells: monocrystalline semiconductor, polycrystalline semiconductor, and amorphous silicon thin-film semiconductor.

Monocrystalline Semiconductor

The **crystallinity** of a material indicates how perfectly ordered the atoms are in the crystal structure. The atoms making up a crystal are repeated in a regular, orderly manner. The uniformity of molecular structure of monocrystalline semiconductor (single-crystal) is ideal for electrons to move efficiently through the material.

An example of a monocrystalline semiconductor is monocrystalline silicon. This is the most widely used type of silicon in wafer-type solar cells because it has the highest efficiency. The drawback is that it is also the most expensive. Typically, the efficiency of monocrystalline Si cells ranges from 14% to 18%, although occasionally you will see 19% to 20% specified. The basic structure of crystalline silicon solar cells were covered in section 3-2.

Another monocrystalline semiconductor is gallium arsenide (GaAs), which is a compound. It is a better absorber of photons than silicon is and generally exhibits higher efficiency. One disadvantage of GaAs is that it is more expensive than silicon to use in PV cells. Its main application is in concentrating PV applications, where a much smaller amount of material is required in a cell. GaAs cells are also commonly used in satellites and other space applications because of their high resistance to heat.

Polycrystalline Semiconductor

This type of semiconductor cell generally has a lower conversion efficiency compared to monocrystalline cells, but manufacturing costs are also lower. The polycrystalline material is composed of numerous smaller crystals so that the orderly arrangement is disrupted from one crystal to another.

A common example of a polycrystalline cell is polycrystalline silicon. Cell efficiency typically is 13% to 15%. Polycrystalline silicon is also widely used because it is less expensive than monocrystalline silicon. A variation on the polycrystalline silicon wafer is ribbon silicon, which is formed by drawing flat thin films from molten silicon. Although less efficient, ribbon silicon cells are less costly because their production does not require sawing wafers from ingots.

Another polycrystalline cell is cadmium-telluride (CdTe) thin-film. It uses the compound CdTe, which is a semiconductor with a band gap of 1.44 eV. This is a good match to the solar spectrum because it utilizes the portion of the solar spectrum with the highest energies. It also has high photon absorption. It is less expensive than silicon and has a higher efficiency than amorphous silicon. Figure 3-8 shows the structure of a basic CdTe photovoltaic cell. The cadmium sulfide (CdS) layer is doped as an *n*-type material, and the thicker CdTe layer is doped as *p*-type and is the main energy absorber. In some variations, the CdTe layer is intrinsic (not doped) and another layer of *p*-doped material, such as zinc telluride (ZnTe), is added below to create an electric field through the CdTe layer from the CdS layer to the ZnTe layer. This is done because *p*-type CdTe exhibits significant resistance, whereas intrinsic CdTe does not.

Another example of a polycrystalline semiconductor is copper-indium-gallium-diselenide (CuInGaSe$_2$ CIGS]). This compound semiconductor material is commonly used in thin-film PVs. It typically has higher efficiency than CdTe, and it has one of the highest absorption ratings of all semiconductors. It can also produce more current per unit area than other thin-film technologies. Figure 3-9 illustrates the basic structure, although several variations are common.

Amorphous Silicon Thin-Film Semiconductor

This type of silicon is noncrystalline and can absorb up to forty times more solar radiation than moncrystalline silicon. **Thin-film photovoltaic** uses layers of semiconductor materials from less than a micrometer (micron) to a few micrometers thick; wafer-type silicon cells can have thicknesses from 100 to several hundred micrometers. Thin-films use much

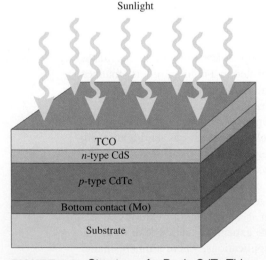

FIGURE 3-8   Structure of a Basic CdTe Thin-Film Photovoltaic

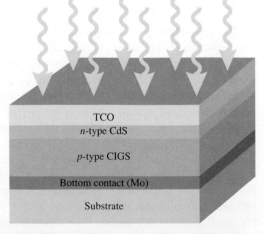

FIGURE 3-9   Structure of a Basic CIGS Thin-Film Photovoltaic Cell

thinner semiconductor layers than wafer-type photovoltaic cells (typically hundreds of times thinner). The advantage of thin-films is that they are much cheaper than crystalline silicon because they use only a fraction of the material and because the manufacturing process is simpler.

Amorphous silicon (a-Si) has a higher band-gap energy (1.75 eV) than crystalline silicon (1.12 eV), which means it absorbs the visible part of the solar spectrum better than the **infrared** portion. Thin-film a-Si solar cells are commonly known as hydrogenated amorphous silicon or a-Si:H. Amorphous silicon is deposited on low-cost fixed or flexible substrates such as plastics, glass, or other substrates using chemical vapor deposition or other methods. Although, amorphous silicon is the least expensive type of silicon photovoltaic material, its conversion efficiency is also the lowest. Typical conversion efficiencies range from 5% to 7% in spite of the higher light absorption. Amorphous silicon, in its pure form, has defects in its atomic bonding that reduce its effectiveness in photovoltaic applications. It is usually alloyed with hydrogen to form hydrogenated amorphous silicon (a-Si:H), which results in significantly fewer defects.

Figure 3-10 shows the cross-section of a hydrogenated amorphous silicon (a-Si:H) thin-film photovoltaic structure. A typical thin-film device doesn't have a metal grid on the top electrical contact like a wafer-type silicon cell does. Instead, it uses a thin layer of a transparent conducting oxide (TCO), such as tin oxide that allows light to pass to the active layers below. These oxides are excellent conductors of electricity and serve as an electrode for the cell. A *p-i-n* structure is usually used for generating and moving the electric charge in which an intrinsic layer (*i*) separates *p* and *n* regions. The intrinsic silicon layer (*i*-type a-Si:H) absorbs most of the photon energy and produces most of the electron–hole pairs. The much thinner *n*-type and *p*-type layers primarily generate the electric field. The free electrons are swept toward the *n*-type layer, and the holes are swept toward the *p*-type layer by the electric field. Some thin-films use a TCO and a reflective layer in place of the bottom contact of molybdenum (Mo) to increase the usable light energy by reflecting it back into the semiconductor layers. A glass protective layer is usually added on the top, as is a layer of anti-reflective material.

Other Types of Photovoltaic Materials

The PV materials previously discussed are all in production, with ongoing research to improve efficiency and lower the cost. Two other types of PV cells are newer and still largely in the research and development stage, but they are beginning to be commercialized for limited applications.

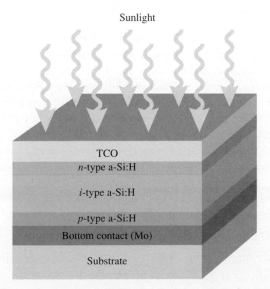

FIGURE 3-10   Structure of a Typical a-Si:H Thin-Film Photovoltaic Cell

Solaronix Dye Cells

(*Source:* David Buchla.)

One area of research for DSSCs is in the production of a dye solar cell on a single substrate using alternating strips of the photoanode, electrolyte, and cathode. Solaronix, a Swiss company, has produced prototype cells this way and is working to solve the problem of containing the electrolyte on the single layer without drying out. If Solaronix is successful, a major drop in cost can be expected.

Dye-Sensitized Solar Cell

The dye-sensitized solar cell (DSSC) is a thin film cell that uses a process that is similar to the one plants use as they absorb sunlight in a dye (chlorophyll) and convert it to chemical energy. In the DSSC, however, the energy is ideally converted to electrical energy. In its simplest form, the DSSC consists of a photoanode, a cathode, and a liquid electrolyte (or mediator solution). The photocathode consists of a thin layer of titanium dioxide (TiO_2) that has been sensitized by a dye and deposited on a glass substrate. When a photon is absorbed in the sensitized layer, an electron is moved to a higher energy level and is injected very rapidly into the conduction band of the TiO_2 semiconductor. (A relatively few electrons recombine within the TiO_2, without providing useful energy to the load.) After passing through the external load, the electrons are passed back to the cell at the cathode, which is composed of an extremely thin (transparent) layer of platinum (Pt) deposited on a glass substrate. Charge balance in the cell is maintained by moving the electrons collected at the cathode back to the dye layer through an electrolyte in a complex oxidation-reduction chemical reaction.

Dye cells have important advantages that offer possibilities for future applications. For one thing, they do not require the expensive high-temperature and high-vacuum fabrication techniques that is required for silicon cells, so they can be fabricated readily without the expense of silicon cells. The TiO_2, a common pigment in paint, is plentiful, inexpensive, and nontoxic. Dye cells work in extremely low light conditions, which may be important in northern climates or indoor applications. Because dye cells can be manufactured at low cost, the payback period (the time required for the cell to produce as much energy as it took to manufacture it) may become better as the technology improves. A disadvantage of dye cells is that they are much less efficient than solid-state cells and they require a liquid electrolyte. Much research is focused on increasing the efficiency and decreasing the cost of these cells by investigating alternatives to the anode, dyes, cathode, and electrolytes of dye cells.

Organic Photovoltaic Cell

Another type of thin-film cell is the organic photovoltaic cell (OPV). In its basic form, the OPV consists of a single layer of active polymer material (the dye) sandwiched between two electrodes. Organic cells are flexible and are very low cost. They can be manufactured in large volume. One drawback is that they are inefficient. Researchers are currently working on improving these cells by using different materials, graded junctions, and production methods.

Neither the DSSC nor the OPV uses the *pn* junction, as do all the previously covered PV technologies.

Environmental Impact

The environmental impact of solar PVs is minimal. It is generally limited to any perceived negative visual impact on the landscape that a solar installation may have, land use issues, effects on plants and wildlife, and concerns about the leaching of certain chemicals used in PV cell manufacturing into the soil and groundwater. Cadmium, which is used in CdTe cells and to a lesser extent in CIGS cells, is a very toxic substance that, like mercury, can find its way into the food chain. Most companies that manufacture PV cells with cadmium have established recycling programs for cadmium-based thin-film PV cells when they reach the end of their useful life. The National Renewable Energy Laboratory (NREL) and others are working on research and development of cadmium-free solar products.

SECTION 3-3 CHECKUP

1. What are three categories of materials used in silicon PV cells?

2. What is a TCO layer and what does it do?

3. What is the advantage of thin-film PV cells over wafer-type cells?

4. What is the purpose of the dye in a DSSC?

3-4 Multijunction Thin-Film

Creating multijunction PV cells is a technique for increasing efficiency. Multijunction PV cells are basically a stack of single-junction PV cells that can absorb a wider band of light energy than any one cell.

A **multijunction** thin-film PV cell is basically two or more individual single-junction cells arranged in descending order of band gap. The main purpose is to increase the amount of energy absorbed and thus increase the efficiency. The band gap of each cell is set by adding or adjusting the amount of semiconductor alloys or varying the type of material. With the highest band gap, the top cell converts the high-energy photons and passes the rest of the remaining photons to be absorbed by the lower band gap cells. The fundamental multijunction concept is shown in Figure 3-11(a) for a three-cell structure.

The spectrum of the sun's radiation consists of ultraviolet (UV) light, visible light, and infrared (IR). The type of radiation depends on the wavelength and frequency. Wavelength is inversely proportional to frequency. IR wavelengths are longer (lower frequency) than visible light. UV wavelengths are shorter (higher frequency) than visible light. Visible light has wavelengths from 400 nm (nanometer) to 700 nm. Red light is at the higher frequency end (longer wavelength) followed by orange, yellow, green, blue, and violet as the wavelength decreases. The energy contained in light is inversely proportional to wavelength or directly proportional to frequency.

A certain level of bright sunlight (sometimes called 1 sun or peak sun) has an irradiance of 1 kW/m^2 at sea level. Of this amount, about 44% is from visible light and about 53% is from IR. About 3% is due to UV. Obviously, much of the energy from the sun lies in the IR part of the spectrum. With multilayered PV cells, energy from a broader range of the spectrum, including visible light and IR, can be captured. Usually a single-cell PV captures only a portion of visible light energy. Figure 3-11(b) shows an arbitrary absorption curve for each of the three cells and, as you can see, the light absorption is spread out over a broad range of the solar spectrum

A typical a-Si (amorphous silicon) multijunction PV cell is shown in Figure 3-12. It has a pin structure for each cell. Light enters through a transparent conducting oxide (TCO). The energy absorbers are intrinsic a-Si and a-SiGe (amorphous silicon-germanium). The *n*-type material is doped a-Si, and the *p*-type material is doped μc-Si (microcrystalline silicon). The germanium is added to amorphous silicon as an alloy to adjust the band gap. The μc-Si is a better conductor of electrons than a-Si and, perhaps more important, it has increased absorption in the red and infrared wavelengths for capturing energy in the sun's infrared part of the spectrum.

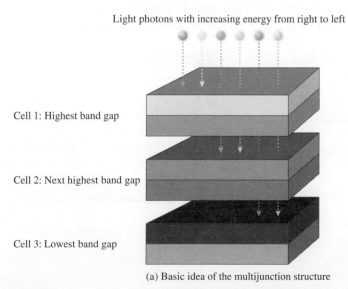

Light photons with increasing energy from right to left

Cell 1: Highest band gap

Cell 2: Next highest band gap

Cell 3: Lowest band gap

(a) Basic idea of the multijunction structure

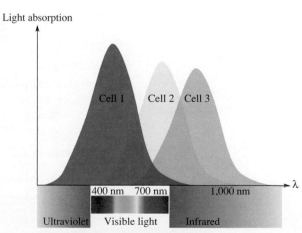

Light absorption

Cell 1   Cell 2   Cell 3

400 nm   700 nm   1,000 nm   λ

Ultraviolet   Visible light   Infrared

(b) Basic idea of spectral coverage

FIGURE 3-11 Fundamental Concept of a Multijunction PV Cell

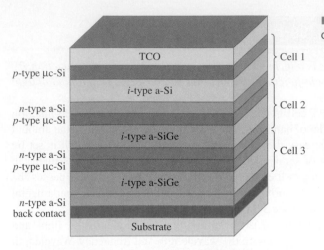

FIGURE 3-12 Basic Structure of a Multijunction a-Si PV Cell

SECTION 3-4 CHECKUP

1. What is the main purpose of a multilayered PV cell?

2. What is the range of visible light in terms of wavelength?

3. Does infrared radiation have a longer or shorter wavelength than visible light?

4. Which has the most energy, red light or violet light?

3-5   Photovoltaic Cell Characteristics and Parameters

PV cell characterization involves measuring the cell's electrical performance characteristics to determine conversion efficiency and critical parameters. The conversion efficiency is a measure of how much incident light energy is converted to electrical energy. The optimum operating point for maximum output power is also a critical parameter, as is spectral response. that is, how the cell responds to various light frequencies. Other important characteristics include how the current varies as a function of output voltage and as a function of light intensity or irradiance.

Current-Voltage Curves

The current-voltage (*I-V*) curve for a PV cell shows that the current is essentially constant over a range of output voltages for a specified amount of incident light energy. Figure 3-13

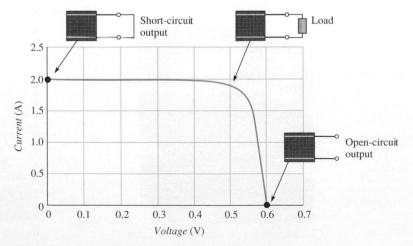

FIGURE 3-13 Typical *I-V* Characteristic Curve for a PV Cell

shows a typical *I-V* curve for which the short-circuit output current, I_{SC}, is 2 A. Because the output terminals are shorted, the output voltage is 0 V. For an open output, the voltage, V_{OC}, is maximum (0.6 V) in this case, but the current is 0 A, as indicated.

Output Power

The output power of the PV cell is voltage times current, so there is no output power for a short-circuit condition because $V_{OUT} = 0$ or for an open-circuit condition because $I_{OUT} = 0$. Above the short-circuit point, the PV cell operates with a resistive load. Between the short-circuit point and the knee of the curve, the output power depends on the voltage because the current is essentially constant. The maximum output power occurs in the knee at the point, called the maximum power point (MPP), where the voltage is approximately 0.52 V. After that point, the curve decreases drastically and the output power declines, as illustrated in Figure 3-14. In the figure, the blue curve is the power. In this particular case, the maximum output power is $P_{OUT(MAX)} = V_{MPP}I_{MPP} = (0.52 \text{ V})(1.85 \text{ A}) = 0.96 \text{ W}$. For maximum power, the load on the cell would have to be:

$$R_L = V_{MPP}/I_{MPP} = 0.52 \text{ V}/1.85 \text{ A} = 0.281 \text{ }\Omega$$

I-V Characteristic as a Function of Irradiance

The output voltage of a PV cell is affected only slightly by the amount of light intensity (irradiance), but the current, and thus the power, decreases as the irradiance decreases. PV cell parameters are usually specified under standard test conditions (STC) at a total irradiance of 1 sun (1,000 W/m²), a temperature of 25° C, and coefficient of air mass (AM) of 1.5. The AM is the path length of solar radiation relative to the path length at zenith at sea level. The AM at zenith at sea level is 1. Figure 3-15 shows typical *I-V* curves for three values of irradiance. Cells vary in the amount of current at a given irradiance.

Energy Conversion Efficiency

The **efficiency** of a PV cell is the ratio of light energy falling on the cell to the light energy that is converted to electrical energy. It is expressed as a percentage, as shown in the following formula:

$$\% \text{ efficiency} = \left[\frac{P_{out(max)}}{E \times A} \right] 100 \qquad \text{Equation 3-1}$$

where

$P_{out(max)}$ is the maximum electrical power output of the cell, in watts (W)

E is the irradiance (light energy) at the surface of the cell, in watts/meter² (W/m²)

A is the surface area of the cell, in meter² (m²)

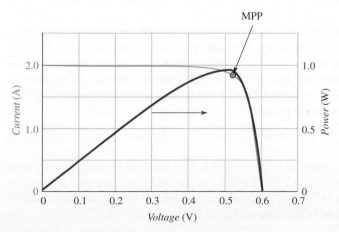

FIGURE 3-14   Power Curve for a Typical PV Cell

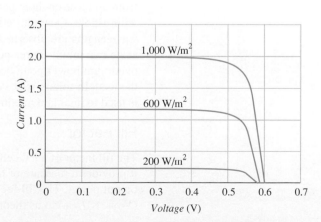

FIGURE 3-15   *I-V* Characteristics as a Function of Irradiance

PV cells are typically square, with sides ranging from about 10 mm (0.3937 inches) to 127 mm (5 inches) or more on a side. Typical efficiencies range from 14% to 18% for a monocrystalline silicon PV cell. Some manufacturers claim efficiencies greater than 18%.

EXAMPLE 3-1

A certain PV cell is illuminated with an irradiance (E) of 1,000 W/m^2. If the cell is 100 mm × 100 mm in size and produces 3 A at 0.5 V at the maximum power point, what is the conversion efficiency?

Solution

$$\text{Surface area} = A = (100 \text{ mm})(100 \text{ mm}) = (0.1 \text{ m})(0.1 \text{ m}) = 0.01 \text{ m}^2$$

The input solar power at the surface is:

$$P_{\text{in(surface)}} = \text{irradiance} \times \text{surface area} = E \times A = (1,000 \text{ W/m}^2)(0.01 \text{ m}^2) = 10 \text{ W}$$

The maximum output power of the cell is:

$$P_{\text{out(max)}} = V_{\text{MPP}}I_{\text{MPP}} = (0.5 \text{ V})(3 \text{ A}) = 1.5 \text{ W}$$

The percent efficiency is:

$$\% \text{ efficiency} = [P_{\text{out(max)}}/(E \times A)]100 = (1.5 \text{ W}/10 \text{ W})100 = \textbf{15\%}$$

> The power output of solar modules can be boosted by 10% just by applying a large transparent sticker to the front. The sticker is a polymer film embossed with microstructures that bend incoming sunlight. The result is that the active materials in the panels absorb more light and convert more of it into electricity.

Factors That Effect Conversion Efficiency

Several factors determine the efficiency of a PV cell: the type of cell, the reflectance efficiency of the cell's surface, the thermodynamic efficiency limit, the quantum efficiency, the maximum power point, and internal resistances. When light photons strike the PV cell, some are reflected and some are absorbed. The nonreflective cell coating minimizes the percentage of photons that are reflected. Certain types of irregular cell surfaces help to decrease the reflection of photons. The angle of incidence of the photons striking the cell surface is also a factor in determining the amount of reflection. The thermodynamic efficiency relates to the photon energy. Photons with energies below the band-gap energy of silicon (1.12 eV) cannot produce electron–hole pairs, and their energy is converted to heat rather than to an electrical output. Photons with energies at or above the band gap create electron–hole pairs and contribute to the electrical output. If the photon energy is greater than the band-gap energy, the excess is transferred to the electron–hole pair and eventually dissipated. Quantum efficiency relates to the number of absorbed photons that contribute to the electrical current for a short-circuit condition. Not all of the electrons that are freed from an electron–hole pair end up as current. Some electrons lose energy and recombine with a hole. Quantum efficiency depends on the wavelength of the light because some wavelengths are absorbed more efficiently than others. Finally, operation of the PV cell at its maximum power point is vital to the conversion efficiency. As you have seen, the maximum power point occurs in the knee of the *I-V* characteristic curve as determined by the load. In solar power systems, a method called maximum power point tracking (MPPT) is used to maintain maximum output power.

Fill Factor

The fill factor of a PV cell is an important parameter in evaluating its performance because it provides a measure of how close a PV cell comes to providing its maximum theoretical output power. The **fill factor** (**FF**) is the ratio of the cell's actual maximum power output ($V_{\text{MPP}} \times I_{\text{MPP}}$) to its theoretical power output ($V_{\text{OC}} \times I_{\text{SC}}$).

$$FF = (V_{\text{MPP}})(I_{\text{MPP}})/(V_{\text{OC}})(I_{\text{SC}}) \hspace{2cm} \text{Equation 3-2}$$

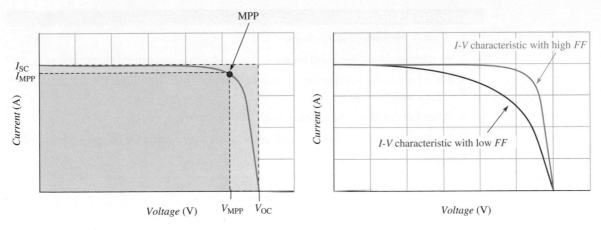

FIGURE 3-16   Fill Factor for a PV Cell

where

FF is the fill factor (dimensionless)

V_{MPP} is the voltage at the maximum power point, in volts (V)

I_{MPP} is the current at the maximum power point, in amperes (A)

V_{OC} is the open circuit voltage, in volts (V)

I_{SC} is the short circuit current, in amperes (A)

Typical commercial solar cells have a fill factor greater than 0.7. During the manufacture of commercial solar modules, each PV cell is tested for its fill factor. If the fill factor is low (below 0.7), the cells are considered as lower grade. Figure 3-16 illustrates the fill factor.

Temperature Dependence of PV Cells

The output voltage and current of a PV cell is temperature dependent. Figure 3-17 shows that, for a constant light intensity, the open circuit output voltage decreases as the temperature increases (due to a change in the band gap) but the current is affected only by a small amount. The important point here is that a PV cell performs better in cooler temperatures.

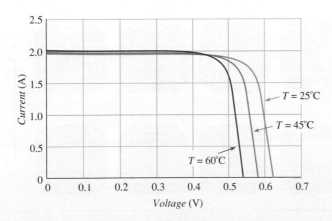

FIGURE 3-17   Effect of Temperature on Output Voltage and Current for a Fixed Light Intensity in a PV Cell

3-6   Solar Modules and Arrays

A single PV cell is impractical for most applications because it can produce only about 0.5 V at the maximum power point. For this reason, multiple PV cells are connected together to form solar modules. One or more modules (sometimes called a panel) are connected to form a solar array. Modules are available in standardized size and are quite rugged; larger panels tend to deliver more watts/dollar. A typical large module delivers about 250 W at noon on a sunny day.

To produce higher voltages, multiple PV cells are connected in series, as shown in Figure 3-18(a). For example, the six series cells will ideally produce 6(0.5 V) = 3 V. Because they are connected in series, the six cells produce the same current as a single cell. For increased current capacity, series cells are connected in parallel, as shown in Figure 3-18(b). Assuming a cell can produce 2 A at the maximum power point, the series-parallel arrangement of twelve cells will produce 4 A at 3 V. The 2 A currents from each of the series connection combine at point P for a total of 4 A through the load.

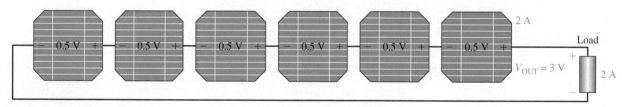

(a) PV cells connected in series

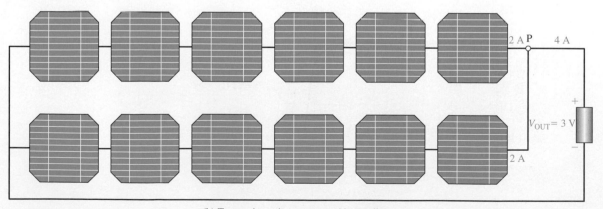

(b) Two series strings connected in parallel

FIGURE 3-18   PV Cells Connected in Series and in Parallel

EXAMPLE 3-2

Six series strings of 12 PV cells are connected in parallel. Assume each PV cell produces a current of 1.5 A and 0.5 V at the maximum power point. Determine the output current, voltage, and power to a load under maximum power point conditions.

Solution

Each series string produces:

$$V = (12)(0.5 \text{ V}) = 6 \text{ V}$$

$$I = 1.5 \text{ A}$$

The total output voltage, current, and power is:

$$V_{OUT} = \mathbf{6 \text{ V}}$$

$$I_{OUT} = (6)(1.5 \text{ V}) = \mathbf{9 \text{ A}}$$

$$P_{OUT} = V_{OUT}I_{OUT} = (6 \text{ V})(9 \text{ A}) = \mathbf{54 \text{ W}}$$

Solar Modules

A combination of multiple PV cells connected to produce a specified power, voltage, and current output is called a **solar module**. Many standard modules consist of 36, 72, or 96 prewired PV cells; however, modules are available with other numbers of cells. Solar modules are generally encapsulated with tempered glass (or some other transparent material) on the front surface and with a waterproof material on the back surface. The edges are sealed for weatherproofing, and a frame of aluminum or some other material holds everything together in a mountable unit. A junction box that provides electrical connections is usually available on the back of the module. A typical solar module is shown in Figure 3-19(a), and

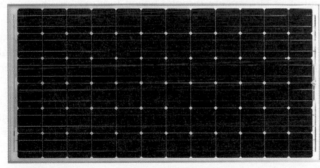

(a) Solar module

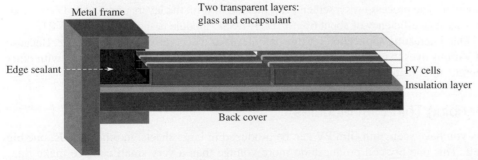

(b) Cutaway of a small segment of a module showing basic construction

FIGURE 3-19 Typical 72-Cell Solar Module and Cutaway View of Typical Construction (*Source:* Part (a), National Renewable Energy Laboratory.)

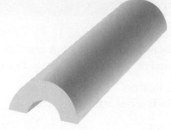

(a) Flat-surface thin-film module   (b) Thin-film sheet mounted on a curved surface

FIGURE 3-20   Example of Thin-Film PV Configurations

a construction view of a small module segment is shown in Figure 3-19(b). The conductive strips on top of each PV cell are connected to the bottom contact of the previous cell by a conductive strip running between the cells, creating a series connection.

Many modules have all series-connected PV cells that produce a specified voltage commonly required in solar power systems. Combinations of series and parallel cells can also be implemented in a module. For example, suppose a certain 36-cell module ideally produces $V_{MPP} = (0.5 \text{ V})36 = 18 \text{ V}$ and a 72-cell module ideally produces $V_{MPP} = (0.5 \text{ V})72 = 36 \text{ V}$. The current output of either module is approximately the same as the current output of a single cell. Assuming the output current at the MMP of the 72-cell modules is 4 A, the rated maximum output power is:

$$P_{OUT} = V_{MPP}I_{MPP} = (36 \text{ V})(4 \text{ A}) = 144 \text{ W}$$

The performance of a PV module is usually rated according to its maximum dc power output (watts) under the standard test conditions (STC) mentioned earlier. The STC uses a coefficient of air mass (AM) of 1.5. AM is the optical path length of the sunlight through earth's atmosphere. As it passes through the atmosphere, light is attenuated by scattering and absorption; the more atmosphere through which light passes, the greater the attenuation. Recall that the term *coefficient of air mass* indicates relative air mass, which is the path length relative to that at the zenith at sea level. At sea-level the coefficient of air mass at the zenith is 1. For the STC, the light source must meet the defined spectral distribution. Because these conditions are not always typical of how solar PV modules and arrays operate in the field, actual performance is usually less than the STC rating.

Thin-film solar cells can be fabricated in large sheets on a rigid substrate and are available in the form of flat-surface modules. They can also be manufactured on large sheets of a flexible substrate that can be mounted on nonflat surfaces, as shown in Figure 3-20.

DSSC Module

A simple process has been developed for making small-area dye-sensitized solar cell (DSSC) modules. The process uses a screen print method to print the layers of the cell. These modules have an efficiency of about 6% to 7%. A DSSC module is shown in Figure 3-21.

One interesting application for dye modules is as a decorative building façade. Because of various dyes that can be used, the colors of the cells can be selected to blend with other architectural features of a building.

Scribing Thin-Film Modules

FIGURE 3-21   Dye-Sensitized Solar Module   (*Source:* National Renewable Energy Laboratory.)

As you have seen, thin-film PV can be produced in large sheets, in other words, one big cell. This one big cell produces no more voltage than a very small cell. To make thin-film practical, a large sheet of PV structure must be divided into many small cells that are all connected in series to produce the required voltage. The technique of *scribing*

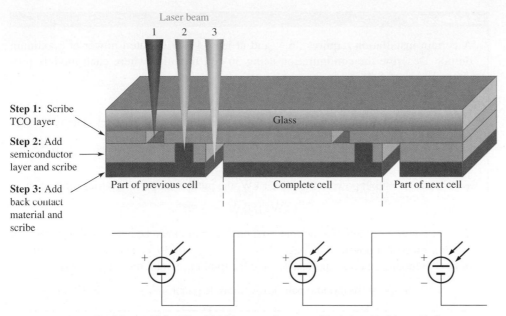

FIGURE 3-22 Scribing a Thin-Film PV Sheet to Create a Module with Many Cells

accomplishes this using a laser beam to make very fine precision cuts (scribe lines) in the various thin-layers. Scribe lines are typically only tens of microns wide. Each layer is targeted separately by varying the type of laser beam and focal points. Scribing accomplishes two goals: It separates the large sheet into many thin-film cells, and it interconnects the small cells in a series arrangement. The basic process is illustrated in Figure 3-22.

Solar Arrays

A **solar array** consists of multiple modules. To increase the rated output power without changing the voltage, two or more modules are connected in parallel in a solar array configuration. This configuration is illustrated in Figure 3-23 for three modules. The outputs have standard connectors for ease of connection with other panels.

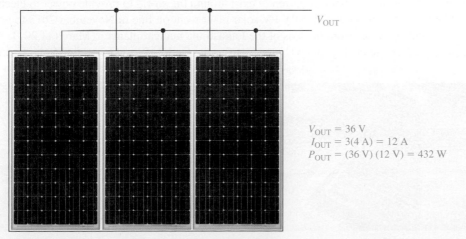

$V_{OUT} = 36$ V
$I_{OUT} = 3(4$ A$) = 12$ A
$P_{OUT} = (36$ V$)(12$ V$) = 432$ W

FIGURE 3-23 Solar Array with Three 72-Cell Modules Connected in Parallel (*Source:* (image) National Renewable Energy Laboratory.)

EXAMPLE 3-3

A certain installation requires 36 V and at least 1 kW of rated power at maximum output. Describe the configuration using 36-cell modules where each module produces 18 V and 5 A.

Solution

To get an output of 36 V, two 36-cell modules are connected in series. The two series modules produce an output power of:

$$P_{OUT} = (36\ V)(5\ A) = 180\ W$$

To achieve an output power of at least 1 kW, the number of pairs of series modules is:

$$1\ kW/180\ W = 5.56$$

We can't have a portion of a module, so 6 module pairs must be connected in parallel to produce an output power of at least 1 kW at a voltage of 36 V. The solar panel consists of twelve modules in this application, and the total maximum output power is:

$$P_{OUT} = (6\ module\ pairs)(5\ A/module\ pair)(36\ V) = \mathbf{1{,}080\ W}$$

It is important to remember that the rated power output of a solar installation is at the standard test conditions. For example, the output power of 1,080 W in Example 3-3 will be less when the panel is shaded by clouds or other objects and the temperature is warmer than 25° C. Also, if the panel does not track the sun's movement, most of the incident light will strike the panel at angles less than perpendicular. Of course, at night there is no output power.

Solar arrays range from very small to very large. Small installations are generally for residential or small business applications or for single-purpose applications such as street or roadway lighting, traffic signals, and so on. Larger arrays are used for industrial applications or to provide power to a small community or to the electrical grid. Figure 3-24(a) shows a typical residential rooftop solar array. Figure 3-24(b) shows a freestanding solar array used to power the visitor center at Mesa Verde National Park in Colorado.

An alternative to placing solar panels or arrays over existing roofing shingle or tiles is the solar-powered shingle. These shingles can replace normal roofing materials as the actual roofing surface.

PV Solar Farms

Large solar array installations currently exist around the world to provide power to electrical grids. In Japan, a huge 70 MW PV solar plant went on line in November 2013 to help resolve power issues from the loss of the Fukushima nuclear plants in March of 2011. An example of a large facility in the United States is the DeSoto Next Generation Solar Energy

(a) Residential roof installation

(b) Solar array

FIGURE 3-24 Solar Arrays (*Source:* Part (a), National Renewable Energy Laboratory; part (b), David Buchla.)

FIGURE 3-25   DeSoto Next-Generation Solar Energy Center in Florida   (*Source:* National Renewable Energy Laboratory.)

Center (see Figure 3-25), which is owned by Florida Power & Light. This installation has over 90,000 solar modules with a total rated capacity of 25 megawatts and produces an estimated 42,000 megawatt-hours (MWh) of electrical energy per year, equivalent to an average output of about 4.8 megawatts. The Mojave Desert and Tehachapi Mountain areas of California are also centers for very large solar farms with very large arrays planned for the future.

Another example of a large solar farm is in Nevada at Nellis Air Force Base near Las Vegas. The Nellis solar energy system generates in excess of 25 million kilowatt-hours (kWh) of electrical energy annually. The peak output power capacity of the system is approximately 13 megawatts.

SECTION 3-6 CHECKUP

1. Does a series connection of PV cells increase the current or voltage?

2. What are the PV cell counts of three standard solar modules?

3. What is a solar module?

4. What is a solar array?

5. Under what conditions does a solar module produce less than its rated output?

3-7   Solar Module Data Sheet Parameters

The most important module parameters include the short-circuit current, the open-circuit voltage, the output voltage, current, and rated power at 1,000 W/m² solar radiation, all measured under STC, as previously discussed. Solar modules must also meet certain mechanical specifications to withstand wind, rain, and other weather conditions.

An example of a solar module data sheet composed of wafer-type PV cells is shown in Figure 3-26. Notice that the data sheet is divided into several sections: electrical data, mechanical data, *I-V* curve, tested operating conditions, warranties and certifications, and mechanical dimensions. Although data sheets vary from one manufacturer to another, most have this type of information. This particular data sheet uses the term *panel* instead of *module.* You will find these terms are sometimes used interchangeably. A CdTe thin-film data sheet is shown in Figure 3-27.

SOLAR PANEL

Electrical Data

Measured at Standard Test Conditions (STC): Irradiance of 1000W/m², AM 1.5, and cell temperature 25°C

Peak Power (+/−5%)	P_{max}	215 W
Rated Voltage	V_{mpp}	39.8 V
Rated Current	I_{mpp}	5.40 A
Open Circuit Voltage	V_{oc}	48.3 V
Short Circuit Current	I_{sc}	5.80 A
Maximum System Voltage	UL	600 V
Temperature Coefficients		
	Power	−0.38%/K
	Voltage (V_{oc})	−136.8mV/K
	Current (I_{sc})	3.5mA/K
NOCT		45°C + /−2°C
Series Fuse Rating		15 A

I-V Curve

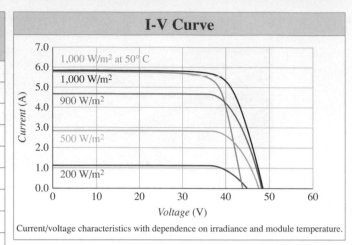

Current/voltage characteristics with dependence on irradiance and module temperature.

Mechanical Data

Solar Cells	72 all-back contact monocrystalline
Front Glass	High transmission tempered glass
Junction Box	IP-65 rated with 3 bypass diodes Dimensions: 32 × 155 × 128 (mm)
Output Cables	1000mm length cables / MultiContact (MC4) connectors
Frame	Anodized aluminum alloy type 6063 (black)
Weight	33.1 lbs. (15.0 kg)

Tested Operating Conditions

Temperature	−40°F to + 185°F (−40°C to + 85°C)
Max load	113 psf 550kg/m² (5400 Pa) front – e.g. snow; 50 psf 245kg/m² (2400 Pa) front and back – e.g. wind
Impact Resistance	Hail 1 in (25 mm) at 52mph (23 m/s)

Warranties and Certifications

Warranties	25 year limited power warranty
	10 year limited product warranty
Certifications	Tested to UL 1703. Class C Fire Rating

Dimensions

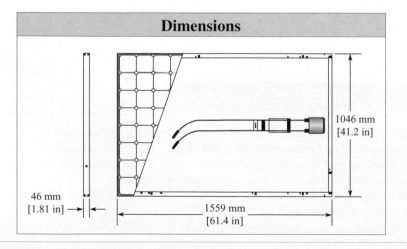

FIGURE 3-26   Typical Solar Module Data Sheet

ELECTRICAL SPECIFICATIONS

Nominal Values		
Nominal Power (+/−5%)	$P_{MPP}(W)$	70
Voltage at P_{MAX}	$V_{MPP}(V)$	65.5
Current at P_{MAX}	$I_{MPP}(A)$	1.07
Open Circuit Voltage	$V_{OC}(V)$	88.0
Short Circuit Current	$I_{SC}(A)$	1.23
Maximum System Voltage	$V_{SYS}(V)$	1000
Temperature Coefficient of P_{MPP}	$T_K(P_{MPP})$	−0.25%/°C
Temperature Coefficient of V_{OC}, high temp (>25°C)	$T_K(V_{OC, high temp})$	−0.25%/°C
Temperature Coefficient of V_{OC}, low temp (−40°C to + 25°C)	$T_K(V_{OC, low temp})$	−0.20%/°C
Temperature Coefficient of I_{SC}	$T_K(I_{SC})$	+0.04%/°C
Limiting Reverse Current[2]	$I_R(A)$	2
Maximum Source Circuit Fuse	$I_{CF}(A)$	10 (2 IEC61730[3])

Nominal Values		
Nominal Power (+/−5%)	$P_{MPP}(W)$	52.6
Voltage at P_{MAX}	$V_{MPP}(V)$	61.4
Current at P_{MAX}	$I_{MPP}(A)$	0.86
Open Circuit Voltage	$V_{OC}(V)$	81.8
Short Circuit Current	$I_{SC}(A)$	1.01

MECHANICAL DESCRIPTION

Length	1200 mm	Thickness	6.8 mm
Width	600 mm	Area	0.72 m²
Weight	12 kg	Leadwire	3.2 mm², 610 mm
Connectors	Solarline 1 type connector		
Bypass Diode	None		
Cell Type	CdS/CdTe semiconductor, 116 active cells		
Frame Material	None		
Cover Type	3.2 mm heat strengthened front glass laminated to 3.2 mm tempered back glass		
Encapsulation	Laminate material with edge seal		

MECHANICAL DRAWING

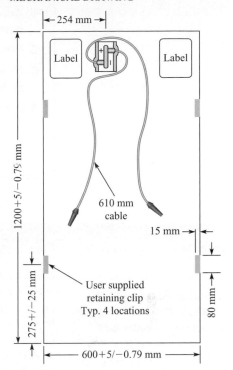

Efficiency at 200W/m²: These PV Modules experience an increase in efficiency of 2% at 200W/m² when compared to the efficiency at 1,000W/m².

* All ratings +/−10%, unless specified otherwise. Specifications are subject to change.

[1] Standard Test Conditions (STC) 1000W/m², AM 1.5, 25°C

FIGURE 3-27   CdTe Thin-Film Module Data Sheet

EXAMPLE 3-4

Refer to the data sheet in Figure 3-26 and determine the fill factor of the solar cell.

Solution

$$V_{MPP} = 39.8 \text{ V}, I_{MPP} = 5.40 \text{ A}, V_{OC} = 48.3 \text{ V}, \text{ and } I_{SC} = 5.80 \text{ A}$$

$$FF = (V_{MPP})(I_{MPP})/(V_{OC})(I_{SC}) = (39.8 \text{ V})(5.40 \text{ A})/(48.3 \text{ V})(5.80 \text{ A}) = \mathbf{0.767}$$

1. According to the data sheet in Figure 3-26, what is the voltage at the maximum power point?

2. What is meant by the term *normal operating cell temperature (NOCT)*?

3. How many modules can be connected in parallel without exceeding the maximum system voltage?

4. According to the data sheet in Figure 3-26, what is the approximate power output at an irradiance of 200 W/m²?

5. What is the approximate decrease in output voltage for the module in Figure 3-27 if the temperature increases from 25° C to 50° C?

3-8   Concentrating Photovoltaics

The Massachusetts Institute of Technology (MIT) is developing a solar concentrator that uses two or more dyes painted onto a pane of glass or plastic. The dyes absorb light across a range of wavelengths, which is then reemitted and transported across the pane to solar cells arranged along the edges of the pane. This technology may lead to windows that are also solar concentrators.

You have probably used a magnifying glass to focus sunlight on a small area and burn a hole in a leaf or piece of paper. This is the same principle as concentrating photovoltaic, except instead of burning a hole in a leaf, the lens causes a large amount of light energy to be focused on and absorbed by the PV cell. Concentrating photovoltaic cells are specialized heat resistant versions of standard solar modules.

A **concentrating photovoltaic (CPV)** system uses lenses or mirrors to focus sunlight on a small area of PV cells. The lenses or mirrors must always be oriented toward the sun so that the sunlight is focused on the PV surface. Therefore, most CPV systems employ single- or double-axis tracking so they can follow the sun's daily and seasonal movement. PV cells operate more efficiently in concentrated light as long as the cell junction temperature is kept cool by suitable methods such as heat sinks. CPV does not operate well in cloudy conditions because an overcast sky creates diffused light, which cannot be concentrated effectively. For the same cell surface area, a CPV can produce many times the power of other types of PV systems. The primary advantage of CPV is the reduction in the amount of PV cell area required, and thus the cost, because the semiconductor material is the most expensive part of any PV system.

CPV generally falls into three categories: (1) low-concentration CPV, which has a solar concentration of from 2 to 100 suns; (2) medium-concentration CPV, which has a concentration of from 100 to 300 suns; and (3) **high-concentration photovoltaic (HCPV)**, which has concentrations of more than 300 suns. HCPV typically incorporates dish reflectors and Fresnel lenses, but cone lenses are also used. Figure 3-28(a) shows the shape of a Fresnel lens versus a conventional convex lens. The Fresnel lens is basically a thin,

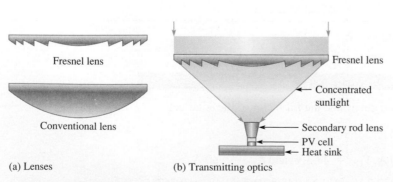

Fresnel lens

Conventional lens

(a) Lenses

Fresnel lens

Concentrated sunlight

Secondary rod lens

PV cell

Heat sink

(b) Transmitting optics

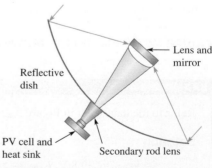

Reflective dish

Lens and mirror

PV cell and heat sink

Secondary rod lens

(c) Reflecting/transmitting optics

FIGURE 3-28   Basic Concepts of HCPV Collectors

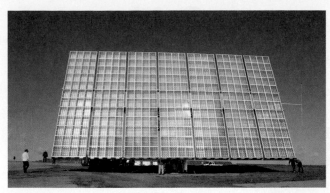

(a) Fresnel array

(b) Reflecting dish

FIGURE 3-29 Examples of HCPV Systems (*Source:* Part (a), National Renewable Energy Laboratory; part (b), Reproduced with permission. © 2011 Solar Systems Pty Ltd.)

flat lens with concentric rings. Each ring is slightly thinner as the rings go out from the center, resulting in light being focused toward the center axis. The PV cells are typically multijunction thin-film, and GaAs is one of the favored materials. The diagram in Figure 3-28(b) illustrates the basic concept of a Fresnel HCPV, which uses transmitting optics (lenses). Figure 3-28(c) shows a dish reflector HCPV, which uses both reflecting (mirrors) and transmitting optics. A typical Fresnel lens array and a reflecting dish are shown in Figure 3-29.

Figure 3-30 is a relative comparison of HCPV to three other types of PV. Each curve represents the relative output power versus time of day, from sunrise to sunset, for PV modules with the same surface area exposed to the same amount of sunlight. Figure 3-31 on page 112 is a typical HCPV data sheet.

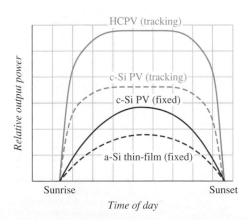

FIGURE 3-30 Comparison of Output Powers for Various Types of PV Systems

---

SECTION 3-8 CHECKUP

1. What do CPV and HCPV stand for?

2. Name two nonsemiconductor components that are used in CPV.

3. What is a Fresnel lens?

4. What is the primary advantage of CPV?

SolFocus.

SF-1136SX (432) Product Specifications

SF-1136SX System

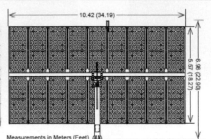

Measurements in Meters (Feet)

System Specifications

Weight (kg) 2735

Note: Inverter and other balance of system components are not included.

SF-1100P Panel (36 panels per array)

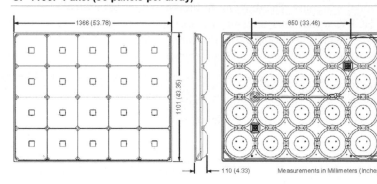

Measurements in Millimeters (Inches)

Panel Characteristics

Rated Power CSTC[1] (Pmp)	432 ±5%
Rated Power CSOC[2] (Pmp)	365 ±5%
Voltage[2] (Vmp)	51.0
Current[2] (Imp)	7.2
Open Circuit Voltage[2] (Voc)	60.5
Short Circuit Current[2] (Isc)	7.8

[1] CSTC (Concentrator Standard Test Conditions): 1000 W/m² DNI, 25°C Cell Temperature, AM1.5
[2] CSOC (Concentrator Standard Operating Conditions): 900 W/m² DNI, 20° Ambient Temperature, AM1.5

Max System Voltage (V)	1000
Power (%/°C)	-0.15
Voltage (V/°C)	-0.09
Multi-Junction Cells	20
By-Pass Diode	20
Length (m)	1.10
Width (m)	1.37
Area (m²)	1.50
Weight (kg)	32
Connector	MC Type 4

SF-1136T Tracker

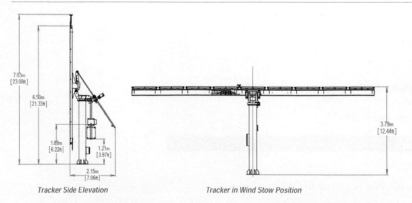

Tracker Side Elevation   *Tracker in Wind Stow Position*

Tracker Controller Characteristics

Azimuth Range – Standard (deg)	±160
Azimuth Range – Tropical (deg)	-315 to 135
Elevation Range (deg)	-5 to 90
Wind Stow Condition (km/hr)	50
Maximum Wind Speed (km/hr)	145*
Communication	Ethernet
Environmental Enclosure	IP65

* Wind stow position

SolFocus, Inc.
1841 Zanker Road • San Jose, CA 95112
Tel: +1 408.850.0300 • Fax: +1 408.850.0301 • www.solfocus.com

FIGURE 3-31   HCPV Module (Panel) and Array Data Sheet   (*Source:* © SolFocus.)

CHAPTER SUMMARY

- Silicon atoms form a crystal structure in which a silicon atom shares an electron with each of its four neighboring atoms, thus creating a covalent bond.

- The doping process adds impurity atoms in very low concentrations to the silicon to create either an excess of free electrons (*n*-type impurity) or a deficiency of electrons (*p*-type impurity),

- The *pn* junction forms a boundary between *n*-type and *p*-type semiconductive materials.

- An electric field exists in the depletion region across the *pn* junction.

- The photovoltaic effect is the basic physical process by which a PV cell converts sunlight into electricity.

- One commonly used process for creating a silicon ingot is called the Czochralski method.

- A wafer-type PV cell consists of a top reflective layer, a grid of conductors, two doped silicon regions, and a back conductive layer.

- The amount of energy required to free an electron from the valence band of a silicon atom is called the band-gap energy and is equal to 1.12 eV (electron volts).

- The output voltage of a PV cell is affected only slightly by the amount of light intensity (irradiance), but the current, and thus the power, decreases as the irradiance decreases.

- PV cell parameters are usually specified under standard test conditions (STC) at a total irradiance of 1 sun (1,000 W/m^2), temperature of 25° C, and a coefficient of air mass of 1.5.

- The efficiency of a PV cell is the ratio of the amount of light energy falling on the cell to the amount that is converted to electrical energy.

- The fill factor is the ratio of the cell's actual maximum power output ($V_{MPP} \times I_{MPP}$) to its theoretical power output ($V_{OC} \times I_{SC}$).

- Two types of PV cells in terms of structure are wafer and thin-film.

- Common types of wafer-type PV cell materials are monocrystalline silicon and polycrystalline silicon.

- Common types of thin-film PV cell materials are amorphous silicon, cadmium-teluride, copper-indium-gallium-disulfide, and gallium arsinide.

- Multijunction PV cells absorb a wider range of wavelength and are therefore more efficient than single-junction cells.

- Laser scribing is used to divide a sheet of thin-film PV material into small series-connected cells.

- CPV uses lenses or mirrors to focus sunlight on a small PV cell.

- HPCV stands for "high-concentration photovoltaics." Two basic types are transmitting (uses lenses) and reflecting (uses mirrors or mirror/lens combinations).

KEY TERMS

band gap   The amount of energy required to free an electron from the valence band of an atom. For silicon, the band-gap energy is 1.12 eV.

concentrating photovoltaic (CPV)   A technology that uses lenses or curved mirrors to intensify sunlight on solar PV cells.

crystallinity   An indication of how perfectly ordered the atoms are in a crystal structure.

doping   The process used to increase the conductivity of a semiconductor in a precise and controlled way.

efficiency   The ratio of light energy falling on the cell to the light energy that is converted to electrical energy.

fill factor (FF)   The ratio of a cell's actual maximum power output to its theoretical power output.

free electron   An electron not bonded to an atom. A conduction electron.

high-concentration photovoltaic (HCPV)   A photovoltaic cell that has concentrations of more than 300 suns. An HCPV typically incorporates dish reflectors and Fresnel lenses, but cone lenses are also used.

hole   A vacancy created in an atomic bond when a valence electron becomes a free electron.

multijunction   A type of thin-film PV cell that is basically two or more individual single-junction cells arranged in descending order of band gap.

photon   Packets of energy in sunlight.

photovoltaic (PV) cell   A device that converts the energy of sunlight directly into electricity using a thin layer or wafer of silicon that has been doped to create a *pn* junction.

photovoltaic effect   The basic physical process by which a PV cell converts sunlight into electricity.

pn junction   The boundary created between an *n*-type and a *p*-type semiconductor.

solar array   A combination of solar modules.

solar module   Combinations of multiple PV cells connected to produce a specified power, voltage, and current output.

thin-film photovoltaics   Type of photovoltaic that uses layers of semiconductor materials from less than a micrometer (micron) to a few micrometers thick.

valence electrons   Electrons in the outer shell of an atom.

FORMULAS

Equation 3-1 $\% \text{ Efficiency} = \left[\dfrac{P_{out(max)}}{E \times A} \right] 100$

Efficiency of a PV cell

Equation 3-2 $FF = (V_{MPP})(I_{MPP})/(V_{OC})(I_{SC})$

Fill Factor for a PV cell

CHAPTER TRUE/FALSE QUIZ

Determine whether each statement is true or false. Answers are at the end of the chapter.

1. PV stands for "photovoltaic."
2. A PV cell and a solar cell are different.
3. PV cell operation is based on the *pn* junction.
4. A PV cell with a larger surface area can produce more voltage than one with a smaller surface area.
5. Output current is increased by connecting PV cells in series.
6. The intensity of the sunlight on a PV cell determines the amount of output current.
7. Output voltage of a PV cell decreases greatly when a cloud passes over a PV system.
8. Groups of PV cells form modules (panels) and groups of modules form arrays.
9. The manufacturing process is the same for both wafer-type and thin-film PV cells.
10. CdTe is used in thin-film PV.
11. Thin-film PV is very rigid.
12. CIGS stands for calcium-indium-gallium-selenide.
13. CPV uses lenses and/or mirrors to focus sunlight onto a PV cell.
14. The Fresnel lens is very thick and heavy.
15. Heat dissipation in the PV cell is a major factor in HCPV systems.
16. GaAs PV cells are commonly used in HCPV.

CHAPTER MULTIPLE-CHOICE QUIZ

Complete each statement by selecting the one correct answer. Answers are at the end of the chapter.

1. Silicon can be doped with
   a. boron
   b. gallium
   c. bismuth
   d. any of these

2. A monocrystalline silicon PV cell consists of
   a. a *pn* junction
   b. a connection grid
   c. a TCO
   d. both (a) and (b)

3. The output voltage of a PV cell increases slightly with
   a. surface area
   b. light intensity
   c. wavelength
   d. none of these

4. The output power of a PV cell increases with
   a. light intensity
   b. surface area
   c. time of day
   d. both (a) and (b)

5. A higher fill factor indicates
   a. higher output voltage
   b. higher output power
   c. larger surface area
   d. more semiconductor layers

6. Compared to a single PV cell, 32 PV cells connected in parallel to a specified load means that the
   a. output voltage increases
   b. output current increases
   c. efficiency increases
   d. output power decreases

7. To achieve 12 V from a PV module with 24 cells, the
   a. cells must be in parallel
   b. light intensity must increase
   c. temperature must decrease
   d. cells must be in series

8. To get more power from a given module,
   a. the cells must be in parallel
   b. the cells must be in series
   c. the cell efficiency must be increased
   d. any of these will work

9. Compared to single-junction cells, multijunction cells are
   a. more efficient
   b. cheaper
   c. more durable
   d. larger in surface area

10. Light with a wavelength of 500 nm is
    a. ultraviolet
    b. visible
    c. infrared
    d. a microwave

11. Light energy per photon increases with
    a. lunar position
    b. frequency
    c. wavelength
    d. cloud cover

12. Laser scribing is used to
    a. test PV cells
    b. inscribe part numbers on modules
    c. separate thin-film PV material into small cells
    d. do none of these

13. Thin-film typically does not use
    a. CdTe
    b. c-Si
    c. CIGS
    d. GaAs

14. The coefficient of air mass relates to
    a. the weight of the atmosphere
    b. the amount of air to cool HCPV cells
    c. the path of sunlight through the atmosphere
    d. none of these

CHAPTER QUESTIONS AND PROBLEMS

1. A certain PV cell is illuminated with an irradiance (E) of 800 W/m^2. If the cell is 200 mm $\times$ 200 mm in size and produces 2.5 A at 0.5 V, what is the conversion efficiency?

2. Assume a conversion efficiency for a certain PV cell is 18%. If the cell is 100 mm $\times$ 100 mm, how much power will it produce at an irradiance of 950 W/m^2?

3. Six series strings of 24 PV cells are connected in parallel. Assume each PV cell produces a current of 2 A and 0.5 V at the maximum power point. Determine the output current, voltage, and power to a load under maximum power point conditions.

4. To obtain 32 V from a PV module, how many individual cells must be used and how should they be connected (assume 0.5 V per cell)?

5. If each cell in the module in problem 4 produces 3A, describe the system required to produce at least 20 A at 32 V.

6. A certain installation requires 36 V and at least 2 kW of rated power. Describe the configuration using 36 cell modules, where each module produces 12 V and 4 A.

7. How would you triple the current capacity of the PV cell array in problem 5 at the same voltage?

8. During one week there was cloud cover for 40% of the time and bright sun for the rest. The irradiance averaged 200 W/m^2 during cloud cover and 1,000 W/m^2 during bright sun. Determine the average irradiance for the week.

9. Assume a 2 A solar cell produces 0.5 V. If a total of 18 V and 4 A is needed, what is the minimum number of solar cells required and how must they be connected?

10. Determine the fill factor for a PV cell with $I_{OC} = 3$ A and $V_{OC} = 0.55$ V if the power at the maximum power point is 1.75 W.

11. If the wavelength of light is changed from 400 nm to 800 nm, how much does the light energy change?

12. Determine the power from the visible, UV, and IR components of sunlight striking the surface of a PV panel with a total irradiance of 500 W/m^2 if the panel is 0.5 m square. What is the energy in kWh if the specified irradiance is on the panel for 10 hours?

FOR DISCUSSION

What specifications are most important to consider when purchasing a solar PV system if space is not a problem? How would your answer differ if space is a problem?

ANSWERS TO CHECKUPS

Section 3-1 Checkup

1. The Bohr model is a planetary model of the atom, with electrons orbiting in discrete orbits about a central nucleus containing neutrons and protons.

2. By doping pure silicon with a pentavalent substance such as phosphorous or arsenic

3. By doping pure silicon with a trivalent substance such as are boron or indium

4. A *pn* junction is the boundary created between an *n*-type and a *p*-type semiconductor.

5. The electric field that forms due to the migration of charge across the boundary

Section 3-2 Checkup

1. In addition to the *n* and *p* regions and the boundary region, there is an anti-reflective coating and conductive grid on the top and a conductive layer on the bottom.

2. The grid allows light to penetrate into the cell.

3. An electron–hole pair is created and a voltage develops across the junction.

4. Band-gap energy is the energy required to free an electron from the valence band.

5. An electron–hole pair is the result when an electron is freed from a crystalline substance and leaves behind a vacancy in the lattice structure.

Section 3-3 Checkup

1. Monocrystalline semiconductor; polycrystalline semiconductor; and amorphous silicon, thin-film semiconductor.

2. A TCO is a transparent conducting oxide. It conducts electric charge and serves as an electrode for the cell.

3. Less costly to produce and the manufacturing process is simpler.

4. The dye sensitizes the photocathode.

Section 3-4 Checkup

1. To absorb more of the incident energy and thus increase the cell efficiency

2. From about 400 nm to 700 nm

3. Longer

4. Violet light has the most energy because its frequency is lower than red.

Section 3-5 Checkup

1. 0 V

2. 0.53 V

3. Energy must be at or above the band-gap energy.

4. Factors that affect efficiency are the type of cell, the reflectance efficiency of the cell's surface, the thermodynamic efficiency limit, the quantum efficiency, the maximum power point, and internal resistances.

5. Better

6. Fill factor is the ratio of the cell's actual maximum power output ($V_{MPP} \times I_{MPP}$) to its theoretical power output ($V_{OC} \times I_{SC}$).

Section 3-6 Checkup

1. Voltage

2. 36, 72, or 96 PV cells

3. Combinations of multiple PV cells connected to produce a specified power, voltage, and current output

4. A combination of solar modules

5. If it is shady (including clouds), the PV cell is not oriented to the sun, or it is night.

Section 3-7 Checkup

1. 39.8 V

2. The NOCT is 45° C $\pm$ 2° C. There is no limit.

3. Reading the graph, $I = 1.2$ A and $V = 37$ V. The maximum power is therefore approximately 44 W.

4. The coefficient is $-0.25\%/°$C for T $> 25°$ C. The output drops $-0.25\%/°$C $\times 25°$ C $= -6.25\%$.

Section 3-8 Checkup

1. CPV = concentrating photovoltaics; HCPV = high-concentration PV

2. A mirror or lens and tracking motors and heat sinks

3. A thin, flat lens with concentric rings

4. The cell surface area is much smaller for the same power.

ANSWERS TO TRUE/FALSE QUIZ

1. T 2. F 3. T 4. F 5. F 6. T 7. T 8. T 9. F
10. T 11. F 12. F 13. T 14. F 15. T 16. T

ANSWERS TO MULTIPLE-CHOICE QUESTIONS

1. d 2. d 3. b 4. a 5. b 6. b 7. d 8. d 9. a
10. b 11. b 12. c 13. b 14. c

Solar Power Systems

CHAPTER OBJECTIVES

- Describe the components and basic block
  diagrams for representative stand-alone solar
  electric systems.
- Explain how to perform an energy audit, evaluate
  electrical requirements, select PV modules,
  and specify cabling and components for a small
  electrical stand-alone system.
- Discuss grid-tied systems from the user
  perspective and describe the basic components
  required for a grid-tied system.
- Discuss advantages, disadvantages, and
  appropriate applications of various types of solar
  concentrators for both users and suppliers of
  electrical power.
- Explain the difference between open-loop hot water
  systems and closed-loop hot water systems, and
  discuss two types of closed-loop systems: the
  pressurized system and the drainback system.

KEY TERMS

*Key terms are shown in bold and color. Definitions for
key terms are provided at the end of the chapter and
in the end-of-book glossary. Bold terms in black are
defined in the end-of-book glossary only.*

- grid-tie system
- charge controller
- inverter
- combiner box
- ground fault protection
  device (GFPD)
- insolation
- depth of discharge
  (DOD)
- transfer switch
- microinverter
- solar concentrator
- Stirling engine
- regenerator
- heliostat
- absorptance
- emittance
- latent heat of
  vaporization
- open-loop
- closed-loop
- drainback system

INTRODUCTION

This chapter introduces solar systems, from the small
to the very large. The basic components of a system
are described, and diagrams of different configurations are given. The chapter begins with a discussion
of systems that are stand-alone, or off grid. These
systems are generally smaller systems for homeowners or remote locations. Grid-tie systems require
a more complex inverter that can synchronize the
output with the utility and take the system offline
(disconnect it) in the event the grid goes down (as a
safety measure for utility workers.) Other solar electric
systems are discussed, including larger systems for
commercial and institutional users and for the utilities
themselves.

In addition to PV modules, electricity can be obtained from the sun by other methods, but these
methods are primarily for utilities and large users.
These methods include Stirling engine collectors and
towers that can obtain extremely high temperatures
for driving steam turbines. Some of these systems will
be discussed. The chapter concludes with another excellent match of a resource and a need: that of solar
hot water. Three different types of solar hot water systems are described.

4-1   Stand-Alone Photovoltaic Solar Power Systems

By definition, a stand-alone system is one that is not designed to send power to the utility grid and thus does not require a grid-tie inverter (but it may still use grid power for backup). Stand-alone systems can range from a simple dc load that can be powered directly from the PV module to ones that include battery storage, an ac inverter, or a backup power supply. They are typically used for low-power applications and are often used where power is otherwise unavailable, such as in certain rural areas and remote locations where the utility grid is not readily accessible.

Basic Systems

Stand-alone solar electric systems do not supply power to the electric utility grid but can use the grid as an input to back up the system. Solar electrical systems can be used to supplement grid power. Grid-free systems do not have any input or output to the grid. By definition, all grid-free systems are stand-alone systems.

Stand-alone systems can have a dc or ac output, which is determined based on the load requirements. In general, stand-alone systems make sense for powering equipment that does not require a huge amount of power, such as lights or small appliances. The most common output voltages for small systems are 12 V or 24 V, with 48 V and higher used with larger systems. As in the case of any electrical system, national electrical codes and general electrical safety rules, including manufacturer's recommendations for wire size, grounding, and required environment for the various components, need to be followed. Depending on the application and the electrical power requirements for the load, most stand-alone systems include a battery for supplying power when there is little or no solar input. For certain basic applications, such as an attic fan or some water-pumping applications, a battery backup is not required, saving the cost and maintenance of batteries.

Table 4-1 shows five configurations for stand-alone systems, in increasing system complexity. Variations of the configurations in Table 4-1 are common, so the table is only representative of these systems. The power output can range from less than 1 W for a small

A remote traffic sign with warning lights is an ideal application for a stand-alone solar power system. The solar input is converted to electricity that charges batteries to operate the lights.
(*Source:* David Buchla.)

TABLE 4-1

Configurations for Stand-Alone Solar PV Systems

System	Components	Typical Applications
1	PV module and dc load.	DC ventilation fans, small water pumps such as circulating pumps for solar thermal water-heating systems, and other dc loads that do not require electrical storage.
2	PV module, dc/dc converter (power conditioning), and dc load.	DC loads that require specific dc voltages but do not require storage, such as a charging station for certain electric vehicles or DC water pumps. This configuration is also useful for miniature applications such as calculators.
3	PV module, charge controller/ battery storage, and dc load.	DC loads that require power even when there is no solar input, such as small yard lights, traffic warning lights, and buoy power, or mobile and remote power for recreational vehicles (RVs) and communication systems.
4	PV module, charge controller/ battery storage, inverter, and dc and ac loads.	AC and dc loads, including appliances such as refrigerators and lights.
5	PV module, charge controller and battery storage (optional), inverter, supplementary generation, and dc and ac loads.	AC and dc systems where there is large seasonal variation in solar input. The supplementary power can be from any other power-generating system such as a wind generator, a gas generator, or the utility grid.

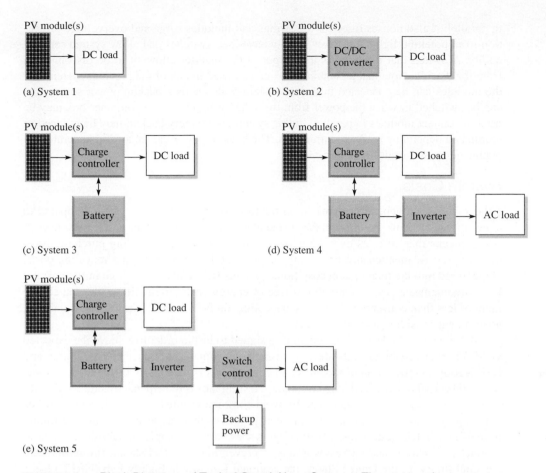

(a) System 1

(b) System 2

(c) System 3

(d) System 4

(e) System 5

FIGURE 4-1 Block Diagrams of Typical Stand-Alone Systems. The systems here are representative of different types; other configurations are possible.

calculator to over 10 kW. Usually systems that are larger than 10 kW are more effective as **grid-tie systems**, which is any electrical generating system that is connected to the electric utility grid (discussed in Section 4-3). Other optional components, such as solar tracking devices or various system-monitoring devices, can be added to any of these systems. System monitoring can provide basic performance data for the system and may include power, energy and possibly service information or advanced diagnostics. Poor performance can be attributed to a bad panel, rodent damage, dirt, leaves, or wiring problems; monitoring can pinpoint the problem area or panel.

Figure 4-1 shows representative block diagrams of the systems listed in Table 4-1. These diagrams are meant only as guides to demonstrate how typical system components are connected together. System 1 represents the simplest system, which is composed of the PV module and a load. A system like this can supply power only when there is solar input, so applications are limited. System 2 adds a dc/dc converter, which allows the designer to match the electrical load or different voltage requirements for better performance. System 3 includes battery backup. With this system, a **charge controller** replaces the dc/dc converter; its main purpose is to regulate and limit charging current to prevent overcharging the batteries. In a solar PV system, the charge controller also prevents draining the batteries back through the PV modules when they are needed for the load. System 4 adds an **inverter**, which converts the dc output to ac for powering small appliances. The inverter is a basic battery-based inverter rather than the more expensive grid-tie inverter, which is required when connecting to a utility grid. (Charge controllers and inverters are covered in detail in Chapter 6.) System 5 is a hybrid system that uses more than one module in parallel, so the outputs are combined in a combiner box (not shown). A **combiner box** is a double-insulated box that allows several strings from modules to be connected together in parallel; it also houses fuses for the strings and includes surge and overvoltage protection from potential lightning strikes. It may have load switches that allow system service-ability. A switch that disconnects the output to the inverter allows it to be disconnected from the dc side. Note that a combiner box can be used in any of the previous systems, or the modules can be connected in series. System 5 also adds a backup power source that can be switched in when the power from the solar system is low; a combiner box may be used to connect modules in parallel. In this system, the battery backup may be reduced or eliminated depending on the requirements. The backup power can be a wind generator, an engine generator, or utility power.

System Costs

Solar systems are generally evaluated on the basis of the cost of the system compared to conventional systems. Frequently, the cost analysis of a system is done as a life-cycle cost, which means that all costs over the expected life of a system, including purchase price, operating costs, maintenance costs, any supplemental energy costs, and recycling costs, are factored into the total system cost. Solar systems frequently have a high initial cost but lower maintenance cost and are almost free of energy costs, so the life-cycle cost can be more or less than conventional systems, depending on factors such as resource availability, government subsidies, and interest cost.

Lifetime costs are less if the product is designed to last in order to achieve the expected yield. Product reliability and safety is ensured with universally accepted standards and certifications; certification of modules and components is mandatory in some parts of the world. The goal of certification from standard testing labs is to help manufacturers identify potential problems with modules and to avoid common failures such as moisture ingress, cracking, ground faults, hot spots, and other problems that can lead to premature failure in the field. Safety testing and certification is viewed as a complementary component to product certification and addresses issues of preventing electrical shock, fire hazards, or personal injury. In the United States, the National Renewable Energy Lab (NREL) has a program to accelerate stress testing in order to predict how and when a test unit might fail. Another program by the US Department of Energy is designed to gather data using simulated environmental stresses to determine reliability.

Computer Simulation of Systems

Solar and other renewable energy systems can be modeled to evaluate costs, design options, and tradeoffs with an NREL program called HOMER.[1] HOMER is a software program that can model both off-grid and grid-connected renewable energy systems and provide the optimum systems for a given situation. The program does hourly calculations on a system for an entire year using data specified by the user for various types of resources (solar, wind, hydro, generators, etc.) and compares the resource with the demand for both electrical and thermal loads. Every system must include a primary load (i.e., a description of the electric demand) or a deferrable load, or be connected to a grid. The user specifies the electrical load in some detail and the cost of various components (capital cost, replacement cost, etc.). Results from HOMER include technical and economic details about each possible system configuration that was selected for evaluation. Figure 4-2 shows a HOMER model of a small grid-tie PV system with battery backup and includes a government rebate or tax incentive in the price of the modules. The model can be modified easily for rate changes, price changes or incentives, product lifetimes, efficiency improvements, and other variables. The program suggests design changes as needed. The four most cost efficient systems, based on net present cost (or life-cycle cost), are displayed under the heading Total NPC. The heading COE indicates the average cost of energy. In this case, the optimum system does not use any renewable energy and has a higher COE that the renewable

---

[1] HOMER can be downloaded from the Homer Energy website. The free download is a good learning tool and is named HOMER Legacy.

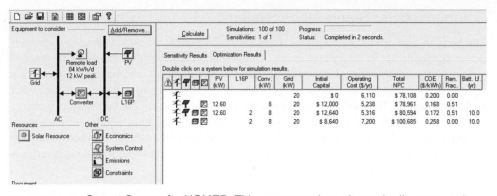

FIGURE 4-2 Output Screen for HOMER. This screen capture shows the list arranged in order of the net present cost for the four best systems of those tested. (*Source:* Reproduced by permission of HOMER.)

system on the second line; however, a small change in module efficiency, service lifetime, or utility price per kWh can change that. Of course, the model does not take into account intangibles such as the desire to be energy-efficient or to reduce carbon dioxide emissions or the value of a backup system when the utility power is not available.

Stand-Alone Systems in Developing Countries

Stand-alone systems are particularly useful for many developing countries because the power grid is either too far away to be practical or it is unreliable. The cost of stand-alone systems is also a factor because it is cheaper to provide electricity if it is not connected to the grid. Most of the cost of solar electricity is the up-front cost of the panels and the installation, and several international organizations can help with these costs.[2] In many developing countries, wood is a traditional fuel source but is in short supply; solar energy is readily available. Frequently, solar electrical systems are supplemented with solar systems for hot water or cooking and refrigeration systems using a solid-absorption ($CaCl_2/NH_3$) cycle. Most of the developing countries are located in very sunny parts of the world, including Africa and large parts of Asia (see the map of the world solar resource in Figure 1-12), so stand-alone solar systems are not only cost-effective, but can help take pressure off remaining wood supplies and other traditional fuel resources, leading to a healthier environment. The solar resource tends to be less dependent on the seasons for locations closer to the equator. Figure 4-3 shows an example of a stand-alone electric system in India at a community health center.

FIGURE 4-3 Stand-Alone Electric System in India. Stand-alone power is important to many developing countries. This solar array provides power to a village health center. (*Source:* National Renewable Energy Laboratory.)

---

[2]One such organization is Solar Energy International (SEI), a group that works with various organizations to help install solar power in developing countries.

Components for Stand-Alone Systems

The heart of a solar electrical system is the PV module, which needs to be able to provide power for the loads in the system and to charge batteries when they are used for backup power. The module selected depends on the load requirements and batteries used. For a 12 V system, the PV module needs to provide about 20 V to charge batteries reliably. For a 24 V system, the PV module should provide 40 V. When battery backup is used, a charge controller is needed. It protects the batteries from overcharging and switches to the battery backup when the PV module power is too low for the load. In cases where there is reliable utility power, it may be used as backup rather than batteries. For ac loads, an inverter is needed that changes the dc to ac.

Codes

Photovoltaic Wire

Photovoltaic (PV) wire is a special type of stranded copper wire that is sunlight-resistant and dedicated for the interconnection of PV modules. In addition, it is designed to operate at elevated temperatures and is rated for wet locations.

Safety Note

Ground faults occur when current finds an alternate return path to the source; they can pose a serious safety hazards. When a ground fault occurs, normally grounded parts can become energized and present a shock hazard to anyone working on them. When troubleshooting a ground fault, it is important to wear protective equipment, including safety glasses and insulating gloves rated for the highest possible voltage in the system. Safety boots are also recommended.

Codes are standards for the building industry and are designed to enforce public health and safety. The **International Building Code (IBC)** is a document that addresses all aspects of the design and installation of building systems, including electrical systems. In the United States, the **National Electric Code (NEC)** is the standards document for electrical work, with specific provisions for solar electric systems. The NEC is a set of codes and standards designed to protect the safety of electrical workers and the public. While neither of these codes are laws, various local jurisdictions often base their requirements on one or more of these standards. The NEC is considered the premier document guiding electrical installations in the United States. Requirements within the NEC include electrical installations in general, wiring and protection circuits, grounding, surge arresters, conduit, boxes, and more. These requirements pertain to solar electric systems. Wiring on roofs must be derated for temperature limitations associated with the current-carrying capacity of conductors. Wiring needs to be protected from sunlight, moisture, and potential damage from rodents, and conduit runs must be able to expand and contract with temperature changes by using expansion couplings and proper support. There are many other provisions for solar electric systems including voltage requirements. In the NEC, high voltage is defined as "more than 600 volts, nominal." For the United States, circuits operating at more than 600 V to ground require special high-voltage equipment and are subject to additional safety guidelines. In Europe, the standard high voltage for PV system is 1,000 V, so products such as cabling, fuses, inverters, and boxes manufactured for Europe may show higher voltage ratings than those in the United States (although utility-scale systems are operating in the US at 1000 V). Another requirement is for a **ground fault protection device (GFPD)** on photovoltaic arrays. The purpose of a GFPD is to reduce the risk of fire associated with a ground fault and to protect personnel that are working if a ground fault should occur. A **ground fault** is current in the ground conductor (rather than in the neutral), which normally should have no current. The detector senses a difference in the current in the hot line and the neutral and trips the breaker if they are not the same. The requirements for a GFPD are to (1) detect the fault, (2) interrupt the current, (3) indicate that a fault has occurred with a visible warning, and (4) disconnect the faulty module. A section of the NEC is devoted to boxes and enclosures, which are always part of a solar electric installation. It is not possible to summarize all of the requirements here, but anyone installing or working on a solar electric system needs to be aware of the code and the special requirements for solar electrical systems.

Another code that is important in solar energy work is the **Uniform Solar Energy Code (USEC).** This code lists plumbing and mechanical requirements, including installation, inspection, and maintenance of solar energy systems. It covers both solar hot water and PV systems, including various components (storage tanks, heat exchangers) and specifies requirements for installing systems in locations where freezing temperatures are possible. Other requirements of USEC include waterproofing when a solar collector is installed on a building structure to avoid water leakage problems.

4-2   Sizing the Stand-Alone System

Designing a solar system requires a systematic approach. The first step in sizing a system is to perform an energy audit, looking for places to save energy. The power requirements are evaluated as part of the audit, and the site is evaluated for the expected solar input. From this, the basic system is designed. In this section, you will go through the steps of the basic process for designing a stand-alone system.

Design Steps for a Stand-Alone System

The following steps provide a systematic way of designing a system:

- Conduct an energy audit and establish power requirements.
- Evaluate the site.
- Develop the initial system concept.
- Determine the PV array size.
- Evaluate cabling and battery requirements.
- Select the components.
- Review the design.

Conduct an Energy Audit and Establish Power Requirements

The load requirements should be the starting point in determining a stand-alone system. Start by calculating the power requirement for each ac electrical load and multiplying by the average number of hours it is powered on each day (Wh/day), which represents energy/day. (Recall that energy is power multiplied by time.) A spreadsheet is a useful tool for developing this analysis. Figure 4-4 shows how a summary of appliances and their usage can give a reasonable estimate of the load requirement; the load is calculated for each month to account for seasonal variations. From this analysis, the average daily use can be determined for a one-year interval. For the month shown, the average daily energy use is 10.79 kWh, which is the energy the solar system will need to provide.

The analysis can also reveal opportunities for conserving energy and/or switching part of the load to other sources, such as propane instead of electricity, and also the savings expected for replacing incandescent bulbs with compact fluorescent bulbs or LED lighting. An often-overlooked load is the standby power used by many electronic devices. Standby power is the power to keep devices ready. In many devices (televisions, computers, printers, phones, etc.), a certain amount of power is used when the device is plugged in but not in service. This power can add up: In the average US household, it is estimated that standby power accounts for a constant load between 40 and 60 W, equivalent to leaving a light on day and night. Removing unnecessary loads like these can reduce the requirements shown in the load analysis and are almost always cost-effective.

AC Load Description	Quantity	Power Rating (W)	Time on per Day (h/da)	Energy Used (Wh/da)
Refrigerator	1	450	8	3,600
Washing machine	1	500	0.5	250
TV	2	100	3	600
Lights (incandescent)	4	60	6	1,440
Lights (fluorescent)	5	30	10	1,500
Toaster oven	1	1,500	0.5	750
Microwave oven	1	1,000	0.4	400
Ceiling fans (medium speed)	3	25	10	750
Computer	2	125	4	1,000
Printer	1	400	0.25	100
Miscellaneous loads	1	200	2	400
			TOTAL =	**10,790**

FIGURE 4-4   Load Requirement Analysis for a Typical Month Using a Spreadsheet

Evaluate the Site

The site should be reviewed to determine the feasibility of a solar system. Look for possible shading problems and potential installation problems. Site data for **insolation** is available from the National Renewable Energy Labs for the United States and from private companies for other locations. The word *insolation* comes from "*in*cident *sol*ar radi*ation*" and is a measure of the energy received on a surface in a specific amount of time. From the data, the average hours of peak sunlight (equal to or greater than $1,000$ W/m^2) can be estimated for each month of the year. The monthly solar irradiance can vary considerably, so this needs to be taken into account when designing a system.

At the specific site, consider possible locations for solar modules and consider the effect of wind or snow loading of arrays and installation issues (roof condition, structural supports and strength, etc.) or future maintenance issues (cleaning, snow or wind loads, etc.). The electrical service is inspected and reviewed for issues that could have an impact on the installation.

Generally modules can be installed on a south-facing roof, a ground framework, or a pole. A single shaded PV module in a series arrangement restricts the current to the other modules and can seriously affect system performance. To analyze a site for possible shading problems, a device like the Solar Pathfinder™ is useful (see Figure 4-5). The Solar Pathfinder™ uses a polished, transparent dome that shows a reflected panoramic view of

FIGURE 4-5   Typical Results from the Solar Pathfinder™. This device can show potential shading problems.   (*Source:* Courtesy of Solar Pathfinder.)

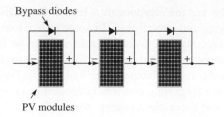

Bypass diodes

PV modules

FIGURE 4-6 Bypass Diodes. Bypass diodes allow current to bypass a shaded module or one that has failed.

the site. Trees, buildings, or other obstacles to the sun are seen as reflections on the surface of the dome. The sun's path is also shown as lines on the dome, so a proposed location can be evaluated for shading problems for the entire year. Using analysis software from Solar Pathfinder™, a custom report can be generated that shows the percentage of shading expected for each month and the expected performance of an array.

To avoid problems with a shaded module that is in series with other modules, or one that has failed, use bypass diodes. Some modules come equipped with one or more bypass diodes. They provide a current path around the module or parts of the module when there is low voltage. When operating normally, the bypass diodes are reverse-biased, effectively removing them from the circuit. If the voltage across a diode is less than about 0.7 V, the bypass diode becomes forward-biased and provides a current path. Figure 4-6 shows how bypass diodes are connected in a group of three series modules, each with one bypass diode. (Diodes were covered in Chapter 2.)

Installation problems include a roof that is not oriented due south or a roof that has a slope that is too steep or is too flat. Ideally, the modules should be perpendicular to the sun's rays, but this is not possible for fixed arrays or most roof-mounted arrays. In general, a flat-plate collector should be tilted at an angle that is equal to the latitude to optimize the greatest *average* radiation, although some people prefer to optimize the winter months by tilting the modules at a higher angle. Ground-mounted and pole-mounted installations allow greater accessibility for making seasonal adjustments to the tilt angle, which can add a few percentage points to the overall energy obtained. For roof installations, mounting brackets can be selected that optimize the tilt angle toward the south, even if the roof is not oriented in that direction (see Figure 4-7).

Develop the Initial System Concept

Having established the resource(s) available and the power requirement, an initial system concept can be developed. The system is evaluated for different operating voltages, which

FIGURE 4-7 Roof Installation with Mounting Brackets. Modules can be oriented south using special mounting brackets.   (*Source:* Joey Kotfica/Jupiter Images/Getty Images.)

affect the size of the cables and the components selected. Consideration is given to system redundancy, battery backup, and overall cost. In some cases, two or more small systems may be preferable to a larger system. System losses need to be accounted for: A typical system (not including batteries) is less than 80% efficient due to losses in cables and connectors, inverter losses, temperature changes from ambient, soiling of the array, and so on. Battery efficiency is another large source of loss because the energy returned from the battery is typically only 80% of the original energy stored. Thus, in a battery-backed system, the overall efficiency may be only 60%.

Determine the PV Array Size

From the projected resource(s), power requirement, and estimated efficiency, the size of the array can be determined. The power requirement for the array, in watts, is estimated for each month by the following equation:

$$P_{array} = \frac{W}{t_{solar}\eta_{sys}}$$

Equation 4-1

where

P_{array} represents the peak power, in watts (W), from the array (based on the manufacturer's module label)

W is the daily energy requirement in, watt-hours (Wh)

t_{solar} is the average hours (h) of peak sunlight for the month

η_{sys} is the overall efficiency of the system

This efficiency does *not* include the inherent inefficiency of PV modules in converting sunlight to electricity but is due to losses in the system, such as inverters, cabling, battery charging losses, and conditions differing from the rated test conditions (such as temperature or irradiance). Notice that less power is required from the array if the system is more efficient, so overall efficiency is important.

EXAMPLE 4-1

A certain site has an average of 5 hours of peak sunlight per day in January. If 11 kWh is required on an average January day from a grid-free system, what peak power is required from the array? Assume the system is 60% efficient, which represents a typical residential system.

Solution

$$P_{array} = \frac{W}{t_{solar}\eta_{sys}}$$

$$P_{array} = \frac{11{,}000 \text{ Wh}}{(5 \text{ h})(0.60)} = 3670 \text{ W} = 3.67 \text{ kW}$$

Choose modules that will provide at least this much power: 18 modules of 215 W is 3.87 kW.

Evaluate Cabling and Battery Requirements

Determine the wire size required from the initial system concept. Recall from Chapter 2 that dc electrical power is the product of current and voltage. For a given power, the current (and hence the required wire size) is smaller as the voltage is higher. In general, smaller diameter wire costs less but is rated for less current and drops more voltage. Note that listed wire capacity needs to be significantly derated for higher temperatures due to wiring on roofs and if the wires are bundled or in conduit. Generally, the size of the wiring is determined so that voltage drops are in the range of 2% or 3%, but relevant codes should be consulted in determining wiring requirements.

EXAMPLE 4-2

Assume that 18 panels (modules) of the type shown in Figure 3-26 are selected for the system in Example 4-1. Their configuration matches system 5 in Figure 4-1. Assume the panels are wired in six parallel groups, with each having three panels in series that are combined at a combiner box, and the output is sent to a charge controller. What minimum wire size is needed for the series groups to the combiner box and the parallel groups from the combiner box? Assume you need to derate the wire so that it carries 50% of its normal rating.

Solution

In the series groups, the current is the same in all panels, so the maximum current is the short-circuit current for one panel, which is specified by the manufacturer as 5.8 A. From Table 2-1 in Chapter 2, the AWG 16 copper wire has a normal current capacity of 15 A. When a 50% derating factor is applied, the current carrying capacity is 7.5 A, which is sufficient. After combining the series groups in parallel, the wiring needs to be able to handle 34.8 A (6 × 5.8 A). The minimum wire size with a 50% derating should normally be able to handle 69.6 A (34.8 A × 2). Choose AWG 4 copper wire (rated for 70 A) to meet the minimum requirement.

After determining the wiring size, you need to evaluate the battery requirement. As you know, batteries are connected in series to increase voltage and in parallel to increase capacity. The battery backup system generally has some combination of a series and parallel arrangement to set the voltage to the required voltage for the system. Batteries can have very high discharge currents, so disconnect switches and overcurrent devices must be selected with the appropriate ratings. Disconnect switches are important to isolate the batteries for maintenance or replacement. Also, batteries need to be in a vented and protected enclosure for safety and to avoid access by children or people unqualified to work with the equipment.

Batteries in solar electric systems tend to be regularly charged and discharged, so they need to be deep-cycle types. Standard automobile batteries have thin plates and are designed for high starting currents, but they are not suitable for solar systems because they are not designed for deep cycling. Golf cart batteries and larger L-16 traction batteries have thick plates, which are suitable for solar systems. Flooded lead-acid batteries are the most common type of deep-cycle battery, but they must be checked regularly for fluid level and specific gravity, cleaned of any corrosion, and refilled as needed with distilled water. Specific gravity is tested with a hydrometer, which indicates the state of charge. A fully charged lead-acid battery has a specific gravity between 1.25 and 1.28. The fill level is important; the electrolyte should be maintained above the minimum and below the maximum level line on the side of the battery.

FIGURE 4-8 Condensed Battery Installation That Is Still Accessible for Maintenance. Batteries should be arranged to prevent accidentally dropping anything on them. A screened vent is needed in the storage area to release any hydrogen gas. (*Source:* Courtesy of Solar Direct.)

Figure 4-8 shows a typical battery installation for a residential installation. The battery array needs to be installed in a cool, well-ventilated area that is not in the living area; a garage or utility room is common (but very high or very cold temperatures decrease battery life). Venting releases any hydrogen gas that builds up as a result of charging; hydrogen is less dense than air, so the vent pipe should be high. Batteries should be fully charged when they are put in service to maximize their life. Charging is complete when the charging current is near zero. Overcharging is dangerous because it can lead to increased hydrogen production and create an explosion hazard.

The capacity of any battery is measured in ampere-hours. The **ampere-hour (Ah)** is a unit of charge, that is, current multiplied by time. From basic math, 1 ampere-hour is equal

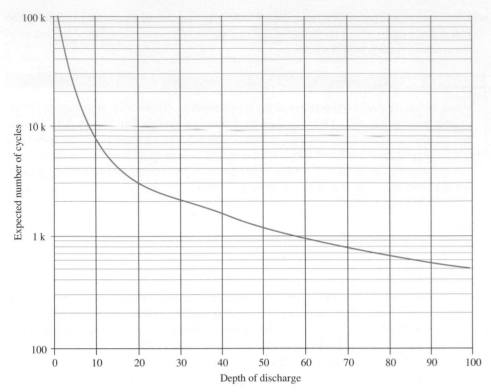

FIGURE 4-9   Expected Life Cycle Versus Depth of Discharge

Safety Note

Flooded lead-acid batteries can produce hydrogen and oxygen, an explosive combination, so they should never be used indoors or near an ignition source and should always be vented to the outside. Electrical equipment (inverters, switches, fuses) should not be in their vicinity. Avoid sparks or open flames near batteries and post "no smoking" signs. Wear eye protection and acid-proof gloves when handling batteries. Have water and baking soda nearby to neutralize any acid spilled from a lead-acid battery and clean up spills immediately.

to a charge of 3,600 coulombs. A typical deep-cycle golf-cart battery is rated for 225 Ah at 6 V. The popular L-16 solar battery is rated for 370 Ah at 6 V. The designer needs to specify the **depth of discharge (DOD)**, which is the ratio of the quantity of charge (measured in ampere-hours) removed from a battery to its rated capacity and can be expressed as a percentage. Although 80% discharge is sometimes specified by manufacturers, continuous deep discharging leads to shortened battery life; many batteries can be damaged if they are deeply discharged. Figure 4-9 shows how battery life for a typical deep-cycle battery depends on depth of discharge. As you can see, the expected lifetime is shortened by deeply discharging the battery.

A reasonable depth of discharge is 50% over a period of days to save on replacement cost (leaving some reserve for emergencies or below-average insolation), but the decision about the number of batteries required depends on various factors, and the depth of discharge varies seasonally. Depending on the projected weather at the site, the batteries are typically sized for three to five days of autonomy in a grid-free system (no solar input). Keep in mind that batteries are expensive and have maintenance issues, so the balance between the number of batteries and the cost of the system is always a compromise. The following formula can be used to determine the required ampere-hours of the batteries:

$$\text{Ah} = \frac{W_{day} \cdot t_{store}}{V B_{dod} \eta_{inv}}$$

Equation 4-2

where

Ah is the required ampere-hours (Ah) from the batteries

W_{day} is the daily energy requirement per day, in watt-hours/day (Wh/d)

t_{store} is the backup time required, in days (d)

V is the dc system voltage to the inverter, in volts (V)

B_{dod} is the battery's maximum depth of discharge, expressed as a fraction

η_{inv} is the efficiency of the inverter and cabling, also expressed as a fraction

EXAMPLE 4-3

Assume that the system described in Example 4-1 is a 24 V system (from the charge controller). Three backup days are required. How many L-16 deep-cycle batteries (rated at 6 V, 390 Ah), are required to provide the three days of backup if the DOD is 50% and the inverter/cables are 95% efficient?

Solution

$$Ah = \frac{W_{day} \cdot t_{store}}{VB_{dod}\,\eta_{inv}}$$

$$= \frac{(11{,}000 \text{ Wh/d})(3 \text{ d})}{(24 \text{ V})(0.5)(0.95)} = 2{,}895 \text{ Ah}$$

The system operates on 24 V, so four 6 V batteries need to be connected in series to give the required 24 V. Each bank of four L-16 batteries will have a 390 Ah rating at 24 V. Eight banks of four series batteries will produce 8×390 Ah = 3,120 Ah, which exceeds the requirement for three days of autonomy. Hence, a total of 32 batteries are required for this system.

Select the Components

Once the system is analyzed, the components are selected. For a stand-alone system, you will generally need the PV modules, combiner boxes, a charge controller, battery backup, and an inverter. In addition, the system will require mechanical and electrical hardware components, which includes mounting hardware, racks, connectors, junction boxes, disconnect switches, fuse holders, contactors, surge arrestors, wiring and conduit, and other parts.

Review the Design

After completion, the design must be reviewed to ensure that the specifications for the system have been met and that all components are operating within their design limits. After completing the design review, plans must be submitted to appropriate government agencies for approvals and permits before any work is done. The electrical utility must approve any connection that is made to the electrical grid.

(*Source:* National Renewable Energy Laboratory.)

Lead-acid batteries are the dominant energy storage technology, but the more expensive lithium ion batteries have important advantages for renewable energy systems that may make them competitive with lead-acid batteries in the future. These advantages include longer lifetime and higher efficiency. They can also be charged and discharged many more times than lead-acid batteries can be. If the cost can be made more competitive, lithium ion batteries may replace lead-acid batteries as the main storage technology in the future.

SECTION 4-2 CHECKUP

1. How does a bypass diode prevent a complete failure if one module in a series string fails?

2. As a rule of thumb, what is the best angle to tilt a fixed flat-plate collector?

3. What measurement unit is used to measure the capacity of a battery?

4. What is meant by the term *depth of discharge*?

4-3 Grid-Tie Photovoltaic Solar Power Systems

Most PV systems are grid-tie systems that work in conjunction with the power supplied by the electric company. A grid-tie system has a special inverter that can receive power from the grid or send grid-quality ac power to the utility grid when there is an excess of energy from the solar system. In addition, the utility company can produce power from solar farms and send power to the grid directly.

Residential and Small Grid-Tie Systems

Grid-tie systems can be set up with or without a battery backup. The simplest grid-tie system does not use battery backup but offers a way to supplement some fraction of the utility power. The major components in this system are the PV modules and an inverter. The modules may be connected in series to the inverter if voltage limits are not exceeded, or a separate combiner box may be used to combine the outputs of various modules in parallel. The inverter must be a special type that can be connected directly to the ac breaker box, so it needs to convert the dc from the PV modules into grid-compatible ac and match the phase of the utility sine wave. It must also be able to disconnect the PV system (using an automatic transfer switch) when the grid is down, so it must be an approved inverter that meets UL standard 1741. A **transfer switch** is an automatic switch that can switch loads between alternate power sources without interrupting the current.

A basic block diagram of a grid-tie system with series PV modules is shown in Figure 4-10. Compared to a system with a battery backup, a battery-free system like this is less expensive, easier to install, and almost maintenance-free. It has the advantage of not having to supply all of the power needed for the home or business; it can offset any fraction of the power and have the utility make up the difference. If the grid is reliable, as it is in most urban areas, then a battery-free system offers the best performance per dollar spent. For many commercial office buildings, stores, and industrial buildings, a battery-free system makes sense. These types of buildings are normally occupied during daylight hours, corresponding to the times when the solar resource is available. Usually the modules can be installed on the roof of the building or a parking structure, so land is not sacrificed for the array. The system can be set up so that any excess power is sold back to the utility, alleviating any concern about weekend or holiday unused capacity.

UL Standard 1741

The Underwriters Laboratories® (UL) is an independent product safety certification organization that writes standards for safety and tests products for compliance. UL standard 1741 lists requirements for inverters, converters, charge controllers, and interconnection system equipment for both utility-interactive (grid-tie) power systems and for non-grid-tie systems. Other UL standards are written for PV modules and junction boxes, cabling, connectors, batteries, and mounting systems. For example, UL standard 1703 specifies standards for PV systems up to 1,000 V. Companies that receive UL certification are allowed to display the UL mark on the product(s).

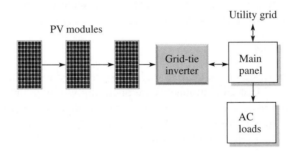

FIGURE 4-10   Simplified Battery-Free Grid-Tie System

Residential and Small Grid-Tie Systems with Battery Backup

Grid-tie systems with a battery backup can continue to supply power any time the grid goes down. The system can switch seamlessly to backup power when an electrical outage occurs. Simultaneously, it disconnects the system from the grid so it doesn't send power out when the grid is down.

A small system with a full battery backup capability is much more expensive than a battery-free system. One way to reduce cost is to split the system into backed-up loads and non-backed-up loads, thus reducing the number of batteries required, saving initial cost, and reducing maintenance and space requirements. This option requires rewiring the service panel and placing non-backed-up loads on a separate dedicated panel from those that are backed up. Essentially, this option is equivalent to having two systems, but rewiring a panel may be a cheaper option than a fully backed-up system. A system with backed-up loads and non-backed-up loads is shown in the block diagram in Figure 4-11. The panels are shown going to a combiner box, but a series arrangement is another option for connecting the modules. A **combiner box** is an electrical connection box for combining the outputs of multiple solar panels into one dc output.

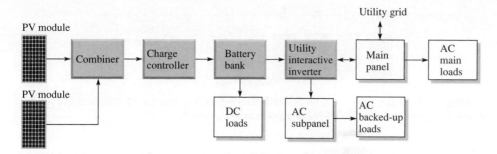

FIGURE 4-11 Simplified Battery Backup System for Part of the AC Load

When the system is in the grid-interactive mode, the inverter takes energy from the sources and sends it to the backed-up loads. The main loads are powered directly from the grid. If there is more energy from the PV modules than is needed by the backed-up loads, the excess is put onto the grid through an internal transfer switch, resulting in a credit for the homeowner (net metering). When the grid is down or out of specification, the transfer switch opens and only the backed-up loads receive power from the inverter. The main loads are solely dependent on the grid, so they will be off until power is restored.

The size of the inverter and battery backup required for a partially backed-up system requires an analysis of the loads that will be put on the backed-up system. To estimate the power requirement for the backup loads, power to each load can be summarized on a spreadsheet like the one shown in Figure 4-4. Motors need more power during starting than during running, so the system must be sized based on starting power. From the results of this analysis, the inverter, including various options, can be selected. One option is to use inverters that can be stacked. The term *stacking* refers to connecting two inverters to provide split-phase 120/240 V outputs. Another option available on some inverters is to provide a backup engine generator input.

The battery bank is sized according to the number of days of autonomy required. The size can be based on historical patterns of time that the grid is down. In general, a system that is backing up the grid is cycled only when the grid is down, so sizing considerations are different than in the grid-free system, which cycles daily. An 80% depth of discharge is appropriate for a system that is cycled infrequently, and the number of days of autonomy is based on grid performance rather than weather patterns. The infrequent cycling means that sealed batteries can be a good choice for a backup system because they require less maintenance than flooded types. The drawback to sealed batteries is that they are more expensive and have a shorter life expectancy than flooded types.

As is the case with other systems, provision must be made for other components such as those listed in Section 4-2. For battery backed-up systems, battery meters that can report the state of charge are useful. These meters show the voltage, current, and percentage of full charge. Another option is a power meter that monitors the performance of the system and alerts the user of fault conditions. Studies have shown that monitoring systems encourage energy conservation and that more detailed information leads to more conservation.

Small Systems with Microinverters

The systems shown previously takes dc to a central inverter and converts it to ac at that point. Another option that is growing in popularity is to use a microinverter at each module. A **microinverter** is a dc to ac converter that is sized to operate with a single solar module. Thus, it can provide maximum power-point tracking for the module and greater efficiency, particularly for situations such as a single shaded module that has reduced output. A basic system is illustrated in Figure 4-12. Each inverter puts out grid-compatible ac that is synchronized to other microinverters in the system. Microinverters are installed in parallel with each other to form a branch circuit. The branch circuits are

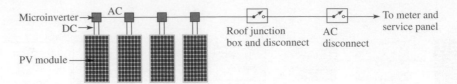

FIGURE 4-12 Basic Microinverter System. DC from each module is converted to ac where it is connected to other microinverters in the system.

often combined at a subpanel. The result is a more modularized system; if a module or microinverter fails, the rest of the system continues to operate (at reduced output) because the other microinverters are connected in parallel and one open source does not affect the operation of the others. The defective module or microinverter can be repaired without taking the rest of the system offline; however, the faulty module may have to be removed for servicing.

Some modules come equipped with a built-in microinverter and circuits to optimize the output. Built-in microinverters do not have access to the dc circuits from the PV module, but they eliminate the dc wiring, connectors, combiner boxes, and so forth. This simplifies installation, making the overall system efficient and cost-effective. It also eliminates high-voltage dc circuits (as much as 600 V), so the microinverter system is safer than high-voltage systems with a central inverter.

Commercial and Institutional PV Systems

Commercial and institutional solar PV systems can offer economies of scale and frequently have the advantage of a relatively lower demand for electricity at night. Most of these systems are designed to reduce the electricity demand for a larger user such as a business, school, or manufacturing facility, so the system is designed to be a grid-tie system. A few systems are designed as off-grid systems for remote applications, such as a PV system that was installed for a marine sanctuary on the Farallones Islands. The marine sanctuary had previously imported diesel to run generators for electricity. In addition to supplementing utility power, another application for commercial and institutional establishments is to provide a solar fuel station for their employees or the public to use. The solar panels are mounted above a parking area, and they supply charging power to electric vehicles, an excellent match of the available resource to the need (charging electric vehicles). Figure 4-13 shows a solar fuel station. Many communities and government entities are providing these stations at public parking facilities to encourage use of electric vehicles and to reduce emissions.

FIGURE 4-13 Solar Fueling Station. The solar modules of this fueling station are used to charge electric vehicles. (*Source:* National Renewable Energy Laboratory.)

Utility Grid-Tie PV Systems

In some areas, utilities have constructed large PV arrays that are designed to feed power to the grid (see Figure 3-25 in Chapter 3). Utilities have much different considerations for implementing solar systems because they are supplying power rather than consuming it. When a utility company is considering adding solar power, the system is first analyzed and modeled to determine the effects, load balancing, equipment loading, and power quality issues. The overall cost, such as any new transmission and distribution systems required, and the impact on existing facilities, such as reduced fuel costs, are evaluated. In some cases, it may be more economical to develop distributed systems using smaller solar arrays deployed on specific feeders to handle additional load and reduce capital costs. Distributed systems can also reduce line-related cost due to power dissipated in transmission lines.

SECTION 4-3 CHECKUP

1. What is the requirement for grid-tie inverters?
2. What are two reasons for having a grid-tie PV system that is not backed up?
3. How does sizing a battery array in a grid-tie system differ from sizing a battery array in a grid-free system?
4. Why is constant system monitoring useful for a grid-tie system?
5. What cost factors should utilities consider for adding solar PV resources that a homeowner does not need to consider?

4-4   Solar Concentrators

A solar concentrator collects light over a certain area and focuses it onto a smaller area. The light can be focused with either a lens or a mirror. For PV systems, the concentrator can increase the amount of electrical power from each cell in the array (as discussed in Section 3-8). An important application for thermal concentrators is to heat a fluid (often water), which can be used for electricity production or for process heat. Electricity production is discussed in this section; specific issues dealing with heating water are covered in Section 4-5.

A **solar concentrator** uses mirrors or lenses to focus solar energy onto a specific area. Concentrators focus direct radiation rather than diffuse radiation, so they work best in locations with high direct solar radiation, such as the southwest United States. Three applications for solar concentrators include (1) enhancing the energy on photovoltaic modules; (2) heating fluids for large electrical power plants; and (3) heating fluids for other applications, including residential hot water, food-processing plants, hospitals, and other commercial applications.

Concentrating Photovoltaic Systems

Concentrating photovoltaic systems (CPVs) put more light energy onto the PV cells using mirrors or lenses (this topic was introduced in Chapter 3). Mirrors, which are more widely used than lenses, can be configured to reflect light onto a module, as shown in Figure 4-14. In this design, flat mirrors have been installed between two PV modules to reflect sunlight onto the modules, thereby increasing the effective solar insolation as long as the assembly is oriented toward the sun. If this assembly is not oriented to the sun, the mirrors cast a shadow, negating any advantage. Thus, the array is designed to track the sun. In this particular collector, the solar energy is nearly doubled at the PV modules by the flat mirrors over the amount that would arrive without the mirrors. The solar concentration ratio C is the ratio of solar intensity at the collector, with mirrors (or lenses) taken into account over

FIGURE 4-14   Concentrating Photovoltaic Collector Using Flat Mirrors   (*Source:* Courtesy of Poulek Solar.)

the incident solar intensity without mirrors or lenses. With other arrangements of mirrors, the concentration ratio can be increased considerably, but heating limits the efficiency and can adversely affect the PV cells, so special cells are used for high concentrations.

PV concentrators have two main drawbacks: the need for tracking the sun and heat buildup. Concentrating collectors require tracking to optimize the solar energy collected. For this reason, they are not as widely used on roof installations, where a fixed flat panel is generally more economical. Tracking collectors are more expensive at the outset, require energy to move them, and require space for the array and between the units (to avoid shadows). As long as space is available, more energy is collected from tracking arrays than from comparable fixed arrays. (Solar trackers are discussed in detail in Section 5-3.)

Heat buildup is a second problem for PV concentrators. Even low concentration ratios can produce heating problems with standard PV cells and reduce their service life. Many manufacturers will not honor warranties if their cells are used with concentrators. Overheating can result in decreased efficiency and even failure. High concentration ratios require high-capacity heat sinks and cooling to prevent thermal destruction and reduce performance losses. Multijunction gallium arsenide (GaAs) PV cells, which were discussed in Section 3-3, are much better at handling heat than normal cells, but they are expensive. They are usually selected with concentrators in cases where the increased cost of the cells is offset by higher concentration ratios. These cells are used in the mirror system shown in Figure 4-15, which has a concentration ratio of 650; thus, each cell sees 650 times more

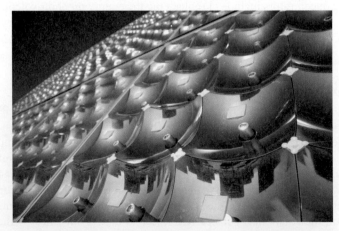

FIGURE 4-15   Concentrating Photovoltaic System That Uses Small Concave Mirrors to Concentrate the Light on the PV Cell   (*Source:* Courtesy of SolFocus.)

FIGURE 4-16   Concentrator Array at Victor Valley College in California   (*Source:* Courtesy of SolFocus.)

light than a cell on a flat-plate collector. The specific cell characteristics determine how much the system can concentrate light.

Large PV systems have important advantages in general. One of the most important advantages is the fact that the only water needed for PV systems is for cleaning purposes. Figure 4-16 shows a large array at Victor Valley College in the California desert. This large array uses a concentrator PV array. The panels have a flat glass cover, which simplifies cleaning. Because it tracks the sun, the system gathers more energy in a day than a comparable flat-panel system with the same collector surface. In addition, most of its components are recyclable when its life cycle is finished, so there is less waste.

Large Power Plant Arrays

Large thermal concentrator arrays located in high solar insolation areas can produce electricity at prices that are nearly competitive with other methods of producing power. Currently, most of these systems are constructed with help from government grants. A key advantage of thermal concentrators using fluids for electrical production over PV modules is that heat can be stored with greater efficiency than electrical energy can be stored in batteries, so the systems can continue to produce power after the sun goes down. The heat is typically stored in molten salts and hot fluids.

For large solar arrays used for commercial power production, the main technologies that have been built or planned and are in the planning phase are parabolic trough mirrors, linear Fresnel mirrors, parabolic dish/Stirling engines, and heliostat mirrors focused on a tower (so-called tower power).

Parabolic Trough Mirrors

Figure 4-17 shows parabolic trough collectors, which are the most common type of concentrating collectors. The parabola mirror focuses light on a long tube positioned at the focal point. The parabola rotates on a single axis to keep the light focused on the tube. The tube carries a special heat transfer fluid (usually synthetic oil) that is heated and the heat is used to power a turbine, which spins an electrical generator. To be efficient, the system must produce high temperatures for the turbine. Steam turbines are generally used so in this case, the system requires a heat exchanger to move the heat from the oil into water, which is directed to the turbine wheel and spins the turbine. Turbines also require cooling to condense the steam and complete the cycle. The most widely used type of cooling system uses water and an evaporative cooling process because it is effective and less costly to install. It

FIGURE 4-17   Parabolic Trough Concentrator Array   (*Source:* National Renewable Energy Laboratory.)

requires a large amount of water, however, which is a problem in arid environments. Cooling water accounts for more than 90% of the water requirements in a water-cooled system. Air cooling is not as efficient as water cooling and is generally more costly to install; however, air cooling may be justified in the desert, depending on the availability of water.

A new type of parabolic trough commercial plant uses direct-steam generation (DSG) to avoid the need for heat exchangers and to power the turbines directly. The concept requires recirculating and injection cooling to control the system temperature and pressure for safety system optimization. Superheated steam is directed to a conventional steam turbine. Several pilot projects using DSG have been tested, and the first commercial operating plant for DSG is now in operation in Malaysia, with a 5-megawatt plant. Other DSG plants are in the planning stage.

Linear Fresnel Mirrors

The concept of the linear Fresnel mirror is basically the same as the parabolic trough mirrors: Direct light is reflected by a line of long parallel mirrors back to a collector that runs along the line of mirrors. A small space exists between the mirrors, which are often referred to as Fresnel mirrors. They are less expensive than parabolic mirrors, and lighter-weight mirrors can be employed because the spacing produces less wind loading. As a result, the structural supports can be smaller, too. Fresnel lenses are also being applied to DSG systems by Solar Power Group, a company specializing in linear Fresnel lens technology. Both parabolic trough mirrors and linear Fresnel mirrors can produce moderately high-temperature heat, which can be used for driving a turbine, heating water, or for other applications such as food processing. Heat storage is also possible with these collectors because the superheated water can be stored directly or the heat can be stored in salts.

Dish/Stirling Engine Collectors

Another type of solar concentrator under consideration by utilities for power production is the Stirling engine system. The **Stirling engine** is a type of heat engine that cools and compresses a gas in one portion of the engine and expands it in a hotter portion to obtain mechanical work. In a solar energy system, the Stirling engine is used with a tracking parabolic dish to focus solar energy on a hot region of the engine. The dish tracks the sun in two axes in order to provide the high temperatures required by the Stirling engine for maximum efficiency. The Stirling engine is a type of external heat engine (unlike

CPV System Using Immiscible Liquids for Tracking

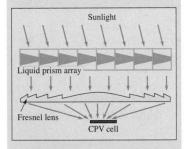

Scientists at the University of North Texas are researching a method of tracking the sun without moving parts. The idea is to use a liquid prism array to bend sunlight toward a Fresnel lens that can then focus the light on a CPV cell. The liquid prism is formed by two immiscible liquids that have different indexes of refraction to bend the light. The amount of bending is determined by the fluid interface angle, which can be controlled with a voltage applied to the sidewalls of the prism. The approach offers a unique method for tracking without the need for mechanical tracking, with its inherent cost and maintenance issues.

FIGURE 4-18 Dish/Stirling Engine Collectors (*Source:* National Renewable Energy Laboratory.)

a gasoline engine) that is air-cooled and does not require any cooling water. This is an important advantage for desert regions and other locations where water is scarce. The only water needed in the system is for washing the mirror. Figure 4-18 shows an array of dish/Stirling engine collectors.

The Stirling engine was invented in 1816 by Robert Stirling and is based on a closed-loop repeated heating and cooling of a sealed working gas (typically hydrogen or helium). There is no exhaust, very few moving parts, no ignition system, and almost no vibration or noise. In the solar dish/Stirling engine, the sun provides the heat source, so there is no combustion. The efficiency of the engines can be quite high (as much as 40%) when there is a significant temperature difference between a hot source and cold sink. The efficiency of a heat engine is proportional to the difference in temperature between the hot and cold reservoirs, so it is an advantage to have a very hot source (hence the large concentrator) and a large radiator surface for cooling. The Stirling engine has a long history, but it has had relatively limited applications because it is slow to warm up and is bulky. But they are very well suited to direct conversion of solar energy to electricity. Another drawback to the dish/Stirling engine system is that it does not lend itself to storing heat for the times when there is no solar input.

The Stirling engine has many variations, but essentially it derives power from a true thermodynamic cycle of the internal gas. When energy is supplied to the gas in the form of heat, the pressure rises and pushes against a piston to produce a power stroke. When the gas is cooled, the pressure drops, which means less work is needed to recompress the gas on the return stroke. The piston drives a mechanical linkage that can do mechanical work as the gas flows between hot and cold heat exchangers. An important part of the engine is a **regenerator**, which is a wire mesh that is located between the heat exchangers and serves as temporary heat storage as the gas cycles between the hot and cold sides. The regenerator increases the efficiency of the engine. The Stirling engine does not have any valves, and the early engines developed rotary motion using an internal crank mechanism connected to two pistons. In 1964, William Beale invented the free-piston Stirling engine, which eliminated the crank mechanism. Another refinement was to use the back-and-forth motion of the piston to drive a linear ac generator (alternator) and produce electricity directly (without the need to convert the piston motion to rotary motion). This is the mechanism used in some dish/Stirling engine collectors. Modern Stirling engines are a model of simplicity, so they tend to be maintenance-free and last for years.

Dish/Stirling engines lend themselves to sloped terrain, so there is minimal preparation work in leveling a site for an array of collectors. They are modular, so a system can easily be expanded if the need arises. Because the engine efficiency is high, the overall system efficiency is also high: About 24% of the incident sunlight power is converted to electricity.

Tower Power

The term *tower power* refers to an array of nearly flat mirrors that are all aimed at a receiver on top of a tower, which collects the energy. The temperature at the collector is the highest of all solar concentrators because of the sheer number of mirrors aimed at one spot. Each mirror is mounted on a device called a **heliostat**, which is a device that moves the mirror in a way that keeps the sun on a specific target. The mirrors need to have precise aim at the tower, so dual-axis control is required to track the sun. The heliostats are individually controlled by a central computer that calculates the direction of the sun from each mirror and sends control signals to motors that keep the sunlight aimed at the receiver on the tower. The receiver absorbs energy and heats a transfer fluid that is used to drive a conventional steam turbine. The heat transfer fluid can be water, a molten salt mixture, or even compressed air.

Figure 4-19 shows the basic block diagram of a solar power tower system, with molten salt as the heat transfer fluid. The heated salt is used to produce superheated steam, which drives a conventional steam-turbine power cycle. Some of the hot molten salt is stored for the system to use in the production of power after the sun is down. Salt was first used in Solar Two, a demonstration solar power tower in Barstow, California, that was operated between 1996 and 1999. At one point, the hot salt at night enabled the plant to have continuous operation for seven continuous 24-hour periods. In arid climates, the condensing can be done by air cooling rather than water.

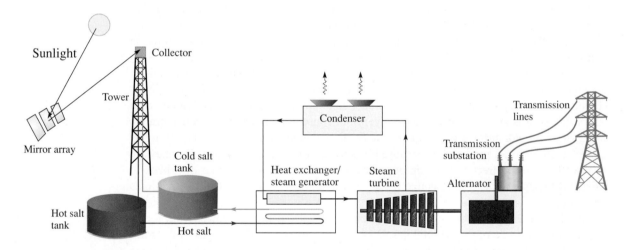

FIGURE 4-19   Block Diagram of a Solar Power Tower System

The largest solar tower power facility in the world is supplying power to California from a dry lake bed in the Mojave Desert and is called Ivanpah Solar Electric Generating System (Ivanpah SEGS). This plant, which was constructed by BrightSource Energy, utilizes the company's proprietary LPT 550 technology. Figure 4-20 shows an installation of the SEGS in Israel. The Ivanpah SEGS consists of three towers, with a combined output of 370 megawatts (nominal). LPT 550 technology uses a closed-loop system that creates superheated steam (over 1,000° F) at the top of the tower and reuses the condensed water to avoid wasting water. Condensing the steam is accomplished with a dry-cooling technology, avoiding the water requirements for cooling that is typical for power plants. The system uses thousands of flat mirrors that track the sun using two-axis tracking. The extremely high temperature means that the turbines can be very efficient; recall that the efficiency is related to the temperature difference between the hot and cold side of the turbine.

FIGURE 4-20 Solar Electric Generating System (*Source:* Courtesy of BrightSource Energy.)

A high-temperature tower that has been constructed in Australia has another application. In addition to electricity, the system is designed to convert natural gas and water into SolarGas, which is a synthetic gas with 25% more energy content than natural gas. The resulting gas is then piped to the user. This fuel has the convenience of natural gas, but with higher energy content. The electricity production from the plant uses a Brayton cycle gas turbine that can operate with no water, a useful feature in Australia's arid environment. The turbines operate at extremely high temperatures; thus, for solar systems, the turbines can be used only with solar towers.

A similar concept has been implemented at Colinga, California. Steam that is generated by solar concentrators is used to enhance oil recovery efforts at a demonstration facility. Figure 4-21 shows the concentrators focused on a tower where steam is produced. The steam decreases the viscosity of the heavy oils at the site. Injection of steam is one of the methods for recovery of heavy oils and may play a role in developing oil sands and oil shale to produce oil in the future.

FIGURE 4-21 Solar Steam Plant for Oil Field Stimulation (*Source:* Courtesy of BrightSource Energy.)

SECTION 4-4 CHECKUP

1. What three types of solar systems use concentrators?

2. Which type of solar concentrator produces the highest temperatures?

3. What type of PV cell is used with PV solar concentrators?

4. What is the advantage of a high temperature difference between the hot source and the cold sink in a heat engine?

5. What is DSG and what does it refer to?

4-5   Solar Hot Water Systems

Water heating is one of the most efficient uses of solar energy and one that has matured over more than 100 years. For home use, a solar hot water heater is a way to reduce energy bills significantly with a proven technology. In general, the payback period for installing a hot water heating system is less than that of a PV electric system. In this section, collectors for hot water systems and representative systems are discussed.

Solar water heating can be divided into passive and active systems. Passive systems are simple systems that do not use auxiliary power such as pumps to operate, whereas active systems require electrical power for external pumps or fans. In some passive systems, the heat collection and storage are separate; others combine the two functions. One of the first solar water heaters used black cans to warm water, an example of a totally passive system but one suited to locations that do not freeze. Systems are still in place in many parts of the world that are essentially no more than an insulated box with a glass cover and an enclosed black tank. This type of hot water system can be quite heavy, depending on the size of the tank, so the structure supporting it needs to be strong enough to carry the weight.

Considerations for choosing a system for heating water include climate, hot water requirements (quantity, time of day, temperature), physical size (storage tanks, etc.), available space for collectors, and cost. The most important factor is climate, which includes the effects of temperature and snow and wind loading. It is important that the system is designed for worst-case conditions, keeping in mind that weather records can be broken!

Collectors

Flat-Plate Collectors

Several different types of collectors are used in hot water heating systems. Choosing the best option for a particular application depends on the climate where the collector is installed and the temperature requirement for the water. In warm climates where there is no chance of freezing, potable water can be circulated through the collector, and a simple flat-plate collector is the most economical solution. Figure 4-22 shows two flat-plate collectors ganged together. They are essentially made from a strong aluminum frame with a copper pipe manifold in a tightly enclosed, tempered glass enclosure. The frame is insulated on the sides and back, and an absorber plate is placed in direct contact with the copper manifold. The absorber plate is made from a good heat conductor that can transfer heat to the pipes; it is coated with a material that has high absorptance and low emittance. **Absorptance** is a dimensionless number that is the ratio of absorbed radiation to incident radiation. **Emittance** is the total flux (radiant energy) emitted per unit area from a material; it is related to the ability of the

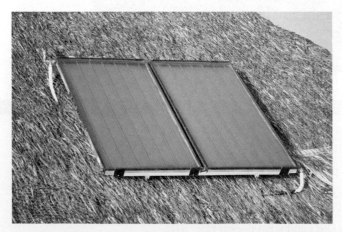

FIGURE 4-22   Two Flat-Plate Collectors Ganged Together   (*Source:* David Buchla.)

material to give off radiant heat. In an insulated flat-plate collector with a good absorber plate, the inside temperature can reach 180° F. The water in this system is circulated to an inside storage tank using a small pump.

Another type of flat-plate collector is the manifold swimming pool heater constructed from unglazed ultraviolet-resistant polymer material. Swimming pool heaters are used in any climate because they are used only when there is no danger of freezing; in the winter, water is diverted from the collectors and the collectors are drained. The basic pool heater is mounted on a south-facing roof; cool water is run into the bottom and warm water emerges from the top. Usually, the water to the pool heater is moved through the system using the pool's pump; an automatic controller and sensors determine when solar heat is available and the pool needs heat. In this case, an automatic diverter valve sends water from the pool to the collectors.

Solar Heat Pipes

A second type of collector consists of an array of evacuated solar heat pipes, which function on the principle of an evaporation and condensation cycle. The cross-section of the heat pipe is shown in Figure 4-23. The basic pipe is a coaxial arrangement with a glass outer tube and a closed copper inner tube that holds a nontoxic fluid. The inner tube has low pressure in it so that the small amount of fluid vaporizes at a lower than normal temperature. Solar insolation striking the assembly causes the fluid to evaporate easily, and the hot gas moves to the top of the inner copper tube, where the heat is transferred to a heat transfer fluid and eventually to the potable water. The heat pipe is very efficient at moving heat because, when compared to liquid, the gas carries energy called the **latent heat of vaporization**; this latent (or hidden) heat is released at the top of the pipe, where it makes contact with a transfer material. The heat of vaporization is the heat absorbed or released during a change of state from a liquid to a gas and is very large compared to the heat absorbed to cause a temperature change in a substance. As the gas moves to the top of the pipe, it cools and releases the heat of vaporization as it condenses back to a liquid. The liquid runs down the tube, completing the cycle. The outer tube contains a hard vacuum, so it eliminates conduction or convection loss from the gas.

For the evaporation and condensation cycle to function properly, heat pipes normally need to be mounted so that they are raised a minimum of 25° from the horizontal (although there are heat pipes designed to be laid flat). Water temperatures are typically between 120° F and 190° F. In freezing locations, the transfer fluid is usually food-grade propylene glycol—not to be confused with poisonous ethylene glycol, which is antifreeze used in radiators but is hazardous for solar water heaters, where it should never be used.

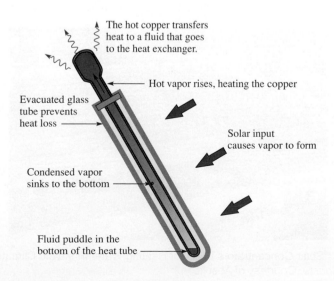

The hot copper transfers heat to a fluid that goes to the heat exchanger.

Hot vapor rises, heating the copper

Evacuated glass tube prevents heat loss

Solar input causes vapor to form

Condensed vapor sinks to the bottom

Fluid puddle in the bottom of the heat tube

FIGURE 4-23   Solar Heat Pipe Construction

FIGURE 4-24   Hot Water Heating System in a Cold Climate (Switzerland) Using Enclosed Heat Tubes   (*Source:* David Buchla.)

Solar hot water heating panels composed of heat tubes are more efficient in cold climates than are flat panels, and they are not affected by outside air temperature or wind because the evacuated glass prevents heat loss by conduction and convection. Figure 4-24 shows heat pipe collectors for a hot water system in Switzerland, where freezing is common.

Concentrating Collectors

Another type of collector that can be used in any climate is the concentrating collector, which is useful for producing very hot water and process heat. The concentrating collector can also be used for electricity production by installing PV cells at the focus. Process heat is useful in a number of industries, including food, chemical, and textile. Concentrating collectors need to have tracking to optimize output, but they can use simple one-axis tracking. For this reason, this type of system is more suited to larger installations. Figure 4-25 shows an Absolicon $\times$10 parabolic trough concentrator with built-in tracking for producing hot water. The trough is covered with glass to help retain heat. In cold climates, nontoxic antifreeze is used to circulate to the collector, and the heated fluid is carried to a separate heat exchanger.

FIGURE 4-25   Solar Concentrators Used for Heating Water in a Cold Climate (Sweden)   (*Source:* Courtesy of Absolicon.)

Open-Loop Systems

In an **open-loop** or direct system, potable water is circulated through the collectors. The simple black-can water heater that was mentioned previously, in which the collector and water storage are integrated into the same unit, can be considered a passive, open-loop type system. These systems can be used in warm climates where freezing is not a problem or in seasonal applications (like campgrounds or summer cabins) where they are drained for the winter months.

Thermosiphons

A more sophisticated open-loop passive system is a thermosiphon, in which hot water is stored in an insulated solar storage tank mounted above the collector. Figure 4-26 shows a basic thermosiphon system that uses a heat pipe collector, an arrangement that works in cold climates. The heat pipes move heat to the solar storage tank, and an internal heat exchanger warms the water in the tank. Cold water is routed directly to the solar storage tank, where it is warmed because of passing through a heat exchanger. When hot water is drawn from the system, it is taken from the backup tank, and preheated water from the solar tank goes to the inlet of the backup water heater. Exposed water pipes in this system should have minimum exposure to the cold and must be insulated to prevent freezing. The insulation for outdoor piping is limited to certain types of insulation that can withstand temperature extremes of the hot water in summer to freezing conditions in winter. In exterior applications, the insulation should have jacketing to protect it from UV radiation, rain, and snow, as well as squirrels, insects, and birds, which find insulation handy to use in nests.

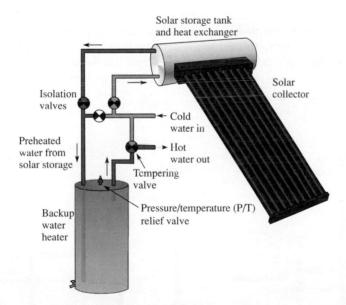

FIGURE 4-26 Thermosiphon Hot Water Heater That Uses Heat Pipes for the Collector

Closed-Loop Systems

In a **closed-loop** water heating system, potable water is never exposed to the outside environment: A separate loop is used with a fluid that is heated. Generally this fluid is a propylene-glycol mixture that is heated and sent to a heat exchanger, where the heat is transferred to the potable water. In addition to protection from freezing, a separate loop has the advantage of protecting the collectors from corrosion and deposits caused by hard water. There are two basic types of closed loop systems: pressurized systems and drainback systems.

Closed-Loop Pressurized System

A closed-loop pressurized system uses a propylene-glycol-water mixture that is circulated to the collector using a recirculating pump. Typically a flat-plate collector is used, but any type of collector will work. At the collector, the propylene-glycol-water mixture is heated and returned to a solar storage tank that contains a heat exchanger. (In some systems, the heat exchanger is an external component.) A common mixture of propylene-glycol and water is a 50-50 mixture, but the particular ratio depends on the climate and the type of system. A basic closed-loop pressurized system is shown in Figure 4-27. The system in the figure has a separate backup water heater that provides hot water if the solar system is unable to do so, but some systems are constructed without the separate tank.

The system is monitored and controlled by a controller, which monitors temperatures to and from the collector and determines when heat is needed and available. In this case, the controller turns on the recirculating pump to move heat from the collector to the solar storage tank. The backup hot water heater is connected so that prewarmed water from the solar storage tank is used in place of cold water. Because temperatures may be hotter than desired, a tempering valve automatically adds cold water as needed to set the final output temperature.

In the propylene-glycol-water loop, an expansion tank is necessary to prevent a system failure that could occur when the fluid expands. As you know, liquids expand when they are heated, and they are not compressible; the expansion tank provides room for the hot fluid. The expansion tank has an enclosed air chamber that is separated from the circulating fluid with a bladder that expands and contracts as the fluid temperature changes.

Drainback Systems

A **drainback system** is one in which the fluid is heated in collectors only during times when there is available heat. A drainback system is a popular method of heating hot water and avoiding freezing problems. The circulating fluid can be pure water or a propylene-glycol-water mix in extremely cold climates or if extra protection from freezing is desired.

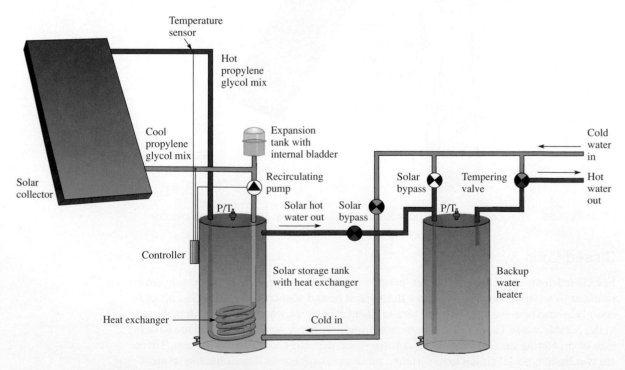

FIGURE 4-27   Closed-Loop Pressurized Hot Water System

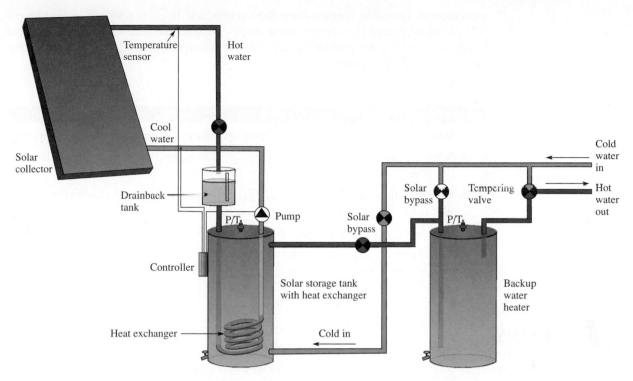

FIGURE 4-28   Drainback Hot Water System

A basic system is illustrated in Figure 4-28; other configurations are possible, but this system shows the basic components. The system pump is on only when heat is available at the collectors and needed at the storage tank; otherwise, it is off. When the pump is off, the collectors drain by gravity back to the drainback tank. While this action should prevent freezing problems, it is possible to see some damage if complete drainage of the system does not occur, so it is important that the system is designed to drain completely and rapidly. The collectors must be installed at an angle and the drain must be set up at the lowest point to ensure drainage. Pipe runs cannot be allowed to sag or collect water when they drain (avoid horizontal pipe runs), and the pump needs to be adequate to ensure reasonable head pressure at the top of the collector. Some system designers prefer to use a propolyene-glycol-water mixture no matter what the conditions, but particularly where conditions can be severe.

Instead of the expansion tank used in a pressurized system, a drainback system requires a reservoir or drainback tank, which is an unvented tank designed to hold all of the recirculating fluid when it is not in the collectors. The tank is located in a protected environment, where it has no chance of freezing, and includes a means of checking the level within the tank. It must also be located below the level of the collectors and within an area safe from freezing so that gravity allows the fluid to drain completely from the collectors and outside plumbing when the collectors are not in use. The drainback tank should also be located as high as possible within the warmer space to reduce the load on the pump.

As in the case of the pressurized closed-loop system, a controller is used. The controller monitors the temperature of the water in the tank and the collector temperature, and turns on the pump if there is enough temperature difference to make it worthwhile to turn it on. One option is to use a dc pump powered by a dedicated PV module.

Failure can occur in any system. Solar hot water systems can be designed to be very reliable, but a leak can occur, or the pump can even become stuck on, in which case, water may be released or water could be exposed to freezing conditions. It is important that the system be checked regularly and maintained as needed. For example, the recirculating fluid

Hybrid Systems

A new trend in solar systems is to combine the generation of electricity with hot water heating. Hybrid PV/thermal systems can represent a cost-effective system for some locations. Flat-plate PV collectors naturally get hot in the sun, so they are a ready source of heat for a thermal water system. One hybrid system, called the Echo system, draws outside air under the PV modules, thus warming the air. The hot air is drawn into the attic of a structure through a roof vent and then passed into a heat exchanger. During the heating season, it can help warm the house (or any structure) as needed; otherwise, it can serve as a preheater for the hot water system.

can evaporate over time, requiring new fluid to be added. In cases where propylene-glycol is used, the propylene-glycol can become acidic over time and need replacement. A visual check of the system for problems such as rodent damage, leaks, or corrosion is a useful exercise in keeping a system active.

SECTION 4-5 CHECKUP

1. What is the difference between an open-loop and a closed-loop solar hot water system?
2. What are the characteristics that make a certain material a good absorber plate in a flat-plate conductor?
3. How does the latent heat of vaporization make a heat pipe more effective at moving heat to the top of the pipe?
4. Why is it necessary to include an expansion tank in a pressurized hot water system?
5. What steps are important to prevent freezing in a drainback hot water system?

CHAPTER SUMMARY

- Stand-alone solar electric systems vary in complexity, but they all have a PV module and a dc load. Most also include a charge controller and battery storage to supply power when there is no solar input.

- Solar electric systems that have a battery backup need a charge controller to prevent overcharging the battery or draining the battery through the PV modules when it is used for the load.

- An inverter is required to change dc from the module to ac for a load.

- Three codes that are important in solar energy work are:
  1. International Building Code (IBC) for design and installation of building systems
  2. National Electric Code (NEC) for electrical work in the United States
  3. Uniform Solar Energy Code (USEC) for plumbing and mechanical requirements for solar energy systems

- A ground fault is a condition where current is in the ground conductor. A ground fault protection device (GFPD) detects the fault, interrupts the current, indicates a fault has occurred, and disconnects the faulty module.

- The steps for choosing the proper size for a residential solar system are: (1) perform an energy audit, showing monthly usage patterns; (2) perform a site analysis including snow and wind loading; (3) develop a system concept; (4) determine the array size needed for the location; (5) determine cable and battery needs; (6) select components; and (7) review the design.

- Grid-tie systems are connected to the utility grid. They may or may not have a battery backup that can be turned on automatically if the grid is down.

- PV arrays and solar towers have been constructed by utilities to supply power to the grid and, in Australia, to increase the energy content of natural gas and create SolarGas.

- Solar water heating methods include passive systems that do not require any electrical power, such as a thermosiphon, and active systems that use pumps to circulate a propylene-glycol-water mixture in the collectors. Two types of active systems are closed-loop pressurized systems and closed-loop drainback systems.

- Collectors for hot water include flat-plate collectors, heat pipes, and concentrators.

KEY TERMS

absorptance   A dimensionless number that is the ratio of absorbed radiation to incident radiation.

charge controller   A device that regulates and limits charging current to prevent overcharging batteries.

closed-loop   (1) In feedback theory, a condition where a portion of the output is returned to the input. (2) In water heating systems, a condition where the potable water is never exposed to the outside environment because a separate loop is used with a fluid that is heated.

combiner box   A double-insulated box that allows several strings from modules to be connected together in parallel; it also houses fuses for the strings and includes surge and overvoltage protection from potential lightning strikes.

depth of discharge (DOD)   The ratio, expressed as a percentage, of the quantity of charge (usually in ampere-hours) removed from a battery to its rated capacity.

drainback system   A solar water heating system in which the circulating fluid is circulated only when heat is available at the collector; otherwise, the collector and exposed plumbing is drained.

emittance   The total flux emitted per unit area from a material; it is related to the ability of the material to give off radiant heat.

grid-tie system   An electrical generating system that is tied to the utility grid.

ground fault protection device (GFPD)   A device that has the following functions: (1) detect a ground fault, (2) interrupt the current in the line, (3) indicate that a fault has occurred with a visible warning, and (4) disconnect the faulty module.

heliostat   A device that moves a mirror in a manner to keep the sun on a specific target.

insolation   From "*in*cident *sol*ar radi*ation*," a measure of the energy received on a surface in a specific amount of time; it can be measured in units of W/m^2.

inverter   A device that converts dc to ac.

latent heat of vaporization   The heat absorbed or released during a change of state from a liquid to a gas.

microinverter   A dc to ac converter that is sized to operate with a single solar module so it can provide maximum power-point tracking for the module and provide greater efficiency.

open-loop   (1) A type of system that does not use feedback to adjust its parameters. (2) A solar water heating system in which the potable water is circulated through the collectors.

regenerator   In a Stirling engine, a wire mesh that is located between the heat exchangers and serves as temporary heat storage as the gas cycles between the hot and cold sides.

solar concentrator   A type of solar collector that collects light over a certain area and focuses it onto a smaller area.

Stirling engine   A type of heat engine that cools and compresses a gas in one portion of the engine and expands it in a hotter portion to obtain mechanical work.

transfer switch   A switch that can switch loads between alternate power sources without interrupting the current.

FORMULAS

Equation 4-1 $P_{array} = \dfrac{W}{t_{solar}\eta_{sys}}$ Power from an array

Equation 4-2 $Ah = \dfrac{W_{day} \cdot t_{store}}{VB_{dod}\eta_{inv}}$ Ampere-hours of batteries

CHAPTER TRUE/FALSE QUIZ

Determine whether each statement is true or false. Answers are at the end of the chapter.

1. A stand-alone system is one that does not have a connection to the utility grid.

2. A charge controller is necessary for all solar PV systems.

3. The purpose of an inverter is to convert ac to dc.

4. The National Electrical Code has specific provisions for solar electric systems.

5. A ground fault is said to be present when there is current in the neutral conductor.

6. An energy analysis can reveal opportunities for conserving energy.

7. Bypass diodes allow current to bypass a module if the module voltage is very low.

8. A grid-tie system always requires a special inverter that is designed to produce grid-quality ac.

9. A battery of standard automobile batteries is an excellent backup for solar electric systems.

10. The depth of discharge can be expressed as a ratio of the ampere-hours removed from a battery to its rated capacity.

11. Generally, flooded batteries are more expensive and have a shorter life expectancy than sealed types.

12. If a module fails in a system using microinverters, the rest of the system will continue to operate (but with reduced output).

13. To estimate power required for a battery backed-up motor load, calculate the power the motor requires based on starting power rather than running power.

14. Concentrating type collectors require tracking to optimize the solar energy collected.

15. Solar concentrators work primarily with diffuse radiation.

16. A Stirling engine is a closed-loop external heat engine that requires a high temperature source to be efficient.

17. Power towers use tracking heliostats to reflect sunlight to a tower collector.

18. Passive solar water heaters use a small pump to cycle potable water from a storage tank to the collector.

19. A good flat-plate collector for hot water systems has an absorber plate with high emittance and low absorptance.

20. Heat of vaporization is the heat absorbed or released during a change of state and is used in heating pipes.

CHAPTER MULTIPLE-CHOICE QUIZ

Complete each statement by selecting the one correct answer.
Answers are at the end of the chapter.

1. A PV solar electric system with an AC output must have a(n)
   a. battery backup
   b. supplementary generator
   c. inverter
   d. all of these

2. Batteries for a PV system should be
   a. automobile batteries
   b. lantern batteries
   c. deep-cycle batteries
   d. lithium ion batteries

3. Special requirements are listed in the NEC for voltages above
   a. 200 V
   b. 600 V
   c. 1,000 V
   d. 10,000 V

4. The output of a group of parallel modules generally goes directly to a
   a. charge controller
   b. dc load
   c. switch control
   d. combiner box

5. The optimum angle for setting a flat-panel collector to obtain the highest average radiation is
   a. 0°
   b. 45°
   c. 70°
   d. equal to the latitude

6. A microinverter is designed to operate with
   a. a single module
   b. two modules
   c. a single string of modules
   d. a group of parallel modules

7. A drawback for a PV concentrator type of collector is
   a. it needs to track the sun
   b. heat buildup
   c. both (a) and (b)
   d. none of these

8. A dish/Stirling engine has the advantage of
   a. no moving parts
   b. high efficiency
   c. heat storage capability
   d. all of these

9. A type of solar system that would be likely to use a steam turbine to generate electricity is a
   a. solar tower
   b. nontracking system
   c. PV concentrator
   d. Stirling engine

10. The simplest type of hot water system for an area that never experiences freezing temperatures is a
    a. closed-loop system
    b. open-loop pressurized system
    c. open-loop drainback system

11. A solar hot water system that is an open-loop passive system is a
    a. pressurized system
    b. drainback system
    c. thermosiphon
    d. swimming pool heating system

12. A solar hot water system that requires an expansion tank is a
    a. pressurized system
    b. drainback system
    c. thermosiphon
    d. swimming pool heating system

13. Heat of vaporization is the heat
    a. required to move a substance from the freezing point to the boiling point
    b. required to change steam at 212° F to superheated steam at 1,000° F
    c. required to melt 1.0 kg of ice
    d. absorbed or released during a change of state from a liquid to a gas

14. In a thermosiphon hot water system that uses heat pipes, a heat exchanger is located in the
    a. the water storage tank
    b. the heated space
    c. the backup storage tank
    d. none of these

CHAPTER QUESTIONS AND PROBLEMS

1. What is the purpose of a charge controller in a battery backed-up solar system?

2. Contrast applications for a stand-alone solar electric system that does not require a battery backup with applications that do need a battery backup.

3. What types of requirements are specified in the Uniform Solar Energy Code?

4. What is meant by life-cycle cost and why is it important in evaluating a solar system?

5. Assume that a solar electric system has a power rating of 1,000 W. If it receives 6 hours per day of peak sunlight, how many watt-hours of energy are delivered during the peak time?

6. How do automobile batteries differ from deep-cycle batteries? Which should be specified for a solar electric system?

7. Answer the following questions for a battery backup system that is used for a 6 kWh/d, 24 V system that has 12 batteries; each battery is rated for 6 V at 370 Ah.

   a. How are the batteries configured in this system?

    b. How many days can the backup system provide backup if the depth of discharge is 50% and the overall efficiency is 95%?

8. How would you change the system in Example 4-3 to provide four days of autonomy using the same batteries?

9. Compare a grid-tie, battery-backed system with one that is not battery backed-up. What are the advantages and disadvantages of each?

10. Compare a utility solar system with a residential or small commercial system. What are important differences in the types of systems for each?

11. What determines the efficiency of a Stirling engine?

12. Describe how a large concentrator array for a solar power tower converts sunlight to electricity.

13. Describe how a heat pipe works.

14. Compare three different types of solar hot water systems. List the advantages and disadvantages of each.

FOR DISCUSSION

For the climate zone you live in, does solar water heating make sense? If it does, what system do you think would work best? If not, explain.

ANSWERS TO CHECKUPS

Section 4-1 Checkup

1. Advantages are that they (1) can provide power in remote locations where the power grid is unavailable, (2) are less expensive in some cases than bringing in power from the grid, (3) can reduce the need to burn limited wood supplies in developing countries.

2. No. An attic fan is needed to cool the attic primarily when the sun is shining, so a battery backup would be unneeded.

3. The goal of certification is to help manufacturers avoid common failures with modules by identifying potential problems.

4. The code enables the evaluation of multiple systems and options, and provides cost analysis of these systems.

5. A combiner box is a junction box that allows several strings of modules to be connected together in parallel; it also includes surge and overvoltage protection from potential lightning strikes.

6. Ground fault protection device

Section 4-2 Checkup

1. A failed module with no output normally would block current from the other modules. A diode can provide a current path around the failure and is active only if the module in question has < 0.7 V.

2. If the tilt is not adjustable, it normally will be set for the latitude or slightly more to optimize the collection in winter.

3. The ampere-hour (Ah)

4. DOD is the ratio of the quantity of charge (usually in ampere-hours) removed from a battery to its rated capacity and can be expressed as a percentage.

Section 4-3 Checkup

1. Grid-tie inverters need to synchronize their output with the utility and be able to disconnect the solar system if the grid goes down.

2. (1) A system that is designed to supplement grid power and not replace it at any time does not need backup, so installation is simplified. (2) Battery backup is expensive, takes up space, and requires regular maintenance.

3. In a grid-tie system, the battery must replace the grid only during outages, so the likelihood and length of outages is the key factor in determining battery size. In a stand-alone system, the key factor in determining battery size is the weather at the location and prospects for long periods of clouds or rain that would prevent the system from operating at its best.

4. System monitoring can provide basic performance data for the system and help pinpoint problems with the system.

5. Some factors that utilities need to consider are: load balancing; equipment loading; power quality issues; overall cost, including any new transmission and distribution systems; as well as many other factors.

Section 4-4 Checkup

1. PV array concentrators, large thermal arrays and towers, Stirling engines

2. Towers

3. Multijunction gallium arsenide

4. The efficiency of the engine is proportional to the temperature difference.

5. DSG stands for "direct-steam generation" and refers to concentrator systems that use water rather than oil in the collectors.

Section 4-5 Checkup

1. In an open-loop system, potable water is circulated through the collectors; in a closed-loop system, a separate fluid, usually a propylene-glycol-water mixture, is sent to the collectors and heat is exchanged with potable water.

2. The plate is constructed from a good heat conductor and it is coated with a material that has high absorptance and low emittance.

3. The heat of vaporization is very large. Because of the change of state, more heat is transferred as the liquid vaporizes.

4. The expansion tank prevents a system failure that could occur when the fluid expands as it is heated in the collectors.

5. The circulating fluid must drain completely when heat is not available at the collector, so the collectors and all plumbing should be checked for proper slope and any sags, and the pump needs to be adequately sized to bring water to the top of the collector with reasonable pressure.

ANSWERS TO TRUE/FALSE QUIZ

1. T   2. F   3. F   4. T   5. F   6. T   7. T   8. T   16. T   17. T   18. T   19. F   20. T
9. F   10. T   11. F   12. T   13. T   14. T   15. F

ANSWERS TO MULTIPLE-CHOICE QUESTIONS

1. c   2. c   3. b   4. d   5. d   6. a   7. c   8. b
9. a   10. a   11. c   12. a   13. d   14. a

Solar Tracking

CHAPTER OBJECTIVES

- Describe the motion of the sun from various points on the earth's surface.
- Explain the effect of the earth's tilt on the diurnal variation of the sun's motion.
- Define terms that are used in locating points on the earth's surface and in space, including latitude, longitude, meridian, ecliptic, synodic day, and so on.
- Compare costs for tracking, including other effects such as space requirements, with the benefits, and discuss which types of solar systems benefit most from tracking.
- Explain how seasonal variations in tilt angle of a solar panel can increase the solar energy collected.
- Compare the advantages and disadvantages of different types of solar trackers, including alt-azimuth and equatorial trackers.
- Compare the costs and benefits of single- or dual-axis trackers for different types of solar collectors.
- Discuss the advantages and disadvantages of permanent magnet dc motors.
- Explain how counter emf causes a reduction in armature current and torque in wound rotor dc motors.
- Compare the torque and power for a motor.
- Discuss the key components of a dc motor.
- Explain why eddy currents in motors are undesirable.
- Describe the magnetic path for the stator field in a dc motor.
- Discuss the advantages for a brushless dc motor and explain how it works.
- Draw the wiring diagram for a series-wound, shunt-wound motor and for the compound motor, and discuss the key differences between them.
- Explain basic troubleshooting steps for a dc motor.
- List three types of stepper motors.
- Explain how a basic stepper motor works.
- Explain how to half-step a stepper motor.
- Describe the process of microstepping a stepper motor.

KEY TERMS

Key terms are shown in bold and color. Definitions for key terms are provided at the end of the chapter and in the end-of-book glossary. Bold terms in black are defined in the end-of-book glossary only.

- latitude
- longitude
- prime meridian
- meridian
- zenith
- synodic day
- celestial equator
- ecliptic
- analemma
- heliostat
- active tracker
- passive tracker
- altitude angle
- azimuth angle
- rotor
- stator
- commutator
- brushes
- end bells
- stepper motor
- holding torque

INTRODUCTION

The amount of power harvested each day by any solar collector can be increased if the collector is oriented to the sun so that the maximum amount of sunlight falls on the collector. To accomplish this, it is necessary to track the sun. Tracking is vital for concentrating collectors as they focus the sun's direct radiation on the collector. Tracking will always increase the power harvested from a given set of collectors, but for some types of collectors it is not always cost-effective to implement.

To understand how tracking works, you need to understand the motion of the sun—both the diurnal (daily) motion and the seasonal changes—and how your position on earth affects this motion. The chapter opens with a discussion of the sun's motion in the sky, and then discusses the costs and benefits for tracking. Single- and dual-axis trackers are discussed. Most trackers rely on motors to move them, but a few use hydraulic systems, and some passive mechanisms have been devised. The chapter closes with sections on dc motors and stepper motors because these are the most common types of motors used with solar trackers.

5-1   Movement of the Sun

To understand how solar tracking devices work, it is necessary to understand the movement of the sun. The sun's apparent motion relative to an earthbound observer is due to the earth's daily rotation, its tilt on its axis, and its annual trip around the sun.

Latitude, Longitude, and the Meridian

The earth is divided into an imaginary grid of circles that are reference lines for measuring any point on the earth. Horizontal lines, called latitude lines, are actually horizontal circles with varying radii that are parallel to the equator. The **latitude** at any given location is the angle in degrees formed by a line that extends from the center of the earth to the equator and another line that extends from the center of the earth to a given point on the globe. At the equator, the latitude is $0°$. Northern latitudes are assigned positive values and southern latitudes are assigned negative values. Thus, the North Pole is located at latitude $+90°$ (or $90°$ N), and the South Pole is located at latitude $-90°$ (or $90°$ S).

Longitude lines are great circles that are perpendicular to the equator and converge at the poles. The reference longitude is defined as a great circle that passes through both poles and the Royal Observatory at Greenwich, England. This longitude is sometimes called the **prime meridian**. Your local **meridian** is a line of longitude that passes through both poles and a point directly over the observer's head, called the **zenith**.

The longitude assigned to any point on earth is defined by the angle formed by a line drawn from the center of the earth to the prime meridian and a line drawn from the center of the earth to the meridian for that point. Positive angles are measured east of the prime meridian; negative angles are measured to the west. Figure 5-1 illustrates latitude and longitude. Longitude can be expressed in degrees or in hours and minutes. Hours and minutes are sometimes used because the earth rotates through $360°$ in approximately 24 hours. Thus, each $15°$ of longitude represents one hour.

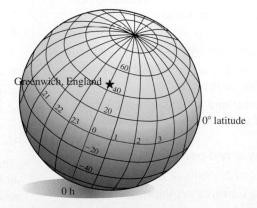

FIGURE 5-1 Latitude and Longitude. The equator is the reference for the latitude. The line crossing Greenwich, England, is the reference for longitude and is defined as the prime meridian ($0°$ or 0 hours as shown).

Motion of the Sun and Stars

The local meridian makes a useful reference for an observer. The earth makes one *rotation* on its axis in almost exactly 24 hours, a time known as **synodic day**. Although we commonly refer to the time from sunrise to sunset as a day, it is not a scientific definition. A precise definition of a day is defined 86,400 s, which is exactly 24 h. The actual time for one rotation of the earth varies a tiny amount from day to day; a leap second is occasionally inserted in the calendar to account for this variation.

The time required for the earth to complete one *revolution* about the sun (with the sun as a reference) is approximately 365.24 days. For the purpose of the calendar, we define a year as 365 days and add an extra day every four years to account (nearly) for the fractional part. The earth rotates eastward, so that the sun appears to rise in the east and set in the west due to this rotation. Because of the earth's path around the sun, the stars appear to move differently than the sun from the vantage point of the earth. If you are following the stars, it is necessary to move the tracking device at a rate that is equal to one rotation in 23 h and 56 m to account for the difference in apparent motion between the sun and the stars. Thus, tracking devices designed to follow the stars (such as on telescope drives) run slightly slower than the rate required to track the sun.

The time for sunrise or sunset depends on your location on the earth and the time of year. The earth's equator is inclined (tilted) 23.37° with respect to the ecliptic. A projection of the earth's equator in the sky is called the **celestial equator**. This inclination of the earth's equator accounts for the seasons and the sun's path in the sky. For an observer on earth, (most of us!), the **ecliptic** is defined as the sun's path in the sky. (It is also the path that the planets follow within a few degrees.)

The position of the ecliptic in the sky as seen by any given observer depends on the location of the observer on the earth's surface and the season. On any summer day at the North Pole, the sun follows the path of a circle without rising or setting. On the summer solstice, the sun is at its highest elevation, which is 23.37° above the horizon, as shown in Figure 5-2(a). At the equator, the sun rises in the east and sets in the west every day. At the equinox (approximately March 21 or September 22), it rises exactly due east, goes through the zenith, and sets due west. Figure 5-2(b) illustrates this case.

Most observers are not at either the North Pole or the equator, so they see that the sun's path on any given day is the projection of a circle at an oblique angle with respect to the horizon. The sun follows a path that is at an angle of 90° − L (where L is the observer's latitude) on the day of the equinox. At other times, the sun's path traces a line that is north or south of this line. Figure 5-3 illustrates the sun's path at latitude 40° N (for example, in the central United States, in Spain, or in Japan). In the northern hemisphere, the path moves north starting at the winter solstice and rises due east and sets due west at the equinoxes.

These definitions can be summarized in a view drawn from the perspective of space. Figure 5-4 shows the idea (obviously not to scale!). Keep in mind that the sun's diurnal

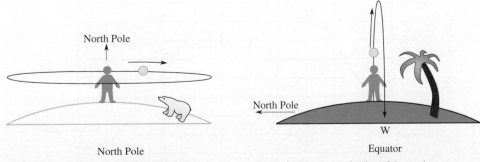

(a) First day of summer at the North Pole   (b) At the equator on the day of the equinox

FIGURE 5-2 Sun's Path (Ecliptic). The ecliptic is shown from two different points on the earth's surface on different days.

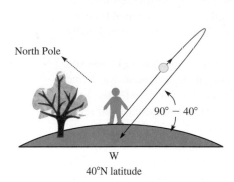

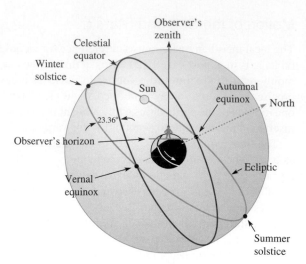

FIGURE 5-3 Sun's Path at a Latitude of 40° N on the Day of an Equinox

FIGURE 5-4 Pictorial Summary of Astronomical Definitions

(daily) motion is due to the earth turning on its axis every 24 h. The sun's journey along the ecliptic is an annual trek, so the sun's movement on the ecliptic is much slower, taking slightly more than 365 d to make one complete cycle. The sun is shown on a fall day (below the celestial equator) and moving slowing (a little less than 1° per day) toward the winter solstice.

A plot of the sun's path from the vantage point of an arbitrary location on earth over the course of a year reveals a sinusoidal shape, with the peaks occurring on the summer and winter solstices. Figure 5-5 illustrates the path for one year for a northern hemisphere observer (in the southern hemisphere, the seasons are reversed). The peaks are $\pm 23.37°$ from the center, which is due to the earth's tilt on its axis. At the midpoint between the peaks, the sun crosses the celestial equator, which is a projection of the earth's equator in the sky. The celestial equator traces a line that is at an angle above the horizon. The angle is equal to $90° - L$, where L is measured from the southern horizon. The crossing points mark the spring and fall equinoxes.

Analemma

While the time for a revolution of the earth is a fixed 24 hours, the orbital speed is not. This is because earth's orbit is slightly elliptical, and the earth speeds up when it is slightly

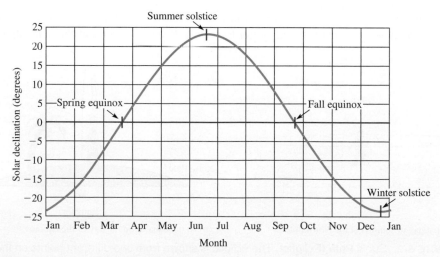

FIGURE 5-5 Path of the Sun for One Year

FIGURE 5-6 Example of the Analemma. The analemma is the figure 8 pattern the sun makes in its annual motion if the position is observed regularly at the same location and time of day. (*Source:* David Buchla.)

closer to the sun (perihelion) and slows down when it is further away (aphelion). The combination of the north–south path of the sun and the seasonal speeding up and slowing down of the orbital speed means that, over the course of a year, the sun will trace out a figure 8 in the sky if it is observed at exactly the same intervals of time and at the same place. (The exact shape depends on your location and observing time.) You may have seen this figure, called the **analemma**, printed on a globe. Figure 5-6 shows the analemma as you would see it if you looked at the same time every week for an entire year. For the sun's apparent motion in the sky, it is important to realize that the sun is not in the identical location in the sky 24 hours later.

SECTION 5-1 CHECKUP

1. How is latitude measured?

2. How is longitude measured?

3. With respect to an observer on earth, the stars move faster than the sun. Explain.

4. Describe the sun's path on the first day of summer at the North Pole.

5-2 Costs and Benefits of Tracking

Many systems are simple roof-mounted panels that are fixed in position. The panels are the most efficient when the sun's rays are perpendicular to the panel, which actually is only during a limited period each day as the sun moves across the sky. By moving the panels to follow the sun, the amount of power harvested each day can be increased by an amount that depends on the site and the season.

Many solar systems are designed to *track*, or follow, the sun's movement across the sky. Tracking devices are categorized by the number of axes (either single-axis or dual-axis) that turn. Both types can follow the daily east–west motion of the sun, but only dual-axis trackers can track the seasonal north–south variations. Compared to single-axis trackers, the initial cost of dual-axis trackers is higher, and they require more space to avoid being shaded by other panels.

Cost Factors

In general, the question of whether a given system should be designed to track the sun boils down to costs versus benefits. The obvious question for the system designer is, "Does the value of the energy gained from tracking offset the added cost of the system?" For concentrating type collectors, the answer is nearly always yes because concentrating collectors need high temperatures to be efficient; they obtain most power from direct radiation, and they are ineffective if they are not pointed toward the sun. Certain concentrating collectors use **heliostats**, which are devices that use mirrors that are mounted so that they reflect the sunlight to a target (such as a tower collector) designed to collect the energy. Heliostats do not point directly to the sun, but they do require tracking devices to keep them moving to the proper orientation, and each one must move independently in order to keep the sun on the target. Because light reflects from mirrors such that the angle of incidence is equal to the angle of reflection, the heliostat is moved in a manner to cause the sun's angle (measured from the normal) and the target angle to be the same (see Figure 5-7).

Tracking systems add to the cost of any system because the tracking hardware requires a control system and a means to move the collectors or mirrors (typically with a dc or stepper motor). In general, they require stiffer structural supports than fixed arrays require. A typical flat-panel tracking collector is mounted on a pole that requires a mechanically strong structural support so that it can move into a given wind load. Tracking mounts tend to be like sails: They need to be capable of withstanding wind loads and heavy snowfall to avoid damage. Long-term maintenance costs need to be analyzed as part of the total life-cycle cost because moving components generally require periodic maintenance.

Roof-mounted flat-panel PV systems are very common because they can be mounted in a mechanically stable arrangement. They are not usually designed to track because of the added cost of a roof-mounted tracking system, appearance considerations on buildings, and the added complexity and maintenance cost of tracking systems. Maintenance of a roof-mounted system is more difficult than a ground-mounted system and may add expense if the roof itself needs repairing or replacing.

In the case of flat panels, the loss in energy with no tracking is primarily in the early morning and late afternoon, when the sun's rays are not perpendicular to the panel. This can reduce the total energy collected by 25% to 40%, depending on the site. By tracking in one axis, most of the energy can be collected, even if the sun is not shining directly on the collector. With single-axis tracking, the effective area of the collector is equal to the cosine of the angle formed between the normal line and the angle of incidence, i, of the sun's rays. (The normal line is a line that is perpendicular to the surface.) Figure 5-8 shows the geometry. For example, if the incident rays are 20° lower than optimum, the incident radiation is reduced only by 6% due to the tilt angle. Many panels are configured so that seasonal

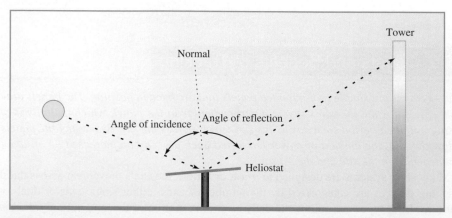

FIGURE 5-7 Heliostat Measurement. Heliostats are positioned to keep the angle of incidence equal to the angle of reflection. The angles are measured with respect to the normal, which is perpendicular to the surface.

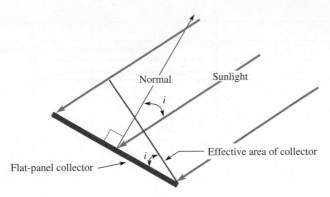

FIGURE 5-8   Flat-Panel Orientation

adjustments to the tilt angle can be accomplished. Changing this angle at least four times per year can increase the total energy from the panel and can help with other considerations such as reducing a snow load.

For tracking modules, another consideration is the additional space required to allow for the movement of the panels and to prevent shading. Both single- and dual-axis trackers require extra space to avoid shadowing effects from one panel to the next, so space may not be utilized as effectively as with fixed panels. For example, available space on a roof may be at a premium, and tracking could take away from the total collector area. A shaded panel also negates the benefit of tracking. Dual-axis trackers need even more space than single-axis trackers because of the larger range of motion. Figure 5-9 illustrates the problem with shading when more than one panel is part of a system.

Benefits

Flat-plate collectors collect more power during the course of a day using tracking compared to an optimum fixed installation. The specific benefit depends on the location and efficiency of the panels and varies with the time of year, the weather, and the type of tracking device. Some locations require more or less site work than others to install trackers. Flat locations that are also unobstructed can be less costly for building purposes than hillsides. More efficient panels are more cost-effective for tracking because an efficient panel can produce more power than a less-efficient one, so the number of panels (and tracking devices) may be reduced. Seasonal variations affect the total energy provided by the panel and have an effect on efficiency.

The ideal curve (clear day) for a PV module is shown in Figure 5-10 for a typical spring day in a northern latitude with and without tracking. In this case, the latitude angle

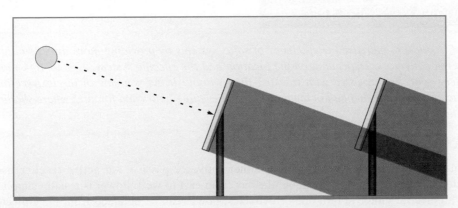

FIGURE 5-9   Shading on One Panel Cast by Another. The shadow cast by one panel on another can create problems as the panels move, particularly when the sun is low in the sky.

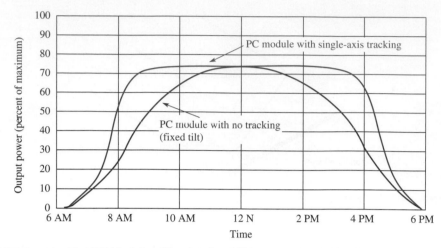

FIGURE 5-10 Flat PV Modules Showing the Difference Between Tracking and Nontracking for a Typical Spring Day

of the tracker is set to the same angle as the fixed module, so the peak power is the same. (Note that if the angles are set differently, the peaks do not coincide.) The total power is related to the area under the curve. You can see that tracking has the greatest impact in the morning and evening hours. Also notice that the 70% of maximum power is available for about 2 hours with a fixed module, but the maximum power extends to over 6 hours with single-axis tracking in this case. On a sunny winter day, the tracking advantage is typically 20% greater power. During the summer, the area under the tracking curve is typically 40% greater than that of the fixed array curve.

SECTION 5-2 CHECKUP

1. Why is it necessary for a group of heliostats to move independently?

2. Cite examples of two expenses, other than the initial hardware cost, that are incurred using a tracking system compared to a nontracking system.

3. Why are most roof-mounted PV systems fixed?

4. What is the advantage of an efficient fixed panel in a tracking system over a less-efficient panel?

5. At what times of the day does a tracker provide the most benefit? Explain.

5-3 Single-Axis and Dual-Axis Solar Trackers

Solar trackers increase the efficiency of solar systems by providing more direct sunlight on the collectors. Depending on the location and the specific system, 25% to 40% more energy can be harvested with tracking than without it. As you saw in the last section, tracking is most beneficial in the morning and evening hours and for sites where shading is not an issue.

Active and Passive Trackers

Solar trackers can be categorized as either active or passive. An **active tracker** uses external power to move throughout the day from east to west. Power is usually supplied to an electric motor to turn one or two axes; power can be obtained from the collector and stored in a battery for continuous tracking. The tracker should position the collector to the east to be ready for the following day, so power needs to be available after the sun

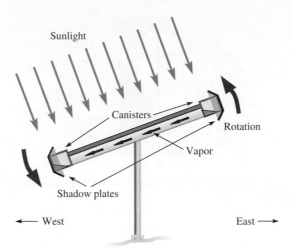

FIGURE 5-11 Passive Tracker. The collector moves as weight shifts between the canisters due to differential heating of the canisters.

is down from a battery or from the utility grid for the reset operation. Another type of active tracker is a hydraulic tracker, which requires electric power for the pump and for the controller. A **passive tracker** does not require external power to turn the axis (usually only one axis is used with this type of tracker). One type of passive tracker uses a liquid, contained in canisters mounted on each side of the module, that can turn easily into a vapor. Figure 5-11 shows the idea. The canisters are connected together by a long tube and depend on the shifting weight of the liquid as they are heated by sunlight. When the sun is shining and if one canister is shaded and the other is not, the liquid in the warmer tube vaporizes and forces more liquid into the shaded tube. The weight shifts and the tracker moves in a way that tends to equalize the temperatures and hence the weight. Passive trackers are simple and do not need any external power, but they do not reset and are left in the evening facing west. It takes about an hour to move them back to the east the following morning in order to produce power, which is time that an active tracker does not waste. Thus, the passive tracker does not harvest quite as much of the available energy as an active tracker.

Single-Axis Trackers

Many solar tracking devices have been developed. Single-axis control means only one axis automatically tracks the daily motion of the sun. The seasonal changes in the sun's path cannot be tracked automatically, but manual adjustments can be made to maximize the sunlight over the course of the year. If an adjustment is done each season, the angle from the horizontal is typically set to the latitude $-15°$ in the summer (flatter orientation) when the sun is high, $0°$ in fall and spring, and $+15°$ in winter (steeper orientation) when the sun is low. The reason for these angles can be seen by referring again to Figure 5-5, which shows the sun's north–south deviation from the celestial equator for one year. Figure 5-12 illustrates these seasonal tilt angles, in which the array faces south for the northern hemisphere and north for the southern hemisphere. (Ignore the negative sign for southern latitudes.)

When this tilt angle cannot be adjusted seasonally, the optimum angle depends on the latitude and climate conditions at the location as well as the user's energy requirements (for example, summer irrigation). In summer, the sun is in the sky longer than it is in the winter, and it is generally more likely to have clear skies; hence, much more energy is available during the summer. For maximum energy intercepted over the course of a year, computer simulations of the weather are used to choose a tilt angle, and they generally favor a larger tilt angle to produce more power during the shorter winter days (December in the northern hemisphere; June in the southern hemisphere).

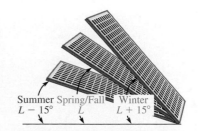

FIGURE 5-12 Optimum Tilt Angle. The optimum tilt angle depends on the season. If the module can be adjusted four times per year, a shift of 15° each season from the latitude (L) is a good choice for maximizing energy.

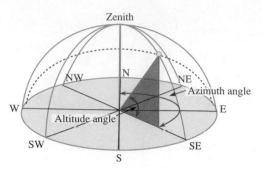

FIGURE 5-13   Altitude and Azimuth Definitions. The dotted line is the path of the sun on a certain day and location on earth.

Altitude-Azimuth Trackers

Most trackers use an altitude-azimuth arrangement with a vertical axis and a horizontal axis. The definition of **altitude angle** is the angle formed between a horizontal plane and an imaginary line pointed to the sun. The **azimuth angle** is a horizontal angular distance from a reference direction, usually north. See Figure 5-13 for an illustration of the definitions for altitude and azimuth angles with reference to the sun.

For a single-axis tracker, the rotating axis is normally the vertical axis which moves the collector east to west during the day. With rotation on the vertical axis, the collector maintains a constant clearance from the ground. The collector is titled from the ground plane by an amount approximately equal to the latitude, and the collector is turned to the east in the morning and slowly turns to face west in the evening. After sunset, active trackers reset to the east to be ready for the following day, but as mentioned previously, passive trackers require sunlight to reset. Figure 5-14 shows an array of single-axis trackers that are driving flat panels. The vertical axis is the only one that tracks.

The path of the sun above the horizon is shortest in winter and longest in summer, so the turning arc is different for the trackers according to the season. In some cases, the tilt for the panels on a single-axis tracker is adjusted manually to account for the seasonal changes in the sun's motion.

Another type of single-axis tracker is shown in Figure 5-15. It uses mirrors to concentrate light on a pipe filled with synthetic oil. In this example, a single actuator can control

FIGURE 5-14   Array of Collectors with Single-Axis Trackers. Notice the heavy support structure.   (*Source:* National Renewable Energy Laboratory.)

(a) Line of collectors at Kramer Junction, California   (b) Hydraulic tracking mechanism

FIGURE 5-15   Single-Axis Tracking for Trough Concentrating Collectors
(*Source:* National Renewable Energy Laboratory.)

an entire line of solar troughs that turn on a common axis. The troughs are aligned along a north–south axis and are rotated to face the sun. During the day, the mirrors rotate from east to west to follow the sun across the sky; they reset back to the east in the evening to await the sun on the following morning. A long array with one drive is an effective arrangement for a single-axis system. This array is a commercial power generation station at Kramer Junction, California. The hot oil is pumped to a heat exchanger that has a water loop on the other side. This water is vaporized to steam that spins a turbine, producing electricity. The same concept of one main driving system can also be applied to a line of flat-panel collectors.

The trough collectors are a form of nonimaging optics that have a wider angle of acceptance of the sun's rays than do the imaging type of concentrators. Nonimaging optics concentrates the radiation energy, but the goals are even illumination on a long target rather than on a small point and good heat transfer to the target. As mentioned, an array like this is ideally suited for single-axis tracking. In this case, both motor and hydraulic drives are used. Figure 5-15(b) shows a hydraulic drive, which is located in the middle of the long trough and rotates the entire trough.

Equatorial Trackers

An equatorial tracker (sometimes called a polar tracker) is a variation of the horizontal-axis tracker where the rotating axis is tilted to be parallel with the earth's axis (pointing due north in the northern hemisphere and due south in the southern hemisphere). In a single-axis equatorial tracker, the panels are mounted so that they are perpendicular to the axis. Figure 5-16 on page 162 shows solar panels mounted on an equatorial mount. The panels are installed to avoid interfering with the motion of the array at sunrise and sunset. The arrangement of the mount shown is similar to the telescope mount on the famous Hale telescope on Palomar Mountain.

Dual-Axis Trackers

Certain concentrating collectors need better accuracy than a single-axis tracker can provide. The higher the concentration ratio, the higher the tracking accuracy that is required. The highest accuracy is required by collectors that are so-called imaging types, such as the parabolic dishes used with Stirling engines. See Figure 5-17 for an example. Imaging concentrating collectors work from direct radiation rather than diffuse radiation, and they are inefficient if they are not pointed directly at the sun. To get the most energy and the highest temperature, dual-axis tracking must be employed.

FIGURE 5-16   Equatorially Mounted Tracker   (*Source:* National Renewable Energy Laboratory.)

As the name implies, a dual-axis tracker can move in either of two directions. Generally, these trackers are set up as alt-azimuth mounts, meaning the axes move vertically (the altitude) and horizontally (azimuth). This reduces the issue of ground clearance, which can be a problem with equatorial mounts, but it requires a controller to make the adjustments in driving two motors. In addition to the benefit for concentrating collectors, dual-axis tracking is sometimes used with flat collectors. A flat collector always receives the maximum amount of solar energy if it is oriented so that the direct radiation is perpendicular to its surface. A dual-tracker moves the collector continuously during daylight hours to meet this condition, thus maximizing the solar energy harvested. Thus, the main consideration when deciding to use single-axis, dual axis, or no tracking boils down to cost.

FIGURE 5-17   Large Dual-Axis Tracker with a Stirling Engine   (*Source:* National Renewable Energy Laboratory.)

Most dual-axis trackers are active types and use two independent motors to turn the axis. The most common method for moving a dual-axis tracker is stepper motors, which can be rotated precisely over some portion of an arc. The motors are controlled by a dedicated controller. Stepper motors and controllers are discussed in Section 5-5. Various tracking algorithms have been written to provide solar coordinates for the controller; among the most accurate is one from NREL that can calculate the solar position over an 8,000-year period with uncertainties of $\pm 0.0003°$. Although this is far more accurate than would ever be needed by any solar tracker, it is useful for certain solar measurements, where a small part of the sun is imaged and tracked.

SECTION 5-3 CHECKUP

1. (a) For a single-axis tracker, what is the optimum angle to set the altitude angle for summer in Phoenix, Arizona (latitude 33.5° N)?

   (b) Would you set a different angle if the angle is fixed and cannot be adjusted seasonally? Explain your answer.

2. How would you orient a single-axis tracker for summer (in December) in Sydney, Australia (latitude 34° S)?

3. Compare dual-axis tracking with single-axis tracking for a flat PV system. What are the primary selection criteria in choosing one or the other?

4. Why does a Stirling engine collector require a dual-axis tracker?

5-4 DC Motors

For many applications in renewable energy, such as tracking, a motor is an integral part of the system. For tracking applications, a motor needs to have high torque. DC motors are suited for this purpose because they can be efficient over a wide range of torque–speed characteristics. In this section, you will learn more about dc motors. In addition to solar tracking, dc motors have other applications in renewable energy systems, such as running water pumps or attic fan motors that use a PV collector for power.

Motors and Magnetic Fields

All motors use the physics principle that a magnetic field is created by a current-carrying conductor and the interaction between magnetic fields produce a force. If you experiment with a magnet and current-carrying wire, you might be surprised to find that this force is fairly small. Motors increase the force by using many turns of wire and strong magnetic fields that have concentrated paths for magnetic field lines.

DC Motors

All motors (and generators) have two basic parts: the rotating part called the **rotor**, and the stationary part called the **stator**. Each part is a magnet, and motors turn by the interaction of magnetic fields. The armature of any rotating machine (motor or generator) is the part from which power is taken. In a motor, the rotor is always the armature because mechanical power is taken from it.

In a dc motor, the stator establishes a magnetic flux field that is constant. It can be established by a permanent magnet or an electromagnet. Recall that a coil of wire becomes a magnet when there is current in it. The polarity of the current determines the location of the north and south magnetic poles in the coil, and the amount of current is related to the strength of the field. The left-hand rule can be used to determine the magnetic polarity of the coil, as shown in Figure 5-18. If the current in the coil is reversed, the poles reverse; this idea is used to reverse the direction of a dc motor.

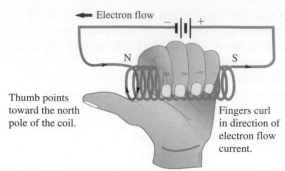

FIGURE 5-18   Left-Hand Rule for Determining the North and South Poles of a Coil

Basic Operation

DC motors can have both a wound rotor and a wound stator (meaning they are electromagnets), or one can be a permanent magnet and the other an electromagnet. With the advent of very strong permanent magnets made from ceramic and rare earth alloys, permanent magnet dc (PMDC) motors for smaller (<5 hp) motors have become popular. They are easier to control (using two wires), and the torque–speed characteristic is more linear. DC motors can be controlled smoothly by adjusting the current to the coils. They can be slowed to zero and then accelerated in the opposite direction.

An advantage of PMDC motors for solar tracking is simplicity and high starting torque: They can use a single source of dc power that is readily available from the PV panel. PMDC motors tend to be more efficient than wound stator motors. Figure 5-19 shows a simplified diagram of a PMDC motor with a permanent magnet stator and a wound rotor. Some parts are not shown for the sake of clarity. The single rotor coil like the one shown does not provide enough torque to be practical, but it explains the operating principles. Because of the slow speed required by a solar tracker, the output speed is reduced through gearing.

Current in the rotor creates a magnetic field that is perpendicular to the rotor winding; the rotor has a north and south pole that is perpendicular to the coil. The rotor moves because of the attractive force between these poles and the stationary stator poles. As the unlike pole of the rotor is attracted to the stator's pole, the polarity of the rotor current is

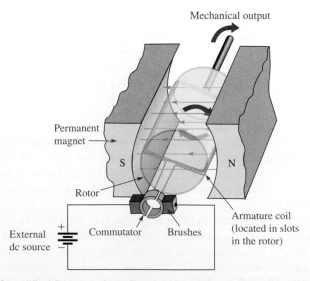

FIGURE 5-19   Simplified Diagram for a Permanent Magnet, Two-Pole DC Motor. Practical motors have many separate windings and a multiple-segment commutator.

suddenly switched, reversing the magnetic poles of the rotor and thus creating a repulsive force. The required switching is done by the commutator. A **commutator** is a series of bars or segments in a dc machine (motor or generator) forming a rotating switch in which the armature coils are connected. The brushes ride on the commutator segments. As soon as it switches, the rotor is repelled from the original stator pole but is now attracted to the one on the opposite side, thus turning on its axis. Essentially the commutator keeps the pattern of current in the rotor the same regardless of its position.

A simple two-pole motor, such as the one illustrated in Figure 5-19, would operate rather roughly and provide very little torque. Additional coils must be added to the motor for it to be practical. With more coils, the commutator needs more segments so that each coil is powered at the appropriate points. The torque from an actual motor depends on its geometry, including the distance from the center of the rotor to the conductor, the length of the conductor exposed to the magnetic flux and the area of the core, and so on. By lumping the geometry factor, including the number of conductors into a constant, K, the torque (in newton-meters) for a dc motor can be expressed as follows:

$$T_q = K\phi I_a \qquad \text{Equation 5-1}$$

where

T_q = torque, in newton meters (n-m)

K = a constant that depends on the number of conductors in the armature, in newton-meters/weber-ampere (n-m/Wb-A)

ϕ = magnetic flux, in webers (Wb)

I_a = armature current, in amperes (A)

Equation 5-1 shows that the torque from a dc motor is proportional to the magnetic flux and the armature current.

Counter EMF

When the armature in a motor begins to rotate, it begins to produce voltage by generator action. A dc motor is constructed like a generator, so it can operate as either a motor or a generator. When the rotor is spun using an external source (such as the wind), it is a generator; recall that when a magnetic field moves across a coil of wire, a current is produced. When electrical energy is applied to it to force it to spin, it is a motor.

When a dc motor is first turned on, the armature resistance is all that limits current, which can be quite high due to the low resistance. In larger motors, fixed resistors are placed in series with the armature to limit current, and these resistors are automatically removed as the motor comes up to speed. As the motor speeds up, a voltage is generated in the armature by generator action that is the opposite polarity to that of the power supply. This voltage is called back voltage, back emf (electromotive force), or counter emf (C_{emf}). This voltage is subtracted from the supply voltage so that the rotor windings see a smaller voltage potential and hence smaller current. Kirchhoff's voltage law states that the algebraic sum of the voltages in a closed path is zero. From this, we can conclude that the applied voltage must be equal to the C_{emf} plus the voltage drop in the armature due to armature resistance. In equation form:

$$V_S = C_{emf} + I_a R_a \qquad \text{Equation 5-2}$$

where

V_S = applied voltage, in volts (V)

C_{emf} = counter emf, in volts (V)

I_a = armature current, in amperes (A)

R_a = armature resistance, in ohms (Ω)

The reduced current as the motor speeds up causes the series motor to lose torque at faster speeds. The load is turned by the motor, so it has inertia, and less torque is required to

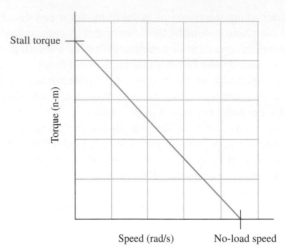

FIGURE 5-20 Torque–Speed Curve for a Small PMDC Motor

keep it turning. The result is less torque as speed increases, as shown in the general plot for a PMDC motor in Figure 5-20. If there is no load on the motor, it accelerates until the C_{emf} is nearly equal to the applied voltage except for a small drop across the armature resistance. In general, for a PMDC motor, the higher the speed, the lower the torque, as illustrated by the line in the figure.

The power delivered by the motor depends on the speed and the torque. There is no power developed if the motor is not turning (stall torque) or if there is no torque (no-load speed). Between these extremes, the motor delivers power to the load.

Stator

For the PMDC motor, the stator consists of a fixed permanent magnet that is separated from the rotor by a very small space. Air has much higher reluctance than iron, so this space, referred to as the gap, is designed to be as small as practical to maximize the magnetic flux yet still allow free rotation of the rotor. The stator magnets are typically made from ferrite, neodymium, or samarium-cobalt materials to provide a very strong magnetic field.

Larger motors have a wound stator, which can provide higher flux and hence can produce more torque. In a wound stator, the windings must be supplied with dc, which can come from a supply that is separate from the rotor supply. This setup offers flexibility in the control of the motor (both speed and torque), but it is generally unnecessary for solar trackers.

Normally with larger motors, the most important rating is power rather than torque. Power and torque are very different quantities. Power is the rate of doing work, whereas torque is a force multiplied by a distance that tends to produce rotation. The relation between torque and power (in SI units) is:

$$T_q = \frac{P}{\omega}$$

Equation 5-3

where

$T_q =$ torque, in newton-meters (n-m)

$P =$ power, in watts (W)

$\omega =$ rotational speed, in radians/second (rad/s)

Although the SI unit for power is the watt, many motors are rated in horsepower, an antiquated unit that was originally developed to compare a steam engine to the power from a draft horse. Today, the accepted conversion equivalency between the two units is 746 W/hp.

EXAMPLE 5-1

Assume a dc motor develops 3.56 n-m of torque at 1,000 rpm. What is the horsepower rating of the motor?

Solution

First, find the number of radians/second represented by 1,000 rpm:

$$\omega = \left(\frac{1,000 \text{ rev/m}}{60 \text{ s/m}}\right)\left(\frac{2\pi \text{ rad}}{\text{rev}}\right) = 104.7 \text{ rad/s}$$

$$P = T_q\omega = (3.56 \text{ n-m})(104.7 \text{ rad/s}) = 373 \text{ W}$$

$$P = \left(\frac{373 \text{ W}}{746 \text{ W/hp}}\right) = \textbf{0.5 hp}$$

Rotor Assembly

Typically, the rotor of a dc motor has a number of low resistance windings, which means that the rotor starting current is high and is limited only by the winding resistance. This creates the high starting torque mentioned previously because torque is proportional to the armature current. As the motor comes up to speed, the current and the torque drop. The high starting torque is why dc motors are useful for the starter motor in cars.

The rotor assembly includes the armature and armature shaft, and the commutator. Windings on the rotor are pressed into slots in the rotor assembly. The slots provide a location for the coil wires while keeping the gap between the rotor and the stator as small as possible and thus providing a low reluctance path for the flux. One end of each coil is connected to a commutator segment so there are two commutator segments for each coil. Protruding from the ends of the rotor is the shaft, which is the connection point to the motor.

The commutator is the mechanical rotating switch located on the rotor that changes the polarity of the current in the rotor coils as the rotor spins. The commutator is segmented; there are two segments for each coil, one for each end of the coil. Thus, an armature with eight coils has sixteen commutator segments. The commutator is made of copper with a thin section of insulation between each segment. This insulation effectively isolates each commutator segment from all others. The commutator segments are used as contact points between the brushes and the rotor. In motors and generators, the **brushes** are electrical conductors that provide a connective path for current from a stationary part to a moving part, in this case, between the electrical source and the commutator.

When a coil of wire is pressed onto the armature, the ends are soldered to a pair of commutator segments. This makes an electrical terminal point for the current from the brushes to the armature. Typically, the armature core is made from laminated steel to prevent the circulation of eddy currents. If the core were solid, the magnetic field could induce these unwanted eddy currents that would circulate in the core material and cause it to heat up. When laminated steel sections are pressed together to make the core, the eddy currents cannot flow from one lamination to another, so they are effectively eliminated. The laminated core also prevents other magnetic losses called flux losses. Flux losses tend to make the magnetic field weaker so that more core material is required to obtain the same magnetic field strengths. The flux losses and eddy current losses are grouped together by designers and are called core losses. The laminated core is designed to allow the armature's magnetic field to be as strong as possible because the laminations prevent core losses.

A typical rotor is shown in Figure 5-21. This rotor has overlapping coils, in an arrangement known as a lap winding. The coils have been pressed into slots along the core. At any given time during the operation of the motor, one conducting coil has one side under one pole of a magnet and the other side under the opposite pole. This combination causes the coil to experience opposite force on each side, which applies a net rotational force to the rotor.

FIGURE 5-21   Rotor Assembly for a DC Motor   (*Source:* Courtesy of Bodine Electric Company.)

Brushes

As mentioned, the brushes rub against the commutator and provide a path for current from the fixed power source to the rotating armature assembly. When the armature is spinning, each commutator segment comes in contact with a positive brush for a short time and is positive during that time. At the other end of the coil, the commutator segment that it is connected to is against a negative brush, completing a path for current in a coil. Typically, several coils are energized at a time. As the armature rotates, different commutator segments come in contact with the brushes. As the armature continues to spin, each commutator segment is alternately powered by positive and then negative voltage, causing the magnetic field polarity to undergo a constant change.

The brushes are made of carbon-composite material or graphite. Usually the brushes have copper added to them when they are formed to improve conduction. Other materials may be added to make them last longer. The end of the brush that rides on the commutator is contoured to fit the commutator exactly for minimum resistance. The process of contouring the brush to the commutator is called **seating**. Whenever a set of new brushes is installed, the brushes should be seated to fit the commutator. The brushes are the main part of a brushed motor that wear out. It is important that their wear be monitored closely so that they do not damage the commutator segments when they begin to deteriorate. Most brushes have a small mark on them called a wear mark or wear bar. When a brush wears down to the mark, it should be replaced. If the brushes begin to wear excessively or do not fit properly on the commutator, they heat up and damage the brush rigging and spring mechanism. If the brushes have been overheated, they can cause burn marks or pitting on the commutator segments and also warp the spring mechanism so that it will no longer hold the brushes with the proper amount of tension.

The brush rigging is an assembly that holds the brushes in place. It is mounted on the rear end plate so that the brushes are accessible by removing the end plate. An access hole is also provided in the motor frame so that the brushes can be checked and adjusted as needed when the motor is initially set up. The brush rigging uses a spring to provide the proper amount of tension on the brushes so that they make proper contact with the commutator. If the tension is too light, the brushes bounce and arc; if the tension is too heavy, the brushes wear down prematurely.

Frame

The armature is placed inside the frame of the motor where the permanent magnet or field coils are mounted. Field windings are mounted on laminated pole pieces called field poles. Similar to an armature, these poles are made of laminated steel or cast iron to prevent eddy current and other flux losses. The field poles are secured so that they will not move when they are attracted or repelled by the armature's magnetic field. Any motion of the poles causes vibration and can damage the outer protective insulation and/or cause a short circuit or a ground condition between the winding and the frame.

The ends of the frame are machined so that the end plates mount firmly into place. An access hole is also provided in the side of the frame or in the end plates so that the field wires can be brought to the outside of the motor, where dc is connected. The bottom of the frame has the mounting bracket attached. The bracket has a set of holes or slots so that the

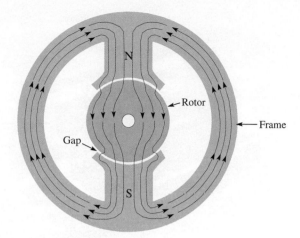

FIGURE 5-22   Magnetic Path for a Motor

motor can be bolted down and mounted securely. There is usually a small amount of adjustment when the motor is used in direct-drive applications so that the armature shaft can turn freely and not bind with the load.

The frame is constructed from steel to provide a low reluctance return path for magnetic field lines and thus allow for stronger magnetic fields than would otherwise be present. Figure 15-22 illustrates the magnetic path for a basic two-pole dc motor. In addition to providing a low reluctance path for the magnetic field, the frame also conducts heat away from the motor due to power dissipated in the motor itself.

End Plates

The end plates (sometimes called **end bells** because of their shape) are mounted on the ends of the motor frame and include the bearings. If the bearing is a ball-bearing type, it is normally lubricated permanently. If it is a sleeve type, it requires a light film of oil to operate properly. In this case, a lubrication tube and wicking material provides lubrication.

The end plates for a sleeve bearing motor are mounted on the motor frame so that the lubricating tube is pointing up. This position ensures that gravity pulls the oil to the wicking material. If the end plates are mounted incorrectly so that the lubricating tube is pointing down, the oil flows away from the wicking and it becomes dry. When the wicking dries out, the armature shaft rubs directly on the metal in the sleeve bearing, which causes it to heat up quickly. In this situation, the shaft seizes to the bearing. For this reason, it is also important to follow lubrication instructions and oil the motor on a regular basis.

Geared DC Motors

For solar trackers and other applications such as telescope drive motors, a direct connection to the rotor would not be practical because of the speed. To slow the shaft, gearing is used. Figure 5-23 on page 170 shows cutaway view of a geared dc motor. In this case, a worm gear is connected directly to the shaft. Worm gearing can slow the shaft speed to a few rpm; additional gearing can slow the shaft speed even more and increase torque. An advantage to dc motors for solar tracking is that they can be reversed, so that they can be returned to the starting position for the following day.

Brushless Motors

Another type of PMDC motor uses a permanent magnet for the rotor, and its stator has coils. Because the rotor is a permanent magnet, there is no need for brushes or a commutator; hence, it is called a brushless PMDC motor. The rotor position is sensed by Hall-effect sensors, which transmit the information to controller logic. The logic sends current with the appropriate polarity to create a rotating magnetic field, which the rotor tries in vain to

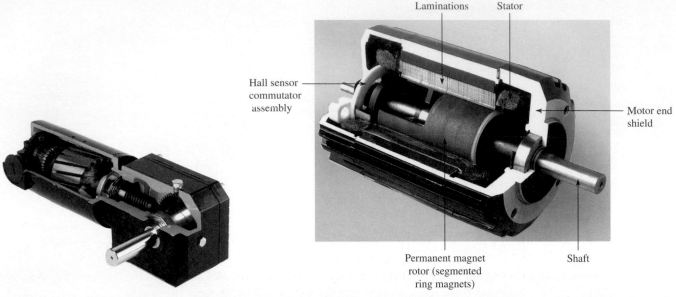

FIGURE 5-23 Cutaway View of Geared PMDC Motor (*Source:* Courtesy of Bodine Electric Company.)

FIGURE 5-24 Cutaway View of a Brushless DC Motor (*Source:* Courtesy of Bodine Electric Company.)

keep up with. Figure 5-24 shows a brushless PMDC motor. The controller also controls the proper voltage to maintain the required speed.

Types of Wound DC Motors

As mentioned, larger dc motors use an electromagnetic field created by stator windings connected to a dc voltage source. The armature and field windings in these motors can be wired in any of three different ways, depending on requirements for torque or speed control. The three basic types of wound stator motors are the series-wound (series) motor, the shunt-wound (shunt) motor, and the compound motor. Figure 5-25(a) illustrates the series motor, and Figure 5-25(b) shows the shunt-wound motor. The compound motor comes in several different arrangements, but the most common type, called a cumulative compound motor, is shown in Figure 5-25(c). Notice that the series field windings are identified as S_1 and S_2, whereas the shunt field windings are identified as F_1 and F_2. You will often see connections to the power supply labeled L_1 and L_2. The choice of which of these configurations is best for a given situation is based on the load requirements. These requirements may include the variation of torque with a change in load and the variation in speed with a change in load. The best choice for high start-up torque is the series-wound motor because it has more torque with a given armature current than the other types. The best choice for regulating speed is the compound motor.

DC Series Motors

A dc series motor provides high starting torque and can move very large shaft loads when it is first energized. From the wiring diagram in Figure 5-25(a), you can see that the field

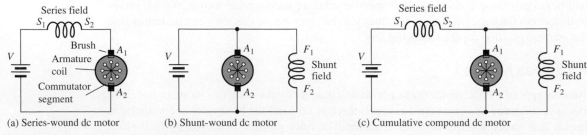

(a) Series-wound dc motor

(b) Shunt-wound dc motor

(c) Cumulative compound dc motor

FIGURE 5-25 Different Types of DC Motor Wiring

winding in this motor is wired in series with the armature winding, which gives the series motor its name. Because the field winding is connected in series with the armature winding, the two windings have the same current, which can be quite large. Series motors that are used to power hoists or cranes may draw currents of thousands of amperes during operation. In most applications, the series-wound motor operates with a high current for only a short time before the current drops as the motor comes up to speed. It is best never to run a series-wound motor without a load or the motor can rotate faster than its rated speed and literally come apart.

Reversing the Direction

The direction of rotation of a series motor can be changed by changing the polarity of either the armature or field winding. If you simply changed the polarity of the applied voltage, you would be changing the polarity of both field and armature windings, and the motor's rotation would remain the same. Because only one of the windings needs to be reversed, the armature winding is typically used because its terminals are readily accessible at the brush rigging.

Troubleshooting the Series Motor

The best way to troubleshoot a series-wound motor is with a voltmeter. The first test should be for applied voltage at the motor terminals. Because the motor terminals are usually connected to a motor starter, the test leads can be placed on these terminals. If the meter shows that full voltage is applied, the problem is in the motor. If the meter shows that no voltage is present, you should test the supply voltage and the control circuit to ensure that the motor starter is closed. If the motor starter has a visual indicator, be sure to check to see that the starter's contacts are closed. If the overloads have tripped, you can assume that they have sensed a problem with the motor or its load. When you reset the overloads, the motor will probably start again, but remember to test the motor thoroughly for problems that could cause an overcurrent situation.

If the voltage test indicates that the motor has full applied voltage to its terminals but the motor is not operating, check for an open in one of the windings or between the brushes and the armature. Each of these sections should be disconnected from each other and voltage should be removed so that they can be tested with an ohmmeter for an open. The series field coils can be tested by putting the ohmmeter leads on terminals + and −. If the meter indicates that an open exists, the motor needs to be removed and sent to be rewound or replaced. If the meter indicates that the field coil has continuity, you should continue the procedure by testing the armature.

The armature can also be tested with an ohmmeter by placing the leads on the terminals marked A_1 and A_2. If the meter shows continuity, rotate the armature shaft slightly to look for bad spots where the commutator may have an open or the brushes may not be seated properly. If the armature test indicates that an open exists, you should continue the test by visually inspecting the brushes and commutator. You may also have an open in the armature coils. The armature must be removed from the motor frame to be tested further. When you have located the problem, you should remember that the commutator can be removed from the motor while the motor remains in place, and the commutator can be turned down on a lathe. When the commutator is replaced in the motor, new brushes can be installed and the motor will then be ready for use.

Another problem is overheating caused by inadequate lubrication or a damaged bearing. The bearing will seize on the shaft and cause the motor to build up friction and overheat. Thermal imaging is useful for checking for heat-related problems. If damage occurs from overheating, the motor may require extensive repairs.

DC Shunt Motors

In the shunt motor, shown in Figure 5-25(b), the field winding is connected in parallel with the armature using smaller gauge wire with many turns. Because the wire for the field winding is relatively small, it can have thousands of turns and can still produce a very strong

Safety Note

When working on motors, be sure to use a lockout after disconnecting power. The **lockout** is a device placed on the handle of the disconnect switch after the handle is placed in the off position. It allows a padlock to be placed around it so it cannot be removed until the work is completed.

(*Source:* kabakoff84/Fotolia.)

In active, solar hot water systems, circulating pumps are required to circulate water or antifreeze between the collector and the storage tank. The pump can be powered with a dc motor that gets power from a solar panel. This is a good match for a solar PV collector with a smaller rated battery backup because most of the time that the pump is needed is precisely the time when electricity from the panel is available.

magnetic field. The shunt motor cannot produce the large current for starting like the series motor; hence, it has low starting torque, which requires that the initial shaft load is small.

When voltage is applied to a shunt motor, the high resistance of the field coil keeps the current and magnetic field low. The armature is similar to the series motor, so it draws a larger current that produces a strong magnetic field. Like the series motor, the armature produces a counter emf (C_{emf}) when it begins to turn. The C_{emf} causes the current in the armature to begin to diminish. The amount of armature current is related directly to the size of the load when the motor reaches full speed. Because the starting load is generally small, the armature current is small.

Unlike the series motor, the shunt motor's speed tends to be constant. When it starts and as it speeds up, the armature begins to produce more C_{emf} until it reaches an equilibrium condition and the motor runs steadily. If at some point the load increases, the armature shaft starts to slow down, and less C_{emf} is produced. This causes the *difference* between the C_{emf} and the applied voltage to become larger. A larger armature voltage increases armature current, which in turn increases the torque. As a result, the speed tends to stay the same. (This situation is an example of negative feedback.) Although it tends to run at constant speed for varying loads, the speed can be varied in two ways: by (1) varying the current in the field windings or (2) varying the amount of current in the armature. Varying the voltage applied to the field is commonly done by adding a rheostat in series with the field windings. The drawback is that the motor's efficiency drops off when it is operated below its rated voltage.

Reversing the Direction

The direction of rotation for a shunt motor can be reversed by changing the polarity of either the armature coil or the field coil, but not both. Usually, the armature current is reversed, as is the case with the series motor. If the motor is connected to a motor starter, the starter usually has contacts for reversing the motor.

Troubleshooting the Shunt Motor

When a DC shunt motor develops a fault, you must be able to locate the problem quickly and return the motor to its normal service or have it replaced. The most likely problems to occur with the shunt motor include loss of supply voltage or an open in either the shunt winding or the armature winding. Other problems may arise that cause the motor to run abnormally hot, even though it continues to drive the load. The motor will show different symptoms for each of these problems.

If the motor will not start, you should listen to see if the motor is humming and trying to start. When the supply voltage has been interrupted due to a blown fuse or a de-energized control circuit, the motor will not be able to draw any current and it will be silent when you try to start it. You can also determine if the supply voltage has been lost by measuring it with a voltmeter at the motor or the motor starter's + and − terminals. If there is no voltage, check the supply fuses and the rest of the supply circuit.

If the motor tries to start and hums loudly, it indicates that the supply voltage is present. The problem in this case is probably due to an open field winding or armature winding. It could also be caused by the supply voltage being too low. Check the windings for continuity. The most likely problem with an open winding is the field winding because it is made from small-gauge wire. Because the windings are connected in parallel, you need to isolate them to read continuity (otherwise you will see continuity when none is present). Because the brushes may be the fault, they should be inspected visually and replaced if they are worn or not seating properly. If the commutator is also damaged, the armature should be removed, so the commutator can be turned down on a lathe. If either the field winding or the armature winding has developed an open circuit, the motor must be removed and replaced.

DC Compound Motors

The dc compound motor is a combination of the series motor and the shunt motor, as shown in Figure 5-25(c). It has two field windings: one that is in series with the armature and one

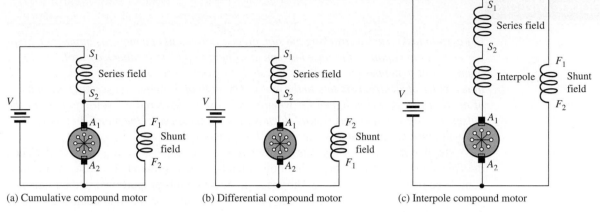

(a) Cumulative compound motor   (b) Differential compound motor   (c) Interpole compound motor

FIGURE 5-26   Examples of Compound Motors

that is in parallel with the armature. The combination of series and shunt windings allows the motor to have the torque characteristics of the series motor and the regulated speed characteristics of the shunt motor.

Variations of compound motor wiring are shown in Figure 5-26. The cumulative compound motor is shown in Figure 5-26(a). It is one of the most common DC motors because it takes the best characteristics of both the series motor and shunt motor, which makes it acceptable for most applications. It provides high starting torque and good speed regulation at high speeds. It is called *cumulative* because the shunt field is wired with similar polarity in parallel with the magnetic field, thus aiding the series field and armature field. When the motor is connected this way, it can start even with a large load and then operate smoothly when the load varies slightly. Recall that the shunt motor can provide smooth operation at full speed, but it cannot start with a large load attached, and the series motor can start with a heavy load, but its speed cannot be controlled.

The differential compound motor is shown in Figure 5-26(b). It uses the same windings as the cumulative compound motor except that the shunt field is connected so its polarity is reversed to the polarity of the armature (compare the F_1 and F_2 positions with the cumulative compound motor). It is considered a short shunt because the shunt field is still connected in parallel with only the armature. It has many of the characteristics of the series-wound motor.

A third type of compound motor is the interpole compound motor (also called a long shunt motor), which is shown in Figure 5-26(c). It prevents a problem with motors called armature reaction. Armature reaction is a flux created around each conductor in the armature by current in the armature. This flux interacts with the main flux from the field coils and tends to shift its position, depending on the armature current. The interpole coils are small coils that are in series with the armature and physically located next to the stator's series coil. They produce a flux that tends to cancel armature reaction by shifting the magnetic field back to its rest position, which helps prevent arcing. As a result, the brushes tend to last longer. The interpoles also allow the armature to draw heavier currents and carry larger shaft loads.

SECTION 5-4 CHECKUP

1. Explain why a series dc motor has high starting torque.

2. How does counter emf control the speed of a shunt motor?

3. Why is it important to keep the air gap small in a dc motor?

4. What is the difference between power and torque?

5. What is the purpose of the commutator in a dc motor?

6. What is the purpose of interpole coils?

5-5   Stepper Motors

For certain tasks, such as tracking the sun, precise movement of a motor is important. Stepper motors are capable of precise movement without requiring feedback control. The basic operation of a stepper motor allows the shaft to move a precise number of degrees each time a pulse of electricity is sent to the motor. The shaft of the motor moves only the number of degrees that it was designed for when each pulse is delivered, so you can control the pulses that are sent and thus control the positioning and speed of the motor. In this section, you will learn more about this important class of motors.

A **stepper motor** moves a rotor in discrete steps in response to a pulse signal. When the frequency of pulses is high, the motion tends to be continuous because each step is completed in a few milliseconds. The speed of the rotation is proportional to the frequency of the pulses, so the stepper motor can position the shaft precisely and control the speed without using feedback.

Stepper motors are available for a variety of motion-control applications. Smaller motors are used in printers and in some computer disk drives and in robots. The important selection criteria for a stepper motor are the number of steps per revolution, the speed, and the torque. Stepper motors are particularly useful for solar trackers because of their need for precise speed control and positioning accuracy. For solar positioning systems, the torque specification is important because the motor must be able to handle any imbalance in holding the collector and must be able to hold the collector for specified wind loading effects. In addition, the type of control system to be used is important in selecting a system because of specific requirements, such as repositioning the system at dusk for the following day.

To implement a stepper motor system, you need the motor, a controller, and a voltage source. A gearbox is optional, but gearing allows the output shaft to make finer steps. The stepper motor requires only a directional signal and a number of pulses of the proper magnitude, corresponding to the number of steps the motor is to move and the direction of motion. Figure 5-27 shows a typical stepper motor and its controller. The controller provides timing and sequencing signals as well as driver circuits that supply current to the coils. The driver circuits are power switching transistors that are designed to supply a voltage a little greater than the motor's rated voltage in order to overcome inductance effects and to improve the performance of the motor. The current is then controlled by a chopper-type amplifier so that the motor sees the rated current.

FIGURE 5-27 Stepper Motor and Controller (*Source:* Photo courtesy of Sanyo Denki.)

Types of Stepper Motors

Three basic types of stepper motors include the permanent magnet motor; the variable reluctance motor; and the hybrid motor, which is a combination of the previous two. In a permanent magnet stepper motor, the magnets are located on the rotor. The advantage of this arrangement is that no brushes are required; however the motor has relatively low torque. Figure 5-28 shows a simplified cross-sectional view of the rotor and stator of a stepper motor. From this figure, you can see that the stator (stationary winding) has four poles, and the rotor has six poles (composed of three complete magnets).

When no power is applied to the motor, the residual magnetism in the rotor magnets causes the rotor to *detent* or align one set of its magnetic poles with one set of magnetic poles of the stator magnets. There are six poles on the rotor and two sets of poles on the stator. This implies that the rotor has twelve possible detent positions. When the rotor is in a detent position, it has enough magnetic force to keep the shaft from moving to the next position. This is what makes the rotor feel like it is clicking from one position to the next as you rotate the rotor by hand with no power applied. While this may appear to be stable, the motor can be rotated if a torque is exerted on it.

When power is applied, it is directed to only one of the stator pairs of windings, which causes that winding pair to become a magnet. In Figure 5-28(a), the rotor is aligned with the poles in the vertical direction because 1A and 1B are the conducting path, while 2A and

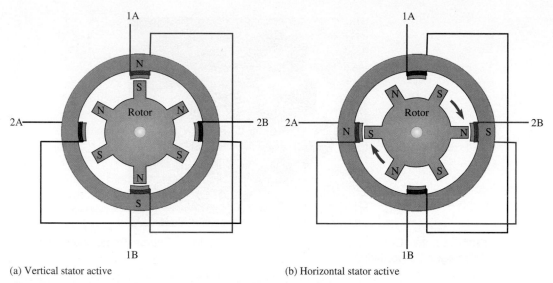

(a) Vertical stator active   (b) Horizontal stator active

FIGURE 5-28   Simplified Permanent Magnet Stepper Motor Diagrams. The motor shown in each part of the figure has permanent magnets located on the rotor.

2B are not. One of the coils for the pair becomes the north pole, and the other becomes the south pole. When this occurs, the stator coil that is the north pole attracts the closest rotor tooth that has the opposite polarity, and the stator coil that is the south pole attracts the closest rotor tooth that has the opposite polarity. When current is flowing through these poles, the rotor has a much stronger attraction to the stator winding, and the increased torque is called holding torque. **Holding torque** is the maximum torque that can be applied externally to a stopped motor without causing it to rotate to the next step position.

If you activate the other stator windings and deactivate the first set, the magnetic field is changed by 90°. However, the rotor will move only 30° before its magnetic fields again align with the change in the stator field. This is shown in Figure 5-28(b). The vertical and horizontal fields are alternately active, and the rotor moves this minimum amount (30° in this case) to align the fields again. The necessity of a complete rotation is why twelve steps are included in this example. In general, a motor with fewer steps can move faster but with lower resolution. For tracking purposes, speed is not important, but resolution and torque are.

The variable-reluctance stepper motor is similar in concept to the permanent magnet stepper motor, but it has a soft iron rotor rather than permanent magnets. The rotor is toothed, and the teeth are attracted to active stator coils, which are positioned on opposite sides of the rotor. Stepping moves the nearest tooth by a small amount to position it under the active magnet pair, minimizing the reluctance of the magnetic path. The teeth are offset on the rotor so that they will not be aligned on the nonactive coils. The number of teeth on the rotor can be quite large, resulting in higher resolution (finer) steps. The amount of torque for this type of motor is small, so it is generally used for small positioning tables and other small loads; it is not useful for higher torque applications such as those required with solar trackers. A problem with this type of motor is low holding torque. When the motor is in a stable position, wind loading or other forces can exert a constant torque on the rotor, causing it to shift position. This causes a step-error (rotor not in the assumed position). If the load torque exceeds the holding torque for the motor, it can lose its position completely.

The hybrid stepper motor is the most widely used and combines the principles of the permanent magnet and the variable reluctance motors. Figure 5-29 shows a cutaway view of a hybrid stepper motor. Four coils are located along the sides of the motor but are not seen in this view (except for the ends). The motor shown is constructed with two laminated multi-tooth rotors. A permanent magnet between each laminated stack pair creates a north pole–south pole orientation along the motor shaft axis. A standard hybrid stepper motor has 200 rotor teeth, which can rotate 1.8° per step. (Other configurations are available.) Hybrid

FIGURE 5-29   Cutaway View of a Hybrid Stepper Motor   (*Source:* Photo courtesy of Kollmorgen Corporation.)

stepper motors have good torque and can be run at higher step than other types, so they are popular for various applications.

Stator Wiring

Stator coils come in two options: unifilar and bifilar. With unifilar wiring, each stator pole has a single coil on it. The motor itself typically has four leads: two for each winding. The driver generally has four outputs that have plus and minus leads for each of two phases. For a four-wire motor, the driver leads are connected to the corresponding leads on the motor.

Bifilar windings have two identical windings, wound in opposite directions, on each stator pole. This simplifies reversing the poles and hence the direction of the motor. Bifilar motors typically have six or eight leads. You need an eight-lead motor for parallel wiring or six leads for series wiring. Series winding allows higher torque at low speed, but lower torque at higher speed. Because low-speed torque is the most important criteria for solar tracking, it is the method generally used. Figure 5-30 shows how to wire either a six-lead or an eight-lead hybrid bifilar motor in series. Notice that an external jumper is added in the eight-wire motor.

Full-Step Switching Sequence

The four-pole stepper motor described in Figure 5-28 uses a four-step switching sequence, with each of the four stator poles becoming a north pole only once and a south pole only once. This is called a full-step sequence and results in the 30° steps previously discussed. This sequence continues through the four steps and then repeats. The four steps cause the motor to rotate one step, or tooth, on the rotor each time.

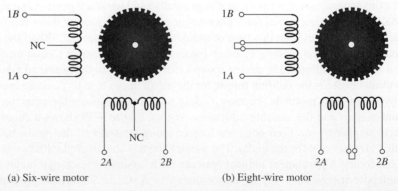

(a) Six-wire motor   (b) Eight-wire motor

FIGURE 5-30   Two Types of Stepper Motor Series Wiring for a Hybrid Stepping Motor

Half-Step Switching Sequence

Another switching sequence for stepper motors is called a half-step sequence. The main feature of this sequence is that you can double the resolution by causing the rotor to move half the distance that it does with the full-step sequence. A standard hybrid motor moves 400 steps instead of 200 steps per revolution. Half-steps can smooth the output motion, but the motor has slightly less torque than it does with full-step mode.

The half-step switching sequence requires a special stepper motor controller. The way the controller gets the motor to reach the half-step is to energize both stator coils at the same time with equal current so that the rotor comes to a stop between the stator poles. This is done every other step so that half steps are inserted into the normal sequence.

Microsteps

The resolution of a stepping motor is limited by the number of discrete positions the motor can assume. In most applications, a step size of one or two degrees is satisfactory. For a few applications, such as printers, a finer step size is desirable. The number of steps can be increased even more by manipulating the current that the controller sends to the motor during each step. The current can be adjusted so that it looks sinusoidal but with discrete current steps, as shown in Figure 5-31. From this figure, you can see that the current in the A and B windings is timed so that there is a constant 90° phase shift between them. The fact that the current to each individual phase increases and decreases in a sinusoidal pattern that is always out of phase with the other current allows the rotor to reach hundreds of intermediate steps. In fact, it is possible for the controller to have 100 or more microsteps for a full-step sequence, which provides more than 20,000 steps for each revolution.

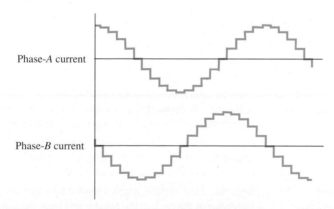

Phase-A current

Phase-B current

FIGURE 5-31   Phase Current Steps for Generating Microsteps

SECTION 5-5 CHECKUP

1. What are three types of stepper motors?

2. What is the disadvantage of a variable reluctance stepper motor as a solar tracking motor?

3. How is half-stepping accomplished on a stepper motor?

4. What is the difference between a unifilar-wound motor and a bifilar-wound motor?

5. What are microsteps?

CHAPTER SUMMARY

- Points on the earth's surface are described by latitude and longitude lines, which form great circles.

- The prime meridian is a great circle of longitude that passes through both poles and Greenwich, England.

- The zenith is a point directly overhead of an observer.

- The ecliptic is a line that describes the sun's path for one year.

- The earth is tilted on its axis by 23.37°, which accounts for north–south variation in the sun's position in the sky over a year.

- Single-axis solar trackers follow the sun's east–west motion in the sky.

- Dual-axis solar trackers can follow the sun's path exactly.

- Imaging collectors, such as Stirling engines, require dual-axis trackers.

- A cost–benefit analysis should be done before implementing solar tracking. Costs and benefits depend on the site, the type of collector, and other factors.

- In general, single-axis tracking can improve the total energy collected on a flat panel from 20% to 40%, depending on the season and other factors.

- Active solar trackers usually use dc or stepper motors to move the collector, but in some cases, hydraulic systems are used.

- Passive trackers can move the collector so that it tracks the sun without requiring electrical energy for tracking.

- DC motors are basic and reliable motors that have high starting torque, which is useful for solar tracking.

- Lower-power dc motors are typically permanent magnet (PMDC) types. They have a single electrical circuit and good torque.

- Torque for a dc motor is proportional to the rotor current and the magnetic flux.

- As a dc motor increases its speed, the counter emf (C_{emf}) increases, reducing the torque.

- In a motor, torque and power are related by the equation $T_q = \dfrac{P}{\omega}$.

- In a wound dc motor, the direction of rotation can be changed by reversing the direction of either the armature current or the rotor current, but not both.

- Wound dc motors are subdivided into series-wound, shunt-wound, and compound motors. The choice of which motor is best depends on the load requirements.

- Stepper motors move in discrete steps in response to a pulsed signal.

- Three types of stepper motors are the permanent magnet motor, the variable reluctance motor, and the hybrid motor.

- The maximum torque that can be applied to the motor without stepping is called the holding torque.

- Stepper motors have either unifilar or bifilar windings. Unifilar means there is one coil on each pole. Bifilar means there are two windings on each pole.

- With special controllers, stepper motors can have half-steps or microsteps.

KEY TERMS

active tracker   A type of solar tracker that uses external power to move throughout the day from east to west.

altitude angle   The angle formed between a horizontal plane and an imaginary line pointed to the sun.

analemma   The figure 8 pattern the sun makes in its annual motion if its position is observed daily at the same location and time.

azimuth angle   A horizontal angular distance from a reference direction, usually north.

brushes   In motors or generators, the electrical conductors that provide a path for current from a stationary part to a moving part.

celestial equator   The projection of the earth's equator in the sky. The celestial equator traces a line that is at an angle above the horizon equal to $90° - L$, where L is the latitude.

commutator   A ring structure with independent conductors that is designed to switch the current in a rotor coil as the structure rotates.

ecliptic   The sun's path in the sky with reference to an observer on earth.

end bells   The end plates of a motor or generator that mount to the frame and include the bearings.

heliostat   A device that uses a movable mirror mounted so that it reflects the sunlight to a fixed target (such as a tower collector).

holding torque   With reference to a stepper motor, the maximum torque that can be applied externally to a stopped motor without causing it to rotate to the next step position.

latitude   The angle in degrees formed by a line that extends from the center of the earth to the equator and from the center of the earth to a given point on the globe.

longitude   Reference circles that are perpendicular to the equator and converge at the poles. Longitude increases in the positive direction in the easterly direction.

meridian   A line of longitude that passes through both poles and a point straight over the observer's head.

passive tracker   A solar tracking device that does not require external power to turn the axis.

prime meridian   A great circle of longitude that passes through both poles and Greenwich, England.

rotor   The rotating part of a motor or generator.

stator   The stationary part of a motor or generator.

stepper motor   A motor that moves a rotor in discrete steps in response to a pulse signal.

synodic day   The time for the earth to make one revolution on its axis. It is almost exactly 24 hours.

zenith   A point directly overhead an observer.

FORMULAS

Equation 5-1	$T_q = K\phi I_a$	Torque for a dc motor
Equation 5-2	$V_S = C_{emf} + I_a R_a$	Applied voltage is back emf plus armature drop
Equation 5-3	$T_q = \dfrac{P}{\omega}$	Torque

CHAPTER TRUE/FALSE QUIZ

Determine whether each statement is true or false. Answers are at the end of the chapter.

1. The angle formed by a line that extends from the center of the earth to the equator and from the center of the earth to a given point on the globe is called the longitude.

2. The first day of summer and the first day of winter is when the sun crosses the ecliptic.

3. A synodic day is almost exactly 24 h.

4. The earth's axis is inclined from the ecliptic by 23.37°.

5. Longitude lines are parallel to the equator.

6. A heliostat should point directly at the sun.

7. It is important to balance the costs of tracking against the benefits.

8. Most of the benefit of tracking with a flat-plate collector is in the middle of the day.

9. Passive trackers require batteries to operate.

10. A single-axis tracker normally tracks just the east–west motion of the sun.

11. An equatorial tracker is aligned so that its axis is parallel to the earth's axis.

12. Stirling engines do not need to track to receive maximum energy.

13. A dual-axis tracker moves the collector continuously during daylight hours.

14. A series dc motor has more torque as it speeds up.

15. The purpose of the commutator is to switch the current in the rotor at just the right time.

16. Permanent magnet dc (PMDC) motors generally have more power than motors that are wound.

17. The counter emf in a dc motor is proportional to the rotor speed.

18. Motors develop maximum torque at the highest rated speed.

19. Stepper motors require a controller.

20. The variable reluctance stepper motor has higher torque than the permanent magnet motor.

CHAPTER MULTIPLE-CHOICE QUIZ

Complete each statement by selecting the one correct answer. Answers are at the end of the chapter.

1. The sun's path in the sky is called the
   a. zenith
   b. celestial equator
   c. prime meridian
   d. ecliptic

2. The factor that is most responsible for the four seasons on earth is the earth's
   a. speed in space
   b. inclination
   c. distance from the sun
   d. distance from the moon

3. At the summer solstice in the northern hemisphere, the sun is at
   a. its furthest point north
   b. its furthest point south
   c. its closest point to earth
   d. the zenith at noon at the equator

4. A great circle drawn through an observer's zenith is called a
   a. latitude line
   b. longitude line
   c. meridian line
   d. horizon line

5. The figure 8 shape you obtain if you photographed the sun at the same time each week for a year is called the
   a. ecliptic
   b. analemma
   c. celestial equator
   d. synodic path

6. A tracking system generally must
   a. be mounted on a roof
   b. have ac power available
   c. be capable of withstanding wind loads
   d. all of these

7. A consideration for deciding if a tracking system is cost effective is
   a. the site
   b. potential shading problems
   c. available power
   d. all of these

8. A alt-azimuth tracker moves the collector so that it tracks the
   a. daily motion of the sun
   b. seasonal motion of the sun
   c. both the daily and seasonal motion of the sun
   d. none of these

9. The optimum tilt angle for a flat-panel collector depends on
   a. the latitude
   b. the season
   c. both the latitude and the season
   d. none of these

10. A flat-panel collector always receives the maximum energy if the normal to the surface is pointed toward the
    a. sun
    b. zenith
    c. equator
    d. meridian

11. The two main parts of a dc motor are the
    a. rotor and commutator
    b. brushes and commutator
    c. rotor and stator
    d. stator and brushes

12. On any motor, the gap is very small to
    a. lower the current in the stator
    b. keep the reluctance small
    c. reduce wear on the brushes
    d. all of these

13. The power from a dc motor is directly proportional to
    a. the gap between the rotor and stator
    b. rotor speed
    c. both (a) and (b)
    d. none of these

14. The brushes in a dc motor are generally made from
    a. stainless steel
    b. plastic
    c. copper wire
    d. carbon composition material

15. The process of contouring the brush to the commutator is called
    a. flashing
    b. seating
    c. slipping
    d. lapping

16. The frame of a motor is generally constructed from
    a. copper to provide a good electrical path
    b. aluminum to keep weight low
    c. fiberglass to save cost
    d. steel to provide a low reluctance path

17. Back emf or counter emf causes
    a. power to increase from a motor
    b. armature current to increase
    c. torque to increase
    d. none of these

18. A variable reluctance stepper motor does not have
    a. stator coils
    b. brushes
    c. a toothed rotor
    d. a soft iron rotor

19. A 200-step hybrid stepper motor rotates
    a. 0.9° per step
    b. 1.8° per step
    c. 3.6° per step
    d. none of these

20. Half-stepping a stepper motor
    a. increases its resolution
    b. increases the torque
    c. both (a) and (b)
    d. none of these

CHAPTER QUESTIONS AND PROBLEMS

1. Explain why summer in northern latitudes corresponds to winter in southern latitudes.

2. Look up your latitude and explain the optimum angle to set a flat-panel collector.

3. The ecliptic is high in the sky at noon in summer. Is it still high in the sky at midnight? Explain your answer.

4. How many degrees per day does the sun move along the ecliptic?

5. How far above the horizon is the sun on the first day of summer?

6. If sunlight strikes a mirror so that the angle of incidence is 40°, what is the angle of reflection?

7. Assume a flat panel is tilted 10° from the optimum. What fraction of the available energy is lost due to the tilt?

8. What time of day does tracking have the greatest impact on the harvested power for a flat panel collector? Explain.

9. Explain how the latitude affects the optimum orientation angle of a flat-plate collector.

10. What is the disadvantage to tracking with a passive tracker compared to an active tracker?

11. Explain how to choose the best angle for a flat plate on an alt-azimuth tracker and how the tracker would move.

12. Explain why a long line of trough collectors is set on a north–south line.

13. What is the advantage of PMDC motors for smaller motor requirements over wound rotor and wound stator motors?

14. Explain why the torque of a dc motor goes down as the motor speeds up.

15. Assume a dc motor develops 1.78 n-m of torque at 800 rpm. How much torque will it develop at 1,200 rpm?

16. At 1,200 rpm, what is the hp rating of the motor in problem 15?

17. Why is the rotor core of a dc motor laminated?

18. How do you reverse the direction of a dc series-wound or shunt-wound motor?

19. What are the functions of a stepper motor controller?

20. What are the differences between a permanent-magnet stepper motor and a variable reluctance stepper motor?

21. What is meant by the term *detent* with respect to a stepper motor?

22. If you see a stepper motor with eight leads, what is the likely wiring configuration?

FOR DISCUSSION

NASA asks you to build a two-axis solar tracking system for a solar collector on Mars. Discuss the factors that you would need to know to accomplish this task.

ANSWERS TO CHECKUPS

Section 5-1 Checkup

1. Latitude is the angle in degrees formed by a line that extends from the center of the earth to the equator and from the center of the earth to a given point on the globe. Positive angles are measured north of the equator; negative angles are measured south of the equator.

2. Longitude is the angle formed by a line drawn from the center of the earth to the prime meridian and a line drawn from the center of the earth to the local meridian. Positive angles are measured east of the prime meridian; negative angles are measured west of the prime meridian.

3. The earth's path around the sun in one year causes the stars to cross the meridian one more time than the sun crosses the meridian. The difference is that the stars appear to move faster than the sun by about 4 minutes per day.

4. The sun will be 23.37° above the horizon because of the earth's tilt and appears to move in a great circle path, never setting.

Section 5-2 Checkup

1. Each heliostat in a group has a different angle between the sun and the target, and the angle changes throughout the day.

2. (1) Tracking systems require maintenance, so the location (rooftop, hillside, etc.) of the system needs to be considered. (2) They require more space than a nontracked system to ensure one collector does not cast a shadow on another. This may affect land cost or limit the size of the collector array.

3. A roof mount is much less expensive for a PV system because the array is out of the way and generally can be installed easily to be angled toward the sun. Using a tracking system negates this advantage; it adds maintenance cost and may make the building unsightly.

4. When space is at a premium, the combination of a more efficient panel and tracking can be cost effective because it can increase the total energy collected.

5. Tracking is most effective during the morning and evening hours because the collectors can be directed toward the sun at those times (see Figure 5-10).

Section 5-3 Checkup

1. (a) Set it to 35.5° unless optimization for a particular season is desired. (b) You can optimize the energy collected by making seasonal adjustments. Set the latitude angle for spring and fall. Set 15° lower in summer and 15° higher in winter.

2. The basic tilt angle would be set for 19° facing north.

3. A dual axis-tracker always collects more energy than a single-axis tracker, but for a flat plate, the additional energy collected may not outweigh the extra expense.

4. The Stirling engine is an example of an imaging-type collector that requires high temperatures to be efficient. This is best accomplished with dual-axis tracking.

Section 5-4 Checkup

1. The rotor resistance is low and the starting current is limited only by this resistance; torque is proportional to the current.

2. An increase in load slows down the shaft, which decreases the C_{emf}. A decrease in C_{emf} causes more armature current, thus increasing the torque and compensating for the load change.

3. A small gap provides a low reluctance path for the flux.

4. Power is the rate of doing work, whereas torque is a force multiplied by a distance that tends to produce rotation.

5. The commutator in a dc motor serves as a mechanical switch to reverse the current in the armature just as the unlike poles are near each other, continuing the rotation of the rotor.

6. The interpole coils produce a flux that tends to cancel armature reaction, which helps prevent arcing. The armature can draw heavier currents and carry larger shaft loads.

Section 5-5 Checkup

1. (1) Permanent magnet motor, (2) variable reluctance motor, and (3) hybrid motor

2. Low holding torque

3. A controller energizes two stator coils at the same time with equal current so that the rotor comes to a stop between the stator poles. This is done every other step, so than half steps are inserted between the normal sequence.

4. A unifilar motor has one winding on each stator pole; a bifilar motor has two identical windings on each stator pole.

5. Very small steps that are generated by a special controller and represent small differences in current in the windings, which have stepped sinusoidal patterns that are phase shifted from each other

ANSWERS TO TRUE/FALSE QUIZ

1. F   2. F   3. T   4. T   5. F   6. F   7. T   8. F   9. F   15. T   16. F   17. T   18. F   19. T   20. F
10. T   11. T   12. F   13. T   14. F

ANSWERS TO MULTIPLE-CHOICE QUESTIONS

1. d   2. b   3. a   4. c   5. b   6. c   7. d   8. a   9. c   17. d   18. b   19. b   20. a
10. a   11. c   12. b   13. b   14. d   15. b   16. d

Charge Controllers and Inverters

CHAPTER OBJECTIVES

- Discuss battery trickle chargers and float chargers.
- Describe the three stages of float charging.
- Describe the PWM and MPPT charge controllers.
- Explain the MPPT process.
- Determine the rating required for a charge controller from given specifications.
- Discuss the basic inverter and types of output waveforms.
- Calculate the power factor given reactive and resistive power and explain how to correct it.
- Explain synchronization.
- Explain anti-islanding in relation to inverters.
- Discuss inverter specifications.

KEY TERMS

Key terms are shown in bold and color. Definitions for key terms are provided at the end of the chapter and in the end-of-book glossary. Bold terms in black are defined in the end-of-book glossary only.

- float voltage
- bulk stage
- absorption stage
- float stage
- pulse width modulation (PWM)
- equalization stage
- maximum power point tracking (MPPT)
- perturb and observe (P&O) algorithm
- harmonic
- three-phase
- power factor
- synchronization
- transfer switch
- islanding
- anti-islanding

INTRODUCTION

In this chapter, the operations of the charge controller and the inverter are discussed. Both are key components in solar power systems, as you learned in Chapter 4. Inverters are used in any residential or commercial system that has an ac output. They are also used in wind power systems, as you will learn in Chapters 7 and 8. The charge controller is used when battery backup is part of the system, and the inverter is used to convert the dc output of a solar module, wind turbine, or batteries to ac power.

6-1   Battery Chargers

The purpose of a battery charger is to charge a battery without overcharging it. The simplest type of charge controller for renewable energy systems monitors the battery voltage and turns the charging current off or reduces it when the battery voltage exceeds a specified level. It turns the charging current back on when battery voltage falls below another specified level; the charge controller turns the circuit on (closed) to resume the charging.

Trickle Charger

The simplest type of battery charger is the continuous trickle charger, which charges the battery at its self-discharge rate by applying a constant voltage and current, regardless of whether the battery is fully charged. Because a simple trickle charger must be turned off manually after a period of time to prevent overcharging the battery, it is not generally used in renewable energy systems, although it could be used in a small home system. The charging current can be preset to the trickle charge requirement for the particular type of battery, which is usually some percentage of the battery's rating. A portion of the total current from the source is diverted through the shunt control, and the rest of the current trickle charges the battery. The current through the shunt control is set at a value that establishes the desired charging current to the battery. This charger provides the same current to the battery regardless of the charge state of the battery. This causes the battery to overcharge and potentially damages the battery once it is fully charged.

The block diagram in Figure 6-1 illustrates one possible configuration of a continuous trickle charger. In this configuration, the source output voltage must be compatible with the battery voltage; if is not, a regulator must be used in series with the source to regulate the module voltage down to a battery-compatible level. The diode prevents the battery from discharging back through the source should the module output drop below the battery voltage. (The diode was discussed in Chapter 2.)

Float Charger

The float charger provides a relatively constant voltage, called the **float voltage**, that is applied continuously to a battery to maintain a fully charged condition.

On/Off Switched Float Charger

In its simplest form, the float charger is a trickle charger with an automatic on/off switch (usually a thyristor or a transistor). This charger senses when the battery voltage reaches a preset reference level (V_{REF1}), which corresponds to full charge or float charge, and shuts *off* the current to the battery. When the battery discharges down to a second preset level (V_{REF2}), it turns the current to the battery back *on*. The sense resistor, R, is used to isolate the battery voltage from the output of the switch so that it can fluctuate independently of the source voltage. The voltage sensing circuit compares the battery voltage to each of the two reference voltages and turns the shunt switch on or off accordingly. This setup overcomes the problem

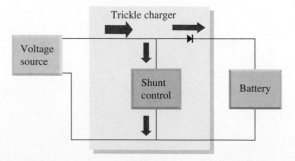

FIGURE 6-1   Continuous Trickle Battery Charger

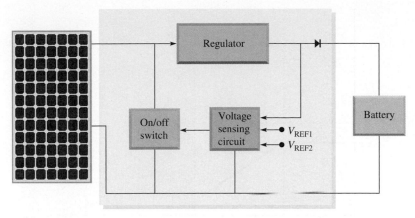

FIGURE 6-2 Switched Shunt Float Charger

with the continuous trickle charge of having to turn it off manually. It is not an ideal way to charge a battery, but it is better than continuous trickle charging that can overcharge the battery. The block diagram in Figure 6-2 illustrates one possible configuration of a float charger.

In another type of float charger, the electronic on/off switch is in series with the source and load. A regulator is used to set the current and voltage. Figure 6-3 shows the concept of a series float charger.

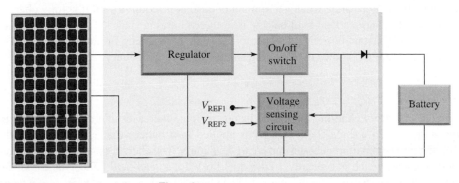

FIGURE 6-3 Switched Series Float Charger

The graph in Figure 6-4 illustrates the idea of series switched float charging. The noncharging portions of the time scale are compressed in order to show more than one on/off cycle. The time during which the battery is not charging (discharge) is typically long compared to the time the battery is charging.

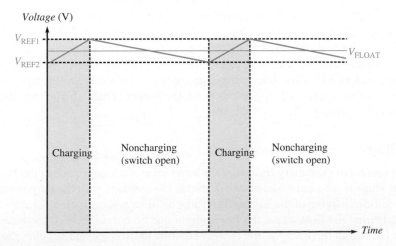

FIGURE 6-4 Typical Series Switched Float Charging Curve

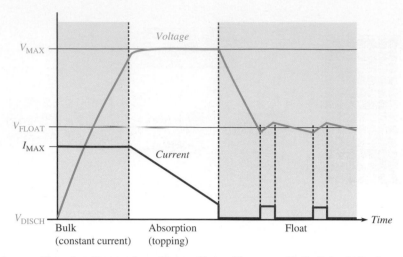

FIGURE 6-5   Charging Curves for a Three-Stage Charger with Switched Float

Three-Stage Float Charger

A characteristic of lead-acid batteries is that you charge them by applying a constant voltage and let the battery draw whatever amount of current it requires until it is fully charged. A good way to charge a lead-acid battery is in three charging stages. These stages are (1) bulk stage (or constant current), (2) absorption stage (or topping or acceptance), and (3) float stage. Figure 6-5 shows the charging stages for a typical battery.

Bulk Stage

The stage of battery charging where the battery voltage increases at a constant rate is the **bulk stage**. When a three-stage charger is applied to a battery that is significantly discharged, there is maximum charge current to the battery. The charger is set to the maximum battery voltage, which is typically 14.4 V to 14.6 V for a lead-acid battery at 25° C. The battery voltage starts from the discharged level (V_{DISCH}) and increases to V_{MAX} at a nearly constant rate during this bulk stage, as shown in Figure 6-5.

Absorption Stage

The stage of battery charging after the battery voltage reaches the maximum and the current through the battery begins to decrease is the **absorption stage**. When the battery voltage is 75% to 80% full, the current through the battery begins to decrease, marking the beginning of the absorption stage. During this stage, the voltage is held at the maximum value while the current decreases. The decrease in current is not limited by the charger but by how much the battery can absorb; getting the correct rate for absorption is important for the longest battery life. This absorption stage continues until current through the battery decreases to a few percent of I_{MAX}. At this point, the battery is fully charged and the battery current is much smaller.

Float Stage

After the current to the battery reaches some lower level, the charger enters the float stage, The **float stage** is the final maintenance or trickle charge stage with the purpose of offsetting any self-discharging of the battery. Typically, the float voltage is from 13.2 V to 13.8 V at 25° C. During the float stage, the float current can be pulsed to keep the battery fully charged. During all three stages, the charger controls the voltage and supplies the current that is required to optimize battery life.

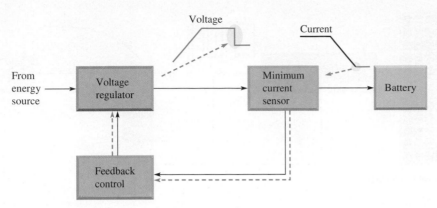

FIGURE 6-6 Function of a Three-Stage Charger

The block diagram in Figure 6-6 illustrates this process for charging the battery. The minimum current sensor issues a signal to the feedback control when the battery becomes fully charged. The feedback control then causes the regulator to decrease its output voltage to the float level.

SECTION 6-1 CHECKUP

1. What is a disadvantage of a trickle charger compared to a float charger?
2. What does *float charging* mean?
3. What are the stages of a three-stage charge controller?
4. What happens during the bulk stage?
5. What happens during the absorption stage?

6-2   Pulse Width Modulation Charge Controller

The pulse width modulation (PWM) charge controller keeps the backup batteries in a solar or wind system charged by switching the energy from the source on and off several times per second. This keeps the battery voltage more constant than continuous charging methods. The PWM controller is generally more efficient than the chargers previously discussed and also operates with the three stages of charge: bulk, absorption, and float. Another stage is sometimes incorporated called equalization.

The pulse width modulation (PWM) charge controller uses pulse width modulation to decrease the power applied to the batteries gradually as they approach full charge. **Pulse width modulation (PWM)** is a process in which a signal is converted to a series of pulses with widths that vary proportionally to the signal amplitude. PWM results in less stress on the batteries and extends the battery life. The controller continuously checks the batteries to determine the rate at which pulses should be applied and the width of the pulses. If the batteries are fully charged with no load, very short duration (width) pulses are applied every few seconds to maintain the batteries in float. If the batteries are discharged, the pulse duration is very long and almost continuous or, depending on the amount of discharge, a constant charge is applied. The controller checks the state of the batteries between each pulse and automatically adjusts the pulse width each time.

Figure 6-7 shows the basic concept of a PWM charge controller. In Figure 6-7(a), the PWM and control circuit produce pulses based on the input from the sampling circuit. The

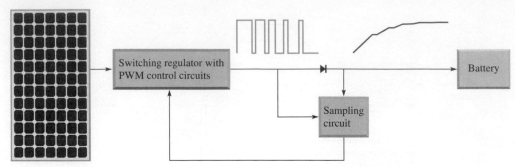

(a) Functional block diagram

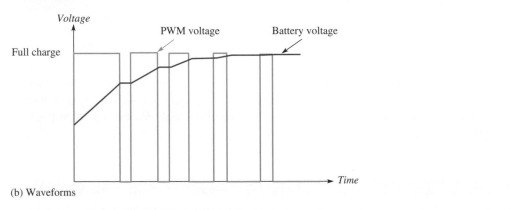

(b) Waveforms

FIGURE 6-7   Three-Stage PWM Charge Controller Waveforms

sampling circuit determines the actual battery voltage by sampling the voltage between pulses. The diode acts as a rectifier to smooth the pulses and also blocks discharge of the battery back through the charger when the source is not generating power. Figure 6-7(b) demonstrates how the battery charges during each pulse and how the width and the time between pulse change as the battery charges.

PWM charge controllers are typically available from about 4 A up to 60 A or more. Figure 6-8 shows typical PWM charge controllers. Higher current rated charge controllers use heat sinks to dissipate excess heat. The black fins in Figure 6-8 are the heat sink.

Equalization

Some PWM charge controllers have an equalization stage in addition to the bulk, absorption, and float stages. These charge controllers are often referred to as four-stage chargers. The **equalization stage** is a charging process where the battery is overcharged

(a)   (b)

FIGURE 6-8   Typical PWM Charge Controllers. The black fins are the heat sink, which is necessary on high current controllers.   (*Source:* Courtesy of Morningstar Corporation®, copyright 2011.)

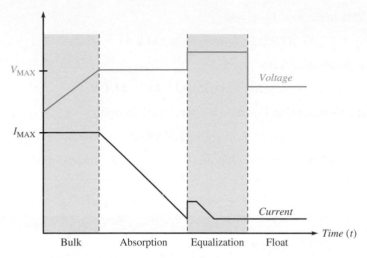

FIGURE 6-9 Ideal Charging Curves, with Equalization

at approximately 1 V above float in order to equalize the charge on all the cells in the battery or battery bank. Some cells may have a charge lower than others, so equalization brings them all to full charge. Excessive overcharging of a battery produces destructive outgassing and water loss. At the float stage, however, the battery can become stratified. Battery stratification occurs when the acid in the electrolyte begins to concentrate at the bottom because it is heavier than the water in the electrolyte. The higher concentration of acid at the bottom causes sulfation (buildup of lead sulfate), which reduces the battery storage capacity and battery life. A controlled overcharge applied periodically prevents stratification by bubbling the electrolyte without overheating. Figure 6-9 shows the ideal charging curves for a battery, with equalization included.

Sizing a Charge Controller

Charge controllers are specified by both current and voltage. A charge controller must match the rated voltages of the solar array and battery bank (usually 12 V, 24 V, or 48 V). Also the charge controller must have sufficient capacity to handle the current from the solar array. A basic formula for determining the minimum current rating is:

$$I_{CH} = 1.56 I_{SC}$$

Equation 6-1

where

I_{CH} is the current rating of the charge controller

I_{SC} is the short-circuit current of the solar array

The derating factor of 1.56 is a result of increasing the charge controller capacity by 25%, which is the same as multiplying by 1.25. A second factor of another 25% (1.25) as specified by the National Electric Code (NEC) for systems in continuous operation (three hours or more) is applied. Multiplying 1.25 by 1.25 gives a factor of approximately 1.56.

EXAMPLE 6-1

A certain solar array consists of six 39.8 V, 215 W modules connected in parallel. The I_{SC} for each module is 5.8 A. Determine the current rating of the charge controller for this application.

Solution

The total I_{SC} for the array is:

$$I_{SC(TOT)} = 6I_{SC} = 6(5.8 \text{ A}) = 34.8 \text{ A}$$

Adding 25% to the total I_{SC} gives:

$$(1.25) I_{SC(TOT)} = (1.25)(34.8 \text{ A}) = 43.5 \text{ A}$$

Applying the derating factor gives:

$$I_{CH} = (1.25)(43.5 \text{ A}) = \textbf{54.4 A}$$

The shorter method using Equation 6-1 gives the following:

$$I_{CH} = 1.56 I_{SC} = (1.56)(40 \text{ A}) = \textbf{54.3 A}$$

The slight difference in this result is due to 1.56 being a rounded value.

SECTION 6-2 CHECKUP

1. What does PWM stand for?

2. What is the purpose of the sampling circuit in Figure 6-5?

3. What is equalization?

4. What two factors are used in sizing a charge controller?

5. How is the factor 1.56 in Equation 6-1 determined?

6-3   Maximum Power Point Tracking Charge Controller

Maximum power point tracking (MPPT) charge controllers eliminate much of the energy loss found in the other types of controllers and produce efficiencies up to 30% over non-MPPT controllers. They are the most widely used type of charge controller, especially in larger systems. The MPPT tracks the voltage and current from the solar module to determine when the maximum power occurs in order to extract the maximum power. The MPPT then adjusts the voltage to the battery to optimize the charging. This results in a maximum power transfer from the solar module to the battery. MPPT charge controllers normally use PWM in their operation.

Maximum power point tracking is the process for tracking the voltage and current from a solar module to determine when the maximum power occurs in order to extract the maximum power. In Figure 6-10, the blue curve is the current-voltage characteristic for a certain solar panel under a specified condition of incident light. The red curve is the power showing where the peak occurs, which is in the knee of the *I-V* curve (blue dot) at I_{MMP} and V_{MMP}. If the incident light decreases, the curves shift down. (These curves were introduced in Chapter 3.)

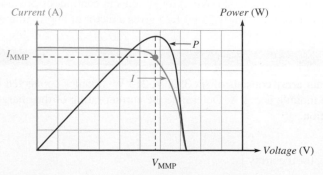

FIGURE 6-10   Solar Module *I-V* and Power Curves

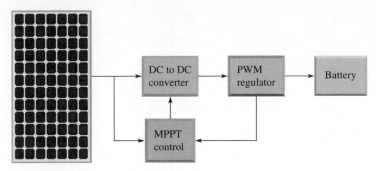

FIGURE 6-11   MPPT Charge Controller

The maximum power point tracking charge controller incorporates PWM and a dc to dc converter. A simplified block diagram of the functional concept is shown in Figure 6-11. The MPPT can be implemented in several ways, so the figure illustrates only the basic functions. The purpose of the dc to dc converter is to isolate the dc input from the dc output so the output can be adjusted for maximum power. The MPPT control typically employs a microprocessor.

The output current of a solar module varies directly with the amount of light (irradiance) as shown in Figure 6-12(a). The maximum power that can be delivered will be greater at a higher irradiance, by reducing the load and maintaining the voltage at a constant level. For changing temperature, the output voltage changes inversely, while the current remains relatively constant, as illustrated in Figure 6-12(b). Notice how the maximum power point (MMP) (blue dot) changes in both cases. The job of MPPT is to keep the operating point of the solar module at the maximum power point as the *I-V* curves change with changes in light or temperature.

MPPT operates using an algorithm, which is basically a series of steps or procedures that is used to accomplish a desired result. Various algorithms are used in MPPT, but we will focus on just one called the perturb and observe algorithm. The **perturb and observe (P&O) algorithm** is a procedure in which a variable is changed (perturbed) and the effect of the change on another variable is monitored (observed). (P&O is also known as the hill-climbing method.) The graphs in Figure 6-13 illustrate points along the *I-V* curve and the power curve that could be used in applying the P&O algorithm.

One method for moving the solar module output voltage along the *I-V* curve is to vary the load on the module incrementally until the maximum power point (MPP) is located. An *I-V* curve is shown in Figure 6-13(a). The MPPT starts by setting a load value and measuring the module output voltage V_1 and the current. The power (P_1) at V_1 is calculated. Next, the load is increased and P_2 is calculated for V_2. P_2 is compared with P_1 and, because P_2 is greater than ($>$) P_1, we are still on the uphill side of the MPP on the power curve shown in Figure 6-13(b). The MPPT then moves the module output voltage to V_3 and calculates P_3. P_3 is greater than P_2, so we are still climbing the power curve toward the MPP. The next

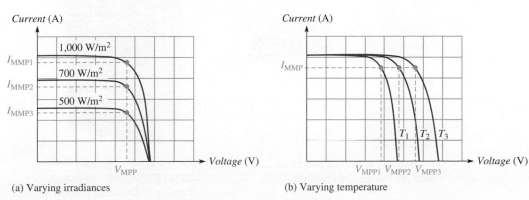

(a) Varying irradiances   (b) Varying temperature

FIGURE 6-12   MMPs for Varying Irradiance and Temperature

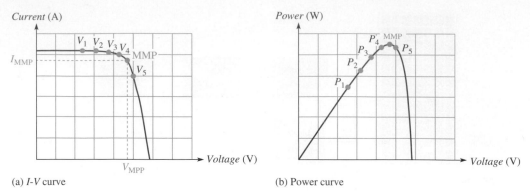

(a) *I-V* curve

(b) Power curve

FIGURE 6-13   P&O Algorithm for MPPT

measurement at V_4 shows that P_4 is greater than P_3. Finally, the measurement at V_5 shows that P_5 is equal to or less than ($\leq$) P_4, which indicates that the MPP has been passed and we are on the downhill side of the power curve. At this point, the MPPT reverses back to V_4 and then goes back and forth (oscillates) between V_4 and V_5 on either side of the MPP. The accuracy of this approach depends on the size of the voltage increments between each measurement. Smaller increments result in more accuracy and give a result closer to the actual MPP. The flow chart in Figure 6-14 further illustrates the simplified P&O algorithm. Inverters come in many sizes and shapes. A typical MPPT charge controller is shown in Figure 6-15.

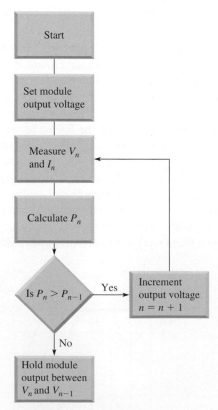

FIGURE 6-14   Simplified Flow Chart for the P&O Algorithm

FIGURE 6-15   Typical MPPT Charge Controller.   (*Source: Courtesy of Outback Power, Arlington, WA.*)

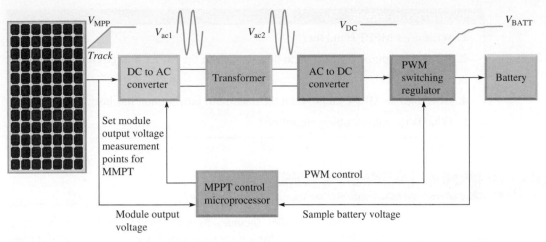

FIGURE 6-16 Operation of an MPPT Charge Controller

Operation of an MPPT Charge Controller

Figure 6-16 is a block diagram of an MPPT charge controller. First, the MPPT microprocessor tracks and sets the solar module output at the maximum power point. The dc to dc converter consists of the dc to ac converter, the transformer, and the ac to dc converter. The purpose of these blocks is to convert the V_{MMP} to ac voltage and transformer-couple the ac voltage to the ac to dc converter, where the ac is converted back to a dc voltage. As you know, a transformer is an electromagnetic device that works only with ac and isolates its input electrically from its output. The reason for the isolation is to allow the output dc voltage to be controlled independently of the voltage from the solar module. The transformer can also step the ac voltage up or down, depending on what is required by the system. The MPPT microprocessor then adjusts the PWM switching regulator to produce the proper voltage required by the battery.

Energy Diversion

In renewable energy systems that use electrical generators, such as wind and small-stream hydroelectric systems, the generator could be damaged if the load was suddenly removed. Without a load, the generator could exceed its speed rating (rpm). A diversion feature is available on many charge controllers that connects a dummy load to the generator when the charger is disconnected during the PWM on/off process. This keeps a constant load on the generator and prevents a rapid buildup in speed that occurs when there is no load. The diversion feature is also used in solar power charge controllers to bleed off and utilize excess energy that the batteries do not need once they reach the float stage. The dummy load is typically a resistive element, similar to those found in water heaters, so that the diverted energy is utilized. Figure 6-17 illustrates the basic idea.

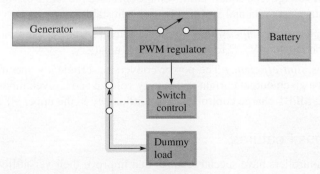

FIGURE 6-17 Diversion in a Charge Controller

6-4   Charge Controller Specifications and Data Sheet

Features of a typical charge controller includes high efficiency; multistage charging, including equalization; PWM; MPPT; display for monitoring performance; and data logging. Typical specifications and a data sheet are presented in this section.

Specifications

Specifications provide the values of operating parameters for a given charge controller. Common specifications are discussed below.

> *Output current* Output current gives the maximum current provided at the output. Some charge controllers have an adjustment feature to limit the output current to a lower value.
>
> *Maximum solar array size* This specification provides the maximum power from a solar array for a given voltage that the charge controller can handle. For example, a 12 V array may be specified for a maximum output power of 800 W, or a 24 V system may be specified for 1,600 W.
>
> *Nominal battery voltage* Most charge controllers are specified for a specific battery bank voltage, such as 12 V, 24 V, 32 V, 36 V, 48 V, 54 V, or 60 V. Some controllers can be programmed for any of these voltages.
>
> *PV open circuit voltage (VOC)* The VOC specifies the maximum voltage that a solar array can produce. Usually the open-circuit voltage is given for the absolute coldest temperature and for start-up and operation at a nominal temperature.
>
> *Standby power consumption* Standby power consumption is the amount of power required by the controller when it is not charging a battery.
>
> *Charging regulation* This regulation specifies the number and type of charging stages for the controller, such as bulk, absorption, float, and equalization.
>
> *Voltage regulation set points* The voltage regulation set points specify the range of voltages for which the output can be regulated by the user.
>
> *Equalization voltage* The equalization voltage specifies the range of voltages above the absorption stage set point and is typically user-adjustable.
>
> *Battery temperature compensation* The battery temperature compensation specifies a compensation rate in mV/°C for each 2 V cell in a battery.
>
> *Power conversion efficiency* The power conversion efficiency specifies the percent efficiency at a given output current and battery voltage and a given input voltage. Typically, a good MPPT charge controller has an efficiency in the upper 90% range.

Miscellaneous Features

Many charge controllers have special features that improve their versatility, in addition to PWM and MPPT.

Voltage step-down This feature allows the controller to charge a lower voltage battery from a higher-voltage PV array.

Status display This screen shows the operational status of the controller, such as stage of operation, voltage, current, and power.

Remote interface This standard interface allows data to be sent to a remote site.

Data logging This feature stores data such as ampere-hours, kilowatt-hours, and float time for each day in a specified period of time.

A data sheet for a typical charge controller is shown in Figure 6-18.

CHARGE CONTROLLER SPECIFICATIONS

Quantity		Typical Specifications
Nominal Battery Voltages		12, 24, 32, 36, 48, 54 or 60 VDC (Single model - selectable via field programming)
Output Current		60 amps maximum with adjustable current limit for smaller systems
Maximum Solar Array Size		12 VDC systems 800 Watts / 24 VDC systems 1600 Watts / 48 VDC systems 3200 Watts
PV Open Circuit Voltage (VOC)		150 VDC absolute maximum coldest conditions / 140 VDC start-up and operating maximum
Standby Power Consumption		Less than 1 Watt
Charging Regulation		Five Stages: Bulk, Absorption, Float, Silent and Equalization
Voltage Regulation Set Points		10 to 80 VDC user adjustable with password protection
Equalization Voltage		Up to 5.0 VDC above Absorb Set Point Adjustable Timer - Automatic Termination when completed
Battery Temperature Compensation		Automatic with optional RTS installed / 5.0 mV per °C per 2V battery cell
Voltage Step-Down Capability		Can charge a lower voltage battery from a higher voltage PV array
Power Conversion Efficiency	Typical	98% at 60 amps with a 48 V battery and nominal 48 V solar array
Status Display		3.1″ (8 cm) backlit LCD screen with 4 lines with 80 alphanumeric characters total
Remote Interface		Proprietary network system using RJ 45 Modular Connectors with CAT 5e Cable (8 wires)
Data Logging		Last 64 days of operation - amp hours, watt hours and time in float for each day along with total accumulated amp hours, kW hours of production
Positive Ground Applications		Requires two pole breakers for switching both positive and negative conductors on both solar array and battery connections (HUB-4 and HUB-10 are not recommended for use in positive ground applications)
Operating Temperature Range		Minimum −40° to maximum 60°C (Power capacity of the controller is derated when above 25° C)
Environmental Rating		Indoor Type 1
Conduit Knockouts		Two $\frac{1}{2}″$ and $\frac{3}{4}″$ on the back; One $\frac{3}{4}″$ and 1″ on each side; Two $\frac{3}{4}″$ and 1″ on the bottom
Warranty		Standard 2 year / Optional 5 year
Weight	Unit	11.6 lbs (5.3 kg)
	Shipping	14 lbs (6.4 kg)
Dimensions (H × W × L)	Unit	13.5 × 5.75 × 4″ (40 × 14 × 10 cm)
	Shipping	18 × 11 × 8″ (46 × 30 × 20 cm)
Options		Remote Temperature Sensor (RTS), HUB and MATE

FIGURE 6-18 Data Sheet for a Typical Charge Controller

SECTION 6-4 CHECKUP

1. What specification identifies the maximum power from a 48 V solar array that can be handled by the charge controller?

2. Can the charge controller in the data sheet in Figure 6-18 charge a 36 V battery bank?

3. What is the maximum output current specified in the data sheet in Figure 6-18?

4. Above what temperature is the charge controller capacity derated?

6-5   Inverters

Recall that the function of an inverter is to convert dc to ac. Most of our electrical needs are for utility-compatible ac. The inverter in most systems is connected directly to either the power source or the backup storage batteries if they are used. A battery-backup inverter is one that includes a built-in charge controller. Although most inverters are for smaller systems and applications, larger ones are used in industrial and commercial operations as well as utility-scale solar farms and some wind machines.

An inverter takes the dc output voltage of the renewable energy system or backup batteries and converts it to ac. In small-scale user systems, the output is typically a standard utility voltage (120 V or 240 V ac in North America) and can be a single-phase output voltage or a three-phase voltage, depending on the system. These inverters are generally rated for less than 100 kW. Figure 6-19(a) shows some smaller inverters used in residential and smaller commercial systems. Microinverters are designed to produce ac only from a single solar panel and may even be integrated into the panel as an embedded module. Many systems today are designed around microinverters, particularly where partial shading is a problem. Microinverters allow monitoring of each panel separately for system performance and to identify any problems.

Some large industrial and commercial inverters are rated to 500 kW, with a few utility-scale inverters rated for over 500 kW. In these cases, the inverter can have an output of several hundred volts. In many newer systems, the voltage from the array is 1,000 VDC. These high-voltage systems reduce wiring costs and the number of connections, so capital cost is less and losses in cables are less during operation due to lower current.

One method for converting the dc from solar panels to ac in a large array is to use a modular approach in which multiple high-voltage inverters are synched together by a master

(a) Small inverter units

(b) Modular grid-tie 150 kW inverters for a large solar array. The entire system is rated at 4.5 megawatts

FIGURE 6-19   Inverter Units   (*Source:* Part (a), National Renewable Energy Laboratory; part (b), courtesy of Nextronex.)

controller. An advantage to this method is that inverters can be added as power increases in the middle of the day and then taken offline as sunset approaches. Thus, only the inverters needed are using power. This makes the conversion process more efficient and minimizes runtime on each inverter to extend its life. It also enables part of the system to be taken offline for maintenance or in case of a problem without taking the entire system offline. Figure 6-19(b) shows 150 kW inverters in a major 4.5 megawatt system that are controlled by a master controller to select inverters depending on the available energy. This particular system has string (series) voltages of up to 1,000 VDC.

Basically an inverter switches the dc output of the energy source on and off and processes the result to create an ac output. The manner in which switching and processing is done is different for different types of inverters, but typically insulated gate bipolar transistors (IGBT) or metal-oxide semiconductor field-effect transistors (MOSFETs) are used for switching (see the discussion in Section 2-6 of Chapter 2). These transistors made it possible to develop new power inverters that are much more efficient (some are over 97% efficient) than older analog switching inverters.

There are three basic types of inverters in terms of the type of output: sine wave, square wave, and modified sine wave as shown in Figure 6-20. The amplitudes of the modified sine wave and the square wave can be designed to have the same root-mean-square (rms) value as that of the sine wave and, as a result, each of the three waveforms can provide the same power to a load.

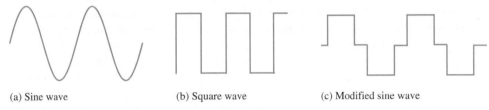

(a) Sine wave   (b) Square wave   (c) Modified sine wave

FIGURE 6-20 Three Types of Inverter Waveforms

Inverters also are available as either grid-tie or non-grid-tie. Grid-tie inverters provide the option of using power from the electrical grid or providing power to the grid. Grid-tie inverters process the switched waveform and produce a low distortion sine wave output that is compatible with the power company sine wave. Grid-tie inverters must not only produce a sine wave within acceptable limits; they must also be synchronized to the grid and include automatic disconnect switches as a safety requirement for utility workers in the event the grid goes down. These inverters use switching circuits, including pulse width modulation, to convert from dc to ac.

In addition to these features, most inverters incorporate readouts for voltage and power levels and communication ports to signal system performance characteristics to a controller or computer. In addition, some inverters are connected to the Internet and can provide a remote user with performance information and diagnostic tools and can even send an email in case of a system error. Non-grid-tie inverters may have either a nonsinusoidal wave or have a sine wave output. They are generally limited to providing power to certain types of loads and are not compatible with the utility company. The square wave inverter is the simplest and least expensive, but it is seldom used today. One drawback to square wave and modified sine wave inverters is that they tend to produce electrical noise (interference) that can be troublesome for electronic equipment.

The harmonic content of a square wave includes a fundamental sine wave at the frequency of the square wave and a series of odd harmonics. In general, **harmonics** are the frequencies contained in a composite waveform, which are integer multiples of the repetition frequency (fundamental). The fundamental is at the same frequency as the square wave, and each odd harmonic is an odd multiple of the fundamental frequency. Figure 6-21 shows a fundamental frequency and two odd harmonics that compose a square wave. (There are more odd harmonics that are not shown.) Available sine wave inverters typically have harmonic distortion less than 3%, which means that the power in harmonics is greatly reduced.

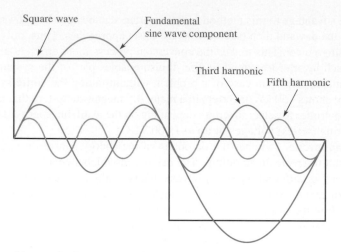

FIGURE 6-21 Harmonic Content of a Square Wave

Square Wave Inverter

A switching circuit is used in the conversion of dc voltage to an alternating (or bipolar) square wave voltage. One method is the use of the inverter bridge (also known as an H-bridge), which is illustrated in Figure 6-22. The switch symbols are used to represent switching transistors (IGBTs or MOSFETs) or other types of electronic switching devices. In Figure 6-22(a), switches S2 and S3 are on for a specified time and S1 and S4 are off. To produce a 60 Hz square wave, each pair of switches is on for half of the period, which is 8.33 ms. As you can see, the direct current is through the load, which creates a positive output voltage. In Figure 6-22(b), opposite switches are on and off. The current is in the opposite direction through the load, and the output voltage is negative. The complete on/off cycle of the switches produces an alternating square wave. The transistors are switched by a timing control circuit, which is not shown for simplicity. The harmonic distortion of a typical square wave output is in the range of 45%, which can be reduced somewhat by filtering out some of the harmonics.

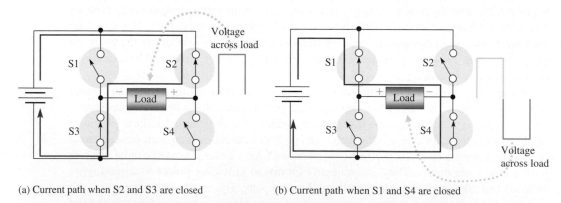

(a) Current path when S2 and S3 are closed   (b) Current path when S1 and S4 are closed

FIGURE 6-22 Inverter Bridge. The inverter bridge (H-bridge) is a method of producing a square wave from a dc voltage.

Modified Sine Wave Inverter

The operation of a basic H-bridge is enhanced to produce the misnamed modified sine wave, which is shown in Figure 6-23. (Perhaps *modified square wave* would be a better name.) The resulting wave is far from resembling a sine wave despite the name and cannot be used for some types of loads. The wave is created from a square wave and adds a short dead time between positive and negatives excursions of the square wave. Although the resulting output is definitely not a sine wave, the harmonic distortion is reduced to 24%

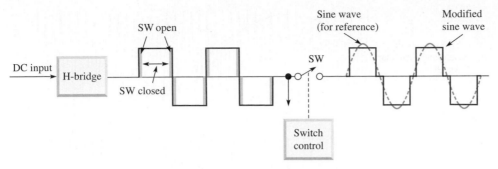

FIGURE 6-23   Concept of a Modified Sine Wave Inverter

compared to a pure square wave's 45%. This is still a significant amount of distortion. Filtering can reduce this distortion; however, filtering also significantly reduces the power to the load and thus reduces the efficiency.

Note that a so-called modified sine wave with only two levels cannot be used for driving many types of loads, including some electronic devices (some computer power supplies are an exception) and cannot be used to send to the grid. Appliances that have electronic speed controls (such as mixers), or those with timers may have problems and may be damaged from the modified sine wave. Also, anything with a motor may not run as efficiently as it would with a pure sine wave. In addition, noise may be generated due to fast transitions in the waveform. This can result in buzzing in speaker systems. As a result of these problems, the modified sine wave has limited use.

By incorporating another level in the two-level modified sine wave, a variation of the modified sine wave is produced that more closely approximates a sine wave, as shown in Figure 6-24. This is accomplished by a more complex switching circuit. The result is a reduced harmonic content that has less distortion than the modified sine wave. This waveform is sometimes referred to as a quasi–sine wave, although this term is also sometimes used to describe the two-level modified sine wave. Other switching arrangements are possible and can produce variations on this waveform.

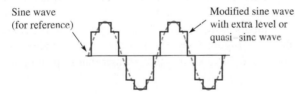

FIGURE 6-24   Three-Level Modified Sine Wave Creating a Quasi–Sine Wave

Single-Phase Sine Wave Inverter

The square wave, modified sine wave, and quasi–sine wave all have a number of harmonics, which, as you know, are sine waves with frequencies that are odd multiples of the fundamental frequency and different amplitudes. Harmonics are especially troublesome in some applications, so high-quality sine wave inverters are the most widely used type. With PWM and low-pass filtering, a nearly pure sine wave can be formed. (Pulse width modulation was introduced earlier in this chapter in relation to charge controllers.) The PWM process produces a pulse waveform with varying pulse widths proportional to the amplitude of a signal (sine wave in this case). In the inverter, a low-power reference 60 Hz sine wave and a higher-frequency triangular wave are used to produce the PWM waveform. The sine wave amplitude values are sampled by the triangular wave to produce the PWM waveform. The PWM signal is passed through a low-pass filter (equivalent to the mathematical process of integrating) and reproduces the sine wave with very little distortion, as shown in Figure 6-25. Distortion can be reduced even further by increasing the frequency of the triangular wave relative to the sine wave, but the addional switching can mean a slight loss of efficiency. (Like most electronics

Transformerless inverters are much lighter in weight due to the lack of a transformer, and they have higher efficiencies than inverters with transformers. They are used in Europe but not as much in the United States because of past NEC requirements. Until 2005, the NEC code required all solar electric systems to be negative grounded. Many electrical engineers voiced concerns about having transformerless electrical systems feed into the public utility grid because of the lack of isolation between the dc and ac circuits.

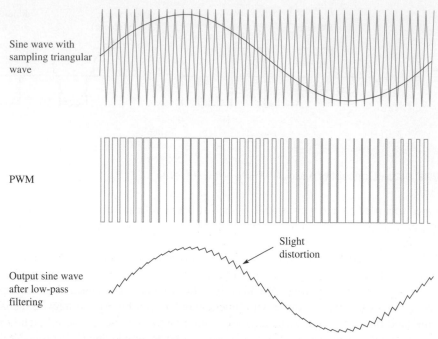

Sine wave with sampling triangular wave

PWM

Slight distortion

Output sine wave after low-pass filtering

FIGURE 6-25 Pulse Width Modulation

systems, there are tradeoffs in design.) A block diagram of a single-phase sine wave inverter is shown in Figure 6-26. Other variations are possible.

Basic Operation

The sine wave inverter uses a low-power electronic signal generator to produce a 60 Hz reference sine wave and a 60 Hz square wave, synchronized with the sine wave. The reference sine wave goes to the PWM circuit along with a triangular wave that is used to sample the sine wave values to produce a PWM control output. This PWM control signal operates an electronic switch that converts the dc input to a PWM voltage. The PWM voltage and the 60 Hz square wave are inputs to the H-bridge. The output of the H-bridge is an alternating PWM voltage, as shown in Figure 6-26. The filter converts the alternating PWM voltage to a utility-quality sine wave that goes to a transformer that steps it up for use by the load.

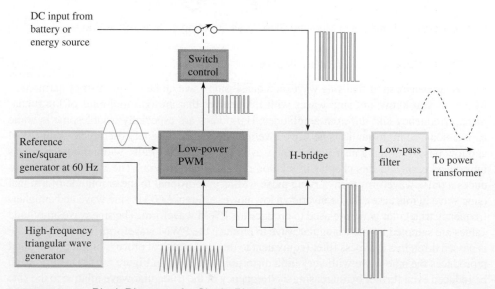

DC input from battery or energy source

Switch control

Reference sine/square generator at 60 Hz

Low-power PWM

H-bridge

Low-pass filter

To power transformer

High-frequency triangular wave generator

FIGURE 6-26 Block Diagram of a Single-Phase Sine Wave Inverter

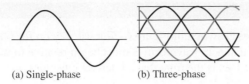

(a) Single-phase   (b) Three-phase

FIGURE 6-27 Comparison of Single-Phase Power and Three-Phase Power

Three-Phase Inverters

Inverters are available that produce a three-phase output rather than a single-phase output. Homes and small businesses generally use single-phase or split-phase power (two opposite phases). Three-phase power is used for distribution over the power lines and for customers using large motors and other high-current loads.

Until this point, our discussion has been limited to a single-phase, sinusoidal output, as shown in Figure 6-27(a). **Three-phase** power has three sine waves that are each one-third of a cycle (120°) apart, as shown in Figure 6-27(b).

Three-phase power is used for electrical distribution because it is very efficient. It is used in industry because three-phase motors and other machines run more smoothly and efficiently and last longer than they would with single phase because the power does not vary over time. With three-phase power, significantly more power can be delivered with three wires as can be delivered with single-phase using two wires (for the same size wires).

The basic difference between a single-phase inverter and a three-phase inverter is the type of inverter bridge used. Figure 6-28 illustrates the concept of a modified inverter bridge used in three-phase inverters. The dc input is processed to produce split-phase PWM voltages that are connected to a three-legged inverter bridge with six transistor switches (mechanical switches are shown for simplicity). The switch control turns each switch on sequentially for 8.33 ms, which corresponds to a half cycle of a 60 Hz signal. The time intervals are all equal; that is, $t_1 = t_2 = t_3 = t_4 = t_5 = t_6 = 8.33$ ms. For example, switches S1 and S2 are turned on and off alternately for 8.33 ms, creating alternating PWM voltage A. Beginning at a precise time during t_3, switch S3 is turned on to create a 120° phase shift between PWM voltage A and PWM voltage B. Then S3 and S4 are turned on and off alternately, thus creating alternating PWM voltage B. Next, at a precise time during t_5, switch S5 is turned on to create a 120° phase shift between PWM voltage B and PWM voltage C. Then S5 and S6 are turned on and off alternately, thus creating alternating PWM voltage C.

The PWM voltages are filtered to produce sine waves, as shown earlier. The final output is a three-phase sinusoidal voltage with 120° phase between each of the sine waves. This output is used to drive a balanced three-phase load such as a three-phase motor.

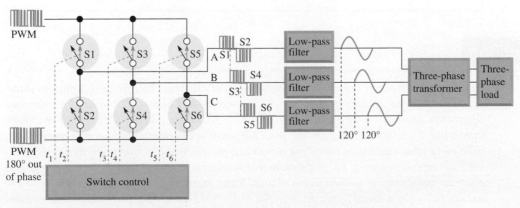

FIGURE 6-28 Using a Modified Inverter Bridge for Three-Phase Generation

EXAMPLE 6-2

Find the time that each switch in Figure 6-28 must stay on for 60 Hz operation. Determine the time (t_{S3}) that t_3 in waveform B must begin relative to waveform A to create a phase shift of 120°. This is the time that switch S3 must turn on.

Solution

The total time for a full 60 Hz cycle is $T = 1/f = 1/60$ Hz $= 16.67$ ms. Therefore, time interval t_1 is 0.5(16.67 ms) $= 8.34$ ms. Once turned on, each switch stays on for 8.34 ms. Using a ratio approach:

$$t_{S3}/16.67 \text{ ms} = 120°/360°$$

$$t_{S3} = 16.67 \text{ ms}(120°/360°) = \textbf{5.56 ms}$$

Another way to create a three-phase output is to use three single-phase inverters and a three-phase sync box, as shown in Figure 6-29. The sync box introduces a phase shift of 120° between each of the three single-phase inputs and produces a three-phase output synchronized with the grid.

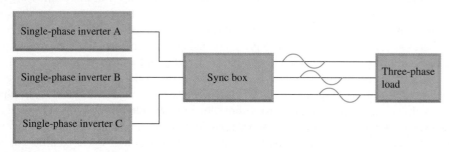

FIGURE 6-29   Three-Phase Output Using Three Single-Phase Inverters

SECTION 6-5 CHECKUP

1. What is the purpose of a dc to ac inverter?
2. Name three types of inverters based on their output.
3. What does an H-bridge do?
4. What is PWM? How is a PWM waveform produced for an inverter?
5. What is the difference between single-phase ac and three-phase ac?
6. What type of ac power is typically used in residences and small businesses?
7. Name one advantage of three-phase over single-phase.

6-6   Inverter Functions

In addition to types of output waveforms, inverters are also classified according to interface as either stand-alone or grid-tie, as you learned in Chapter 4. Another inverter configuration, known as a battery backup inverter, is simply an inverter with a built-in charge controller. Stand-alone inverters are used for small applications such as powering a specific piece of equipment or providing power to a private residence or small business. Grid-tie inverters, also known as grid-interactive inverters, are used mainly in larger systems that can supply power to the electrical grid. Some inverters are used in both modes where excess power in a small application is fed to the grid for credit. Only simplified systems are shown here to illustrate the functions of an inverter. Refer to Chapter 4 to review the various PV solar system configurations.

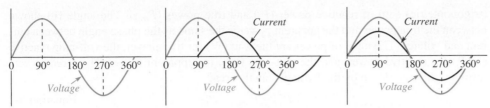

(a) Angular measurement   (b) Current 90° out of phase with voltage (c) Current and voltage in phase

FIGURE 6-30 Sine Wave Phase Relationships

The stand-alone inverter is used in applications where all of the output power is used for a specified load, such as an appliance with an ac motor, and is independent of the electrical grid. The grid-tie inverter is used in applications where all or part of the output power can be sent to the electrical grid. For example, a home solar power system may give excess power not used in the home to the power company for credit if net metering is available. A large solar power system is usually devoted entirely to producing power for the electrical grid. The power company provides single-phase sinusoidal voltage and current to a residence or small business at a voltage of 240 V rms and at a frequency of 60 Hz. The 240 V is split into two 120 V lines for most home and small business applications. Some large appliances, such as clothes dryers and stoves, require the full 240 V.

Power Factor Correction

A single-phase sine wave inverter (both stand-alone and grid-tie) produces one output voltage. One complete cycle of a voltage sine wave goes from 0° to 360°, as shown in Figure 6-30(a). In Figure 6-30(b), current and voltage are 90° out-of-phase; that is, a phase difference of 90° exists between them. This is the case with a pure inductive load and no actual power is delivered to the load. In Figure 6-30(c), the current and voltage are in phase; that is, no phase difference exists between them. The output current and voltage of an inverter are designed to be in phase for a purely resistive load.

Power Factor

The phase difference between current and voltage determines the **power factor**. If the load is pure capacitance or pure inductance, the voltage and current are 90° out of phase with each other and the power factor is 0. In this case, all the power is reactive power, which is returned to the source and no true power is delivered to the load. If the load is pure resistance, the voltage and current are in phase with each other and the power factor is 1, so all of the power is delivered and dissipated in the load. When the load is a combination of resistance and reactance (which is usually the case), the power factor is somewhere between 0 and 1. If the power factor is less than 1, power factor correction is necessary to increase the power factor to as close to 1 as possible. For power factor correction, one type of reactance is added to cancel the effects of the opposite type of reactance in the load, making the load appear to be purely resistive.

The diagram in Figure 6-31 is a common method to illustrate power factor. If there is inductance or capacitance in the load, there will be reactive power plotted on the y-axis. The resistance in the load accounts for the true power. Apparent power (P_a) is the vector

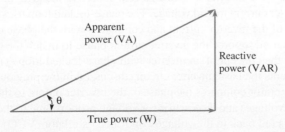

FIGURE 6-31 Power Diagram. The sides of the right triangle are related by the Pythagorean theorem.

(trigonometric) sum of reactive power (P_r) and true power (P_{true}). The angle (θ) shown between the true power and the apparent power is the same as the phase angle between current and voltage. The units of power are the watt (W) for true power, the volt-amp reactive (VAR) for reactive power, and the volt-amp (VA) for apparent power. The three powers form a right triangle, so by the Pythagorean theorem:

$$P_a^2 = P_{true}^2 + P_r^2 \qquad \text{Equation 6-2}$$

The power factor (PF) is expressed as:

$$PF = \cos(\theta) \qquad \text{Equation 6-3}$$

Cos is an abbreviation for "cosine," which is a trigonometric function. On the scientific calculator, simply enter the phase angle and press the COS key.

Most reactive loads tend to be inductive. Appliances with motors, such as air conditioners, refrigerators, washing machines, and dryers, are typical inductive loads. For these types of loads, power factor correction is achieved by the addition of capacitors in parallel with the connected motor circuits. The amount of capacitance required depends on the type of motor and the type of inverter, including the type of waveform. In general, it is important to check with the user manual or the inverter manufacturer before attempting any external power factor correction with an inverter. An incorrect value of capacitance can damage an inverter.

EXAMPLE 6-3

(a) If the true power is 50 W and the reactive power is 10 VAR, determine the apparent power.

(b) If the phase angle is 20°, determine the power factor.

Solution

(a) Take the square root (use the $\sqrt{x}$ key) on both sides of Equation 6-2:

$$P_a = \sqrt{P_{true}^2 + P_r^2} = \sqrt{(50\ \text{W})^2 + (10\ \text{VAR})^2} = \textbf{51 VA}$$

(b) $PF = \cos 20° = \textbf{0.94}$

Synchronization with the Grid

A grid-tie inverter must synchronize its output voltage with the grid voltage in terms of frequency, phase, and amplitude. Ideally, the grid maintains a power factor of 1; this number can vary within specified limits, but it is usually very close to 1. This means that the current and voltage from the grid are ideally in phase, as are the current and voltage from the inverter. When a grid-tie inverter is synchronized, it is at the same frequency and phase as the grid. Also, the voltage amplitude of the inverter must be a bit higher than the grid for the inverter to supply current to the grid. The two basic **synchronization** methods are phase-locked loop and zero-crossing detection.

Phased-Locked Loop

A phase-locked loop can be used to accomplish synchronization of the inverter current to the grid voltage. Because the inverter acts as a current source to the grid, this provides a direct correlation of inverter current to grid voltage. The phase-locked loop uses a phase detector to measure the phase of the inverter current and compare it with the phase of the grid voltage. A feedback loop is used to force the inverter current phase to match the grid voltage phase. Figure 6-32 shows the concept of an inverter with phase-locked loop synchronization. The green block represents the basic inverter circuits discussed in the previous section.

The phase comparator compares the phase of the inverter output to the phase of the grid voltage (reference voltage) and issues an error voltage proportional to the phase difference. The error voltage is fed back to the voltage-controlled oscillator (VCO) to move the phase of the inverter output closer to the phase of the grid voltage. When the phase difference

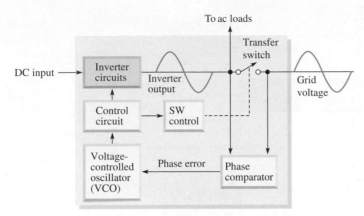

FIGURE 6-32   Phase-Locked Loop Synchronization in a Grid-Tie Inverter

between the grid voltage and the inverter voltage is zero, the error voltage is zero (or near zero) and an in-phase condition is established. During this synchronization process, the transfer switch is open. The **transfer switch** is a switch used to connect a source to the grid. Once the voltage and current are in phase, the transfer switch closes and the inverter is then connected to the grid.

Zero-Crossing Detection

An alternate method for verifying that the inverter and grid are synchronized is to check the zero crossings of each wave. Zero volt crossings occur when a sine wave switches polarity: at 0°, 180°, and 360°. A sine wave crosses the zero volt axis in the positive direction at 0° and at 360° and in the negative direction at 180°. The crossing times of the grid are compared to the crossing times of the inverter output, and the difference in the two waves are used by control circuits to synchronize the two waves. During the synchronization process, the transfer switch is open to avoid connecting the inverter to the grid. Once synchronization is established, the transfer switch is closed and the inverter is connected to the grid.

Anti-Islanding

When utility workers are working on a power line, most utility operating procedures, including the *National Electrical Safety Code* (NESC), require that the line be isolated from all generating sources. **Islanding** occurs when a grid-tie renewable energy source (called a distributed generation [DG] resource), continues to operate and provide power to a certain location and remains connected to the electrical grid after the grid no longer supplies power. DGs can be any type of renewable energy source, such as a solar or wind system that is connected to the grid. The term *islanding* comes from the idea that, when a portion of the grid goes down and thus causes a blackout, a location that still has power from a DG is isolated and surrounded by darkness (no power) like an island surrounded by the sea. An island can be as small as a single residence or as large as an entire community. Figure 6-31 illustrates the basic idea.

Anti-islanding is a protective feature of a grid-tie inverter that detects a power outage by monitoring parameters of the ac voltage at the point of common connection (PCC) and disconnecting from the grid with the transfer switch that is often part of the inverter. The transfer switch can switch from the grid to the distributed generator if it is used for backup only. If the DG is used in conjunction with the grid on a continuous basis, the transfer switch simply disconnects the DG from the grid.

As stated before, the reason for anti-islanding is to protect utility company personnel who are working on the power lines as a result of the grid outage. If renewable energy systems remain connected to the grid when an outage occurs, the power supplied by these systems to the grid is hazardous to the workers and can also damage any equipment connected to the grid at other distribution points. The standards for anti-islanding protection are UL1741 and IEEE1547.

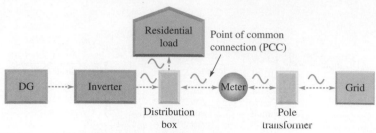

(a) Grid-tie renewable energy system operating normally

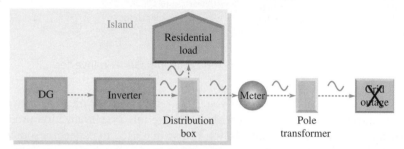

(b) Islanding: grid goes down and DG remains connected to the grid, presenting a safety hazard

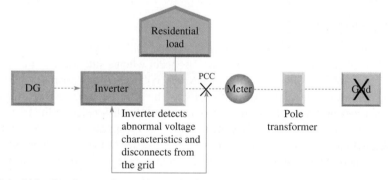

(c) Anti-islanding: inverter detects abnormal voltage at the PCC and disconnects from the grid

FIGURE 6-33   Islanding and Anti-Islanding Detection

Two factors sometimes make detection of islanding difficult. One is that the voltage generated by the inverter is identical to the grid voltage. The other is a motor that is part of a load on the grid may continue to spin and initially act as a generator with a frequency close to the original line frequency (although it will eventually slow down). It may be hard to distinguish either of these sources from the grid voltage.

Three basic types of detection are used for sensing islanding for grid-tie inverters: passive detection, active detection, and utility notification.

Passive detection Generally, most inverters detect the islanding condition by detecting a sudden change in system frequency, voltage magnitude, phase, or power. One or more of these parameters are monitored, and the inverter is shut down within a specified time when these parameters vary from specified limits.

Active detection In this method, a disturbance is injected and the response is measured. For example, the inverter current may be altered and the response to this change depends on the type of load to which the inverter is connected. Typically, a grid-tie inverter is connected to the user load and the grid. If the grid becomes disabled, the load changes and causes a predictable response to the disturbance.

Utility notification For utility notification, the power company knows when it removes power and issues a signal to the inverter to disconnect from the grid.

Maximum Power Point Tracking

Maximum power point tracking (MPPT) was discussed in relation to the charge controller. When an inverter is used in a system that has no backup batteries and therefore no charge controller, the MPPT is built into the inverter. Because the inverter input comes directly from the solar array, the MPPT in an inverter operates the same as one for a charge controller.

SECTION 6-6 CHECKUP

1. What is power factor?
2. What is the purpose of synchronization in an inverter?
3. What is islanding and why must it be avoided?
4. What are three ways to detect islanding?

6-7   Inverter Specifications and Data Sheet

As you have learned, the basic function of the inverter is to convert dc power to ac power because most of our electrical needs are for ac. The inverter is connected directly to either the power source (solar PV array or wind turbine) or the charge controller, depending on whether backup storage batteries are used. Also, some manufacturers offer a single unit containing a charge controller and an inverter.

Specifications

Specifications provide the values of operating parameters for a given inverter. Common specifications are discussed below. Some or all of the specifications usually appear on the inverter data sheet.

Maximum ac output power This is the maximum power the inverter can supply to a load on a steady basis at a specified output voltage. The value is expressed in watts or kilowatts.

Peak output power This is also known as the surge power; it is the maximum power that an inverter can supply for a short time. For example, some appliances with electric motors require a much higher power on start-up than when they are running on a continuous basis. Common examples are refrigerators, air-conditioning units, and pumps.

AC output voltage This value indicates to which utility voltages the inverter can connect. For inverters designed for residential use, the output voltage is 120 V or 240 V at 60 Hz for North America. It is 230 V at 50 Hz for many other countries.

Peak efficiency The peak efficiency is the highest efficiency that the inverter can achieve. Most grid-tie inverters have peak efficiencies above 90%. The energy lost during inversion is, for the most part, converted into heat. It's important to note what this means: In order for an inverter to put out the rated amount of power, it will need to have a power input that exceeds the output. For example, an inverter with a rated output power of 5,000 W and a peak efficiency of 95% requires an input power of 5,263 W to operate at full power.

California Energy Commission weighted efficiency This value is established by the California Energy Commission (CEC). This value is an average efficiency and is a better representation of the inverter's operating profile than is the peak efficiency.

Maximum input current This is the maximum direct current that the inverter can utilize. If a solar array or wind turbine produces a current that exceeds this maximum input current, the excess current is not used by the inverter.

Maximum output current This is the maximum continuous ac that the inverter supplies. This value is typically used to determine the minimum current rating of the protection devices (breakers and fuses) and disconnects required for the output circuit.

Peak power tracking voltage This is the dc voltage range in which the inverter's maximum power point tracker operates.

Start Voltage This value is the minimum dc voltage required for the inverter to turn on and begin operation. This is particularly important for solar applications because the solar module or modules must be capable of producing the voltage. If this value is not provided by the manufacturer, the lower end of the peak power tracking voltage range can be used as the inverter's minimum voltage.

NEMA rating This rating indicates the level of protection the inverter has against water intrusion. Most outdoor inverters are rated as National Electrical Manufacturers Association (NEMA) 3 for most weather conditions.

Total harmonic distortion The total harmonic distortion (THD) is an indication of the purity, or the harmonic content, of the sinusoidal output of an inverter. Most filtered sine waves still contain some harmonics that distort the waveform to a minor degree.

Miscellaneous Features

Inverter features vary from one model to another and from one manufacturer to another. Common features found on many inverters are as follows.

Weatherproof enclosure Most inverters, especially grid-tie inverters, are designed to be installed outdoors and have weatherproof enclosures.

AC/DC disconnects Some inverters have built in ac/dc disconnects for safety and to facilitate removing the inverter if it needs to be serviced.

Ground fault protection The National Electric Code (NEC) requires that roof-mounted solar electric systems must be grounded. Most inverters have built-in ground fault protection.

Maximum power point tracking Tracking the peak power point of a solar panel array is important for maximizing energy obtained from a PV module or array. If a system does not have a charge controller that performs this function, the inverter is connected directly to the PV source and requires MPPT.

Transfer switch A transfer switch is also known as a transfer relay. Grid-tie inverters usually feature a built-in load transfer switch for backup emergency power applications. As long as utility power reaches the inverter's ac input side, the transfer switch passes the ac grid power directly through the inverter to the load. If the utility grid power is interrupted, the transfer relay automatically switches to the battery backup input to the inverter.

Generator start switch Some inverters are available with a separate generator start switch and include a second ac input for an ac generator that is used for backup. The generator can be programmed to start when the utility grid fails or when a low battery charge is detected. In some applications, where the inverter has a built-in battery charger, the generator operates long enough to recharge the batteries. If there is no battery backup, the generator is used as the only power source until the grid is operating again.

Display panel A remote display panel option is available for many inverters to indicate the system status. This feature is particularly useful if the inverter and battery bank are located in an area that is difficult to access. A standard interface allows data to be sent to a remote site.

Data Sheet

A data sheet for a typical inverter is shown in Figure 6-34.

INVERTER SPECIFICATIONS

SPRx
HIGH EFFICIENCY INVERTERS   ELECTRICAL SPECIFICATIONS

Model	SPR-3300x	SPR-3300x-208	SPR-4000x
Maximum AC Power Output	3300 W	3300 W	4000 W
AC Output Voltage (nominal)	240 VAC	208 VAC	240 VAC
AC Voltage Range	211-264 VAC	183-228 VAC	211-264 VAC
AC Frequency (nominal)	60 Hz	60 Hz	60 Hz
AC Frequency Range	59.3-60.5 Hz	59.3-60.5 Hz	59.3-60.5 Hz
Maximum Continuous Output Current	13.8 A	15.9 A	16.7 A
Current THD	< 3%	< 3%	< 3%
Power Factor	> 0.9	> 0.9	> 0.9
DC Input Voltage Range	195-600 VDC	195-600 VDC	195-600 VDC
Max DC Current	18.5 Adc	18.5 Adc	22.1 Adc
Peak Power Tracking Voltage Range	195-550 VDC	195-550 VDC	195-550 VDC
Peak Inverter Efficiency	95.3%	94.6%	95.7%
CEC Efficiency	94.5%	94.0%	95.0%
Night Time Power Consumption	< 1W	< 1 W	< 1W
Output Overcurrent Protection	20 A	25 A	25 A
Grounding	Positive ground		

MECHANICAL SPECIFICATIONS

Operating Temperature Range	−13F to 149F (−25° C to + 65° C)
Enclosure Type	NEMA3R (outdoor rated)
Unit Weight	49.0 to 51 lbs
Shipping Weight	57 to 59 lbs
Shipping Dimensions (H × W × D)	34.1 × 20.4 × 10.3″ (86.6 × 51.8 × 26.2 cm)
Inverter Dimensions (H × W × D)	28.5 × 15.9 × 5.7″ (72.4 × 40.3 × 14.6 cm)
Mounting	Wall Mount (mounting bracket included)

FEATURES

PV/Utility Disconnect	Eliminates need for external PV (DC) disconnect. Complies with UL and NEC requirements
Cooling	Convection cooled, no fan required
Display	Backlit, 2-line, 16-character Liquid Crystal Display provides instantaneous power, daily and lifetime energy production, PV array voltage and frequency, time online "selling" today and fault messages
Communications	RS 232 and Two Canbus RJ45 ports
Wiring Box	PV, utility, ground, and communications connections. Inverter can be separated from the wiring box
Warranty	10 years

FIGURE 6-34 Typical Inverter Data Sheet

EXAMPLE 6-4

Determine the power that a solar module array must provide to achieve maximum power from the SPR-3300x inverter specified in the data sheet in Figure 6-34.

Solution

Because $P_{OUT} = (\text{efficiency})(P_{IN})$, $P_{IN} = P_{OUT}/\text{efficiency}$. Using peak efficiency, the input power to the inverter must be

$$P_{IN} = P_{OUT}/\text{peak efficiency} = 3{,}300 \text{ W}/0.953 = \textbf{3,463 W}$$

Using the CEC efficiency, the input power to the inverter must be

$$P_{IN} = P_{OUT}/\text{CEC efficiency} = 3{,}300 \text{ W}/0.945 = \textbf{3,492 W}$$

Inverter Classes

Inverters can be classed according to their power output. The following information is not set in stone, but it gives you an idea of the classifications and general power ranges associated with them. These ranges may vary from one manufacturer to another. Inverters may also be found with output power specifications falling between each of the ranges listed.

Small residential inverters Small residential inverters are in the 1,800 W to 2,500 W range, with single-phase power.

Large residential inverters Large residential inverters are in the 3,000 W to 6,000 W range, with single-phase power.

Small commercial inverters Small commercial inverters are in the 13 kW to 15 kW range and can include three-phase power.

Large commercial inverters Large commercial inverters are in the 60 kW to 100 kW range. Inverters can be combined to provide up to or above 1 MW (1,000 kW) of three-phase power.

SECTION 6-7 CHECKUP

1. What determines the required input power to an inverter so that it achieves a specified output power?
2. What is CEC weighted efficiency?
3. What does THD stand for?
4. What indicates the level of protection against water intrusion for an inverter?

CHAPTER SUMMARY

- The simplest type of battery charger is the continuous trickle charger.
- Float voltage is the relatively constant voltage that is applied continuously to a battery to maintain a fully charged condition.
- The three stages of float charging are bulk, absorption, and float.
- The PWM charge controller uses pulse width modulation to decrease the power applied to the batteries gradually as they approach full charge.

- In the equalization process, the battery is overcharged to approximately 1 V above float in order to equalize the charge on all the cells in the battery or battery bank.
- The MPPT charge controller incorporates PWM and a dc to dc converter.
- MPPT operates using an algorithm, which is basically a series of steps or procedures for accomplishing a desired result.

- The inverter takes the dc output voltage of batteries or other dc source and converts it to a standard ac voltage.

- Three types of inverter output are the square wave, a stepped wave called a modified sine wave, and a pure sine wave.

- The harmonic content of a square wave includes a fundamental sine wave at the frequency of the square wave and a series of odd harmonics.

- The stand-alone inverter is used in applications where all of the output power is used for a specified load, such as an appliance with an ac motor, and is independent of the electrical grid.

- The grid-tie inverter is used in applications where all or part of the output power is provided to the electrical grid.

- The phase difference between current and voltage determines the power factor.

- A grid-tie inverter must synchronize its output voltage with the grid voltage in terms of frequency, phase, and amplitude.

- Islanding occurs when a grid-tie renewable energy source continues to operate and provide power to a certain location after the electrical grid no longer supplies power.

- Anti-islanding is a protective feature of a grid-tie inverter that detects when there is a power outage and disconnects from the grid with a transfer switch.

- Three-phase power has three sine waves that are 120° degrees apart.

KEY TERMS

absorption stage The stage of battery charging after the battery voltage reaches the maximum and the current through the battery begins to decrease. The voltage is held at the maximum value while the current decreases.

anti-islanding A protective feature of a grid-tie inverter that detects when there is a power outage and disconnects the renewable energy source from the grid.

bulk stage The stage of battery charging where the battery voltage increases at a constant rate.

equalization The charging process in which the battery is overcharged at approximately 1 V above float in order to equalize the charge on all the cells in the battery or battery bank.

float stage The final maintenance or trickle charge stage with the purpose of offsetting any self-discharging of the battery

float voltage A voltage supplied to a battery to maintain it at full charge and prolong its life.

harmonics The frequencies contained in a composite waveform, which are integer multiples of the repetition frequency (fundamental).

islanding The situation when a grid-tie renewable energy source continues to operate and provide power to a certain location and

remains connected to the electrical grid after the grid no longer supplies power.

maximum power point tracking (MPPT) The process for tracking the voltage and current from a solar module to determine when the maximum power occurs in order to extract the maximum power.

perturb and observe (P&O) algorithm A procedure in which a variable is changed (perturbed) and the effect of the change on another variable is monitored (observed).

power factor The cosine of the phase angle between current and voltage.

pulse width modulation (PWM) A process in which a signal is converted to a series of pulses with widths that vary proportionally to the signal amplitude.

synchronization The process of producing a fixed phase relationship between two or more waveforms.

three-phase Three ac voltages that have the same magnitude and frequency but are separated by 120°.

transfer switch A switch used for connecting or disconnecting a source from the grid.

FORMULAS

Equation 6-1	$I_{CH} = 1.56 I_{SC}$	Recommended current rating of the charge controller
Equation 6-2	$P_a^2 = P_{true}^2 + P_r^2$	Power relationships
Equation 6-3	$PF = \cos(\theta)$	Power factor

CHAPTER TRUE/FALSE QUIZ

Determine whether each statement is true or false. Answers are at the end of the chapter.

1. Overcharging can potentially damage a battery.

2. Three stages of battery charging are blast, absorption, and float.

3. PWM stands for "pulse width modulation."

4. The main purpose of a charge controller is to control the charging of batteries.

5. MPPT stands for "mid-point power tracking."

6. P&O is an MPPT algorithm and stands for "perturb and observe."

7. An inverter converts ac to dc.

8. Inverters can produce one of three types of output waveforms.

9. Anti-islanding occurs when an inverter disconnects from the grid when the grid goes down.

10. The main purpose of anti-islanding is to protect the homeowner.

11. A power factor of 0 indicates that the voltage and current are 90° out of phase.

12. Synchronization occurs when the power factor is between 0 and 1.

13. The output of a grid-tie inverter must be synchronized with the power grid.

14. Three-phase power is typically used for residential applications.

CHAPTER MULTIPLE-CHOICE QUIZ

Complete each statement by selecting the one correct answer. Answers are at the end of the chapter.

1. The relatively constant voltage that is applied continuously to a battery to maintain a fully charged condition is called the
   a. absorption voltage
   b. bulk voltage
   c. float voltage
   d. equalization voltage

2. During the bulk stage of charging, the battery voltage
   a. stays the same
   b. increases at a constant rate
   c. decreases at a constant rate
   d. increases at an exponential rate

3. Charge controllers are specified by
   a. current
   b. voltage
   c. power
   d. all of these

4. A PWM charge controller adjusts its output voltage as the battery approaches full charge by
   a. reducing it
   b. increasing it
   c. keeping it level
   d. turning it off

5. Stratification in a battery is caused by
   a. lead sulfate
   b. acid settling to the bottom
   c. overcharging
   d. both (a) and (b)

6. MPPT stands for
   a. maximum power point trigger
   b. maximum power point tracking
   c. mean positive power transient
   d. maximized power plot track

7. The P&O algorithm used in MPPT is also known as
   a. power operation
   b. power optimization

   c. power option
   d. hill climbing

8. One method used in inverters to obtain a square wave from a dc voltage is the
   a. Wheatstone bridge
   b. H-bridge
   c. twin-T bridge
   d. all of these

9. A modified sine wave can be produced from a square wave by
   a. amplification
   b. attenuation
   c. switching
   d. diversion

10. One cycle of a sine wave is
    a. 180°
    b. 90°
    c. 270°
    d. 360°

11. Power factor is determined by
    a. the phase angle between current and voltage
    b. the amount of power delivered to a load
    c. the amount of power from the grid
    d. none of these

12. When the load current and voltage are in phase, the power factor is
    a. 0
    b. 1
    c. 0.5
    d. none of these

13. A transfer switch is used to disconnect the
    a. batteries
    b. PV module
    c. house
    d. grid

14. In three-phase power, the sine waves are separated by
    a. 120°
    b. 90°
    c. 180°
    d. 360°

CHAPTER QUESTIONS AND PROBLEMS

1. What is the difference between a basic trickle charger and an on/off switched charger?

2. What is the purpose of equalization in the charging process?

3. If the charging current to a battery is 10 A during the bulk stage, approximately how much current is there at the end of the absorption stage?

4. A certain solar array consists of four 12 V, 250 W modules connected in parallel. The I_{SC} for each module is 8 A. Determine the current rating for the charge controller in this application.

5. During the MMPT P&O process, how does a charge controller determine when the MMP has been reached?

6. What does the diversion feature of a charge controller do?

7. What are the frequencies of the harmonic content of a 60 Hz square up to the fifth harmonic?

8. What is the purpose of the low-pass filter in a sine wave inverter?

9. If the true power is 10 kW and the reactive power is 1 kVAR, determine the apparent power.

10. If the reactive power is 150 VAR and the apparent power is 300 VA, determine the true power.

11. If the phase angle between current and voltage is 10°, determine the power factor.

12. What is the power factor when the current and voltage are in phase?

13. What are the three types of detection used in anti-islanding?

14. How can single-phase inverters be used to produce a three-phase output?

15. A certain inverter has a maximum power output of 5,000 W and a CEC efficiency of 92%. How much power must be obtained from a solar array for the inverter to operate at maximum output?

FOR DISCUSSION

Lead-acid batteries are the most common way to store energy in a small system, but they have drawbacks. Lithium-ion batteries may eventually replace lead-acid batteries in renewable energy systems. How would this affect how charging is done?

ANSWERS TO CHECKUPS

Section 6-1 Checkup

1. The trickle charger must be turned off manually when the battery is charged or it may overcharge it.
2. It is charging a battery in a manner that keeps it at its fully charged condition.
3. Bulk, absorption, and float
4. During the bulk stage, the battery is charged at its maximum rate.
5. During the absorption stage, the battery continues to charge but at a rate determined by battery capacity.

Section 6-2 Checkup

1. Pulse width modulation
2. It determines the actual battery voltage by sampling the voltage between pulses.
3. A periodic controlled overcharging of a battery to prevent stratification of the electrolyte
4. The rated voltage of the battery and the short-circuit current of the solar array
5. It is the product of two 25% (same as multiplying 1.25 × 1.25) derating factors.

Section 6-3 Checkup

1. Maximum power point tracking
2. It isolates the dc input from the dc output, so the output can be adjusted for maximum power.
3. Perturb and observe; an algorithm that changes a parameter and observes the results
4. The load is varied and the power is monitored to find the highest power.
5. This is the same as the perturb and observe algorithm.

Section 6-4 Checkup

1. Maximum solar array size
2. Yes, as long as current limit is not exceeded (60 A)
3. 60 A
4. 25° C (77° F)

Section 6-5 Checkup

1. The inverter takes the dc output voltage of the batteries or of the energy source and converts it to a standard utility voltage or a three-phase voltage.
2. Sine wave, modified sine wave, and square wave
3. An H-bridge converts dc to an alternating or bipolar square wave.
4. PWM stands for "pulse width modulation." It produces a pulse waveform with pulse widths proportional to the amplitude of a sine wave by comparing a reference sine wave with a high-frequency triangle wave.
5. Single-phase has one ac voltage; three-phase has three ac voltages separated by one-third of a cycle each.
6. Single-phase and split-phase; split-phase has two waveforms that are 180° out of phase with each other.
7. Three-phase is more efficient for driving large motors and does not require separate starting windings or other starting method.

Section 6-6 Checkup

1. The phase difference between current and voltage determines the power factor, which is the cosine of the angle between the current and the voltage or between the true power and the average power.
2. Synchronization makes the inverter voltage the same frequency and phase as the grid voltage.

3. Islanding is when a grid-tie renewable energy source continues to operate and provide power to a certain location and remains connected to the electrical grid after the grid no longer supplies power. Islanding must be avoided to protect utility company personnel who are working on the power lines as a result of the grid outage.

4. Passive detection, active detection, and utility notification

Section 6-7 Checkup

1. The peak or weighted efficiency

2. The average efficiency of an inverter

3. The total harmonic distortion, which is an indication of the purity, or the harmonic content, of the sinusoidal output of an inverter

4. The NEMA rating

ANSWERS TO TRUE/FALSE QUIZ

1. T   2. F   3. T   4. T   5. F   6. T   7. F
8. T   9. T   10. F   11. T   12. F   13. T   14. F

ANSWERS TO MULTIPLE-CHOICE QUESTIONS

1. c   2. b   3. d   4. d   5. b   6. b   7. d   8. b
9. c   10. d   11. a   12. b   13. d   14. a

Wind Power Systems

CHAPTER OBJECTIVES

- Explain what causes the wind to blow and causes the direction and speed of the wind to vary.
- Describe Betz's law and how it relates to wind turbines.
- Identify factors that cause long-term variations in wind speed.
- Discuss the selection criteria for locating a wind turbine.
- Explain how a wind turbine converts wind energy to electricity.
- Identify the basic parts of a horizontal-axis wind turbine.
- Explain the terms *lift, drag,* and *angle of attack* with respect to wind turbines.
- Given basic parameters for a wind turbine and a given wind speed, calculate the energy delivered in a certain time from that wind turbine.
- Identify the advantages and disadvantages of single-bladed, two-bladed, and three-bladed wind turbines.
- Discuss the benefits of large wind farms and the advantages and disadvantages of offshore wind farms.

KEY TERMS

Key terms are shown in bold and color. Definitions for key terms are provided at the end of the chapter and in the end-of-book glossary. Bold terms in black are defined in the end-of-book glossary only.

- millibar
- barometric pressure
- pressure gradient
- power curve
- Betz's law
- wind farm
- lift
- drag
- angle of attack
- stall
- lift-to-drag ratio
- tip speed ratio (TSR)
- horizontal-axis wind turbine (HAWT)
- pitch
- yaw
- vertical-axis wind turbine (VAWT)

INTRODUCTION

Wind generators are viable sources for providing electricity from renewable energy. Many are small units that provide power directly to a residence or small commercial establishment, and they can be connected to the grid. Others are large units connected together in wind farms to provide a large amount of power to the grid. Wind farms are located in favorable locations for wind, which include many offshore sites.

This chapter opens with a general overview of wind, including the science of the wind, the power in the wind, and the wind power curve. The important Betz's law shows that not all of the available energy can be extracted by a wind machine. Blades and how wind energy can turn them are discussed. Horizontal- and vertical-axis wind machines and their basic components are introduced and discussed in detail.

7-1   Power in the Wind

For hundreds of years, the wind has been studied scientifically, primarily to predict weather and its impact on human activities. In the last fifty years, advances in technology and the understanding of wind have been applied to the development of wind turbines. This understanding is important to developing and using the wind resource efficiently.

Ancient Wind Power

(*Source:* Brad Pict/Fotolia.)

Wind energy has been harnessed for centuries to grind grains (hence the name *windmill*) and pump water. The Greeks were known to use "watermills" called *hydraletēs* as early as the first century BC. The photo shows an old Greek windmill that used sails to turn; these windmills can still be seen on some of the Greek Islands.

Physics of the Wind

The scientific study of the wind has helped scientists better understand wind and the causes of wind. This knowledge has led to better predictions of where the strongest and most continuous winds occur and to enable better placement of wind turbines. Wind varies from day to night, from season to season, and with environmental conditions such as ocean temperatures and cloud conditions. The ability to predict these effects in detail has proven to be challenging, even for the supercomputers of today.

Recall that in Section 1-4, global wind patterns were shown to be the result primarily of differential heating and the earth's rotation. Like all gases, air expands when heated and contracts when cooled. When the air is warmed, it rises and cools at high elevation, becoming denser. The rising and falling air currents cause large circulating cells of air. Superimposed on these large circulating currents of air are variations due to many other factors. Pressure variations in the atmosphere are one of the most important predictors of wind. To understand the energy in wind, it is helpful to review some physical science.

Pressure is defined as force per unit area and can be calculated by the following formula:

$$p = F/A \qquad\qquad \text{Equation 7-1}$$

where

p is pressure in newtons per square meter or pascals (Pa)

F is force in newtons (n)

A is area in square meters (m^2)

These units, which are perhaps unfamiliar, are derived units in the SI metric system. Several other pressure units are in common use, depending on application. The **pascal (Pa)** is a very small unit of pressure, so it is common in meteorology (weather science) to use the unit of the **millibar**. One millibar is equal to 100 Pa. A common pressure unit in the English system is the pound per square inch (psi). Each psi of pressure translates to 6,895 Pa. Another pressure unit is the **atmosphere (atm),** which is the pressure exerted by the atmosphere under specified conditions at sea level. One atm is defined as 1.013×10^5 Pa.

EXAMPLE 7-1

There are 4.448 newtons in one pound and 39.37 inches in a meter. Express atmospheric pressure in psi. (The number of Pa in 1 atm is 1.013×10^5.)

Solution
Set up the conversion so that units cancel:

$$1.013 \times 10^5 \text{ Pa} = 1.013 \times 10^5 \left(\frac{n}{m^2}\right)\left(\frac{1b}{4.448\ n}\right)\left(\frac{m}{39.37\ in}\right)^2 = \mathbf{14.7\ \frac{1b}{in^2}}$$

In addition to the pressure units given above, meteorologists frequently use millimeters of mercury (mmHg) or occasionally inches of mercury (in Hg). These units represent the pressure exerted by the atmosphere on a column of mercury contained in an evacuated tube. The height of the mercury column is proportional to the atmospheric pressure. This value is often referred to as **barometric pressure**, which is the pressure exerted by the

FIGURE 7-1 Basic Compass Face. This basic compass face shows the cardinal directions and directions between these points. (*Source:* Disenador/Fotolia.)

atmosphere on the earth. Atmospheric pressure varies from day to day and decreases with elevation. For reference, standard atmospheric pressure is taken for a typical reading at sea level to be 760 mm Hg (29.9 in Hg), which is 101 kPa or 14.7 psi. In Denver, at an elevation of 1,609 m (5,280 ft), the barometric pressure is typically 25.7 in Hg or 653 mm Hg.

Wind Direction

Wind direction is specified by compass readings and can be stated either as a cardinal direction or in degrees. For example, 0° is north, 90° is east, 180° is south, and 270° is west, and 360° is north again. Notice that north is defined as both 0° and 360°. Directions between the four cardinal directions are defined as northeast (NE), northwest (NW), southwest (SW), and southeast (SE). Wind direction *always* refers to the direction from which the wind is blowing. Thus, a north wind indicates the wind is moving from north to south. Figure 7-1 shows a basic compass face.

Causes of Wind

As storms move across land, the air pressure changes. Low pressure is associated with storms or storm fronts. If a high-pressure area is developing, weather forecasters predict better weather, including clear skies. The lowest air pressure reading ever recorded at sea level was taken in the middle of a hurricane, which measured just 25.69 in Hg, or about 12.5 psi. The highest air pressure reading ever recorded was about 32 in Hg, or approximately 15.6 psi. Extreme high and low pressure measurements are not common, but small fluctuations from day to day are important because they help determine the direction and velocity of the wind. Wind direction and velocity are related directly to these differences of air pressure. When the wind blows, it is moving from a region of high air pressure to a region of lower air pressure. The larger the air pressure difference between the two points, the stronger the wind blows. A variation in air pressure is called a **pressure gradient**. Weather forecasters use maps called pressure gradient maps to connect points of constant pressure and to predict the direction and velocity of wind. Figure 7-2 shows a pressure gradient map for North America on a given day. Where lines are close together, the pressure gradient is highest.

Wind does not always blow in a straight line inside a low-pressure or high-pressure area. For example, in the northern hemisphere, winds inside a low-pressure area blow in a counterclockwise direction as the front moves through an area. This means that the wind may blow from one direction as the beginning of the front moves through and then slowly change its direction until it is blowing in the opposite direction as the front moves out. The wind velocity inside the front varies according to the size of the pressure area directly ahead of the front and that of the next weather front that is coming behind it. In the spring in the northern hemisphere, several fronts may line up and come through an area very quickly,

Flying Electric Generators

(*Source:* Courtesy of Sky WindPower Corporation.)

High-altitude winds are also of special interest to researchers looking at flying electric generators that are tethered to the earth. The concept of a flying electric generator is shown in this artist's sketch. The energy in high-altitude winds is consistently high, and promising research has been done using models. In spite of some potential problems (such as aircraft interference), the payoff is huge for this technology because of consistency and the strength of the winds.

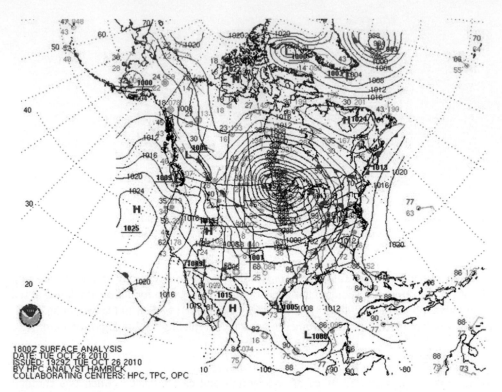

FIGURE 7-2 Pressure Gradient Map. Each line represents constant pressure in millibars. When lines are close together, the pressure change is more pronounced. (*Source:* National Oceanic and Atmospheric Administration Weather.)

within 24 hours. What this means in terms of the wind energy available to a wind turbine is that the velocity and wind direction may vary dramatically over a 24-hour period. In other regions, the wind may come from the same direction and at roughly the same velocity from day to day.

Weather forecasters are interested in the wind velocity and direction at very high altitudes as well as winds at lower altitudes. Technicians and engineers who work with wind turbines are interested mainly in low-altitude winds, less than 60 m from the ground, because these winds are used to turn wind turbine blades. One of the natural phenomena of the wind is that it blows at stronger speeds at higher levels due to friction effects with the ground. A smooth hard terrain or calm water has very low friction, but urban areas or rough terrain can have significant friction. For example, the wind speed at ground level may be 10 mph, but at 10 m elevation, the wind speed may be 20 mph. Keep in mind that the energy in the wind varies as the cube of its velocity, so a doubling of wind speed means eight times more energy is available. This is why larger wind turbines today are mounted on taller towers. Typically, wind turbines have been mounted at heights below 60 m, but research is currently being conducted on some wind turbines that are much taller. Because wind turbines in the future may need to be mounted above 100 m, winds at higher elevations need to be understood.

Boundary Layer and Gradient Wind

Boundary-layer wind is wind in the lowest part of the atmosphere, where turbulence and friction play a role in its behavior. The behavior of boundary-layer wind is influenced by the friction caused by irregularities in the earth's surface and other factors such as temperature changes and moisture. Boundary-layer wind has a large influence on the way a wind turbine can harvest electrical energy from wind blowing over its blades.

Gradient wind blows at a constant speed and flows parallel to imaginary curved isobars just above the earth's surface, where friction from irregularities such as mountains, trees,

FIGURE 7-3 Wind Turbines on the Coast of San Clemente Island   (*Source:* National Renewable Energy Laboratory.)

and buildings cause changes in the flow. Gradient wind causes the wind speed to change when high- and low-pressure fronts move through an area, and as the earth changes temperature from day to night and from season to season.

Sea and Land Breezes

A **sea breeze** is created whenever there is a difference between the water temperature and the land temperature. When the sun is high in the sky during the summer, it shines on the land and the sea and adds heat to both. The land warms more quickly than the water, however, so each day, the land may warm to perhaps 27° C (80° F), while the water remains around 21° C (70° F). When this occurs, a sea breeze is generated from the water and moves over the land, making coastal regions of the world productive areas for wind turbines. Figure 7-3 shows several wind turbines located on San Clemente Island, which is off the coast of California.

Sea breezes occur more often in the spring and summer, when the land warms more quickly than do bodies of water. If the land cools so that it is cooler than the water, then the process reverses and a **land breeze** develops, which means that the wind blows from the land out toward the open water. Farther away from open water, this effect is much less and the wind speed is very light. Land breezes are more common in the fall and winter, when the land begins to cool more than do bodies of water.

Variations in Wind Speed

One of the challenging problems for installation of wind turbines is to determine factors affecting the wind speeds in the area where the turbine is to be located. For example, the seasons that cause temperature changes in winter and summer also bring wide variations in wind speed. In some locations, the variation of the wind speed between day and night may also be high.

Laminar and Turbulent Flow

Winds can be variable and create problems for turbines. When air flows in a straight line, it is called **laminar flow.** When a gas or liquid is disturbed such that the flow is no longer in a straight line, it is called **turbulent flow.** Figure 7-4 shows an example of laminar flow and turbulent flow in a smoke column. Wind can exhibit turbulent flow when it does not blow continually at the same speed (referred to as swirling winds) or encounter obstacles. Swirling winds do not produce the same constant energy as strong straight-line winds. Other turbulence can be caused by storms. The turbulence in a storm is caused by sudden changes in the wind velocity and wind direction, which causes the wind forces on the turbine blades to change rapidly

Wind Turbine Wake

Another type of wind turbulence occurs after the wind passes the blades of a wind turbine and swirling winds are created in the wake, which can affect downstream turbines. These

FIGURE 7-4 Example of Laminar and Turbulent Flow. As the smoke rises, it changes from laminar flow to turbulent flow. (*Source:* Fotolia.)

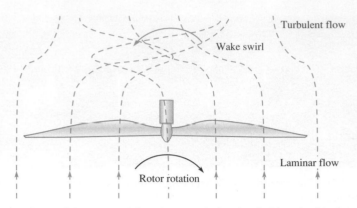

FIGURE 7-5 In front of the rotor, airflow is smooth (laminar). After the blades, the air is turbulent and swirls in a direction that is opposite to the rotation of the blades.

winds swirl in a direction that is opposite to the rotor's rotation. Turbulence is a problem if multiple wind turbines are located on the same site and the swirling winds from one turbine get in the airstream of the next turbine. This may not occur at all wind speeds. The choice of an arrangement for multiple wind turbines is affected by the issue of turbulent wakes. Figure 7-5 shows turbulent and laminar flow around a blade.

To understand the wake behind a wind turbine, consider the fact that a turbine harvests kinetic energy from the wind. In other words, the air behind the turbine must move slower that the air in front because the air will have less energy. (Recall that the kinetic energy is given by $W_{KE} = \frac{1}{2}mv^2$). Because the amount of air entering is the same as the air leaving, the air behind the rotor must occupy more space, hence it expands and air pressure drops behind the rotor. As the slower air mixes with faster air from the surrounding space, a turbulent region forms, which is illustrated in Figure 7-6.

Extreme Wind Speeds

High wind speeds are a problem for upwind horizontal wind turbines because the high winds tend to bend the tips of the turbine blades into the tower. To withstand these high winds, the turbine blades must be designed properly. The design compromise required may make the blades less efficient. Blades are connected to a shaft that enters the nacelle. The **nacelle** is an enclosure or housing on a wind turbine for the generator, gearbox, and any other parts of the wind turbine that are on the top of the tower. In some upwind horizontal wind turbines, the nacelle is designed so that it sits further from the tower with a slight upward tilt so that the blades are farther from the tower even if the blades flex during high winds, the blade tips cannot touch the tower. Another way around this problem is to use a

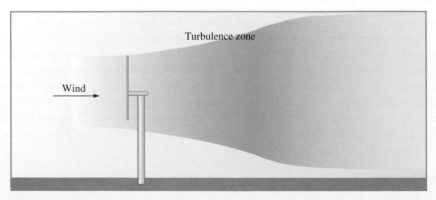

FIGURE 7-6 Turbulence Behind a Wind Turbine. The slower-moving air shown in the darker color mixes with faster air surrounding it, creating turbulence in the wake area.

downwind horizontal wind turbine to allow winds past the tower before hitting the blades. In this case, high winds tend to bend the blades outward and away from the tower.

SECTION 7-1 CHECKUP

1. What are three units for measuring pressure?

2. How is wind direction specified?

3. Refer to the map in Figure 7-2. What is the pressure reading represented by the first gradient line shown off the east coast of the United States?

7-2 Wind Power Curve

In this section, you will learn several ways to measure the peak performance of a wind generator. One way to measure peak performance is to use a table or graph of a power curve. Another way is to measure the amount of usable energy (power produced over time) that the wind system produces in the wind conditions at a site.

Wind Turbine Peak Performance

Wind turbine peak performance occurs when the output of the wind turbine generator is at or above its rated output. One way to measure peak performance is to use a graph of a power curve. A **power curve** is a graph that shows the wind speed and the output power of the wind turbine over a range of wind speeds from zero to the maximum wind speed for which the wind turbine is designed. Figure 7-7 shows a graph of a power curve for a wind turbine. On this graph, the wind speed is shown on the *x*-axis of the graph from 2 to 21 m/s. The output of the generator is shown on the *y*-axis on the left side, and it indicates power in kilowatts from 0 to 70 kW. According to this power curve, the wind turbine produces its maximum output of 63 kW to 65 kW when the wind is about 17 m/s. Although the power in the wind increases considerably as wind speed goes up, the turbine is designed not to exceed its rated power after some point. Each wind turbine manufacturer tests its models and produces a power curve for prospective buyers. This information allows a comparison with similar models from different manufacturers.

The test to create the official power curve for a specific wind turbine is designed to keep as many factors constant as possible so that comparisons of power output among different models of wind turbines can be made. When a wind turbine is put on a specific site, other issues change the power output of the wind turbine. Examples of such issues are icing of the blades in cold weather, dirt on the blades, and even insect impacts that have not been

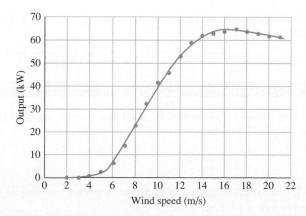

FIGURE 7-7   Generator Output for a 65 kW Wind Turbine

TABLE 7-1

Wind Speed and Generator Output Data for Calculating a Power Curve for a Wind Turbine

Wind Speed (m/s)	Wind Speed (mph)	Power (kW)
2	4.5	0
3	6.7	0
4	8.8	0.7
5	11.2	2.2
6	13.4	6.5
7	15.6	13.8
8	17.9	22.7
9	20.1	32.1
10	22.4	41.3
11	24.6	46.5
12	26.8	52.8
13	29.1	58.7
14	31.3	61.7
15	33.5	62.9
16	35.8	63.8
17	38	64.7
18	40.2	63.5
19	42.5	62.6
20	44.7	61.7
21	46.9	61.2

removed by rain or cleaning for long periods of time. Each of these issues may change the actual power output of the wind turbine so that it is lower by 1% to 2% from the rating indicated on the published power curve.

The data from the graph can also be shown in a table. Table 7-1 shows the data from the graph shown in Figure 7-7. The wind speed is shown in m/s in the first column and the wind speed is shown in mph in the second column. The power output is shown in kW in the last column.

EXAMPLE 7-2

Use Table 7-1 to determine the amount of electrical power the wind turbine produces when the wind speed is 10 m/s. What is the speed of the wind in mph when the wind blows at 10 m/s?

Solution
From Table 7-1, the power the generator produces at a wind speed of 10 m/s is 41.3 kW. The wind speed in mph is 22.4 mph.

EXAMPLE 7-3

Use Table 7-1 to determine the amount of electrical power the wind turbine produces when the wind speed is 38 mph. What is the speed of the wind in m/s when the wind is blowing 38 mph?

Solution
From Table 7-1, the power the generator produces at a wind speed of 38 mph is 64.7 kW. The wind speed in m/s is 17 m/s.

A second way to measure peak performance includes air density. **Density (ρ)** is the mass (*m*) per volume (*Vol*) of a substance, and is written in equation form as follows:

$$\rho = \frac{m}{Vol}$$

Equation 7-2

where

ρ is the density in kg/m^3

m is the mass in kg

Vol is the volume in m^3

Air density (ρ_a) is a measure of how much mass is contained in a specific volume of air. The amount of moisture (humidity) in the air, the air pressure, and the temperature all affect the air density. Air is denser when it is cooled; it also changes with humidity. When the air is dense, the output of the wind turbine increases.

Application of Wind Power Curves

The wind power curve indicates how much power a wind turbine should produce at any given wind speed. The maximum value from the wind power curve may be used in marketing wind turbines and for comparisons between competing models, so the values are sometimes higher than the actual output. If you are using power curves as part of a purchasing decision, you may want to request actual performance data from several installed wind turbines. Some manufacturers use projections and calculations to determine the values on the wind power curve, which may make their projections higher than the output is in real-world applications.

The rated power is used to compare similar wind turbines under standard conditions. **Peak power** is the amount of electrical power the wind turbine can produce at the highest rated wind speed. The wind turbine may not be able to produce power on a continual basis at the peak power rating because wind speeds that high do not normally occur on a continual basis, and the wind turbine generator and gearbox may not able to handle that exceptional load for any sustained period of time.

Testing by Independent Laboratories

Power curve testing should be done by independent testing laboratories as well as by manufacturers. If the data is a set of testing criteria, the same type of data must be collected for different models. Testing agencies are set up throughout the world for these tests. In the United States, one government agency that performs testing is the National Renewable Energy Laboratory (NREL) of the US Department of Energy (DOE).

NREL provides tests in conjunction with other agencies such as the National Aeronautics and Space Administration (NASA). NASA has a large wind tunnel located at Moffett Field in California. A wind tunnel is a testing laboratory designed specifically to create large wind flows under controlled conditions. Originally, wind tunnels were used to test aircraft wings and aircraft stability under operating conditions. The wind tunnel is a large dome filled with compressed air; the air is slowly released to pass through a tunnel where the wind turbine blades and rotor are positioned. The release of the air into the wind tunnel creates a realistic simulation of wind blowing at the height and level where a wind turbine would normally be located on top of its tower. Because the wind turbine is mounted near the ground in the tunnel, it can easily have a large number of sensors connected to it to provide vast amounts of data when it is under load. The wind tunnel at Moffett Field is 24.4 m × 36.6 m (80 ft × 120 ft) in area, and it can produce low- and medium-wind speeds to test wind turbines. The data from these tests is very important because it is collected by a scientific laboratory that is entirely independent from any manufacturer.

In 2008, NREL begin testing small wind turbines with outputs of less than 100 kW. Mostly small companies manufacture these wind turbines, and comprehensive testing is

Wind Turbine Testing

(*Source:* Photo courtesy of US Department of Energy.)

NASA used funding from the National Science Foundation (NSF) and the Energy Research and Development Administration (ERDA) to test a wind turbine for wake turbulence. The test in the photo used smoke from the tip of the turbine to study wake turbulence downwind from the blades.

not cost-effective for these small companies. Until recently, it had been difficult to get data from independent sources. NREL has established test parameters so that data for different wind turbines can be compared.

Measuring Generator Output

Another major part of determining the performance of a wind turbine is to measure the efficiency of the generator. The output of the electrical generator changes with the shaft rotational speed. The field current of some generators can be adjusted so that their efficiency remains fairly constant over a range of speeds. When the generator efficiency varies with speed, the generator speed may not be optimized for a given turbine speed, which makes the overall efficiency of the wind turbine less than its rating. A double peak in the output power may occur at different wind speeds: the first when the turbine blades reach their maximum efficiency and the second when the generator reaches its maximum output.

SECTION 7-2 CHECKUP

1. What is a wind power curve?

2. List two independent testing centers for testing wind turbine power curves.

3. Explain the importance of using actual measured data to compare the efficiency of two wind turbines rather than using projected data.

4. What causes the output of a generator to change? What can be adjusted to make the generator efficiency fairly constant?

7-3   Betz's Law

When wind turbines harvest wind energy, the amount harvested is always less than the wind energy. The German physicist Albert Betz studied this process and developed an equation that relates energy in the wind to the maximum amount that can be harvested.

Work of Albert Betz

In 1919, Albert Betz studied wind energy and determined that the best efficiency that a wind turbine can achieve is 59.3% for an ideal wind turbine (one with no hub, weightless rotors, no drag or friction, etc.). In 1926, Betz developed an equation, now known as **Betz's law**, that can be applied to any turbine with a disk like rotor. The 59.3% limit is an ideal or maximum theoretical efficiency (n max) of a wind turbine (also referred to as the power coefficient.) This is the ratio of maximum power obtained from the wind to the total power available in the wind. The factor 0.593 is also known as Betz's coefficient, which is usually abbreviated as C_p.

Efficiency of the Gearbox and Generator

The overall efficiency of a wind turbine includes the efficiency of the mechanical system, which is the rotor, gearbox, shafts, and generator. Betz's law indicates that the maximum efficiency a wind turbine can have is 59.3%, but in reality all wind turbines have lower efficiency because of various losses (friction, etc.). Wind turbines are designed to be as efficient as possible, so each part of the mechanical assembly is tested to determine its efficiency and to reduce friction and other losses. If a component experiences problems during operation that affect the efficiency (such as loss of lubrication), the monitoring sensors can alert operators to the difficulty and maintenance can be performed immediately. On larger systems that generate 2 MW and up, a small increase in efficiency produces a fairly large increase in the energy harvested.

Calculating Power from a Wind Turbine

The formula for total *theoretical* power in the wind for a given rotor is derived from our basic kinetic energy equation, which finds the kinetic energy that the air passing the turbine blades has in 1 second. (Recall that power involves the *rate* at which energy passes the turbine.) The equation for wind power is derived by first finding the kinetic energy of the mass of air that passes through the rotor. The kinetic energy is:

$$W_{KE} = \frac{1}{2}mv^2$$

To find the power in air, divide W_{KE} by time:

$$P = \frac{\left(\frac{1}{2}mv^2\right)}{t} = \frac{mv^2}{2t}$$

The mass of air that passes the rotor each second (m/t) is the volume of air per second multiplied by its density. The volume of air per second is the sweep area times the velocity of the wind, v ; that is,

$$m/t \,(\text{kg/s}) = v\,(\text{m/s}) \times A\,(\text{m}^2) \times \rho_\alpha(\text{kg/m}^3)$$

Substituting into the power equation gives the *theoretical* power in the wind:

$$P_w = \frac{1}{2}A\,\rho_a v_w^3$$

where

P_w = wind power, in W

A = area swept by turbine, in m^2

v_w = velocity of the wind, in m/s

ρ_a = density of the air in the wind, in kg/m^3

The actual power that is harvested by a wind turbine is much less than the theoretical number. The actual power from a turbine is found by multiplying the theoretical power in the wind by the efficiency of the turbine. In general, the turbine efficiency ranges from 0.25 to 0.45 and is called the maximum power coefficient, which is dimensionless. From this, the power from a turbine in watts can be expressed as follows:

$$P_{\text{tur}} = \frac{1}{2}A\,\rho_a v_w^3 C_{\text{tur}} \qquad \text{Equation 7-3}$$

where

P_{tur} = power out of turbine in W

C_{tur} = the efficiency of the turbine (typically 0.25 to 0.45)

An important point to notice about Equation 7-3 is that the power out is proportional to the *cube* of wind speed. Thus, if the speed of the wind doubles, the power out will be eight times larger (if all other factors remain the same). It is also proportional to the *area* swept by the blades. Thus, blade length is related to power by the square. A blade that is twice as long as another blade sweeps out four times the area and thus produce four times more power (if all other factors remain the same).

EXAMPLE 7-4

Determine the output power of a wind turbine whose blades are 5.0 m long, when the wind speed is 5 m/s (11.2 mph), air density is 1.2 kg/m^3, and the maximum power coefficient for this wind turbine is 0.30.

Solution

$$P_{tur} = \frac{1}{2} A \, \rho_a v_w^3 \, C_{tur}$$

The area swept by a 5 m long blade is 78.5 m². By substituting, we get:

$$P_{tur} = \frac{1}{2} A \, \rho_a v_w^3 \, C_{tur}$$
$$= \frac{1}{2}(78.5 \text{ m}^2)(1.2 \text{ kg/m}^3)(5.0 \text{ m/s})^3(0.3) = 1767 \text{ W} = \mathbf{1.77 \text{ kW}}$$

Variable-Area Wind Turbine Blades

(*Source:* Courtesy of Frontier Wind, LLC.)

Energy Unlimited, Inc. has developed variable-length wind turbine blades. The blades are designed with a tip that automatically extends outward in response to light winds and retracts in stronger winds. The turbine can capture more energy in low-wind conditions and retract in high-wind conditions. The increase in efficiency associated with this innovative development can increase production by as much as 25%.

Calculating Sweep Area

To calculate how much power is available at any given wind speed, you must know the amount of sweep area (also called the swept area) for the blades. The **sweep area** is the amount of area the wind turbine blades cover when they rotate one revolution. You can calculate the blade sweep area using the length of one blade as the radius (ignoring the small contribution of the area of the hub). The following example illustrates this calculation.

EXAMPLE 7-5

Determine the sweep area of a wind turbine with blades that are 12 m long. Express the answer in m² and in ft².

Solution

$$A = \pi(r)^2 = \pi(12 \text{ m})^2 = \mathbf{452 \text{ m}^2}$$

There are 3.28 ft/m. The length of the blades in feet is:

$$(12 \text{ m})\left(\frac{3.28 \text{ ft}}{\text{m}}\right) = 39.4 \text{ ft}$$

$$A = \pi(39.4 \text{ ft})^2 = \mathbf{4870 \text{ ft}^2}$$

The swept area increases as the square of the radius. Table 7-2 shows the area swept for different length rotor blades.

Two spreadsheet tables illustrate how changing wind speed and diameter affects the power in a wind turbine. In Table 7-3, the blade length changes from 1 to 12 meters, and the output power is plotted for a constant 10 m/s wind speed and a coefficient of 0.36. Notice the power from the turbine goes up by a factor of 4 when the blade length doubles.

TABLE 7-2

Rotor Sweep Areas for Wind Turbines

Length of Rotor (m)	Length of Rotor (ft)	Sweep Area (m²)	Sweep Area (ft²)
1	3.281	3.14	34
2	6.562	12.6	135
4	13.124	50.3	541
6	19.686	113	1,218
8	26.248	201	2,164
10	32.81	314	3,382
12	39.372	452	4,870

TABLE 7-3
Power Out at Constant Velocity Wind as a Function of Blade Length

$V_w = 10$ m/s $= 22.4$ mph		$C_{tur} = 0.3$	
Length in m	Area in m^2	P_w (kW)	P_{tur} (kW)
1	3.14	1.88	0.57
2	12.6	7.56	2.27
4	50.3	30.18	9.05
6	113	67.80	20.34
8	201	120.60	36.18
10	314	188.40	56.52
12	452	271.20	81.36

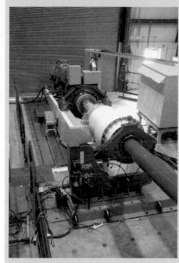

Wind Turbine Dynamometer Testing

(*Source:* National Renewable Energy Laboratory.)

A wind turbine generator and gearbox are connected to a dynamometer to test the output of the system under various conditions. These test are used to determine the actual efficiency of all of the components in the wind turbine drive train under various speeds at which the blades turn the rotor.

These tables do *not* take into account the need to reduce power when the maximum power is reached. The required flattening of the wind power curve depends on the power ratings for the specific turbine.

Table 7-4 illustrates the effect of wind velocity on a wind turbine with a 6 m blade and a coefficient of 0.36. The wind velocity is cubed in this formula, so the power increase as a function of wind speed rises dramatically as the wind speed increases. The effect of curtailing power at higher wind speeds is not included in the spreadsheet in Table 7-4.

Although the calculation of wind power illustrates important features about wind turbines, the best measure of wind turbine performance is annual energy output rather than power out. If it is possible to measure the actual amount of energy produced by a wind turbine over several years, you can get a good idea of how much energy a nearby wind turbine can produce.

A **wind farm** is a group of wind turbines in one area. When a location is being considered for a wind farm, the distribution of wind speed has to be evaluated. Using an average speed to calculate energy available in a given period can lead to grossly inaccurate results.

TABLE 7-4
Power Out at Constant Blade Length as a Function of Wind Velocity

$I = 6$ m (19.7 ft)		$C_{tur} = 0.3$	
v_w (m/s)	v_w (mph)	P_w (kW)	P_{tur} (kW)
2	4.48	0.54	0.16
4	8.96	4.34	1.30
6	13.44	14.64	4.39
8	17.92	34.71	10.41
10	22.4	67.80	20.34
12	26.88	117.16	35.15
14	31.36	186.04	55.81

EXAMPLE 7-6

A site has a wind that blows at a steady 10 m/s (22.4 mi/hr) for 12 hours and 5 m/s (11.2 mi/hr) for 12 hours. Assume a wind turbine with 10 m blades has an efficiency of 0.3. What is the total energy harvested?

(a) Calculate the energy for each 12-hour period and the sum.

(b) Repeat the calculation for the average velocity of 7.5 m/s for 24 hours.

(c) Compare answers and (a) and (b) and explain the difference.

Solution

(a) The swept area is:

$$A = \pi(r)^2 = \pi(10\ \text{m})^2 = 314\ \text{m}^2$$

The power for the first 12 hours is:

$$P_{tur} = \tfrac{1}{2} A\, p_a v_w^3\, C_{tur}$$
$$= \tfrac{1}{2}(314\ \text{m}^2)(1.2\ \text{kg/m}^3)(10\ \text{m/s})^3(0.3) = 56.5\ \text{kW}$$

The energy, W, for this 12-hour period is:

$$W = Pt = (56.5\ \text{kW})(12\ \text{h}) = \mathbf{678\ kWh}$$

The power for the second 12 hours is:

$$P_{tur} = \tfrac{1}{2} A\, p_a v_w^3\, C_{tur}$$
$$= \tfrac{1}{2}(314\ \text{m}^2)(1.2\ \text{kg/m}^3)(5\ \text{m/s})^3(0.3) = 7.06\ \text{kW}$$

Notice that the power is 1/8 as much. The energy, W, for the second 12-hour period is:

$$W = Pt = (7.06\ \text{kW})(12\ \text{hr}) = \mathbf{84.8\ kWh}$$

The total energy for the 24-hour period is the sum:

$$678\ \text{kWh} + 84.8\ \text{kWh} = \mathbf{763\ kWh}$$

(b) The average wind speed is 7.5 m/s. The power from this speed is:

$$P_{tur} = \tfrac{1}{2} A\, p_a v_w^3\, C_{tur}$$
$$= \tfrac{1}{2}(314\ \text{m}^2)(1.2\ \text{kg/m}^3)(7.5\ \text{m/s})^3(0.3) = 23.8\ \text{kW}$$

From this, the energy in 24 hours is:

$$W = Pt = (23.8\ \text{kW})(24\ \text{hr}) = \mathbf{572\ kWh}$$

(c) The total energy as calculated in part (a) is 33% larger than in part (b), illustrating that using the average wind speed leads to erroneous results. The reason is that the average does not account for the cubic relation of wind speed and energy.

Testing Wind Turbines

When a wind turbine is designed, it must be tested to verify the amount of electrical power it can produce at every given wind speed. Its cut-in and cut-out wind speed must also be determined and verified so that it can be published in the turbine data sheets. Some components of the wind turbine, such as the blades, rotor, gearbox, and generator, can be tested separately in a laboratory. This type of testing allows the engineers and technicians to determine the efficiency of the individual components before they are used as a complete system.

A wind turbine generator can be tested by connecting it to a dynamometer or a load bank, which allows the output power from the generator to be used. When the output load is varied as in real conditions, the generator shows the efficiency it will have in converting mechanical to electrical energy. The load on a dynamometer can be designed to run at a specific constant value over a specific period of time, or it can increase and decrease its load to show the effects of changing load conditions.

Finite-Element Analysis for Blade Design

The complex design of the turbine blade is calculated with computer programs that analyze blade design and efficiency. A popular computer program uses a technique called **finite-element analysis (FEA),** which employs complex mathematical algorithms to ensure its accuracy. Basically FEA breaks up a complex structure into a large number of small inter-acting zones, with the physical behavior of the zones described by mathematical models. The analysis can calculate stresses and loading in complex structures such as turbine blades. In addition to identifying any weaknesses in the blade, the program can simulate a destructive failure. The program can put specified loads on the turbine blade to calculate the detailed effects of changing wind speeds and the blades' response. Designers can make changes to the design of the blade and compare the efficiency of the blade and failure points before it is constructed.

Over time, FEA programs have become more and more accurate and have made the testing process more reliable. When combined with actual data from the field, FEA makes it easy to predict the efficiency of additional wind turbines of similar size.

Supervisory Control and Data Acquisition Verification

Today it is possible to collect a wide variety of data on any operating wind turbine with a **supervisory control and data acquisition (SCADA)** system. The data from the SCADA system provides information on wind velocity, wind direction, the amount of electrical power that is produced, and other performance data. When this data is gathered over a long period of time and stored in a computer databank, it can be analyzed so that very accurate projections can be developed. This method of projecting wind turbine efficiency is very accurate, especially when multiple wind turbines are erected on a similar site, such as on a wind farm. SCADA has allowed projections and calculations on wind turbine efficiencies to become more and more accurate, which makes it easier to determine the economic return on investment (ROI). SCADA is discussed further in Chapter 8.

SECTION 7-3 CHECKUP

1. What is Betz's law and how does it relate to wind turbines?

2. What is the relationship between wind speed and power from a turbine?

3. Identify the four parameters used to determine the amount of power that a wind turbine can produce.

4. Explain why finite element analysis (FEA) is used to design wind turbine blades.

5. Explain the benefit of SCADA data for developing wind turbine performance predictions.

7-4   Blade Aerodynamics

*The turbine blade on a wind generator is an **airfoil**, as is the wing on an airplane. By orienting an airplane wing so that it deflects air downward, a pressure difference is created that causes lift. On an airplane wing, the top surface is rounded, while the other surface is relatively flat, which helps direct air flow. The blade on a wind turbine can be thought of as a rotating wing, but the forces are different on a turbine due to rotation. This section introduces you to important concepts about turbine blades.*

A turbine blade is similar to a rotating wing. Differences in pressure cause the blades to both bend and rotate. In normal operation, the rounded front portion of the blades is oriented in the direction of rotation and the flat portion faces the wind. The front of the blade is referred to as the **leading edge** and the back is referred to as the **trailing edge,** as illustrated in Figure 7-8(a).

Lift and Drag

Lift is a component of an aerodynamic force exerted on a body that is perpendicular to a fluid (such as air) flowing past it. For an airplane wing, it is the force that lifts the plane,

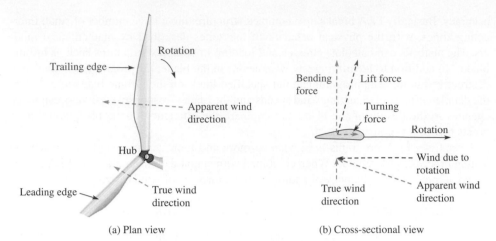

FIGURE 7-8   Air Moving Past a Turbine. (a) The rounded leading edge is oriented in the direction of rotation. (b) A lift force is created by pressure differences that are perpendicular to the apparent wind direction. This force tends to bend the blades and create a smaller rotational force.

hence the term *lift*. In a wind turbine, the term *lift* is a bit of a misnomer because it does not lift the blade; rather, it is a force exerted in a direction that is perpendicular to the *apparent* wind direction rather than the true direction. See Figure 7-8(b). In this case, lift is shown related to the airflow rather than the wing, as would be the case for an airplane wing. Assume the flat part of the blade is facing the true wind. As the blade turns, air that flows across the leading edge appears as a separate component of the wind; thus, the apparent wind direction is shifted to oppose the direction of rotation. The rotation of the blade causes a lift force that is perpendicular to the apparent wind direction. A small portion of this force goes toward turning the blade.

The lift force rotates with the blades so it constantly changes direction. The motion of the blades is opposed by the force required to spin the generator, friction in the system, and **drag**. The drag force is friction caused by air, which opposes the motion. This force is made as little as possible, so that as much of the lift as possible can go into useful work (turning the turbine).

Drag is expressed in terms of the drag coefficient, which is a dimensionless number. Typically, the only area of a wind turbine blade used in the calculation of drag is the front area (leading edge) of the blade. Design engineers aim for the smallest amount of drag. The smaller the drag, the more efficient the turbine is in harvesting wind energy. To reduce drag, blades are made relatively narrow.

A typical drag coefficient for wind turbine blades is 0.04; compare this to a well-designed automobile with a drag coefficient of 0.30. Even though the drag coefficient for a blade is fairly constant, as the wind speed increases, the amount of drag force also increases. The lower the drag coefficient number, the better the aerodynamic efficiency.

Angle of Attack

The angle at which the wind strikes the turbine blade is called the **angle of attack**. When the wind blows at a low angle over a blade, as shown in Figure 7-9(a), the blade has a certain amount of lift, as indicated by the vertical arrow. As the angle of attack increases, the lift also increases, as shown in Figure 7-9(b). At a steep angle of attack, turbulence begins, reducing lift and increasing drag; this point is called **stall** and is shown in Figure 7-9(c). Depending on the wind speed and blade shape, a **critical angle of attack** is reached, at which point the lift is at a maximum. At steeper angles, the turbine blade begins to lose its ability to convert energy from the wind.

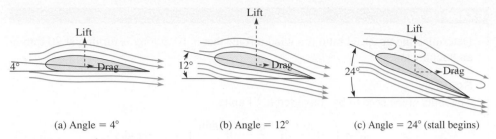

(a) Angle = 4°   (b) Angle = 12°   (c) Angle = 24° (stall begins)

FIGURE 7-9 Angle of Attack. As attack angle increases, lift increases until the airfoil begins to stall at a high angle of attack. At this point, turbulence begins, causing lift to decrease and drag to increase. Stall begins at this point.

Lift-to-Drag Ratio

The **lift-to-drag ratio** is a ratio of the lift force to the drag force, and it varies across the blade. The higher the lift-to-drag ratio, the more efficient the turbine blade is at converting wind energy into torque, which produces more electricity from the generator. Turbine blades have the highest lift-to-drag ratio near the tip of the blade. The blade has more material with very high strength near the hub because of the higher stresses in that region, and less material near the narrow tip. For this reason, lift is increased and drag is reduced near the tip.

Various factors affect drag, including the materials used to construct the blade, wind speed, air density, and air temperature. Even dirt and bugs on the blade affect drag. Drag is also affected by how the blades are oriented.

Stall occurs at very high angles to the wind and the blade no longer has lift. A wind turbine is subjected to the highest and lowest winds that flow at its location. When high winds occur, the turbine blades increase their speed, and the output of the generator may increase to the point at which the generator becomes overheated and damaged. Also, high winds may damage the turbine blades and the tower if the generator is allowed to increase its output at an uncontrolled rate.

Turbine Blade Tip Speeds

The tip of the turbine blade travels at the highest speed of any part of the turbine blade when it is rotating. Because of this speed, the tip passes more air as it travels and hence generates more lift. **Tip speed** is defined as the speed at the blade tip as it rotates through the air. Because the tip is rotating at the highest speed, it comes under considerable stress caused by centrifugal force. Blades are specified for a maximum tip speed and they are tapered to reduce lift at the ends because the faster-moving tip can still generate sufficient lift.

High tip speed is defined as speeds between 65 and 85 m/s, which is about 145 to 190 mph. High tip speeds are needed to make the turbine blade more efficient. At very high speeds, the turbine blade may receive too much stress, which can cause deterioration due to microfracturing. A turbine blade must be designed to withstand the maximum stress. A specification that is important is the ratio of the tip speed to the wind speed, or the **tip speed ratio (TSR)**. Tip speed can be determined from the rotational speed, which is ωR, where ω is the rotational speed in radians per second and R is the radius of the turbine in meters. The optimal tip speed ratio depends on the number of blades and is lower when there are more blades. For three blades, a TSR of 6 to 7 is optimal. If it is less, not all the available energy is captured; if it is more, the blades move into an area of turbulence from the last blade and are not as efficient. Because the speed of the blade is faster near the tip, the tip speed ratio is only valid at the radius used in the calculation. To optimize the angle of attack all along the blade, the blade is twisted from root to tip. The following example shows how to calculate the TSR from a known wind speed and rotational speed.

EXAMPLE 7-7

Determine the tip speed ratio if a wind turbine that is 10 m long is turning at 20 rpm in an 8 mph wind.

Solution

The units given need to be converted to SI units:

$$\omega = \left(20\,\frac{\text{rev}}{\text{min}}\right)\left(\frac{2\pi\,\text{rad}}{\text{rev}}\right)\left(\frac{\text{min}}{60\,\text{s}}\right) = 2.09\,\text{rad/s}$$

The tip speed is $\omega R = (2.09\,\text{rad/s})(10\,\text{m}) = 20.9\,\text{m/s}$.

$$8\,\text{mi/hr} = 3.58\,\text{m/s}$$

$$\text{TSR} = \frac{\text{tip speed}}{\text{wind speed}} = \frac{20.9\,\text{m/s}}{3.58\,\text{m/s}} = \mathbf{5.84}$$

Twist

Because of the difference in speed along the blade, the optimum angle for the tip is not the same as the optimum angle of the main part of the blade. A twist is added along the length of the blade to optimize the amount of energy harvested. Typically, 10° to 20° of twist is included, with the twist at the tip being the highest. This produces a change in the apparent wind direction across the blade. Recall that if the rotational speed is zero, the true wind direction and the apparent wind direction are the same.

SECTION 7-4 CHECKUP

1. What is lift?
2. How is drag different on a wind turbine than on an airplane wing?
3. What is the lift-to-drag ratio?
4. How do you calculate tip speed?
5. What is TSR?

7-5   Horizontal-Axis Wind Turbine

This section introduces the horizontal-axis wind turbine (HAWT), which is by far the most common type of wind turbine. Horizontal-axis wind turbines may produce less than 100 kW for basic applications and residential use, or as much as 6 MW for offshore power generation. Even larger turbines are on the drawing board.

How Wind Turbines Create Electricity

The **horizontal-axis wind turbine (HAWT)** is a wind turbine in which the main rotor shaft is pointed in the direction of the wind to extract power. The principal components of a basic HAWT are shown in Figure 7-10. The rotor receives energy from the wind and produces torque on a low-speed shaft. The low-speed shaft transfers the energy to a gearbox, high-speed shaft, and generator, which are enclosed in the nacelle for protection. Notice how the blades are connected to the rotor and to the shaft. This shaft is called the low-speed shaft because the wind turns the rotating assembly at a leisurely 10 to 20 revolutions per minute (rpm) typically.

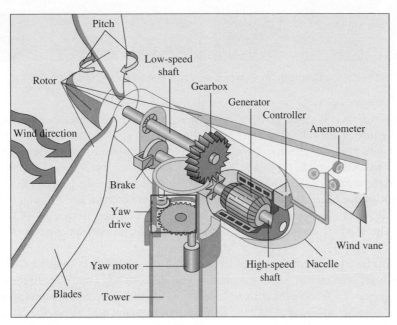

FIGURE 7-10   Basic Parts of a Horizontal-Axis Wind Turbine   (*Source:* US Department of Energy.)

The low-speed shaft connects to the gearbox, which has a set of gears that increase the output speed of the shaft to approximately 1,800 rpm for an output frequency of 60 Hz (or a speed of 1,500 rpm if the frequency is 50 Hz). For this reason, the shaft from the gearbox is called the high-speed shaft. The high-speed shaft is then connected to the generator, which converts the rotational motion to ac voltage. This speed is critical if it is used to turn the generator directly because the frequency of the ac from the generator is related directly to the rate at which it is turned.

Almost all horizontal-axis wind turbines have similar components to those discussed in this chapter, but there are some exceptions. For example, direct-drive wind turbines do not have a gearbox, and they usually have a dc generator rather than an ac generator. These may or may not include a converter to ac (which can be located at the tower base). In commercial turbines, a computer or programmable logic controller (PLC) is the controller. The controller takes data from an anemometer to determine the direction the wind turbine should be pointed, how to optimize the energy harvested, or how to prevent overspeeding in the event of high winds.

Controlling the output frequency and keeping it constant despite varying winds can be done in one of three ways. One way is to control the speed at which the generator shaft turns, which can be accomplished by adjusting the pitch and yaw. **Pitch** is the rotational angle of the blades on a wind turbine; **yaw** is the direction the wind turbine blades and nacelle are facing. Pitch and yaw can be adjusted so that a high-speed shaft runs at a constant rate to produce the required output frequency (typically 50 Hz or 60 Hz) from the generator.

HAWTs may also use a gearbox or set of gears, which changes the slow rotation of the blades into a faster rotation for the generator. The optimum blade rotation is generally between 10 and 20 rpm, and the gear ratio can be used to make the high-speed shaft rotate at the speed the generator requires.

The second method for controlling the frequency is to allow the turbine to run freely at any speed that is within its ratings and send the voltage to a **power electronic frequency converter.** This method is also used with vertical-axis wind turbines (VAWTs). When a frequency converter is used, the rotational speed of the turbine is not controlled until the maximum speed is reached, at which point speed controls take over. The frequency converter consists of the features shown in Figure 7-11.

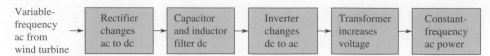

| Variable-frequency ac from wind turbine | Rectifier changes ac to dc | Capacitor and inductor filter dc | Inverter changes dc to ac | Transformer increases voltage | Constant-frequency ac power |

FIGURE 7-11   Block Diagram for Power Electronic Frequency Converter

The inverter accepts single-phase or three-phase ac to its input circuits within a specified range of frequency and voltage level. The ac is filtered and converted to dc by the rectifier and smoothed with passive filters to remove any trace of the input frequency. The next section has an inverter that converts the dc voltage back to single-phase or three-phase ac voltage at the precise frequency and phase required by the grid. This method has the advantage of having a wider range of operating conditions without requiring more complicated gearing.

A few applications can use pure dc, which can be obtained from a point before the inverter. The internal parts for a commercial power electronic frequency converter are shown in Figure 7-12.

The third way to control the output frequency of the generator is to use a double-feed, inductive-type generator in which the ac field current is tightly controlled to the required output frequency by feeding the current through an electronic circuit that produces an exact frequency. Another option is to take the field current from the grid. When the field current is a specific frequency, the generator output frequency is exactly the same. This method is discussed in more detail in Chapter 13.

Towers

The tower for a HAWT may be 40 to 100 m (approximately 130 to 328 ft) high so that it is tall enough to position the turbine blade into the strongest wind flow. Most sites have the strongest winds well above ground level. Today, most towers for larger wind turbines used to produce electrical power for utilities are in the range of 65 to 100 m tall. The Encore E126, recently installed in Germany, has a tower that is 138 m (453 ft) high. The blades are located on the main shaft, on a rotor at a considerable distance in front of the tower, so they are far enough out to clear the tower when the blades are rotating.

World's Largest Horizontal-Axis Wind Turbine

(*Source:* Courtesy of Vestas Wind Systems A/S.)

Vestas has plans for the world's largest wind turbine. The blades for this wind turbine will be 164 meters (538 feet) in diameter and will have a rated capacity of 8 megawatts. The new wind turbine will be an offshore wind turbine located near Aberdeen Bay in Scotland.

FIGURE 7-12   Power Electronic Frequency Converter   (*Source:* Tom Kissell.)

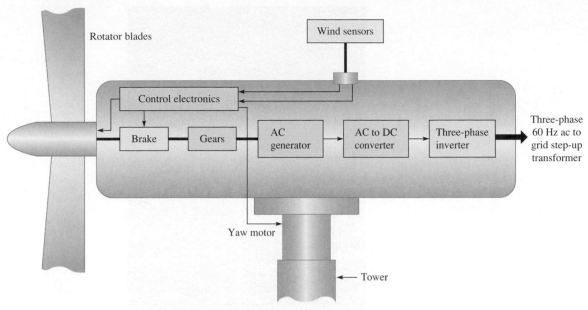

FIGURE 7-13 Typical Horizontal-Axis Wind Turbine (HAWT)

Controller

The blade pitch and the direction the turbine faces have already been described as functions monitored by the controller. The controller also uses sensors to measure the generator output (voltage and frequency), turbine blade speed, vibration, turbine and drive train parameters, and other parameters such as number of complete rotations of the vertical axis (yaw control). Some systems limit the number of full rotations made by the turbine yaw in one direction before reversing directions. The data from these sensors is usually stored for operators to review as necessary. Figure 7-13 shows all the parts in the horizontal-axis wind turbine (HAWT).

Number of Blades

HAWTs may be designed with one, two, three, or more blades. The fewer blades a wind turbine has, the faster the blades must turn to harvest the same amount of energy as a wind turbine with more blades. For example, a three-blade wind turbine does not have to turn as fast as a two-blade wind turbine to harvest the same amount of energy. Therefore, the tip speed ratios of a two-blade wind turbine and a three-blade wind turbine are different.

Smaller, residential-size units are designed for cost efficiency and the size of the electrical load of the home. Turbines used for commercial production of electric power may be two-blade, three-blade or five-blade, all of which are designed for much larger energy loads. The vast majority of horizontal-axis wind turbines used in the commercial production of power for utility companies are three-blade turbines.

Single-Blade Turbines

Single-blade wind turbines are used in a few limited applications, but they are the least used of all the HAWTs. To rotate smoothly, single-blade turbines must have one or two counterbalances. Figure 7-14 shows a single-blade wind turbine with two counterbalances. The advantage of this type of wind turbine is the lower cost because of the use of only one turbine blade (and the small weight savings), but single-blade turbines must run at much higher speeds to convert the same amount of energy from the wind as two-blade or three-blade turbines with the same size blades. Because the single-blade turbine must run at higher speeds, more wear and fatigue are generated on the blade and bearings in the mounting mechanism, which in turn means higher maintenance costs over the life of the turbine. Single-blade turbines also require extensive setup procedures to ensure that the

FIGURE 7-14 Single-Blade Horizontal-Axis Wind Turbine with Two Counterbalances. (*Source:* Courtesy of Powerhouse Wind LTD. Image courtesy of Pat Wall.)

FIGURE 7-15   Typical Two-Blade Wind Turbine   (*Source:* Tom Kissell.)

blade is mounted perfectly and is balanced to limit oscillation and vibration. Because of these problems, very few single-blade turbines are in use today.

Two-Blade Wind Turbines

Compared to three-blade turbines, two-blade wind turbines have the advantage of saving on the cost and the weight of the third rotor blade, but they have the disadvantage of requiring higher rotational speed to yield the same energy output. This is a disadvantage in terms of both noise and wear of critical bearings, shafts, and gearboxes. Two-blade turbines have experienced high-fatigue failures of the blade and other mechanical parts, so they have limited application. Figure 7-15 shows a two-blade wind turbine.

Another way to improve the efficiency of the two-blade turbine is to make the two blades thicker and wider than traditional turbine blades so that the two blades can convert more wind energy. The thicker blades also mean that the blades are stronger and better able to resist fatigue problems. New composite materials allow the increased size without adding substantial weight to each blade. These materials also allow the blade to be produced at a lower cost. Even with these more efficient blades, however, the two-blade turbine is still slightly less efficient than the three-blade turbine.

One advantage to a two-blade turbine is that it is faster and safer to install than the three-blade version. The two-blade turbine can be lifted into position after the turbine blades have been mounted while it is still on the ground because the blades can be mounted in a horizontal position and easily lifted as a unit. A three-blade turbine always has one blade pointing downward if it is raised as a unit, so it is more difficult to get the larger wind turbines off the ground as a unit for mounting.

Three-Blade Wind Turbines

The majority of large horizontal-axis wind turbines use three blades, with the rotor position maintained upwind by the yaw control. Figure 7-16 shows a three-blade wind turbine. The three blades provide the most energy conversion while limiting noise and vibration. The three blades provide more blade surface for converting wind energy into electrical energy than a two-blade or single-blade wind turbine.

FIGURE 7-16   Three-Blade Wind Turbine   (*Source:* David Buchla.)

The blades for the larger horizontal-axis wind turbines are so large they must be transported individually by a truck and trailer. This also means that one or more very large cranes are needed to set the tower and turbine in place. The tower to hold the larger three-blade turbine must also be larger and reinforced to support the weight and to withstand the increased wind power that is harvested to produce its maximum output. The blades on larger three-blade wind turbines are typically installed one at a time after the nacelle is mounted on the tower. On smaller three-blade turbines, the blades can be mounted to the rotor while the rotor is on the ground. Then the entire rotor assembly is lifted with a crane and attached to the shaft after the nacelle is mounted on the tower.

Five-Blade Wind Turbines

A few wind turbines have five blades to produce electrical energy efficiently from low-speed winds. Figure 7-17 shows a five-blade wind turbine. A five-blade wind generator normally has narrower and thinner blades, which creates issues with strength. While they are excellent in low-speed winds, they become inefficient in high-speed winds and they are noisier. The tower and base are mounted into the roof of the building, which is a concrete-reinforced building. This type of five-blade wind turbine needs a very strong base and tower to hold the wind turbine in the wind. Notice the thickness of the tower and the cowling around the blades, which helps direct wind directly into the blades.

Comparison of Blade Types

Wind turbine blades can be compared in a number of ways, such as by size, weight, material, and the way they are manufactured. Wind turbine blades can be made from a variety of materials, from wood for smaller blades to aluminum and other metals for small and medium-size blades. Turbine blades must be stiff enough to prevent the blade tips from being pushed into the tower by high winds, yet agile enough to convert wind power into electricity efficiently.

FIGURE 7-17   Five-Blade Wind Turbine   (*Source:* Tom Kissell.)

The largest commercial wind turbine blades are made of composite materials (carbon composition, plastics, and fiberglass), which makes them lighter in weight yet strong enough to hold up in high winds. The core may be filled with plastic foam or other light-weight substance to add rigidity. A typical fiberglass blade for a 100 kW wind turbine is 9 m (30 ft) long; a typical blade for a 2 megawatt wind turbine is 45 m long. Blade Dynamics is a wind turbine developer in the UK that is developing a blade that will measure between 80 and 100 m long! The blade will be made from carbon fiber and assembled from smaller pieces. It will be used for future turbines in the 8-10 MW range. Table 7-5 summarizes advantages and disadvantages of single-, two-, and three-blade wind turbines.

TABLE 7-5

Advantages and Disadvantages of Single-, Two-, and Three-Blade Wind Turbines

Type of Wind Turbine	Advantages	Disadvantages
Three-Blade Turbine	1. Quietest of the three types of turbines 2. Least amount of vibration 3. Available blade pitch control allows blade to catch maximum amount of wind 4. Lowest energy cost when compared to other turbines with similar size blades	1. Heavier than single- and two-blade turbines 2. Most capital-expensive of the three types 3. Requires active yaw control to make blades face into the wind 4. Requires the largest cranes to construct 5. Requires the largest and heaviest tower 6. Larger blades are more difficult to transport to the tower site
Two-Blade Turbine	1. Initial cost and weight are lower and they are simpler to mount 2. Produces more energy than single-blade turbine	1. Noisier than the three-blade turbine 2. Produces less energy than the three-blade turbine (when blade size and speed are the same)
Single-Blade Turbine	1. Least expensive 2. Easiest to erect because of its light weight and because the blade can be mounted while it is on the ground 3. Requires the smallest and lightest tower	1. Noisier than the three-blade turbine 2. Must run at highest speed to produce the same amount of electrical power 3. Most prone to vibration at the blade

1. What are the major parts of a horizontal-axis wind turbine?

2. If a one-blade rotor and a two-blade rotor of the same diameter are producing the same power with a certain wind speed, will there be a difference in noise level? Explain.

3. What do you think is the primary reason that three-blade rotors are more widely used than other types?

4. Why does a single-blade wind turbine need one or more counterbalances?

5. Identify three ways a wind turbine can provide voltage at the correct frequency for the grid.

7-6   Vertical-Axis Wind Turbine

Vertical-axis wind turbines come in one of two basic types: the Darrieus wind turbine, which looks like an eggbeater, and the Savonius turbine, which uses large scooped cups. Vertical-axis wind turbines were tested and used more extensively in the 1980s and 1990s because they were quieter and could operate without requiring yaw controls, regardless of the wind's direction. This section will explain the operation of vertical-axis wind turbines and discuss their advantages and disadvantages.

The **vertical-axis wind turbine (VAWT)** is a wind turbine that has its main rotational axis oriented in the vertical direction. VAWTs were innovative designs that have not proven as efficient in general as HAWTs, but they have a few good features, including quiet operation. Because they are not as efficient as HAWTs, they are rarely used in large units. Most VAWTs are smaller units that can be located in residential and commercial locations because they are much quieter than the horizontal-axis turbines.

The two types of vertical-axis wind turbines are the **Darrieus wind turbine,** which turns a shaft using lift forces, and the **Savonius wind turbine**, whose cups are pushed by direct wind forces. Vertical-axis wind turbines can produce electrical power at lower speeds and at a variety of changing speeds. Because they vary widely in speed, the ac generators they use do not produce a constant output. Usually the output goes to an inverter that converts it to standard ac (either single-phase or three-phase). Another option is to use dc as the output.

Darrieus Vertical-Axis Wind Turbine

Figure 7-18 shows a typical Darrieus vertical-axis wind turbine. The physical appearance of the Darrieus wind turbine looks like a large egg beater. The blade is mounted on a large monopole, and the generator is located at the bottom of the blade. The top of the pole has a number of guy wires that hold the pole in place when the force of the wind causes the blade to rotate. Figure 7-19 shows the internal parts of the Darrieus wind turbine.

When the Darrieus blade is operating, it is moving through the air in a circular path. The oncoming airflow generates a net force pointing obliquely forward and is projected inward past the turbine axis at a certain distance, giving a positive torque to the shaft. This helps the blade to rotate in the direction it is already traveling. The action of this blade is similar to the aerodynamic principles used in helicopters, and it makes the operation of this type of wind turbine quieter than a horizontal-axis wind turbine of the same size. Because there is less friction on the blade, the blade can rotate with equal torque regardless of the wind's direction.

A problem with the Darrieus wind turbine is that it is not self-starting, so it uses its generator as a motor to get the rotor started. As the wind increases the blade speed, power to the motor is turned off and it begins working as a generator. Darrieus wind turbines were installed on early wind farms, but most of them have been taken out of use in commercial applications because they are not as efficient as HAWTs and they require constant maintenance.

Smaller Vertical-Axis Wind Turbines

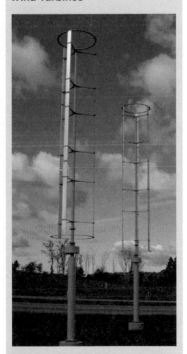

(*Source:* Tom Kissell.)

Smaller vertical-axis wind turbines operate well in urban environments, where they provide quiet, vibration-free operation. In urban areas, wind speed and directions are frequently changing, and wind speeds tend to be lower because of buildings and other objects that create wind shadows. Vertical-axis wind turbines can generate voltage at low wind speeds, and they do not have to change direction to catch usable wind.

FIGURE 7-18   Darrieus Wind Turbine   (*Source:* US Department of Energy.)

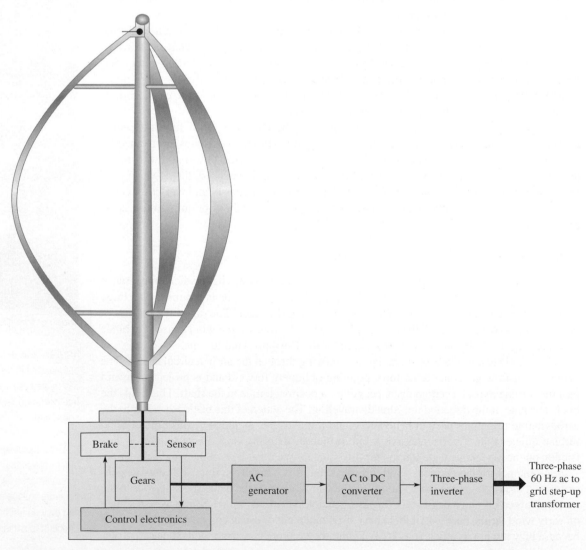

FIGURE 7-19   Internal Parts of the Darrieus Wind Turbine

FIGURE 7-20 Quietrevolution QR5 Vertical-Axis Wind Turbines   (*Source:* National Renewable Energy Laboratory.)

Quietrevolution VAWTs

Figure 7-17 shows another type of vertical-axis wind turbine called the Quietrevolution QR5. This unit, a variation on the Darrieus wind turbine, quietly produces enough energy to supply a home or small office even at lower or variable wind speeds. The unit requires a three-phase electrical connection, which may not be available at some smaller commercial or residential sites. At the present time, it is used only in the United Kingdom. The rotor is approximately 5 m (16.4 ft) tall and 3.1 m (10.17 ft) in diameter, and it weighs approximately 450 kg (990 lb). The blades are on a 3.4-m (11.15-ft) shaft where height restrictions are a concern; the blades are on a larger one when permitted.

The QR5 turbine uses a direct-drive, permanent magnet synchronous generator integrated into the base of the rotor. The peak power with wind speeds of 14 m/s is 7.4 kW and 6.2 kW dc. The British Wind Energy Association (BWEA) rated power is 8.5 kW aerodynamic power, which is rated at a wind speed of 16 m/s. After converting to electrical power, the output is 7.0 kW dc or 6.5 kW grid quality ac. The QR5 turbine can provide between 5,000 and 11,000 kWh per year, depending on the amount of wind at the site. The aerodynamic helical blade results in a very smooth, nearly silent operation.

Savonius Vertical-Axis Wind Turbine

The Savonius vertical-axis wind turbine uses cups, called scoops, instead of blades to capture wind power. Figure 7-21 shows an example of a Savonius vertical-axis wind turbine. When the wind blows, it creates a positive force in the scoop and a negative force on the back side of the scoop. This difference in force pushes the turbine around. In a typical Savonius turbine, the wind comes from the front of the cylinder, causing rotation. However, wind also strikes the back of the other scoops, tending to slow the rotor.

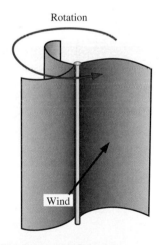

FIGURE 7-21   Savonius Vertical-Axis Wind Turbine

Advantages and Disadvantages to VAWTs

Table 7-6 on page 242 lists the advantages and disadvantages of vertical-axis wind turbines.

SECTION 7-6 CHECKUP

1. What are the two types of vertical-axis wind turbines?

2. Explain how a Darrieus wind turbine is different from a Savonius wind turbine.

3. Why are vertical-axis wind turbines used in cities more often than are horizontal-axis wind turbines?

TABLE 7-6

Advantages and Disadvantages of VAWTs

Advantages	Disadvantages
Quieter and less vibration than horizontal-axis wind turbines	Requires power and a starting motor to start the Darrieus wind turbine
Does not need a yaw control because it can produce electricity regardless of the direction the wind is blowing	Need guy wires to ensure the pole stays vertical so blade rotation is smooth
Produces electrical energy at very low wind speeds	Not as efficient as horizontal-axis wind turbines
Slower blade speeds because the blades are closer to the axis of rotation.	
Can be installed in urban environments and residential and commercial areas because they operate at low noise levels	
More eye pleasing than some larger horizontal wind turbines; can be coated with colors to match buildings and surroundings because the smaller turbines are mounted low to the ground near buildings	
Vertical-axis towers are much shorter than horizontal-axis wind turbines.	
Generator is generally mounted closer to the ground, so a crane is not needed for servicing.	

7-7   Wind Farms

Some large wind farms have several hundred wind turbines. Grouping wind turbines in close proximity to each other has many advantages, which are discussed in this section.

Figure 7-22 shows a portion of a large wind farm. Large wind farms take full advantage of a favorable wind location. An economic advantage to a wind farm is that several wind

FIGURE 7-22   Part of a Large Wind Farm. Notice the transmission lines, which are dwarfed by the huge turbines.   (*Source:* David Buchla.)

turbines may share transmission lines or a large transformer that is needed to increase the voltage to the grid. When large wind farms are in the planning stage, the substation, transformers, and capacity of the grid are all important design considerations for designing the system. The grid itself is basically the transmission and distribution system. (Transmission and distribution systems are covered in Chapter 14.)

Another advantage of wind farms is that the turbines can have shared periodic inspections. Thus, a single large crane can be used to service all of the turbines at the same time, which reduces maintenance cost. Many times, the cost of transporting a crane is over half the cost of maintenance. In some cases, a dedicated crane can be purchased and maintained on the wind farm, so that it is readily available. A full-time maintenance crew may be assigned to the wind farm. If the wind farm is smaller and has a few wind turbines, the equipment to maintain them is not always available, and the maintenance personnel are generally contracted for short periods of time.

The trend is toward the establishment of larger wind farms. The largest installations can provide power that is comparable to a coal-fired plant. These wind farms produce sufficient power to make additional construction of other types of electrical power plants unnecessary. Some of the larger energy producers, such as BP, General Electric, and Siemens, are involved in large wind projects in the United States and other locations worldwide. Large wind farms, including those offshore, have been proposed by other energy companies throughout the world. One offshore wind farm in Scotland is planned to produce 1.5 gigawatts (GW) of power, about the same as a large nuclear power station.

The world's two largest wind farms, each with over 1 terawatt of power, are in the United States and India (Jaisalmer Wind Park). In the United States, the Alta Wind Energy Center in the Tehachapi Pass of California is the largest operational wind farm in the world. Many wind farms with nameplate capacities over 300 MW are located principally in the United States and China. The midwestern United States is one of the world's best wind regions (refer to the map in Figure 1-15, in Chapter 1). Offshore wind farms tend to be smaller; the largest one currently is Greater Gabbard in the United Kingdom, with over 500 GW listed as the power output. Because of the rapid growth in wind energy (20% to 30% annually), expect larger wind farms in the future. Several large ones are already under construction.

> **Wind Turbine Capacity**
>
> The rated capacity of a wind turbine is the amount of electrical power its generator could possibly produce if the maximum amount of wind was blowing across its blades every hour of the day, every day of the year. Because the average wind speed is much less than the maximum amount for much of the time, the amount of electrical power that the wind turbine can provide to power a home is much less than its generator's rating.

Wind Energy Production

New wind farms are coming online around the world and old ones are expanded. Energy production from wind is growing very rapidly; worldwide growth has been about 25% per year. For any given year, the actual statistics are available in the Statistical Review of World Energy, an annual detailed report from BP. The top wind-producing nations are China, the United States, Germany, Spain, and India, with China leading all other nations in production.

In the United States, government incentives to build wind farms and more efficient wind turbines have dramatically increased the wind energy produced from wind farms in recent years. Typically growth in new capacity has been 20% to 30% per year.

Small Community Wind Farm Projects

Many small communities have control over their electric utility company and have choices in where they purchase their electrical energy. In these communities, two options are to install wind turbines or lease land to companies that install and operate wind turbines so that the electrical energy from them can be used by the community. This energy offsets the electrical energy the community would otherwise have to purchase off the grid. Figure 7-23 shows this type of installation: four 1.8 megawatt, utility-scale wind turbines installed in Bowling Green, Ohio. The wind turbines provide enough electricity to power approximately 1,800 homes. An advantage to small localized wind farms like this is that they can provide backup power to the community when the grid goes down.

FIGURE 7-23   Four Wind Turbines at a Small Wind Farm in Bowling Green, Ohio.   (*Source:* Tom Kissell.)

Nysted Wind Farm, Denmark

(*Source:* Courtesy of Siemens AG Energy Sector.)

The Nysted Wind Farm is a joint Danish–Swedish wind farm that was installed in 2003 near Lolland, Denmark. It has 72 Siemens wind turbines, each producing 2.3 megawatts. The 72 wind turbines give a total generating capacity of 166 megawatts. Lolland is the fourth largest island of Denmark and is located in the Baltic Sea. The personnel who work on these wind turbines travel to and from them in boats.

World Offshore Wind Farms

England, Scotland, Denmark, and Germany have large numbers of **offshore wind farms** that have been in production for over 20 years. An offshore wind farm is a group of wind turbines that are all located offshore in the ocean or large body of water. Newer and larger wind turbines are being installed on offshore wind farms to provide additional electrical energy for these countries.

Offshore wind turbine locations are generally considered to be at least 10 km (6 miles) or more from land. The offshore wind turbines have several advantages over those located on inland areas. For example, they are less obtrusive at sea than are turbines on land because their apparent size and noise is limited due to the distance they are located from shore. The average wind speed at sea is usually higher, and the winds last longer than they do over land. Offshore towers can generally be taller than onshore towers because there are no height restrictions as there are on land, so they can reach for higher wind speeds.

A disadvantage with offshore installations is that they are more expensive than onshore installations, primarily because of the requirements for the underwater structures and installation costs (such as a crane). In the North Sea, four of the monopod structures weigh 750 tons each and are driven 20 m (77 ft) into the seabed. Some will be in water as deep as 50 m (164 ft). Saltwater causes substantial and continuous corrosion, which raises the overall cost of maintenance (painting, etc.). Because of maintenance issues, it is more practical to install the largest possible wind turbines to harvest the most energy with the fewest number of turbines. Power transmission from offshore turbines is also an issue; it is transmitted through underwater cables, often using high-voltage dc for longer distances. Other issues include effects on marine life and bird migration, as well as hazards to navigation.

Offshore Installations in the United States

The United States has excellent wind resources offshore on the Pacific Coast, Atlantic Coast, Gulf of Mexico, and the Great Lakes (see Figure 1-14 for a US map). Offshore locations are excellent for larger wind turbines (3 megawatts and larger). Some 5 mega-watt turbines are being constructed in the North Sea. The United States has plans to test wind turbines in Texas that would be located in the Gulf of Mexico, 8.5 miles from

Galveston. When planners are determining the locations for offshore wind turbines, they must consider weather problems such as hurricanes, ice in winter, and the depth of the water for the towers.

Spacing Wind Turbines on Wind Farms to Limit Turbulence

Wind turbines on larger wind farms typically have rotor diameters of about 100 meters (328 feet). In the past, their towers have been spaced about seven rotor diameters (700 m, or 2,300 ft) apart. New research suggests placing wind turbines approximately 15 rotor diameters (1,500 m, or over 4,900 ft) apart to allow each wind turbine to generate more electrical power and to increase its efficiency. When the wind blows past wind turbine blades, it creates wind turbulence in its wake. Wind turbines located in the wake of one in front intercept turbulence, which decreases the amount of energy available. Turbulence can also be caused by terrain and obstacles such as trees, buildings, or other structures.

A factor in the placement of wind turbines on wind farms is the height of the towers. Modern wind turbine towers on wind farms are 80 m (262 ft), and some of the newest ones are 100 m (328 ft) tall. The taller towers allow wind turbines to have larger and larger blades. Larger blades allow them to harvest more wind energy that is converted into electrical power. Researchers have found that the taller towers also create a turbulence that draws powerful kinetic energy from higher altitudes when the wind is blowing. Because they are located close to one another, it is easy to connect multiple wind turbines in a wind farm to substations and the grid.

SECTION 7-7 CHECKUP

1. What are the advantages of large wind farms over smaller ones?

2. What are the best areas for wind farms in the United States?

3. What are some advantages and disadvantages to offshore wind farms?

4. What causes turbulence? How can its effects be avoided?

Wind Turbine Blade Damage from Lightning Strikes and Insects

(*Source:* National Renewable Energy Laboratory.)

Wind turbine blades, towers, and generators can be damaged by lightning. Lightning protection and grounding systems are used on wind turbines to protect them from lightning strikes. If a blade is damaged by lightning, it must be repaired and its fiberglass structure must be returned to its original form. Wind turbine blades can also be damaged by insect strikes on their leading edge. Special coatings have been created for wind turbine blades to protect them from surface damage and to allow them to be cleaned more easily.

CHAPTER SUMMARY

- A pressure gradient is the change in air pressure as an air mass moves across a given distance.

- Boundary-layer wind is wind that is close to the surface of the earth. A sea breeze is created whenever there is a difference between the water temperature and the land temperature. In the summer, when the sun is high in the sky, it shines on the land and the sea and adds heat to both. The land warms more quickly than the water, however, so each day, the land may warm to perhaps 80° to 85° F while the water remains around 75° F.

- When the wind blows, it is moving from a region of high air pressure to a region of lower air pressure.

- In the United States, the prevailing winds blow from the West Coast to the East Coast.

- Wind turbulence is a condition in which the wind does not blow in a straight line and does not blow continually at the same speed. Sometimes these winds are called swirling winds.

- The data collected through the supervisory control and data acquisitions (SCADA) system is put into a data bank where it can be analyzed.

- A power curve is a graph that shows the wind speed and the output power of a wind turbine over a range of wind speeds from zero to the maximum wind speed for which the wind turbine is designed.

- Betz's law states that the ideal efficiency of a wind turbine with a disk like rotor is 59.3%. Practical wind turbines will always have a lower efficiency than this.

- A computer program uses a technique called finite-element analysis (FEA) to design and test wind turbine blades, and it employs complex mathematical algorithms to ensure its accuracy.

- The turbine blade on a wind generator is an airfoil.

- The angle at which the wind strikes the turbine blade is called the angle of attack.

- Drag is the force that opposes the motion of the airfoil as it moves through the air.

- The lift-to-drag ratio is a ratio of the value of lift to the value of drag. A higher lift value and a lower drag value provide a higher lift-to-drag ratio.

- Tip speed is defined as the measured speed at the blade tip as it rotates through the air.
- Three bladed horizontal-axis wind turbines are the most widely used type and have the lowest cost for the energy harvested.
- The two basic types of vertical-axis wind turbines are the Darrieus wind turbine, which looks like an eggbeater, and the Savonius turbine.

- A wind farm is a group of wind turbines that may consist of two or more turbines located in one area. Some large wind farms have more than 100 wind turbines, all of which are placed close to each other so they can share the electrical grid connection, transformers, and maintenance equipment.

KEY TERMS

angle of attack   The angle at which the wind strikes turbine blades.

barometric pressure   The pressure exerted by the atmosphere on the earth. Barometric pressure at sea level is 760 mm Hg, 101 kPa, or 14.7 psi.

Betz's law   A formula originated by Albert Betz that states that the highest possible efficiency that a wind turbine can achieve is approximately 59%.

drag   The force that opposes the motion of the airfoil as it moves through the air.

horizontal-axis wind turbine (HAWT)   A wind turbine in which the main rotor shaft is pointed in the direction of the wind to extract power.

lift   A component of an aerodynamic force exerted on a body that is perpendicular to a fluid (such as air) flowing past it.

lift-to-drag ratio   The ratio of the value of lift force to the value of drag force

millibar   A unit of atmospheric pressure equal to 100 Pa.

nacelle   An enclosure or housing on a wind turbine for the generator, gearbox, and any other parts of the wind turbine that are on the top of the tower.

pitch   The rotational angle of the blades on a wind turbine.

power curve   A graph that shows the wind speed and the output power of a wind turbine over a range of wind speeds.

pressure gradient   The change or variation in atmospheric pressure per unit of horizontal distance in the direction in which the pressure changes most rapidly. It is expressed in units of pressure per unit length.

stall   A reduction in the lift force as the angle of attack increases beyond some point.

tip speed ratio (TSR)   The ratio of the tip speed to the wind speed.

vertical-axis wind turbine (VAWT)   A wind turbine that has its main rotational axis oriented in the vertical direction.

wind farm   A group of wind turbines in one area.

yaw   Direction that the wind turbine blades and nacelle are facing.

FORMULAS

Equation 7-1	$p = F/A$	Definition of pressure
Equation 7-2	$\rho = \dfrac{m}{Vol}$	Definition of density
Equation 7-3	$P_{tur} = \frac{1}{2}A \times \rho_a \times v_w^3 \times C_{tur}$	Turbine power

CHAPTER TRUE/FALSE QUIZ

Determine whether each statement is true or false. Answers are at the end of the chapter.

1. Pressure gradient force is one of the main forces acting on the air to make it move as wind.

2. The turbine blade on a wind generator is called an airfoil.

3. A higher lift value and a lower drag value provide a lower lift-to-drag ratio.

4. Generally, the pressure gradient force is directed from high-pressure toward low-pressure zones, which causes wind to move from high-pressure areas toward low-pressure areas.

5. Wind turbulence is a condition in which the wind blows in a straight line and blows continually at the same speed.

6. A computer program called finite-element analysis (FEA) employs complex mathematical algorithms; it is used to design and test wind turbine blades.

7. The angle at which the wind strikes the turbine blade is called the angle of attack.

8. Stall occurs when the turbine blade no longer has lift.

9. Tip speed is defined as the measured speed at the blade tip as it rotates through the air.

10. Drag is the force that opposes the motion of the airfoil as it moves through the air.

11. Single-blade wind turbines are the most used of all the horizontal-axis wind turbines.

12. Two-blade wind turbines yield the same energy output as three-blade wind turbines (when blade size and speed are the same).

13. When the wind turbine blade rotates, the tip of a wind turbine blade travels at the same speed as the base of the blade where it is connected to the rotor.

14. Gradient wind blows at a constant speed and flows parallel to imaginary curved isobars just above the earth's surface.

15. The Darrieus wind turbine is a horizontal-axis wind turbine.

CHAPTER MULTIPLE-CHOICE QUIZ

Complete each statement by selecting the one correct answer. Answers are at the end of the chapter.

1. What is the angle of attack in reference to a wind turbine blade?
  a. The speed at which the turbine blades are turning
  b. The number of turbine blades divided by the speed at which they are turning
  c. The angle at which the wind strikes the blade

2. Betz's law and the efficiency of energy conversion for a wind turbine indicates that the maximum amount of wind energy that an ideal wind turbine can harvest is
  a. approximately 50%
  b. approximately 59%
  c. approximatcly 75%
  d. approximately 99%

3. Small residential wind turbine systems may need an electronic inverter if
  a. the wind turbine generator is used to charge batteries in the home, and all the loads in the home require ac voltage
  b. the wind turbine generator produces dc voltage, which needs to be converted to ac voltage
  c. the wind turbine blade speed is usually not controlled and can turn the generator shaft at any speed; an inverter can control the frequency of the ac voltage output
  d. all of these are true

4. What is the difference between active and passive stall control of wind turbines?
  a. Active stall control uses the design of the blades to create the stall condition; passive stall control uses a hydraulic or mechanical system to change the pitch and thus cause the stall condition.
  b. Active stall control uses a hydraulic or mechanical system to change the pitch; passive stall control uses the design conditions of the blade to create the stall condition.
  c. Both active and passive stall control use hydraulic or mechanical systems to cause the stall condition.

5. A wind power curve shows the
  a. wind speed in miles per hour or meters per second and the amount of electrical power that is produced at that speed
  b. speed of the blades and the frequency for the output at any speed
  c. torque on the blades at any given wind speed

6. Which type of wind turbine blades need counterbalances?
  a. Three-blade wind turbines
  b. Two-blade wind turbines

  c. Single-blade wind turbines
  d. None of the horizontal-axis wind turbines

7. Lift is produced on the wind turbine blade
  a. when the wind blows across the blade from the leading edge to the trailing edge
  b. when the wind blows across the blade from the trailing edge to the leading edge
  c. when the blade is set to the furled condition
  d. any time the wind blows, regardless of the direction from which it blows against the turbine blade

8. When the wind speed increases across a wind turbine blade, the
  a. lift and drag forces increase
  b. lift and drag forces decrease
  c. lift increases and the drag decreases
  d. lift decreases and the drag increases

9. Atmospheric pressure is approximately
  a. 100 psi at sea level
  b. 14.7 psi at sea level
  c. 20 psi at sea level

10. A sea breeze
  a. blows in toward the land
  b. blows away from the land and toward the sea
  c. may blow away from the land or toward the land at night

11. Which way do the prevailing winds blow in the United States?
  a. from east to west
  b. from north to south
  c. from west to east
  d. from south to north

12. Wind turbulence occurs when
  a. the wind blows in a straight line
  b. wind swirls and does not blow in a straight line
  c. wind blows more strongly in the winter

13. The three parts of the formula for determining the amount of power a wind turbine can create are
  a. wind velocity, the density of the air in the wind, and the diameter of the turbine blades
  b. wind direction, the density of the air in the wind, and the diameter of the turbine blades
  c. wind velocity, the humidity of the air in the wind, and the diameter of the turbine blades

14. What does SCADA stand for?
  a. supervisory control and data acquisition
  b. service center and data acquisition
  c. supervisory control and data applications

CHAPTER QUESTIONS AND PROBLEMS

1. What is the difference between turbulent and laminar flow?

2. What is boundary-layer wind?

3. Explain how a sea breeze is created.

4. Explain what causes a land breeze.

5. Explain how lift is created when wind blows across the blade of a wind turbine.

6. What is the lift-to-drag ratio?

7. Name two advantages of locating wind turbines over water.

8. What happens to wind power when the density of the air increases?

9. What is the effect on a wind turbine when the air becomes denser?

10. What is a power curve for a wind turbine and how is it used?

11. What is Betz's law and why is it important?

12. How is the energy in the wind related to wind speed?

13. How is available energy in the wind related to the length of the blades in a wind turbine?

14. Why is the constant rotational speed of the generator shaft important for certain wind turbines?

15. Why does the addition of an inverter or power electronic frequency converter allow a wind turbine to rotate at various speeds?

16. What is blade tip speed and why is it important?

17. What is the function of the gearbox in a horizontal-axis wind turbine?

18. What is an advantage of a vertical-axis wind turbine for a residential area?

19. Determine the sweep area of a wind turbine whose blades are 60 ft in diameter.

20. Determine the sweep area of a wind turbine whose blades are 40 m in diameter.

21. Determine the power output of a wind turbine whose blades are 12 m in diameter and when the wind speed is 6 m/s, the air density at sea level is about 1.2 kg/m^3, and the maximum power coefficient for this wind turbine is 0.35.

FOR DISCUSSION

Several proposals have been offered for harnessing winds at high altitudes with flying generators. Discuss the merits and problems with these proposals. Do you favor this approach?

ANSWERS TO CHECKUPS

Section 7-1 Checkup

1. Pascals, psi, atm

2. From cardinal directions (N, E, S, W) or by compass readings specifying the direction from which the wind is blowing

3. 1,016 millibars

Section 7-2 Checkup

1. A curve that relates wind speed to output power

2. In the United States, the National Renewable Energy Laboratory (NREL) and National Aeronautics and Space Administration (NASA)

3. Many variables affect performance; measured data allows potential buyers to compare turbines.

4. Variations in the shaft speed change the generator output; it can be made more efficient by adjusting the field current for the generator.

Section 7-3 Checkup

1. It is an equation developed by Albert Betz that indicates that the maximum theoretical efficiency (*n* max) of a wind turbine is 59.3%.

2. Power increases by the cube of wind speed.

3. Swept area, velocity of the wind, density of the air, and efficiency of the turbine

4. FEA can calculate turbine performance accurately and predict effects before expensive construction and testing is performed.

5. The SCADA data is developed over a long period of time, so it can make accurate projections of future performance.

Section 7-4 Checkup

1. It is a force on a blade or wing created by a pressure difference between two surfaces.

2. Drag is a friction force that always opposes motion. On an airplane wing, it is oriented toward the rear; on a wind turbine blade, it is a rotational force that is directed away from the blade motion.

3. The ratio of the lift force to the drag force and it varies across the blade

4. Tip speed $= \omega R$, where ω is the rotational speed in radians per second and R is the radius of the turbine in meters.

5. Tip speed ratio; it is the ratio of the tip speed to the wind speed.

Section 7-5 Checkup

1. The tower, rotor and rotor blades, low-speed shaft, gearbox, high-speed shaft, generator, and controller; there may also be electronic frequency converters

2. Yes. Because they are both producing the same power, the one-blade rotor is turning twice as fast as the two-blade rotor and is therefore noisier.

3. Three-blade turbines produce more energy for their investment than other types.

4. To prevent vibration

5. Three ways are (1) control the speed of the turbine using pitch and yaw control, (2) allow the turbine to free run and control the output frequency with an electronic converter system, and (3) use a double-feed inductive-type generator in which the ac field current is tightly controlled.

Section 7-6 Checkup

1. The Darrieus and the Savonius

2. The Darrieus turns a shaft using lift forces; the Savonius is pushed by direct wind forces.

3. They are quieter.

Section 7-7 Checkup

1. They can share facilities and maintenance equipment such as cranes.

2. Offshore and in the Midwest

3. Advantages: generally better wind speed, no obstructions, less obtrusive. Disadvantages: more expensive to install and maintain, maintenance issues with saltwater, and need for underwater cables.

4. Turbulence occurs as wind changes velocity past turbine blades and mixes with other air. Its effects are avoided by consideration of spacing.

ANSWERS TO TRUE/FALSE QUIZ

1. T   2. T   3. F   4. T   5. F   6. T   7. T   8. T   9. T
10. T   11. F   12. F   13. F   14. T   15. F

ANSWERS TO MULTIPLE-CHOICE QUESTIONS

1. c   2. b   3. d   4. b   5. a   6. c   7. a   8. a
9. b   10. a   11. c   12. b   13. a   14. a

Wind Turbine Control

CHAPTER OBJECTIVES

- Explain the purpose of a gearbox that is used on many horizontal-axis wind turbines.
- Describe the yaw control for a horizontal wind turbine.
- Explain the difference between and cite the advantages and disadvantages of upwind and downwind turbines.
- Identify the shafts and components in the drive train of a horizontal wind turbine.
- Explain the operation of hydraulic brakes on a wind turbine.
- Identify the basic parts of a closed-loop control for the blade pitch of a wind turbine.
- Calculate the current required in a 4 to 20 mA current loop required to rotate the blades a given amount.
- Calculate the gear ratio for a wind turbine given the input revolutions per minute (rpm) and the required output rpm.
- Explain why the speed of the generator in some wind turbines must be controlled.
- Explain where you will find brakes in a horizontal-axis wind turbine.
- Identify the parts of a direct-drive wind turbine.

KEY TERMS

Key terms are shown in bold and color. Definitions for key terms are provided at the end of the chapter and in the end-of-book glossary. Bold terms in black are defined in the end-of-book glossary only.

- passive stall control
- active stall control
- upwind turbine
- downwind turbine
- drive train
- drive train compliance
- direct-drive wind turbine

- anemometer
- wind direction indicator
- yaw brakes
- rotor brakes
- high-speed shaft brakes
- fail-safe brakes
- dynamic braking

INTRODUCTION

This chapter discusses control of wind turbines, including details of pitch and yaw control, active and passive stall control, the controller, braking, and related topics. Pitch control and yaw control were introduced briefly in Chapter 7, but both topics are covered in more detail in this chapter. Yaw control is used only with horizontal-axis wind turbines; vertical-axis wind turbines do not need yaw control. The yaw position is changed so that the blades are moved out of the wind and into a furled position when the wind speed is too strong. When the wind speed is strong, a governing system is activated to avoid damage due to stress or heating of bearings and other turbine elements. The chapter concludes with a discussion of braking systems for wind turbines. Wind turbines have brakes on the yaw drive, rotor, and high-speed shaft, and the location and function of each are explained.

8-1   Pitch and Yaw Control

Wind turbines require pitch and yaw control to optimize conditions for maintaining speed or for decreasing speed when necessary in high winds. These controls are also used to ensure that the wind turbine is never allowed to turn faster than its rated speed, thus avoiding damage to the blades. Small wind turbines may not use these control systems, but all commercial high-power units do.

Recall that the pitch of a turbine blade was defined in Chapter 7 as the angle formed between the direction in which the blades rotate and the tilt of the blades. Changing the pitch changes the attack angle and the force that winds apply to the blades, so controlling the pitch can be used to adjust the speed of the turbine and optimize output power. Pitch control is one of the principal ways of adjusting the speed. Speed can also be controlled by generator load control. In high wind conditions, pitch control can be used to turn the rotor to a large angle of attack, which starts to stall the blade, thus increasing drag and decreasing lift (refer again to Figure 7-9 in Chapter 7 for an illustration of stall). This action slows the rotor to prevent overspeeding.

The pitch requires constant adjustment to keep the rotational speed constant in many wind turbines. This adjustment is done with a controller that uses sensors to determine rotational speed and other parameters. Using automated systems, the pitch can be adjusted several times each second. Depending on the type of system, the controller may adjust the pitch to keep constant rotational speed or maximize the energy harvested. The optimum setting for the wind conditions and load is determined by the controller that sends data to the **pitch control,** which sets the blade angle to this optimal position.

Yaw was previously defined as the direction the wind turbine blades and nacelle are facing; the **yaw control** turns the nacelle to face the wind. The yaw control assembly consists of the yaw drive, which is a large gear, and the yaw motor, which has a gear mounted to its shaft that engages the larger yaw gear. The gear ratio between the smaller gear on the yaw motor and the larger gear on the yaw drive gives the motor sufficient power and torque to rotate the nacelle. All of these parts are mounted securely inside the nacelle.

In general, the purpose of the yaw drive is to point the turbine into the wind, but yaw control can also turn the rotor out of alignment with the wind when a given wind speed is exceeded or to help control the speed. Small wind turbines can use passive yaw control. **Passive yaw control** uses the power in the wind itself to orient the nacelle. A common arrangement for a small turbine is to use a tail fin at the back (like a weather vane) to orient the turbine; however, this tail fin is not satisfactory for large wind turbines because the wind cannot exert sufficient torque to turn the heavy nacelle.

In **active yaw control,** a yaw motor is coupled through a hydrodynamic coupling arrangement that introduces damping into the system to prevent the system from responding to small changes in wind direction, thus reducing tower oscillations and fatigue. The hydrodynamic coupling is connected to a yaw gearbox and drive pinion that causes the yaw ring on the nacelle to rotate the nacelle on the top of the tower. Figure 8-1 illustrates the basic motions for the pitch and yaw controls.

Wind turbines can be characterized as constant-speed turbines or variable-speed turbines. In both types, pitch control is used to adjust the speed. The constant-speed turbine is the most basic type of wind turbine. The goal with a constant-speed rotor is to keep the generator turning at constant speed and thus maintain the generator's frequency at the same frequency as the grid by constantly and rapidly adjusting the pitch for the conditions.

Most constant-speed turbines can withstand a little variation in the rotor speed (a few percentage points) without affecting the grid synchronization. As the wind turbine starts to spin the rotor, the induction machine acts as motor until it reaches a certain speed. This action helps start the rotor in low-wind conditions because wind turbines have low starting torque. At the synchronous speed, the stator supplies power back to the grid, at the grid frequency. Because small variations in speed can be tolerated, the constant-speed turbine can absorb gusts of wind that may change the speed slightly. In practice, a special type of induction generator (called a doubly-fed induction generator, which is covered in Chapter 13)

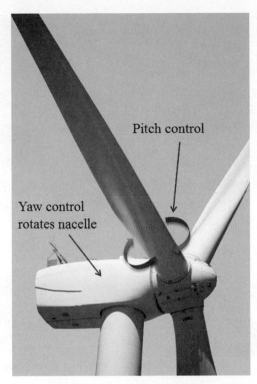

FIGURE 8-1 Pitch and Yaw Control. Pitch control rotates the blades at the point they are connected to the rotor to adjust the turbine speed. Yaw control moves the nacelle into or out of the wind; the motor and gear are internal to the tower.   (*Source:* National Renewable Energy Laboratory.)

is usually used because this type of induction generator can allow the rotor to turn above or below synchronous speed, which is a big advantage in wind turbines.

Variable-speed turbines have the ability to capture more of the wind energy, but they initially have variable ac output or a dc generator. In either case, the output needs to be converted to grid-quality ac before connecting it to the grid. Generators and control algorithms are different for variable-speed turbines than they are for constant-speed turbines, but the basic idea of a variable-speed turbine is to allow the turbine to operate with the optimum power efficiency over a wider range of wind speeds than an equivalent constant-speed turbine. For wind turbines that can produce power over a continuous range of rotor speeds, the tip speed ratio can be optimized by changing the pitch for various wind speeds, resulting in more efficient operation. (Recall from Chapter 7 that tip speed ratio is the ratio of the tip speed to the wind speed.)

If the generator produces a varying ac, it must be converted to the fixed-frequency, fixed-voltage ac required by the power grid using additional electronics. In general, the payoff warrants the additional cost, so most new turbines are variable-speed types. On average, a variable-speed turbine collects 10% more power over its lifetime than an equivalent constant-speed turbine.

In addition to adjusting the turbine blades for the optimum speed, pitch can be used to feather the blades. Feathering means turning the edge of the blades into the wind to reduce their angle of attack and minimize wind forces. In addition to feathering, brakes may be set to stop the blades completely. This is an important safety measure for maintenance or when wind speeds exceed the maximum rated number for safe operation. Figure 8-2 shows a wind turbine with feathered blades. Compare the blades in Figure 8-2 to the blades of a turbine producing power in Figure 8-3.

Hydraulic Pitch Control

A common way to control blade pitch rapidly and precisely is by adding hydraulic actuators on the rotor hub. The in-and-out motion of the hydraulic cylinder is converted to rotational

FIGURE 8-2 Wind Turbine with Blades Feathered for Maintenance   (*Source:* Tom Kissell.)

FIGURE 8-3 Wind Turbine That Is Producing Power. Notice that the blades are turned to present the flat side toward the wind. (*Source:* Tom Kissell.)

FIGURE 8-4 Hydraulic Cylinder Control of Blade Pitch (*Source:* Courtesy of Bosch Rexroth Group.)

motion for the blades. Figure 8-4 illustrates a hydraulic actuator. When the cylinder rod fully extends, the blade is rotated approximately 90°. This pitch control system uses a proportional hydraulic valve that controls how far the hydraulic cylinder is extended or retracted, which in turn controls the angle of the blade pitch down to 1° increments. The proportional hydraulic valve is usually controlled by a current signal from the controller. A common method is called a 4–20 mA loop, which uses an analog signal that ranges continuously between 4 mA (lowest level) to 20 mA (highest level) to make adjustments in the blade pitch. (A zero current indicates a loop failure.) The 4–20 mA signal loop tends to be unaffected by noise, so it is ideal in the noisy electrical environment of the wind turbine. The signal loop can have multiple receiving devices (but only one transmitter). The value of current sent by the controller (transmitter) determines the rotation of the hydraulic valve; the following example illustrates this idea.

EXAMPLE 8-1

Assume that a hydraulic proportional valve receives a 4–20 mA signal to control the hydraulic cylinder, which extends or retracts to set the position of the blade pitch between 0° and 90°.

(a) What current needs to be sent to the proportional valve to set the blade pitch to 45°?

(b) What current will set the blade pitch to 15°?

Solution

The number of mA in the full range of the 4–20 mA signal is 20 mA − 4 mA = 16 mA.
    The number of mA per degree of rotation is 16 mA/90° = 0.178 mA per degree.

(a) In 45°:

$$0.178 \text{ mA/degree} \times 45 \text{ degrees} = 8.0 \text{ mA}$$

To determine the current corresponding to 45°, add 8 mA to the 4 mA offset:

$$8.0\,\text{mA} + 4.0\,\text{mA} = \mathbf{12.0\,mA}$$

(b) In 15°:

$$0.178\,\text{mA/degree} \times 15° = 2.67\,\text{mA}$$

To determine the current corresponding to 15°, add 2.67 mA to the 4.0 mA offset:

$$2.67\,\text{mA} + 4.0\,\text{mA} = \mathbf{6.67\,mA}$$

Electrical Pitch Control

Another type of pitch control is electrical pitch control, which use batteries or capacitors to supply electricity. Typically, the batteries are located in the hub and have a limited life, thus requiring replacement. Electrical pitch control systems have certain advantages, such as avoiding any issues with hydraulic fluid (potential leaks or problems with viscosity in very cold climates). In addition, they use less power because hydraulic systems require a pump to be running constantly.

Mechanical Pitch Control

Large wind turbines use hydraulic power to change the blade pitch. In some smaller wind turbines, a mechanical system can be used to change the pitch of the blades using a strong spring. When the wind speed is less than the maximum rating for the wind turbine, the wind force operating against the spring causes the pitch to self-adjust so that the blade is harvesting the maximum amount of wind energy. When the wind speed increases and exceeds the maximum rating for the wind turbine, the spring tension changes the blade pitch to stall and thus slow down.

Passive Stall Control

Passive stall control is a design feature of turbine blades that causes them to create turbulence on the back of the blades in high winds, thus reducing the lift force that drives the rotor. This type of control is known as passive control because there are no moving parts to fail and no expensive controls are needed. The blades are attached to the hub at a fixed angle. This stall prevents the normal lift force from acting on the blades, thus slowing them. You may have noticed the twist in turbine blades; this twist induces stall gradually to reduce dramatic stresses on the blade.

Active Stall Control

Active stall control is a method of controlling power in large turbines by moving the pitch angle in and out of stall condition as necessary. In normal pitch control, the system is designed to avoid stall conditions. Active stall control regulates power by increasing pitch to match the stall threshold once rated power is achieved. The power is maintained at a constant level by moving the blades in and out of stall condition, and the blade design allows the onset of stall to be more gradual. At the stall point, lift is reduced and drag is increased, thus reducing power. The system can respond to short-term events (wind gusts) with the active stall control, while the pitch control system holds the average power constant. Active stall control makes it easier for the wind turbine to be brought under control if an emergency stop is necessary.

Individual Blade Pitch Control

In larger wind turbines, the latest type of pitch control is called individual blade pitch control, which allows maximum efficiency at all wind levels. The controller uses the sensors in this system to allow the pitch of each blade to be adjusted in small increments several times during each revolution. This means that the pitch of one blade may be adjusted slightly more or less than the other blades. During any one rotation, each blade may be adjusted separately to provide the best overall efficiency for a given wind speed. Data about blade motion is transmitted to the controller and displayed on a server on the ground. Individual control of the blades is especially useful when winds are very gusty and the wind changes direction and speed frequently.

Yaw Mechanism

The yaw mechanism on a wind turbine rotates the nacelle to control the direction in which the nacelle and blades are aimed at the wind. The yaw drive is generally moved by high-torque motors. The nacelle can be rotated so that the blades face directly into the wind when the turbine is trying to harvest the maximum amount of wind, or it can move the blades out of the wind when the winds are too strong and could damage the wind turbine.

The yaw ring gear is a fixed gear at the top of the tower. It has gear teeth that mesh with the yaw drive motors. When the drive motor shaft rotates, the entire nacelle rotates. Figure 8-5 shows the yaw mechanism, with the motors shown in blue. The yaw control can use up to seven yaw motors to move the yaw ring on the largest wind turbines because the nacelle and blades on these turbines are so large. A small wind turbine may have only one yaw motor.

Yaw Drive Motors

The yaw drive can be powered by an electric or a hydraulic motor. Ideally, it should move the nacelle slowly and avoid constant small adjustments that increase wear and stress. An electric yaw drive motor includes a gear reducer to produce high torque and thus move the nacelle easily but relatively slowly. Permanent magnet motors work well as yaw drives because they produce very high torque at low speed. The hydraulic motor is an excellent motor for producing very high torque at low speed, which allows the yaw motor to move the heavy nacelle with minimum energy expended.

FIGURE 8-5 Yaw Motors and Ring Gear for a Large Turbine. The gear on each drive motor engages the fixed yaw ring gear to rotate the nacelle. (*Source:* Courtesy of Bosch Rexroth Group.)

Yaw Control

As you learned in Chapter 7, yaw control sends signals to the yaw drive motors to move the nacelle to point into the wind. When the wind speed is above the maximum safe rating, the yaw control sends signals to move the blades out of the wind. The yaw drive motors are controlled by the programmable logic controller (PLC) or computer controller. The controller uses onboard instrument data to determine the pointing direction.

Figure 8-6 shows two large wind turbines with their blades pointing in slightly different directions. This difference in direction is due to variation in wind direction and the way the control system in each turbine adjusts the pitch and direction of the blades through the yaw control to maximize the wind energy harvested. These two wind turbines are located within a few hundred meters of each other, and the wind direction and speed is varying enough at this distance to cause the yaw mechanism control on each wind turbine to move its respective turbine to a slightly different position.

Wind turbines have cables to transmit power from the rotating nacelle to the stationary tower. Most turbines use a yaw counter to ensure that the nacelle makes only four to six complete revolutions in any direction before it is forced back in the opposite direction to avoid winding the cables inside. If the nacelle were allowed to rotate continually in one direction, it could cause the electrical cables to twist and possibly break. On systems in which a PLC or computer controls the yaw position, the computer control sends a signal to reverse the rotation several times in order to put slack back into the cable from the generator.

Another way to compensate for the yaw rotating more than 360° is to use slip rings. Slip rings allow a fixed electrical connection to be made to a rotating assembly by using brushes to make contact with rotating rings. Specialized slip rings for wind turbines are designed for up to several hundred amps at 600 V. Other types of slip rings pass data signals from the turbine to the ground.

Slip rings can wear and generally require checking and minor maintenance continually throughout the life of the turbine. One problem with slip rings is that they can arc. Arcing roughs up the surface with minor pitting, which increases the electrical resistance. To avoid arcing problems, redundant brushes can be employed to prevent one brush from skipping at the same time as another. Because the slip rings are at the top of the tower, where the turbine is mounted, checking them requires a technician to climb the tower and perform the maintenance.

FIGURE 8-6 Two Wind Turbines in Close Proximity. These two wind turbines are located close to each other, but their yaw mechanisms are pointing the blades of the two turbines in slightly different directions.   (*Source:* Tom Kissell.)

1. How is the pitch adjusted for maintenance?

2. Why is pitch variable on large wind turbines?

3. What is active stall control?

4. What is passive yaw control?

5. Why are yaw motors designed to be low speed and high torque?

8-2 Turbine Orientation

The most common type of wind turbine is the horizontal-axis wind turbine, and the most common configuration is three blades. The horizontal-axis wind turbine is mounted on the top of a pole or tower. The wind turbine can be designed so that the wind blows over the blades and then over the generator nacelle, or the wind can blow over the nacelle first and then through the blades.

Upwind Turbine

An **upwind turbine** is designed so that the wind blows over the blades and then over the nacelle and tower, as shown in Figure 8-7. The rotor blades need to be rather rigid and placed sufficiently far in front of the tower to avoid striking the tower in high winds. Large upwind turbines require a yaw mechanism to point the turbine into the wind. On smaller wind turbines, the yaw control can be a simple rudder. By far, the majority of turbines are upwind types.

Direction of wind for upwind turbine

FIGURE 8-7 Upwind Turbine. The wind blows over the blades first and then over the nacelle and tower.   (*Source:* Tom Kissell.)

The advantage to the upwind design is that there is less effect from the tower on the wind because the turbine is upstream from the tower. Still, there is a small effect as the wind starts to bend the blades away from the tower. This effect creates a drop in power as the blades pass the tower; however, the problem is less pronounced in the upwind configuration. The effect of the tower has another more subtle problem associated with it. As blades pass the tower, they flex slightly. The flexing can cause potential damage, and it can produce a period of oscillation. For example, if a three-blade turbine turns so that a blade passes the tower every second, it can set up a vibration of 1 Hz. If this frequency corresponds to a resonant structural frequency, damage can result. Engineers must account for these possible structural resonances and avoid them.

The upwind turbine has a rather rigid and inflexible rotor, and it is placed some distance in front of the tower so that the blades can safely flex somewhat in stronger winds. If the blades are not mounted far enough in front of the tower, strong winds can make the blades flex back into the tower and possibly cause damage. Some earlier upwind machines did not account for flexing of the blades in stronger winds, and the blades flexed into the tower and were severely damaged.

Two approaches to increasing the tip clearance from the tower are to (1) move the rotor further from the tower or (2) tilt the nacelle up so that the bottom clearance of the blades increases. Both of these methods are used to some extent. The problem with extending the rotor from the tower is that the yaw bearing must be larger, which makes it more costly. Tilting the nacelle loses some efficiency, but a few degrees results in a minor amount of power loss (a 5° tilt loses about 0.4% of the available power.)

Wind turbine blades for large upwind turbines need to be designed with less flex than blades for downwind turbines. To keep the blades light and yet keep them rigid, they are made from fiber-reinforced plastics, fiberglass, or carbon fibers. Carbon fibers are stronger and stiffer, but they tend to be much more expensive and so are limited in use. Except in small turbines, metals and wood are generally not suitable materials for turbine blades because they do not have a high strength-to-weight ratio. Blades are hollow and contain internal structure, made with spar caps and shear webs, to stiffen them. Figure 8-8 shows two structures using spar caps and shear webs to stiffen the blades. Designers must trade blade efficiency, which calls for thin lightweight blades, with structural strength requirements, which are better with thick heavy blades.

In addition to the internal consideration in blade design, the shape of the blade and width changes from the root (the part closest to the hub) to the tip. The root needs to be strong enough to support the blade. The blade's strength can be reduced further from its root, but keep in mind that the speed at the tip is much faster than at points closer to the hub. Thus, the blade is twisted along the length to keep the angles correct. The calculation of the tradeoff in blade design requires a computer that uses finite element analysis to complete a design.

Blades need to be tested for various potential failure modes and more. In the United States, the Wind Technology Testing Center (WTTC), located in the Charleston Harbor district of Boston, Massachusetts, can test blades up to 90 m in length. Blade testing is required as part of turbine certification for deploying wind turbines to avoid catastrophic failures. Two incidents in which very large blades have broken loose and fallen to the ground occurred in 2013, so this type of testing continues to be important.

Blade Bending

(*Source:* Tom Kissell.)

In high winds, the tips of the blades are bent toward the tower. Damage is possible if the blades strike the tower. The blades on large wind turbines must be set far enough out to avoid touching the tower in the strongest winds. The blades are also reinforced so that they do not bend so much at the tip ends. The constant forces on the blades produce stress, which illustrates the need for regular inspections to detect cracks.

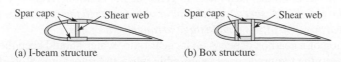

(a) I-beam structure   (b) Box structure

FIGURE 8-8   Cross-Sectional View of the Internal Structures of Wind Turbine Blades

FIGURE 8-9 Downwind Turbine. The wind blows over the nacelle first and then the blades.   (*Source:* Tom Kissell.)

Downwind Turbine

In the **downwind turbine**, the wind blows over the nacelle first and then over the rotor and blades. Thus, the blades and rotor are on the downwind or lee side of the tower. In this case, any flexing of the blades moves the tips of the blades away from the tower. Figure 8-9 shows an example of a downwind turbine. On some smaller downwind turbines, the wind itself adjusts the yaw so that the blades remain directed into the wind, which is another type of passive yaw control.

In general, wind shading from the tower exacerbates the problems mentioned with upwind turbines. The tower can create turbulence behind it that makes the blades less efficient. The blades tend to flex and can have resonances; however, the blades do not tend to bend toward the tower, which is an advantage. Some smaller downwind turbines are designed so that they orient to the wind direction without a yaw drive, saving the cost of a yaw drive motor and gearing. In this case, yaw rotation must be controlled or limited because of the electrical cables, as mentioned previously. Another option is to use slip rings, as mentioned previously, but this option is best only for smaller turbines. Slip rings are not useful for very large turbines with high current output.

SECTION 8-2 CHECKUP

1. What is the difference between an upwind and a downwind turbine?

2. What problems are created by tower shading?

3. What are the tradeoffs that designers must consider when designing a turbine blade?

4. What is one advantage of downwind turbines?

8-3   Drive Train Gearing and Direct-Drive Turbines

In some larger wind turbines, the blades and rotor are designed to rotate fairly slowly, but the generator needs to rotate much faster to generate the required frequency. It is common to use a gearbox to change speed from the input of the rotor to the requirement for the generator. The gearbox typically is a multistage arrangement of gears.

The selection of a gearbox depends on its gear ratio, which should keep the output shaft speed as close to the required speed as possible when the rotor turns at its design speed. The required rpm for the generator depends on the grid frequency and the number of poles in the generator (this topic is discussed in Chapter 13). For many wind turbines, the rotor is designed to turn very slowly (about 12 to 34 rpm), but the generator may have to turn at 1,500 or 1,800 rpm to generate the correct frequency for the grid. If you are converting between rpm and Hz, keep in mind that rpm is revolutions per *minute* and hertz is cycles per *second*, so there is a factor of 60 to account for in the conversion.

EXAMPLE 8-2

What is the rotor frequency, in Hertz, of a wind turbine rotor that turns at 18 rpm?

Solution

Set up the units to cancel:

$$\left(18\frac{\text{revolutions}}{\text{minute}}\right)\left(\frac{\text{minute}}{60\text{ seconds}}\right) = 0.30\text{ Hertz}$$

The gearbox is an integral component that is manufactured and assembled as a subsystem prior to being installed on the wind turbine. Figure 8-10 shows a large gearbox. This gearbox has its own lubrication and cooling system integrated into it so that it can operate as a self-contained system. The gearbox can have liquid cooling or air cooling, depending on its size and load. Lubrication is important, and the gearbox requires regular oil changes, just like a car. The process of changing oil, particularly on a tower located in the ocean, is a complicated and expensive procedure, so it is important to use high-quality lubricants to extend their life.

The gearbox ratio is the same as the ratio of the required generator speed to the rotor speed; that is, generator speed/rotor speed = gear ratio. The application of this formula is demonstrated in Example 8-3.

EXAMPLE 8-3

Determine the gear ratio for a wind turbine transmission if the generator needs to turn at 1,800 rpm and the rotor turns at 20 rpm.

Solution

Use the formula for determining the gear ratio: generator speed/rotor speed = gear ratio.

$$1,800\text{ rpm}/20\text{ rpm} = 90{:}1$$

The gear ratio is **90:1.**

FIGURE 8-10 Cutaway View of a Wind Turbine Gearbox   (*Source:* Courtesy of Bosch Rexroth Group.)

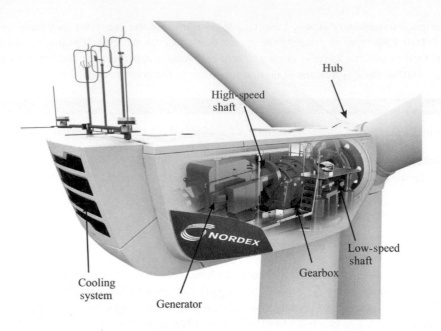

FIGURE 8-11   Drive Train of a Horizontal Wind Turbine   (*Source:* Courtesy of Nordex SE.)

The gearbox can have a large rotor plate that can be used with a mechanical braking system. The mechanical brakes provide friction pads that come into contact with this rotor plate and cause the gear system to slow down and stop the rotation when the system is locked down for maintenance or when it is in danger of becoming damaged from overspeeding. Braking is discussed in more detail in Section 8-5.

Drive Train

The **drive train** in a wind turbine consists of the turbine blades, rotor, low-speed shaft, gearbox, high-speed shaft, and generator. This set of components allows the wind turbine blades to harvest wind energy and transfer it through the shafts and gearbox to the generator, where it is converted to electrical energy. Figure 8-11 shows all of these components.

Drive Train Mounting Arrangement Options

The drive train components are mounted inside the nacelle on the nacelle bedplate. The nacelle bedplate consists of a very large metal plate that has the strength to support all the components and keep them in alignment. In many systems, the rotor, gearbox, and generator are mounted on the bedplate and are aligned precisely in a shop. After they are started up and tested, they are then lifted into the wind turbine nacelle as an assembled unit. This process ensures that the components are initially aligned and tested as an assembled unit, that much of the vibration is removed, and that other adjustments are made while the system is in the shop. Some turbine manufacturers use a test stand to test these components under operating conditions and while they are fully loaded in the shop so they can make any adjustments while the unit is on the ground. Figure 8-12 shows a typical gearbox and generator mounted as a unit, ready to be tested.

Drive Train Compliance

Drive train compliance is the correct alignment of all the components in the drive train. If any of the components do not align to a very tight tolerance, vibrations and premature wear on the components can result. Any misalignment of the shafts can cause them to rotate in an elliptical pattern, causing vibration and premature failure due to wear. Misalignment can be amplified where energy is transferred from one component to the next.

FIGURE 8-12   Gearbox and Generator Mounted as a Unit. The gearbox and generator are assembled together and tested as a complete unit prior to installation in a wind turbine nacelle.   (*Source:* National Renewable Energy Laboratory.)

Special equipment is used to ensure that the alignment is nearly perfect and also to measure vibration due to any imbalance. Vibration analysis is used to determine the source of any detected vibration. Adjustments can be made to alignments and/or weights added to points on the shafts to eliminate vibrations. Vibration monitors are used in some systems to detect any unwanted vibration as soon as it occurs. In systems without vibration monitors, the drive train must be checked periodically after a certain amount of runtime.

Figure 8-13 shows a wind turbine with its nacelle opened so that the shafts, gearbox, and generator can be inspected. When you work on a wind turbine, you may need to remove and replace the gearbox or the generator, and this work must be done through an open nacelle.

Direct-Drive Systems

A **direct-drive wind turbine** has its rotor connected directly to the generator; it does not have a gearbox. Figure 8-14 shows an example of a direct-drive wind turbine. The output frequency of the generator for a direct-drive wind turbine is constantly changing as the wind changes the rotational speed of the turbine blades. Because the output frequency is constantly changing, the generator output is generally converted to dc, which is sent to an

FIGURE 8-13   Nacelle Open for Maintenance. The top of the nacelle is open so the gearbox and generator can have work done.   (*Source:* Courtesy of Siemens AG Energy Sector.)

Direct Drive Wind Turbines in Iceland

(*Source:* David Buchla)

Landsvirkjun is Iceland's largest producer of electricity and has significant hydropower and geothermal plants. They are currently testing two 900 kW direct drive turbines on the inland plains in Iceland, a location with almost constant winds. In particular the studies include wind measurements as a function of height, and the effect of icing, snow, and volcanic ash on turbine operation in the severe Iceland climate. They will also study the effect on birds and other wildlife. The direct drive turbines produce electricity at lower speed and thus produce less stress, which is expected to increase the years of service. In addition they require less maintenance, a significant advantage in Iceland. Results so far have been encouraging.

FIGURE 8-14   Direct-Drive Wind Turbine   (*Source:* Courtesy of Emergya Wind Technologies, www.ewinternational.com. Copyright 2010.)

electronic inverter for conversion back to ac. The inverter output is a constant frequency for the grid at the required frequency (50 Hz or 60 Hz). In the past, direct-drive wind turbines were not practical because the inverters were not large enough. Today inverters are large enough to handle all of the electrical power the direct-drive wind turbine produces. In some cases, the inverter can be on the ground to reduce the weight on the tower. Figure 8-15 shows a labeled diagram of a direct-drive wind turbine.

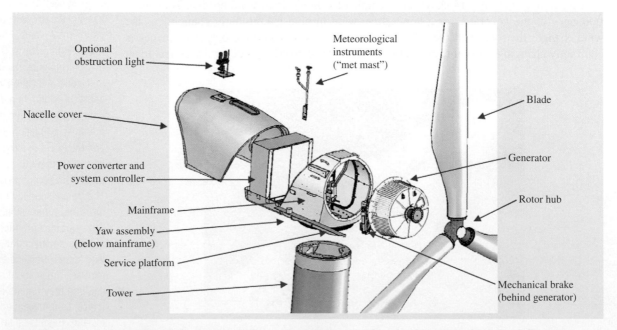

FIGURE 8-15   Labeled Diagram of a Direct-Drive Wind Turbine   (*Source:* Courtesy of Northern Power Systems Inc.)

The advantage of a turbine with no gearbox is a savings in moving parts that require regular maintenance and lubricants. Lower maintenance costs and fewer lubricants are particularly useful features for offshore turbines, where maintenance is more difficult. Several very large offshore turbines now use direct drive. The disadvantage is that a multipole synchronous generator is required to accommodate the slow speed. This type of generator generally uses permanent magnets that are made from costly rare earth materials. Thus, the initial capital cost is higher and the generator weighs more, but these costs are balanced by higher efficiency and lower maintenance costs. Experience with direct-drive units has now been accumulating, and they have proven to be as reliable as units with gears.

One compromise between geared and direct-drive turbines is a single-stage gearbox, which enables the generator to be smaller and lighter in weight while still offering many of the advantages of direct drive. Essentially it is a transmission that increases the speed at which the generator turns. The transmission is less complex than the multistage gearbox it replaces, but it still provides an increase in speed for the generator shaft.

SECTION 8-3 CHECKUP

1. What is the purpose of the gearbox in a wind turbine?
2. Name the parts in the drive train of a wind turbine.
3. Why is alignment of the drive train critical?
4. What is the fundamental difference between a direct-drive turbine and other turbines?
5. What are the advantages of direct-drive turbines?

8-4 Wind Measurement

Wind turbines have a variety of data requirements, such as wind speed, wind direction, generator voltage and current, power production, blade pitch, and maintenance issues such as the number of hours the blades have been rotating. This information is transmitted to a PLC or dedicated computer controller. The instruments measuring the wind are located on the nacelle.

Wind Turbine Instruments

Data is constantly acquired about the winds, and the information is gathered and transmitted through a data acquisition system connected to a network link. Figure 8-16 shows an example of a data access point link that takes data from the wind turbine and sends it over a telephone, Internet, or Ethernet connection. All of this data is sent to a central location and archived so that it can be retrieved and analyzed at a later date. Over time, a continuous record of the wind is created and can be matched to the turbine performance. This information also indicates which months produced the best winds and whether other wind turbines should be located nearby.

Some of the data that are gathered can pinpoint maintenance issues. For example, the total number of hours a generator produces electrical power, and the amount of voltage and current that is produced, can be tracked. This information is compared to maintenance records to determine the life of bearings, gears, shafts, and other mechanical parts of the wind turbine. The data collected from the wind turbine can be checked in real time or stored for later analysis in a spreadsheet or other program. Technicians can use this information to determine any problems with the system by comparing the wind data to the amount of electrical power produced.

Anemometer

The **anemometer** is an instrument that measures wind speed; it is mounted on the top of the nacelle, usually near the back. Figure 8-17 shows a typical arrangement. The signals

Ultrasonic Wind Measurement Instruments

(*Source:* Courtesy of R. M. Young Company, www.youngusa.com.)

An ultrasonic wind measurement instrument measures wind speed, wind direction, and air temperature. The ultrasonic wind instrument also has an integrated heating circuit that melts snow and ice during the winter months to ensure that it operates accurately. The ultrasonic sensor has multiple sensor tips, and an ultrasound signal is sent between the tips. The speed of the ultrasonic signal is constant when there is no wind; when the wind begins to blow, it causes the ultrasonic pulse to be changed, and this information is converted to wind direction and wind speed.

FIGURE 8-16   Data Access Point That Sends Data from the Wind Sensors (*Source:* Tom Kissell.)

from the anemometer are sent to the controller, where they are used to set blade pitch to the optimum angle.

Anemometers come in one of two types. In a rotational anemometer, the rotation of an element serves to measure wind speed. The second type, called a cooling-power anemometer, uses heat transfer from an object to air at elevated temperature to determine wind speed.

When the anemometer is connected to the controller, it can sample wind speeds several times a second and provide an average as needed. One type of rotational anemometer uses a simple permanent magnet dc generator, which is driven by spinning cups mounted on a shaft. When the wind blows into the cups, they rotate in proportion to the wind and generate an output voltage that is proportional to the wind speed.

FIGURE 8-17   Anemometer on the Back of a Wind Turbine Nacelle   (*Source:* David Buchla.)

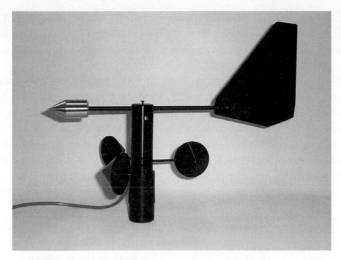

FIGURE 8-18   Cup Anemometer. A cup anemometer measures wind speed, and its wind vane shows the direction of the wind in a single package. This sensor uses an electronic amplifier.   (*Source:* Courtesy of NovaLynx Corporation.)

Another type of rotational anemometer uses the spinning cups to catch and sample the wind, but instead of turning a small generator, the anemometer produces a series of electrical pulses that are proportional to wind speed. The pulse stream is converted into wind speed data for recording or display. The anemometer is calibrated to within 0.5 mph so that it can provide accurate wind speed data. Figure 8-18 shows a cup anemometer with a wind direction instrument on the same shaft. The outputs of the instruments are connected to an electronic circuit that converts the signals to a 4–20 mA current loop in this case, but other types of outputs are used (including wireless transmission of the data.)

Wind Direction Indicator

The **wind direction indicator** (also known as a wind vane) is a blade on a shaft that is rotated by the wind until it is pointed in the direction from which the wind is blowing. Typically it is mounted on the same shaft as the anemometer, as shown in Figure 8-18. In this case, the shaft is connected to the wiper of a 20-kΩ potentiometer specifically selected for linearity. The movement of the shaft can change the resistance from 0 to 20 kΩ within one revolution of the shaft (360°). A signal is developed across the potentiometer that is proportional to the wind direction. The signal is sent to the PLC analog input module as a 4–20 mA current signal, where it can be used to determine the correct direction of the nacelle. In addition, the wind data can be stored for later analysis.

Supervisory Control and Data Acquisition System

The supervisory control and data acquisition (SCADA) system was introduced in Chapter 7. It has been used in industry for many years to monitor important production information. The SCADA system for a wind turbine collects data on a single wind turbine or on a large number of wind turbines for a wind farm. The system can be used to start, stop, or reset wind turbine generators remotely, either individually or in groups of wind turbines in a wind farm.

The SCADA system can collect information on vibration, generator power output, low- and high-speed shaft data, diagnostic data for the generator and data from the gearbox. Wind speed and direction data are also collected and correlated with the power output of the generator. The data from the SCADA system can also track maintenance problems for comparison with similar wind turbines, as well as track maintenance that has been performed on each wind turbine.

After a few years of data gathering, the data can be used for production predictions and load planning. When a large number of wind turbines on a wind farm are connected to the grid, the SCADA system can help with grid loading and integration with coal-fired and nuclear-powered electrical generation plants. SCADA systems have been used for coal-fired and nuclear-powered electrical generation stations over the years, and much of the electrical production and generator data are similar to the data collected from wind turbines. Security information, which can include video camera or fire protection information for each wind turbine, is also provided through the SCADA system.

The SCADA data serves two purposes. Data routed to the PLC or computer operate and control the wind turbine. A second purpose is to archive performance information over a long period of time. The SCADA system can use telephone, Internet, Ethernet, RS232 serial system, Profibus, wireless, and other network protocols to send data between wind turbines, their PLCs or computers, and the network between wind turbines on a wind farm. Because some of the data are proprietary, basic network protection such as firewall software is used to protect the information from outside intrusion. The SCADA system also has a SCADA server, which is a dedicated computer on the network that gathers and stores the data.

SECTION 8-4 CHECKUP

1. Name two types of anemometers.

2. What is the purpose of an anemometer on a wind turbine?

3. What is wind direction data used to control?

4. What is a supervisory control and data acquisition (SCADA) system? What is the data collected by a SCADA system used for?

8-5 Braking

Wind turbines require control over the blade rotation speed for several reasons. Any time maintenance is to be done on the blades, gearbox, or generator, the blades must be secured with brakes so they do not rotate. Dynamic braking is used to help keep loading constant by using electrical resistive loading. Brakes are applied at three different places on larger wind turbines: the yaw drive, the rotor, and the high-speed shaft. All three are discussed in this section.

Wind turbines need to be stopped for maintenance or for controlling the load. Brakes are used on wind turbines in three locations: the yaw ring, the rotor hub, and the high-speed shaft. These three locations for brakes are shown in Figure 8-19.

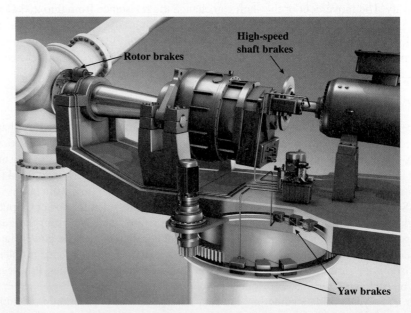

FIGURE 8-19 Drive Train and Yaw Drive Brakes (*Source:* Courtesy of Bosch Rexroth Group.)

Yaw Brakes

Yaw brakes are a series of caliper brakes located on the yaw ring gear to secure the yaw mechanism so that it cannot move. When the yaw motion mechanism needs to move, the brakes are released. The yaw brakes on some smaller wind turbines are caliper brakes that are normally set. These caliper brakes are essentially two pads mounted in either side of a moving disk. The brakes are set by spring tension, and the hydraulic system overrides the spring tension on the brakes to open the pads and allow the yaw mechanism to move. When the yaw mechanism has moved to the desired position, the hydraulic pressure is released from the calipers and the spring tension puts them back into the set position, where they make contact with the yaw ring. When the brakes are set, they take some of the strain off the yaw motion system.

Rotor Brakes

Rotor brakes are caliper brakes located directly behind the rotor hub at the front of the wind turbine to keep the rotor from turning. Dynamic braking is used to slow the rotor before the mechanical brakes are applied. (Dynamic braking will be explained later in this section.) Figure 8-20(a) shows the caliper brakes on the rotor plate of a smaller wind turbine. The calipers close and make friction contact with the rotor plate, just as the disk brakes on the front wheels of a car do. The rotor disk provides a very large surface, where the brake pads can provide pressure and create sufficient friction to make the shaft stop rotating. Figure 8-20(b) shows the large hydraulic cylinder that extends its rod upward to move the caliper mechanism to the open position. The brakes are kept in the set position by heavy springs, and when the calipers need to open to allow the rotor disk to rotate, the hydraulic cylinder extends and makes the caliper pads open.

(a) Hydraulic caliper brakes for a rotor on a small wind turbine.

(b) A hydraulic cylinder that extends to make the brake pads open. Strong springs make the brakes pads press tightly against the rotor when the cylinder is retracted.

FIGURE 8-20 Caliper Brakes (*Source:* Tom Kissell.)

Magnetic Tip Brakes on Wind Turbine Blades

(*Source:* Tom Kissell.)

The magnetic tip brake is provided on many small and midsize wind turbines. When the turbine blades are subjected to wind speeds that are so high they could damage the turbine, the tip brakes are extended by centrifugal force and they cause the blade to unload energy. When the wind speed slows to a safe level, the permanent magnet pulls the brake tips back to their running position.

High-Speed Shaft Brakes

High-speed shaft brakes are caliper brakes that are mounted around a rotor plate on the high-speed shaft to prevent the high-speed shaft from rotating. The rotor plate is attached to the shaft. (Recall that the high-speed shaft is between the gearbox and the generator.) When the wind turbine must be stopped, both the rotor brakes and the high-speed shaft brakes are set. The high-speed shaft brakes may also be set whenever the gearbox transmission is set to idle, to lock the generator shaft.

Hydraulic Brakes

Some wind turbines use hydraulic systems to control the brakes, and these brakes are a type of fail-safe brakes because they use a spring to keep the brakes in the set position and open with hydraulic pressure. **Fail-safe brakes** are a general type of brake designed to stop motion in the event of a loss of power or hydraulic pressure.

In a wind turbine, the hydraulic system causes the brakes to open, whereas a hydraulic failure causes the springs to set the brakes automatically. When the brakes need to be released or opened, hydraulic fluid causes a cylinder rod to extend and move the brake caliper mechanism a small distance and thus make the two pads of the disk brakes open. When the pressure is removed, strong spring tension closes the brakes against the rotor surface again, creating friction that makes the rotor stop rotating. When the pressure from the hydraulic fluid is turned off for any reason, spring pressure causes the disk brakes to return the pads to a position where they are touching the rotor and creating enough friction to stop the rotor. The distance between the brake pads and the rotor is kept to a minimum, usually less than one half inch. Because the distance is very short, the brakes can be applied and the rotor can be stopped very quickly.

The hydraulic system uses a directional valve, flow control, and pressure control to ensure that the right amount of hydraulic pressure is applied to keep the brakes open far enough to allow the rotor to turn. The springs keep pressure on the calipers to maintain friction on the rotor. The hydraulic pressure can be controlled by a proportional or servo hydraulic system, which makes its application very accurate.

Mechanical Brakes

Small wind turbines may not have a hydraulic system and may activate their brakes through mechanical means. These brakes are normally a fail-safe type. The springs in the brakes apply pressure all the time, unless the turbine blades are allowed to turn. When the wind begins to blow enough for the blades to turn, a cam activates the brake release and causes them to release their grip on the rotor. If the wind speed drops below the minimum needed to turn the blades, the cam moves back and allows the springs to apply the brakes again.

If any problems occur with the blades, the springs always cause the brakes to be applied and stop the blades from turning. This feature also protects against overspeeding.

Mechanical Parking Brakes

Wind turbines need a parking brake to lock the blades in position when the blades must be secured for maintenance or if the blades need to be stopped to protect them from high winds. The **mechanical parking brake** may be a separate brake or it may be part of the braking system. In the United States, the federal Occupational Safety and Health Administration (OSHA) requires that a parking brake be installed in the wind turbine and that it can be activated to stop the blade from rotating when necessary. The parking brake can also be set as fail-safe control to ensure that the blade can be stopped quickly or brought under control to ensure a safe working area.

Dynamic Braking

Another method of controlling the speed of a wind turbine is through dynamic braking. **Dynamic braking** increases the load on the generator to slow it down. The dynamic braking system puts electrical resistance called a load bank across the generator output, which causes the generator to experience a reverse torque and slow down. A similar method is used in locomotives to slow a train as it goes downhill. The advantage of the system is there is no wear and tear on brake pads and rotors. The reverse torque slows the entire drive train, including the main rotor. The force of the wind energy turning the blades and the force from the load on the generator form a balanced system, in which the generator load is changed as needed to control the speed.

Dynamic braking is used to control the speed of the wind turbine through a wide variety of conditions. For example, if the wind is blowing very hard but the electrical load is at its minimum, the back torque the generator produces is minimized and the speed of the blades begins to increase. This action could allow the blades to spin above their safe speed. When this occurs, the dynamic braking system can be employed to control the speed. As the regular electrical load picks up again, the resistive load bank is removed, allowing all the electrical power to go directly to the regular electrical load. When the load banks are put across the generator, a large amount of heat is produced as the load bank dissipates excess power. If dynamic braking is used, a method to cool the resistors in the load bank must be provided. If the wind turbine is used to supply voltage and current to the grid, this type of load problem very rarely occurs because the load to the grid is fairly constant. Dynamic braking is typically used prior to applying mechanical rotor brakes.

SECTION 8-5 CHECKUP

1. Name three types of brakes on a horizontal-axis wind turbine.
2. Why are hydraulic brakes designed to be closed with large springs?
3. Why are brakes needed on wind turbines?
4. What is meant by the term *dynamic braking*?

CHAPTER SUMMARY

- Blade pitch control is used to hold constant speed, optimize power output, or to feather the blades.
- The turbine blades are turned with the flat side toward the wind when producing power and with the front edge to the wind when not producing power.
- A common method of pitch control is the use of hydraulic actuators on the rotor hub.
- Passive stall control uses the design of the turbine blades to help slow the turbine down in the event of winds that become too strong and might damage the blades.
- Active stall control is used to control power in some large turbines by rotating the blade toward the stall condition.
- Individual blade pitch control allows the adjustment of the pitch of each blade independently.
- The yaw mechanism on a wind turbine rotates the nacelle to control the direction in which the nacelle and blades are pointing. Yaw is controlled by driving a yaw gear with high-torque, low-speed drive motors.
- The yaw control moves the yaw position so that the blades can harvest the maximum amount of energy when the wind speed is slow. It can also be used to move the blades out of the wind if the wind speed is too high.

- An upwind turbine is designed so that the wind blows over the blades and then over the generator nacelle; the majority of all HAWTs are upwind turbines because there is less wind shading from the tower.
- Upwind turbines require stiffer blades, which are made using internal structures called spar caps and shear webs. Blades are designed to be lightweight and stiff.
- A downwind turbine is one in which the rotor is placed on the downwind or lee side of the tower. Downwind turbines can be constructed without a yaw drive, but they suffer from more tower shading than upwind designs.
- In some larger wind turbines, the blades and rotor rotate slowly and a gearbox is used to increase the speed to the generator so that the turbine is useful for generating the proper frequency for the grid.
- The drive train generally includes the turbine blades, rotor, low-speed shaft, gearbox, high-speed shaft, and generator.
- Drive train compliance is the correct alignment of all the components in the drive train. Any components out of alignment can cause major vibrations, which in turn causes failures and/or premature wear on the components. Vibration monitors are used in some systems to detect any vibration before it can do damage.

- The direct-drive wind turbine does not use a gearbox but requires a multipole generator to produce asynchronous ac, which is converted to regulated ac usually by first converting to dc and then to utility-compatible ac with an inverter. Direct-drive turbines have been used on very large offshore installations.

- Direct-drive wind turbines require less maintenance and lubricants than those with a gearbox.

- An anemometer is mounted on the top of the nacelle, usually near the back, where it can gather wind speed information. Most anemometers for wind turbines use rotational anemometers and transmit data to a controller using a 4–20 mA current loop.

- The wind direction indicator is an instrument mounted on the back of the nacelle and transmits wind direction information to the controller. It can use a specially designed precision potentiometer to indicate the direction.

- The supervisory control and data acquisition (SCADA) system can be used to start, stop, or reset wind turbine generators remotely, either individually or as a group of wind turbines in a wind farm.

- Information the SCADA system can collect includes vibration information, data about generator performance, data from the low- and high-speed shafts, and the gearbox. Data about the wind speed and direction are also collected and correlated with the power output of the generator.

- Larger wind turbines can have three braking systems: the yaw brakes, the rotor brakes, and the high-speed shaft brakes.

- On wind turbines that use hydraulic systems to control the brakes, the brakes are fail-safe because spring pressure keeps the brakes in the set position unless the brakes are released by hydraulic pressure.

- A method of controlling the speed of a wind turbine is through dynamic braking. Dynamic braking is a process that sends energy produced by the generator to a resistive load that increases generator loading, causing it to slow down.

KEY TERMS

active stall control   A method of controlling power in large turbines by moving the pitch angle in and out of stall condition.

anemometer   An instrument that measures wind speed.

direct-drive wind turbine   A type of wind turbine in which the rotor is connected directly to the generator; it does not have a gearbox.

downwind turbine   A type of wind turbine that is designed so that the wind blows over the nacelle first and then over the rotor and blades.

drive train   The turbine blades, rotor, low-speed shaft, gearbox, high-speed shaft, and generator in a wind turbine.

drive train compliance   The correct alignment of all the components in the drive train of a wind turbine.

dynamic braking   A method of braking that increases the load on the generator to slow it down and control speed. The dynamic braking system puts an electrical load bank across the generator output, as required for control.

fail-safe brakes   A type of braking system that is normally set to stop motion in the event of a loss of power or hydraulic pressure.

high-speed shaft brakes   Caliper brakes that are mounted around a rotor plate on the high-speed shaft to prevent the high-speed shaft from rotating.

passive stall control   A design feature of turbine blades that causes them to create turbulence on the back of the blades in high winds, thus reducing the lift force that drives the rotor.

rotor brakes   Caliper brakes that are located directly behind the rotor hub at the front of the wind turbine to keep the rotor from turning.

upwind turbine   A type of wind turbine that is designed so that the wind blows over the blades and then over the nacelle and tower.

wind direction indicator   A blade on a shaft that is rotated by the wind until it is pointed in the direction from which the wind is blowing. Also known as *wind vane*.

yaw brakes   A series of caliper brakes located on the yaw ring gear to secure the yaw mechanism so that it cannot move.

CHAPTER TRUE/FALSE QUIZ

Determine whether each statement is true or false. Answers are at the end of the chapter.

1. The speed of a wind turbine is controlled primarily by the yaw motor.

2. The words *pitch* and *yaw* are synonymous.

3. A major advantage to a variable-speed turbine is that it can operate with the optimum efficiency for producing power.

4. The direct-drive wind turbine uses a transmission and gears between the blades and the generator.

5. Generally, direct-drive turbines use a dc generator.

6. Passive stall control is important for very large wind turbines.

7. Smaller downwind turbines do not require a yaw drive.

8. The maximum speed that a wind turbine rotates can be controlled with active stall control.

9. An advantage of a downwind turbine is that it does not have as much tower shading as an upwind turbine.

10. Blade stiffness is more important for upwind turbines than downwind turbines.

11. The gearbox ratio is the ratio of generator speed to the low speed shaft speed.

12. The anemometer is a device that measures barometric pressure.

13. Designers need to be concerned with mechanical resonance, which should be avoided.

14. The supervisory control and data acquisition (SCADA) system can be used to start, stop, or reset wind turbine generators remotely, either individually or for groups of wind turbines in a wind farm.

15. Larger wind turbines have three basic places that brakes are applied: the yaw brakes, the rotor brakes, and the high-speed shaft brakes.

16. Dynamic braking is a process that uses rotors and brake pads for yaw brakes.

17. On wind turbines that use hydraulic systems to control the brakes, the brakes are set by hydraulic pressure.

18. The yaw mechanism on a wind turbine rotates the blades to control the speed at which the blades are rotating.

CHAPTER MULTIPLE-CHOICE QUIZ

Complete each statement by selecting the one correct answer. Answers are at the end of the chapter.

1. Pitch control determines the
   a. speed at which the turbine blades are turning
   b. direction the nacelle is pointing
   c. output voltage from the generator
   d. all of these

2. Pitch is adjusted to
   a. slow the turbine when the wind is too strong
   b. feather the blades for maintenance
   c. optimize turbine speed
   d. all of these

3. Active stall control uses
   a. brakes to lock the blades
   b. a hydraulic or mechanical system to change the pitch
   c. an electrical motor to change the yaw
   d. none of these

4. A gearbox is not used on
   a. upwind turbines
   b. downwind turbines
   c. direct-drive turbines
   d. small turbines

5. An advantage of a direct-drive turbine is that it
   a. requires less maintenance than other types of turbines
   b. rotates much faster than other types of turbines
   c. has a self-lubricating gearbox
   d. all of these

6. The drive train does not include the
   a. rotor
   b. yaw motor
   c. gearbox
   d. generator

7. Drive train compliance
   a. minimizes vibrations
   b. maximizes power output
   c. synchronizes the ac
   d. does all of these

8. An anemometer is used to determine
   a. wind speed
   b. wind direction
   c. both (a) and (b)
   d. none of these

9. The SCADA system can
   a. collect data for a single turbine
   b. remotely start, stop, or reset turbines
   c. track maintenance issues
   d. do all of these

10. In a direct-drive wind turbine, the rotor is connected to the
    a. high-speed shaft
    b. gearbox
    c. generator
    d. inverter

11. The yaw drive motor needs to be
    a. high-speed, low-torque
    b. low-speed, high-torque
    c. high-speed, high-torque
    d. low-speed, low-torque

12. The primary purpose of hydrodynamic coupling of a yaw motor is to avoid
    a. vibration
    b. unneeded movements
    c. motor stall
    d. alignment issues

13. An upwind turbine is designed so that the wind blows over the
    a. blades and then the nacelle
    b. nacelle and then the blades
    c. sometimes the blades first and sometimes the nacelle first

14. The direct-drive type wind turbine
    a. has blades, a rotor, a low-speed shaft, a gearbox, a high-speed shaft, and a generator
    b. uses a gearbox between the blades and the rotor and generator
    c. has the rotor connected directly to the generator

15. The purpose of the inverter in a direct-drive turbine is to convert
    a. unregulated ac to regulated dc
    b. unregulated dc to regulated dc
    c. unregulated dc to regulated ac
    d. none of these

16. Three brake systems on a turbine are the
    a. rotor, yaw, and high-speed shaft brakes
    b. rotor, pitch, and low-speed shaft brakes
    c. pitch, low-speed shaft, and yaw brakes
    d. low-speed shaft, high-speed shaft, and gearbox brakes

CHAPTER QUESTIONS AND PROBLEMS

1. Why is constant rotational speed of the generator shaft important for many wind turbines?

2. Explain how pitch control regulates the speed of a turbine.

3. What is the advantage of using a 4–20 mA current loop for control signals?

4. Assume that a hydraulic proportional valve receives a 4–20 mA signal to control the hydraulic cylinder, which extends or retracts to set the position of the blade pitch between 0° and 80°. What current will set the pitch to 65°?

5. The hydraulic proportional valve receives a 4–20 mA current loop signal to control the hydraulic cylinder, which extends or retracts to set the position of the blade pitch between 0° and 90°. What is the value of the electrical signal that needs to be sent to the proportional valve to set the cylinder rod to extend to 75% so the blade pitch is set at 67.5°?

6. What happens to the air over a wing or turbine blade when it is positioned to stall?

7. How does active stall control adjust power in high winds?

8. What are two problems with tower wind shading?

9. Explain why blades are tapered and twisted.

10. Identify the places on a horizontal wind turbine where brakes are used.

11. Explain the difference between active stall control and passive stall control.

12. Explain why some smaller wind turbines do not need yaw control.

13. Explain why the yaw position of a wind turbine may need to be changed.

14. What data are important for analyzing wind turbine performance?

15. What are slip rings? Where are they used on a wind turbine?

16. Explain how an anemometer determines the speed at which the wind is blowing.

17. What happens if all the components in the drive train are not in compliance?

18. Describe the components you expect to find inside the nacelle of a typical horizontal-axis wind turbine and explain the function of each.

19. Explain how a direct-drive type wind turbine can provide electrical power at the correct frequency to match the grid.

20. Explain how SCADA can be used for a wind farm.

FOR DISCUSSION

For many seaside communities, the prospect of a wind farm located offshore is unpopular, yet the community uses electrical energy for their homes and businesses. What standards should be in place for locating wind turbines in visually appealing areas like many coastal regions?

ANSWERS TO CHECKUPS

Section 8-1 Checkup

1. During maintenance, pitch is adjusted to feather the blades.

2. Variable pitch allows constant adjustment to control speed and optimize power output.

3. Active stall control is a method of controlling power in large turbines by moving the pitch angle in and out of stall condition.

4. Passive yaw control uses the power of the wind to orient the nacelle. Typically, a fin causes the nacelle to point into the wind.

5. The position of the turbine should be moved slowly to avoid wear and stress. The motors need to be high-torque to move the heavy nacelle easily.

Section 8-2 Checkup

1. In an upwind turbine, the air flows over the blades and then over the nacelle and tower; in a downwind turbine, the blades are behind the tower and the air flows past the tower and nacelle first.

2. Tower shading can cause the blades to flex and increase wear, and it may set up resonances within the tower structure. In addition, there is a slight drop in power due to shading.

3. Efficiency is optimized with thin, lightweight blades, but structural strength is better with heavy, thick blades.

4. Blades tend to bend away from the tower, so they can be positioned closer to the tower; smaller wind turbines do not need a yaw motor.

Section 8-3 Checkup

1. The gearbox changes the low rotor speed into a higher speed for the generator.

2. The turbine blades, rotor, low-speed shaft, gearbox, high-speed shaft, and generator

3. Any misalignment of the drive train can cause vibrations and premature wear on the components.

4. Direct-drive turbines do not require a gearbox, and they use a large number of poles on the generator.

5. The weight of the nacelle is lower; there are fewer moving parts, and less maintenance and fewer lubricants are required.

Section 8-4 Checkup

1. The rotational anemometer and the cooling-power anemometer

2. To measure the wind speed for the controller

3. The yaw drive

4. The SCADA system controls the turbine and collects information over a long period of time to track wind turbine performance and to determine if maintenance is needed.

Section 8-5 Checkup

1. Three brakes are the yaw brakes, rotor brakes, and high-speed shaft brakes.

2. This forms fail-safe brakes that apply the brakes when there is no hydraulic pressure.

3. To keep the turbine from turning during maintenance or to help remove the load from the generator.

4. Dynamic braking puts electrical resistance across the generator output to create a reverse torque and slow down the generator.

ANSWERS TO TRUE/FALSE QUIZ

1. F 2. F 3. T 4. F 5. T 6. F 7. T 8. T 9. F   17. F 18. F
10. T 11. F 12. F 13. T 14. T 15. T 16. T

ANSWERS TO MULTIPLE-CHOICE QUESTIONS

1. a 2. d 3. b 4. c 5. a 6. b 7. a 8. a 9. d
10. c 11. b 12. a 13. a 14. c 15. c 16. a

Biomass Technologies

CHAPTER OBJECTIVES

- Discuss the carbon cycle.
- Describe several sources of biomass.
- Describe how ethanol, biodiesel, and green diesel are produced.
- Explain how energy is obtained from algae.
- Discuss anaerobic digestion and its applications.
- Explain the types of combined heat and power (CHP) systems.

KEY TERMS

Key terms are shown in bold and color. Definitions for key terms are provided at the end of the chapter and in the end-of-book glossary. Bold terms in black are defined in the end-of-book glossary only.

- biomass
- photosynthesis
- respiration
- carbon cycle
- carbon reservoir
- algae
- combustion
- thermal conversion
- chemical conversion
- ethanol
- fossil energy replacement (FER) ratio
- petroleum replacement ratio (PPR)
- anaerobic digestion
- combined heat and power (CHP)
- organic Rankine cycle ORC

INTRODUCTION

In this chapter, biomass as a source of renewable energy is presented, and several processes for the production of energy from biomass (organic materials) are introduced. Biomass sources are discussed. Fuels such as ethanol, biodiesel, green diesel, and methane are products of biomass. These fuels are discussed and the production processes described in this chapter. The use of algae as a renewable biomass source is covered, and anaerobic digestion in sewage and landfill treatment is introduced. Also, the methods for combined heat and power (CHP) systems are explained in terms of the Rankine cycle, the organic Rankine cycle, and gasification. Recall that the basic Rankine cycle was discussed in Chapter 1.

9-1   The Carbon Cycle

The biosphere includes the atmosphere, waters, and earth's crust that support life. The carbon cycle describes the movement of carbon as it is transported throughout the biosphere. The carbon cycle is made up of a sequence of events that are essential for sustaining life.

Carbon is a naturally abundant nonmetal element with an atomic number of 6. All organic materials (living or once living) contain carbon; it is the basic element of life. (Some inorganic materials also contain carbon.) **Biomass** refers to organic (carbon-containing) materials that have energy stored as carbohydrates, which can be used as feedstock for the production of renewable energy. **Photosynthesis** is the process whereby plants, algae, and some species of bacteria capture energy from the sun and use it to convert carbon dioxide and water into carbohydrates that are stored in the plant for food. In the photosynthesis process, plants use carbon dioxide and water to produce oxygen, water, and carbohydrates (glucose). The chlorophyll in leaves is the agent that converts water and CO_2 into the carbohydrates and oxygen. Chlorophyll also contributes the green color in leaves. The amount of energy collected by plants worldwide is estimated to be about six times more than the total energy used by humans. The process of photosynthesis in plants is illustrated in simple form in Figure 9-1.

The chemical reaction of photosynthesis occurs in the leaf. The stomata in a leaf are pores where the CO_2 enters the leaf and the oxygen exits the leaf during photosynthesis. Most of the stomata are on the underside of the leaf and are opened and closed by cells called guard cells. When there is no sunlight, the stomata closes. Photosynthesis can be expressed in words as follows:

$$\text{Sun energy} + \text{carbon dioxide} + \text{water} \rightarrow \text{oxygen} + \text{carbohydrate}$$

The corresponding chemical reaction is

$$\text{Sunlight} + 6CO_2 + H_2O \rightarrow 6O_2 + C_6H_{12}O_6$$

Respiration is the process in which oxygen is used to break down organic compounds into carbon dioxide (CO_2) and water (H_2O). The process of respiration can be stated as

$$\text{Food} + \text{oxygen} \rightarrow \text{energy} + \text{carbon dioxide} + \text{water}$$

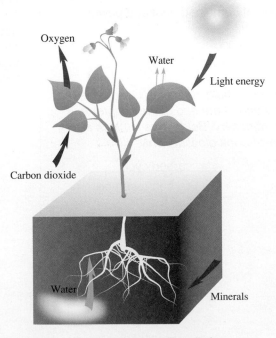

FIGURE 9-1   Photosynthesis   (*Source:* designua/Fototlia.)

Glucose, derived from carbohydrates, is normally used as a food supply. Respiration occurs at the cellular level in animals and plants. In animals, breathing supplies oxygen for the respiration process, but it is not the same as respiration. Animals, plants, and decomposing materials all respire, but only animals breathe. Decomposition is also a significant source of CO_2. Decomposers are microorganisms that feed on the rotting remains of plants and animals and release carbon dioxide into the atmosphere.

The **carbon cycle** is the process by which carbon atoms are endlessly cycled around the biosphere (earth and atmosphere). When plant material is burned, the energy stored is released in the form of heat. The net effect is a balance between the capturing and releasing of carbon dioxide. The **carbon reservoir** is any natural feature that stores carbon. Examples include living organisms, the atmosphere, the ocean, soil, certain rocks and fossil fuels. Carbon in the atmosphere, in the form of CO_2, is that part of the carbon reservoir that plants utilize in the process of photosynthesis. Also, the carbon in the form of carbohydrates stored in plants and eaten by animals or humans for energy is returned as CO_2 to the atmosphere via respiration. When plants and plant residues are burned or processed for energy, the CO_2 also returns to the atmosphere. Once the carbon cycle is completed, it starts again and continues endlessly. Fossil fuels that are burned and natural events such as volcanic eruptions release CO_2 into the atmosphere. Also, the oceans absorb CO_2 from the atmosphere through photosynthesis by plant-like organisms (phytoplankton) and simple chemistry, and then release some of it back to the atmosphere. Unfortunately, the oceans absorb far more CO_2 than they release, which has tended to increase the acidity of the oceans, as mentioned in Section 1-1. This can have a serious effect on marine life, particularly shellfish and corals. The carbon cycle is illustrated in Figure 9-2.

Although biomass is one of the oldest sources of energy used by humans, its use decreased for many years because the energy density in fossil fuels is much higher and fossil fuels have historically been inexpensive. The drawback to the burning of fossil fuels is that it adds carbon to the normally neutral cycle because carbon from fossils has been buried and unavailable for a very long period of time (a process called sequestering). When the CO_2 from fossil fuel burning is released, it causes an imbalance in the carbon cycle, as do other events such as volcanic eruptions. The net effect has been a continued increase in atmospheric CO_2, a greenhouse gas.

Efforts are being made to slow the increase of CO_2 in the atmosphere. The most effective way is to reduce burning fossil fuels through conservation. Switching to renewable

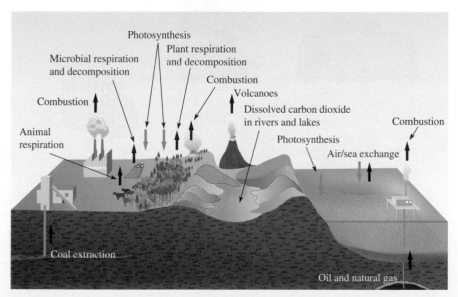

FIGURE 9-2 Carbon Cycle. Emission is shown with black arrows; absorption is shown with green arrows.

energy sources and fuels with lower carbon content is another way to reduce the level of CO_2 in the atmosphere. Research into carbon dioxide capture and sequestration is trying to discover how to capture and store CO_2 in deep underground reservoirs under an impermeable seal. One proposal from Lawrence Livermore National Laboratory is to trap CO_2 as it is emitted in flue gas using a seawater/limestone gas scrubber to form dissolved calcium bicarbonate. The calcium bicarbonate could be released in the ocean, where it might benefit marine life by offsetting the acidification of the ocean.

SECTION 9-1 CHECKUP

1. What is the carbon cycle?

2. What are four sources of CO_2?

3. What are two major processes in the carbon cycle?

4. Where does photosynthesis occur?

5. Is respiration the same as breathing? Explain.

9-2   Biomass Sources

Biomass for renewable energy consists of wood and its by-products, unused parts of food crops, grasses, and livestock manure and tissue can be sources of energy. Some of the household garbage that you discard each day can also be used to create energy. Landfills are tapped for methane gas, which is given off as garbage decomposes. Ethanol from corn is one example of a biomass fuel. A wide range of biomass resources are beneficial because their use will reduce overall carbon emissions and provide other benefits, such as reducing our dependence on fossil fuels.

Feedstock is any raw material used to create an energy product, such as electricity, heat, or fuel. Biomass feedstock includes forests and grasslands, agriculture, trash, sewage, animal waste, and industrial waste, as shown in Figure 9-3.

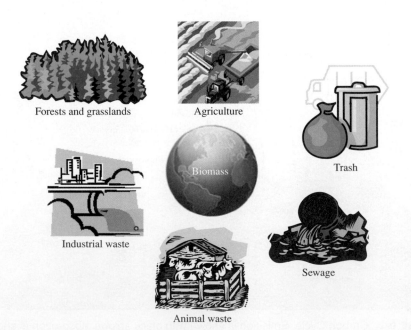

Forests and grasslands   Agriculture

Biomass

Trash

Industrial waste

Animal waste

Sewage

FIGURE 9-3   Examples of Biomass Sources

Forest Biomass

Stem Wood

Generally, in logging operations, trees with stems (trunks) from 5 inches to 9 inches in diameter are used for paper and some wood and chemical products. Those with stems greater than 9 inches are generally harvested for lumber and other building products. Lower-quality and smaller stem trees and trees with defects are often culled for bioenergy use. Smaller trees to be used for bioenergy must be harvested at the same time that the larger trees are logged to make the logging operation cost-effective.

Branches and Tops

Trees are harvested for the stem wood. The tops and branches are removed from the stem and are considered logging residue. If they can be gathered and chipped efficiently, this logging waste can be used in bioenergy.

Thinning

Thinning operations are done in young timber stands to improve tree growth and reduce the risk of wildfire damage. The small trees and underbrush can be used for bioenergy. Traditional forestry equipment along with chippers can be used during the thinning operation. Also, some types of small trees and undergrowth can be baled to reduce transportation costs.

Mill Residues

Saw mills produce residues during their manufacturing process. These residues include sawdust, wood chips, shavings, and bark. Almost 100% of these residues can used to produce additional products, and almost half of the residues can be used for bioenergy.

Urban Plant Waste

Several sources of waste in metropolitan and other developing areas can be used as bioenergy. These sources include construction waste, wood from building demolition, wooden pallets, and others. Also, large amounts of shrub and tree trimmings as well as leaves are usually collected during the fall season by municipalities, and this material can be used for bioenergy. Use of any of these materials in urban areas also saves landfill space.

Agriculture Biomass

Food Crops

Food portions of crops such as corn, grains, beets, and sugarcane are processed to produce ethanol. Soybeans and sunflowers are refined into biodiesel fuel. The disadvantage of using these food crops as biomass is that it takes away from food production and tends to increase the cost of these foods in the marketplace.

Crop Residues

The parts of food plants left over after harvesting can be used as biomass. Examples are the stalks, leaves, and cobs from corn harvesting and processing. The husks, shells, stems, and leaves of other crops such as rice, wheat, barley, and nuts are also useful. Care must be taken to avoid removing too much crop residue because a certain amount must be left in the fields to help prevent erosion and to provide soil nutrients.

Energy Crops

Certain types of grasses and trees can be planted in ways that don't affect food production. They can be grown on marginal land or pastures, or they can be planted in rotation with food crops. For example, switchgrass is a hearty species that can be grown in many locations.

Biofuel Plant for Forest Management

(*Source:* David Buchla.)

A 2 megawatt biomass plant is under construction in Northern California that will produce electricity using a gasification technology. This technology converts organic slash in the forest to gas that runs a turbine. The project will reduce air pollution from burning the slash in prescribed burns and reduce wildfire fuel, which helps preserve wildlife habitat. The plant is being constructed near the forest to avoid transportation costs.

Algae

Algae is a type of energy crop that has great potential as a biomass source for biofuels. **Algae** are simple organisms that have no roots, leaves, or other elements like plants do. Like plants, most algae use the sun's energy to make their own food with the process of photosynthesis; they capture more of the solar energy and produce more oxygen than all other plants combined. However, algae lack the roots, stems, leaves, and embryos typical of true plants, so they are generally classified separately. Algae are found in diverse habitats and in greatly varying types and sizes. Microscopic algae (microalgae), called phytoplankton, live in lakes, ponds, and oceans. The largest type of algae is seaweed, including kelp. Algae can also grow on trees, soil, animals, and rocks.

Trash Biomass

Trash in landfills is a major source of bioenergy. Energy can be extracted from a landfill either by incinerating combustible trash or by collecting the methane gas given off as garbage decomposes. In the United States, the Environmental Protection Agency's Landfill Methane Outreach Program states that approximately 425 landfill gas energy projects are currently in operation or under development in North America. We can expect that number to increase.

Sewage Biomass

Sewage sludge results from the treatment of municipal wastewater. Most of the dry matter content is nontoxic organic compounds, but the sludge also contains some inorganic material and some toxic elements. This biomass can be processed using anaerobic digestion, incineration, and co-incineration to produce energy in the form of heat, electricity, or biofuels.

Animal Biomass

Livestock Manure

Waste from farm animals such cattle, hogs, goats, and poultry is a biomass source that can be used to produce biogas such as methane. Dry manure has provided heating and cooking fuel for rural societies for centuries. If the water content is low enough, dry manure can be burned directly. Wet manure produced from livestock in confined areas produces biogas, which contains a high amount of methane. Liquid manure is created when livestock pens are flushed with water to clean them. This manure can be collected and broken down by bacteria in a process called anaerobic digestion to create methane. Anaerobic digesters are used on a small scale on farms across the United States.

Animal Tissue

Disposing of livestock carcasses has always been a major concern of farmers. Huge amounts of animal tissue are created when livestock die due to natural disasters, common mortality, and diseases. Also, the butchering of livestock for food leaves a certain amount of waste tissue and bone. One relatively recent method of energy generation from animal tissue is co-firing with coal to reduce the amount of coal used.

Industrial Biomass

Industrial biomass can be considered a subset of forest biomass, but it is treated as a separate category because the materials are created as a result of a production process.

Wood Residues

Scraps of wood and particle board, along with the shavings and sawdust, are a biomass feedstock. For example, a furniture factory creates much scrap lumber and other by-products.

The building trades also create lots of wood scraps. In addition, industries and businesses use large quantities of cardboard boxes and shipping crates.

Pulp and Paper Mill Sludge

Sludge from the papermaking process is solid waste material. Sludge consists of pulp residues and inorganic additives. Because of high carbohydrate content, pulp mill sludge can be used as biomass feedstock for conversion to various forms of energy.

Biomass Energy Processes

Energy from biomass is produced by burning (combustion) or by thermal, chemical, or biochemical conversion to fuel. Several processes can convert biomass into energy.

Combustion

Burning is the simplest process for converting biomass to usable energy. **Combustion** is the burning of a substance; it is a rapid exothermic reaction that is the oxidation of a fuel (combining fuel with an oxidant (usually oxygen)). Combustion generates heat that can be used directly, such as burning wood in a stove or fireplace, or used indirectly for the production of electricity by a steam turbine.

Co-Firing Process

In co-firing, biomass is used to replace from 15% to 20% of the coal in a coal-fired power plant. Generally, in addition to animal tissue, switchgrass is used but the method is somewhat inefficient.

Thermal Conversion

Thermal conversion is any of several processes that use heat to extract energy from biomass through noncombustion chemical reactions. The processes for biomass fuels are as follows:

Gasification This thermal conversion process uses high temperatures, oxygen, and steam to convert biomass fuels into syngas, (Gasification is also common for fossil fuels such as coal).

Pyrolysis When biomass is heated without any oxygen, the most common end product is charcoal.

Torrefaction Heat is used to remove oxygen and moisture from the biomass in a thermochemical process. Biocoal is generally the end product and is made into briquettes or pellets for heating.

Chemical conversion **Chemical conversion** is the direct conversion of the biomass into various forms of fuel without combustion. Because most biomass contains carbohydrates, it can be reduced to various chemicals that are useful fuels, such as methane. For example, corn, sugar cane, and soybeans can be converted into liquid fuels that can replace diesel and gasoline. Biodiesel can be produced using recycled cooking oils. When fermentation and anaerobic digestion are included in these processes, they are known as biochemical conversion.

SECTION 9-2 CHECKUP

1. What is biomass?
2. What are the major biomass feedstock sources?
3. What is one biofuel made from corn?
4. What is a disadvantage of using a food source as biomass?
5. What is the major energy product of livestock manure?

9-3   Biofuels: Ethanol

Biofuels are produced from living organisms or from organic waste. They are derived from the photosynthesis process, which gets its energy from the sun. In a sense, biofuels and biomass in general can be considered to be derived from solar energy. Two classes of biomass sources are used for the production of ethanol: starch-based biomass, which includes various food crops such as corn, and cellulose-based biomass, which includes wood, wood residues, crop residues, and grasses. Algae also promise to be sources of significant feedstock for many types of biofuels, although much research has yet to be done to make it commercially practical.

Ethanol is the most common biofuel around the world. It is an alcohol made by fermentation of sugars derived from corn, wheat, rice, sugar beets, sugar cane, sorghum, potatoes, and other starchy food sources. Trees, grasses, corn stalks, and other materials containing cellulose can also be used as feedstock for cellulosic ethanol production.

Most internal combustion gasoline engines can run on blends of ethanol and gasoline called E10 (10% ethanol). Because of its smaller energy density, more ethanol is required to produce a certain amount of energy compared to gasoline. Flexible-fuel vehicles (FFVs) can run safely on E85, which is 85% ethanol. Corn is currently the most common food feedstock for ethanol production, although it has a negative impact on food supplies and food prices. As a result, cellulosic ethanol has been developed using plants containing cellulose, such as wood, wood residues, switchgrass, and other nonfood plants. Although cellulosic ethanol is not commercially competitive with corn ethanol at the present time, it has significant potential to replace corn-based ethanol.

Fossil Energy Replacement Ratio

Ethanol yield can be measured as a ratio of the energy delivered to the consumer in the form of ethanol (or other biofuel) to the amount of all fossil fuel energy used in the process. This ratio is called the **fossil energy replacement (FER) ratio** and is the ratio of the energy delivered to the consumer to the fossil energy used at the production site. The FER indicates the amount of reliance on fossil energy for the production of ethanol and other biofuels. A higher FER means that less fossil energy was used in producing the biofuel. The FER ratio can be expressed as

FER = energy delivered to consumer/fossil energy used

For example, fossil fuels (gasoline and diesel) are used for growing, transporting, and processing, and include fuel for tractors, equipment, and trucks for transportation of the feedstock to the ethanol plant and for delivery of the final product to the consumer. The ethanol production process requires electricity and a significant amount of heat, generally from a fossil fuel source. The bar graph in Figure 9-4 shows typical FER ratios for several fuels. For example, cellulosic ethanol delivers over five times the amount of energy than that contained in the fossil fuels used in the production process.

Petroleum Replacement Ratio

The **petroleum replacement ratio (PRR)** is a ratio of the energy delivered to the consumer in the form of biofuel compared to the petroleum energy used in the process. PRR differs from the FER ratio because it includes only petroleum and not other fossil fuels such as coal. In the United States and other parts of the world, a large percentage of petroleum is imported. A higher PRR means that less petroleum energy was used in producing the biofuel. The PRR can be expressed as

PRR = energy delivered to consumer/petroleum energy used

Figure 9-5 shows typical PRR values for several fuels. For example, cellulosic ethanol delivers about twenty times the amount of energy than that contained in the petroleum used in the production process.

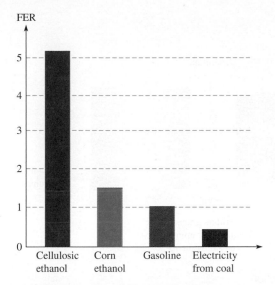

FIGURE 9-4 Fossil Energy Replacement (FER) Ratios

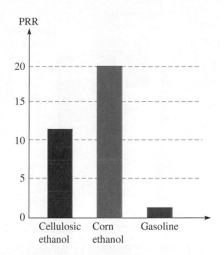

FIGURE 9-5 Petroleum Replacement Ratios (PRRs)

EXAMPLE 9-1

If you measure energy in British thermal units (BTUs), and if 15,000 BTUs of fossil energy are used to produce corn ethanol, how many BTUs will the resulting corn ethanol contain? How many BTUs will cellulosic ethanol contain? How many kWh of energy does the corn ethanol contain?

Solution

Refer to Figure 9-4. The FER ratio is approximately 1.5:

$$\text{BTU in corn ethanol} = \text{FER (fossil energy)} = 1.5\,(15{,}000\ \text{BTU}) = \textbf{22,500 BTU}$$

For cellulosic ethanol, the FER is approximately 5.1:

$$\text{BTU in cellulosic ethanol} = 5.1(15{,}000\ \text{BTU}) = \textbf{76,500 BTU}$$

One BTU equals 0.00029 kWh. Thus, the corn ethanol contains:

$$22{,}500\ \text{BTU} = (0.00029\ \text{kWh/BTU})(22{,}500\ \text{BTU}) = \textbf{6.525 kWh}$$

Starch-Based Ethanol Production

Starch-based ethanol can be produced via two processes: dry milling and wet milling.

Dry Milling

Corn or other starchy food plants are most commonly processed to produce ethanol using **dry milling.** A diagram of the dry milling process is shown in Figure 9-6.

The grinder in the dry milling process reduces the corn kernels or other starchy source to a course flour or meal. The meal is slurried with water and sent to the cooker. The cooker uses high temperatures to reduce the bacteria levels and liquefy the corn mash (liquefaction). Enzymes are then added to the mash to convert starch (complex carbohydrates) into simple sugars by a process called **saccharification.** After cooling, the mash is sent to the fermenter, where yeast is added. During fermentation, the mash is kept cool and agitated to activate the yeast and the conversion of sugar to ethanol and CO_2 occurs. The CO_2 is captured by the scrubber for use as beverage carbonation or for dry ice production. After fermentation, the result, called beer, is sent to the distillation columns, where the ethanol

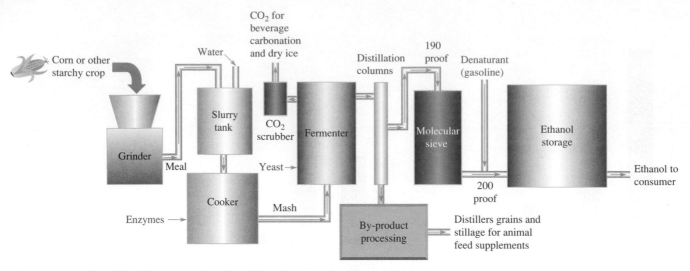

FIGURE 9-6   Simplified Diagram of the Dry Milling Process for Ethanol Production

is separated from the stillage (residue grain), and the ethanol is dehydrated to 190 proof by conventional distillation. It is further dehydrated by the molecular sieve to create 200 proof ethanol. To avoid alcohol tax and to render it undrinkable, the ethanol is blended with approximately 5% gasoline. The final product is then sent to the ethanol storage for shipment to consumers. Figure 1-24 in Chapter 1 showed an overview of an ethanol plant. Figure 9-7 shows the distillation column of a typical ethanol plant.

Wet Milling

Wet milling is the process in which a feedstock is soaked in water usually with dilute sulfurous acid to soften it before further processing. Wet milling is not as widely used as dry milling for converting corn to ethanol. For corn, the first step in wet milling is soaking the whole kernels in water and diluted sulfurous acid to soften the kernels. This process makes it easier to separate the kernel into the starch, fiber, germ, and protein, thus allowing a variety of by-products to be made, such as corn syrup, biodegradable plastics, food additives, cooking oil, and livestock feed. Other than this first wet step, much of the rest of the processing to obtain ethanol is similar to the dry milling method. For this reason, wet milling is not covered in detail. Deriving the by-products, however, requires additional processes.

FIGURE 9-7   Distillation Column of a Corn Ethanol Plant   (*Source:* National Renewable Energy Laboratory.)

Cellulose-based Ethanol Production

The production of ethanol from cellulosic biomass is an alternative to starch-based sources. One advantage is that it doesn't use the edible portion of food crops as an energy source. The main cellulosic biomass sources are wood; wood residues such as sawdust and shavings; leaves; grasses such as switchgrass and others; and food crop residues such as corn stalks, corn cobs, stems, peelings, pits, shells, and so on.

Although the cellulosic parts of plants are not edible, they contain sugars that can be fermented. The process of breaking down plant cellulose and hemicellulose into sugars is difficult because they are locked into complex carbohydrates, and the process requires large amounts of energy. Cellulose has very stable crystalline structures and is not soluble in water or in normal solvents, which compounds the difficulty in separating the sugars. Another problem is the gathering and transportation of the cellulosic materials. When gathering the waste from food crop harvesting, such as corn, care must taken to leave a sufficient amount of the cellulosic waste on the ground for soil enrichment and erosion prevention. One type of cellulosic feedstock that holds great promise is switchgrass. The entire plant can be utilized, it can be planted on soils not suitable for food crops or planted as a rotation crop, and it can be harvested easily. A field of switchgrass is shown in Figure 9-8.

A cell wall surrounds the plasma membrane in all plant cells. The cell wall in cellulosic biomass is primarily composed of cellulose, hemicellulose, and lignin. Cellulose contains only anhydrous glucose, but hemicellulose contains many different sugars. Lignin fills the spaces in the cell wall between cellulose and hemicellulose, and binds the cellulose fibers together. Figure 9-9 shows a simplified diagram of the process for converting cellulosic biomass into ethanol. The two key steps in the process are hydrolysis and fermentation.

Handling The first step in the production of cellulosic ethanol is reducing the size of the biomass to facilitate processing and increase efficiency. This step is called biomass handling and includes grinding, chipping, or shredding, depending on the type of biomass. The agricultural residues such as corn stalks, stems, and cobs are ground into small pieces; wood is converted into chips of similar size.

Pretreatment In the pretreatment step, dilute sulfuric acid is mixed with the biomass, and the hemicellulose portion of the biomass is broken down into simple sugars by hemicellulose hydrolysis. These sugars are called pentoses (5-carbon sugars), the most

FIGURE 9-8 Switchgrass (*Source:* National Renewable Energy Laboratory.)

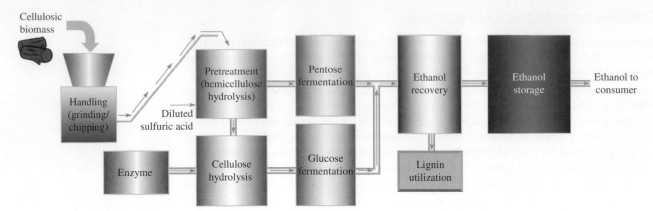

FIGURE 9-9   Production of Ethanol from Cellulosic Biomass

common of which is xylose. Also, a small amount of the cellulose is also converted to glucose in this step.

Pentose fermentation These hemicellulosic sugars then go to the pentose fermentation stage, where they are fermented by hydrolysis.

Cellulose hydrolysis The remaining cellulose then goes to the cellulose hydrolysis stage, where enzymes are added to separate the glucose. Cellulose hydrolysis is also saccharification because the end result is breaking down the starches into simple sugars.

Cellulose fermentation Cellulose fermentation consists of a process that converts glucose to ethanol. Yeast or bacteria feed on the sugar and turns it into ethanol broth while giving off CO_2, which can be captured and used as a by-product or released into the atmosphere.

Ethanol recovery In the recovery step, the ethanol is separated from the ethanol broth that resulted from glucose and pentose fermentation, and final dehydration removes any water from the ethanol.

Lignin utilization The lignin from the cellulose biomass is processed and utilized as a fuel to produce electricity required for the ethanol production process or to be sold to the utility grid.

The conversion of cellulosic biomass to ethanol is very expensive compared to using food crops. However, some manufacturing plants have already been constructed, and much research is being done to improve efficiency and costs.

SECTION 9-3 CHECKUP

1. What two classes of biomass are used for making ethanol?

2. What is the major food crop used in the production of ethanol?

3. What is one advantage of cellulose-based biomass in the production of ethanol?

4. What are the two methods for producing starch-based ethanol?

5. What is the FER ratio?

9-4   Biofuels: Biodiesel and Green Fuels

Biodiesel is a fuel derived from vegetable oils such as soybean oil or recycled cooking oils, animal fats, and recycled greases. An important by-product that results from converting oils to biodiesel is glycerin, a syrupy liquid that is used in a number of other products, including household and medical products. Green diesel generally uses the same feed-stocks as biodiesel, but it uses different processing methods.

Biodiesel

Transesterification is a chemical process in which an organic ester reacts with an organic alcohol. The reaction is often catalyzed with an acid or a base catalyst. (A **catalyst** is a chemical that speeds up the reaction but emerges from the reaction unchanged.) Transesterification is the main process used in converting biomass oils to biodiesel fuel. In the biodiesel process, certain feedstocks must be pretreated by a process called acid esterification before they can go through the transesterification process. These feedstocks include inedible animal fat and recycled grease, which have more than 4% free fatty acids. Feedstocks with less than 4% free fatty acids do not require acid esterification. These feedstocks include vegetable oils such as soybean oil and some food-grade animal fats. Figure 9-10(a) is a simplified illustration of the biodiesel process; Figure 9-10(b) shows a typical biodiesel plant.

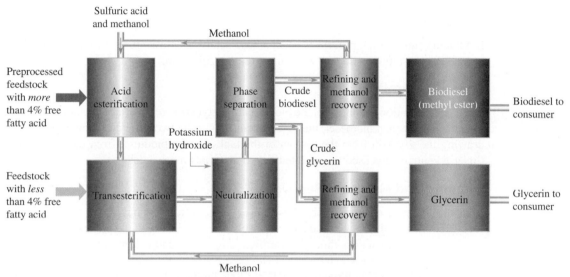

(a) Simplified diagram of biodiesel production process

(b) Typical biodiesel plant

FIGURE 9-10 Biodiesel Production (*Source:* Part (b), Mauricio Simonetti/Getty Images.)

Acid Esterification

As mentioned, feedstocks containing more than 4% free fatty acids must go through a chemical process called acid esterification to improve the yield of biodiesel. Water and contaminants in these feedstocks are removed before they enter the esterification process. Sulfuric acid is added as a catalyst; it is dissolved in methanol, mixed with the pretreated feedstock, and heated. This process converts the free fatty acids to biodiesel, which is then dehydrated and sent to the transesterification stage.

Transesterification

Feedstocks containing less than 4% free fatty acids are also preprocessed to remove water and contaminants. The feedstocks then go to the transesterification stage and are mixed with the product from the acid esterification stage. Potassium hydroxide, a catalyst, and methanol, a simple alcohol, are then added to the mix.

Neutralization

If an acid esterification process is used, a base catalyst, potassium hydroxide, is added to neutralize the sulfuric acid that was part of the esterification.

Phase Separation

Once neutralization is complete, the biodiesel and glycerin are separated. Because glycerin is much denser than biodiesel, the glycerin sinks, and the two can be separated by simply drawing the glycerin from the bottom of the tank and the biodiesel from the top. Sometimes, a centrifuge is used to separate the two materials more quickly.

Refining and Methanol Recovery

Methanol is removed from the biodiesel and glycerin after they have been separated. The recovered methanol is treated and recycled back to the beginning of the process to be reused.

The biodiesel then goes through a purification process and is sent to storage. Biodiesel can be used as pure diesel (B100) or it can be blended with petroleum diesel to form B2 (2% biodiesel), B5 (5% biodiesel), or B20 (20% biodiesel). Other blends are also sometimes used. The use of biodiesel in a conventional diesel engine results in substantially lower emissions of unburned hydrocarbons, carbon monoxide, and particulate matter compared to emissions from petroleum diesel.

The glycerin goes through an additional process where certain components and contaminants are removed. Glycerin is used in certain soaps, preservatives, lotions, and certain pharmaceutical products. Although not explosive itself, glycerin is also used in the explosive nitroglycerin.

Green Diesel

As you have seen, biodiesel is produced using the process of transesterification. Unlike biodiesel, which is an ester and has different chemical properties from petroleum diesel, green diesel is composed of long-chain hydrocarbons like petroleum diesel. Green diesel (a more descriptive name might be petroleum-like biodiesel) normally uses the same feedstocks as biodiesel but uses different processing methods. One method involves partially combusting a biomass source to produce carbon monoxide and hydrogen. A special chemical process is used to produce complex hydrocarbons. This process is commonly called the biomass-to-liquids (BTL) process. The other method is hydroprocessing, which may occur in the same facilities used to process petroleum. Figure 9-11 shows how biodiesel and green diesel differ.

Hydroprocessing

Hydroprocessing is a general chemical engineering term for various chemical processes that react feedstock (oils) with hydrogen at a high temperature and pressure to alter the

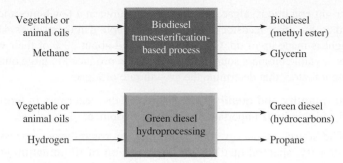

FIGURE 9-11   Comparison of Biodiesel and Green Biodiesel

chemical properties of the feedstock. In renewable energy systems, hydroprocessing is focused on the catalytic conversion of feedstocks such as algae to produce fuels. A by-product of hydroprocessing a biomass oil is propane (not glycerin, which is a by-product of biodiesel production).

The product of hydroprocessing a biomass feedstock is a fuel called green diesel that is similar to petroleum diesel. The main advantages of green diesel over biodiesel is that green diesel has a greater heating content, it has better cold weather properties, the propane by-product is generally preferable over glycerin, and production costs are generally lower.

SECTION 9-4 CHECKUP

1. What is the process for producing biodiesel?
2. What is the difference between biodiesel and green diesel?
3. What is the by-product of biodiesel production?
4. Where is the biodiesel by-product used?

9-5   Biofuels from Algae

Algae are any of various aquatic, photosynthetic organisms that range in size from single-cell organisms to giant kelp. Algae convert carbon dioxide and sunlight into energy in the form of an oil. They can produce up to fifteen times more oil per acre than other plants used as biofuel sources. Algae oil is a sustainable feedstock alternative to soybean, canola, and palm oils and can be used for making biofuels such as biodiesel, biojet, methanol, and ethanol.

Comparison of Biomass Feedstock

Algae has great potential compared to most other biomass feedstocks in terms of the amount of oil that can be produced. Table 9-1 lists several oil-producing feedstocks and their typical oil output.

TABLE 9-1

Potential Feedstock Oil Output

Feedstock	Gallons per Acre per Year
Corn	18
Soybeans	48
Sunflower	102
Palm	635
Microalgae	5,000–15,000*

*Some sources indicate that between 50,000 and 100,000 gallons is achievable.

Under the right conditions, algae can double its volume in a few hours and can be harvested every day; the other feedstocks take longer to replenish themselves. Up to 59% of the algae weight is made up of oil. Oil-palm trees yield about 20% of their weight in oil. For algae to be a viable biomass source, they have to be produced in large quantities. Here are the important factors that determine the growth rate of algae:

- **Nutrient quantity and quality** Composition of the water mix in terms of nutrient contents and salinity is important to proper development of the algae.
- **Light** The source of energy that drives the photosynthesis process in algae is light. Intensity, spectral quality, and the duration of illumination are important considerations.
- **pH** The pH range for most algae species is between 7 and 9.
- **Aeration and mixing** Mixing is used to prevent sedimentation of the algae, to ensure that all cells are equally exposed to light, and to prevent temperature stratification.
- **Temperature** Typically, most algae can tolerate temperatures between 16° C and 27° C. Temperatures below this range slow the growth of the algae; temperatures above this range can be lethal to the algae.
- **Type of algae** Different types of algae have different growth rates.

Algae Production and Processing

Microalgae can be cultivated in open ponds, which can be large shallow ponds, artificial ponds, tanks, or raceway ponds. Algae can also be cultivated in closed-loop systems such as the **photobioreactor (PBR),** which is a bioreactor that incorporates a light source. The two main production methods are the raceway pond and the photobioreactor. The basic stages of production (cultivation) and processing of the algae for oil are shown in Figure 9-12. The algae are grown in water in a container (pond or photobioreactor) with sunlight, CO_2, and nutrients added. The algae are then separated from the solution, and the oil is extracted using one of several methods.

Raceway Pond

The **raceway pond** is a large, shallow, open water configuration that resembles a racetrack. The water, with its algae and nutrients, is moved around the track by a rotating paddle assembly. The disadvantages of this type of facility are that it is exposed to weather, it is susceptible to outside contamination, and the conditions are difficult to control. The advantage is that the racetrack system is relatively inexpensive to construct and operate. Figure 9-13 illustrates a raceway pond for algae production.

Photobioreactor

A disadvantage of the pond system is that the growth of algae at the surface tends to block sunlight from penetrating deeper levels and thus reduces the efficiency. The open pond is also subject to unwanted algae strains. The algae photobioreactor (PBR) uses a closed system consisting of aboveground transparent tubes or thin tanks through which the algae liquid is moved. The PBR system functions essentially like the raceway pond but allows a significant increase

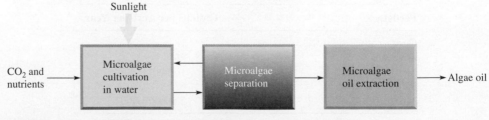

FIGURE 9-12 Basic Stages in Algae Cultivation and Processing

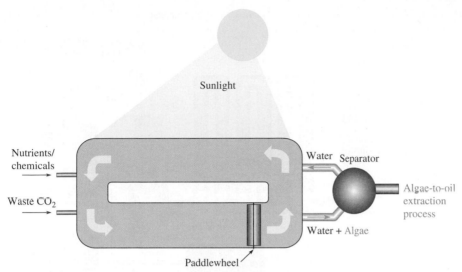

(a) Diagram of an algae raceway pond

(b) Large microalgae raceway facility

FIGURE 9-13 Algae Raceway Pond (*Source:* Part (b), Peter Ginter/Getty Images.)

in sunlight penetration. The enclosure protects against intrusion of unwanted contaminants, organisms, and bugs. Also, the temperature and other parameters can be better controlled in the closed PBR system. Figure 9-14 is a process flow diagram for algae oil production.

The feedstock consists of water; algae, which has been developed in incubation tanks; CO_2; and various nutrients to enhance algae growth. This feedstock is put into the feeder vessel, and the mix is pumped into the PBR. The algae multiply rapidly when exposed to the sunlight through the transparent bioreactor tubes. The algae then go into a separator where much of the water is removed, leaving slurry. The slurry is dried to remove more of the liquid and then sent to a mechanical press, where the oil is squeezed from the algae. The oil is processed through a centrifuge to remove the remaining water and stored in a holding tank. The algae biomass (solids) left from the press operation is collected and can be burned to produce electricity and/or heat. The water recovered during the process is recycled into the system. Ideally, the PBR can be located near a power generation facility so

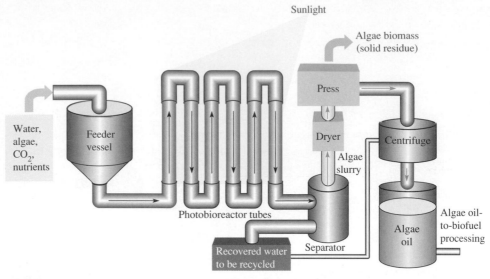

FIGURE 9-14   Process Flow Diagram for Production of Algae Oil with a Photobioreactor (PBR)

that the waste CO_2 can be used. Sensors and computer monitoring control various parameters, such as flow rate, temperature, CO_2 concentration, pH, nutrient levels, and algae growth rate. Figure 9-15 shows a small PBR that uses a tank instead of tubes.

Separation Methods

Several techniques are available for separating algae from the water mix in which they are grown.

Filtration This method uses a microstrainer or a reverse-flow vacuum pump to separate the algae from the liquid medium.

Centrifugation A centrifuge causes the algae to settle to the bottom of a tank where it can be collected.

Flotation In this method, algae are made to float on the surface of the medium, where they can be recovered easily. One flotation approach is to adjust the pH,

Biofilm-Based Algae Growth Systems

(*Source:* Courtesy of Bioeconomy Institute, Iowa State University.)

The Bioeconomy Institute (BEI) of Iowa State University is conducting research with various projects involving algae, including a method that can be used to separate algae by allowing the algae to attach itself to the surface of a material. The material alternates between a carbon dioxide–rich gas phase and a nutritious growth-medium phase that has been found to increase algae growth rates and simplify harvesting the algae. When ready for harvesting, the algae is easily scraped off the surface of the material.

FIGURE 9-15   Algae Photobioreactor (PBR) at Iowa State University Bioeconomy Institute (BEI). The flat panel tanks are an alternate method to the photobioreactor tubes illustrated in Figure 9-14. The flat tank allows the maximum amount of sunlight to reach the algae. Air is bubbled in from the bottom of the tank to stimulate growth. (*Source:* Courtesy of Bioeconomy Institute, Iowa State University.)

bubble air through columns, and collect the froth of algae that accumulates above the liquid level.

Flocculation This approach to separating algae from the medium uses chemicals to cause the algae to form lumps.

Extraction of Oil from Algae

Mechanical Methods

One method of extracting oil from algae is the expression/expeller press. In this method, the algae is dried and then pressed to extract the oil. Another method, which is the simplest, is mechanical crushing. A third method is ultrasonic-assisted extraction, which uses ultrasonic waves and liquid jets to break the algae cell walls and release the oil content.

Chemical Methods

Several methods with different chemicals are used. In each case, the chemical forces the algae to release its oil. Chemical methods are generally less desirable than mechanical methods because of the problems and risks of handling and disposing of the chemicals.

SECTION 9-5 CHECKUP

1. Why are algae an excellent feedstock?

2. Name two general methods for extracting oil from algae.

3. Name two processes for algae production.

4. What are four methods for separating algae from a water mix?

9-6   Anaerobic Digestion

Anaerobic digestion is used for waste and sewage treatment. Anaerobic digestion is also used as a source of renewable energy if it produces a biogas consisting of methane and carbon dioxide. The gas can be used directly in combined heat and power (CHP) systems or it can be upgraded to biomethane.

Biogas

Municipal sewage, agricultural animal waste, and landfills are good sources of methane gas produced by **anaerobic digestion**, which is a bacterial fermentation process in which microorganisms break down biodegradable material in the absence of oxygen. The direct product of anaerobic digestion is **biogas,** which consists primarily of methane (CH_4) and carbon dioxide (CO_2), with small amounts of ammonia (NH_3) and hydrogen sulfide (H_2S). Hydrogen sulfide is a deadly gas that needs to be removed. Compost is also a product of anaerobic digestion.

Biogas typically contains 60% methane and 29% CO_2, with traces of H_2S and water vapor, but these numbers vary widely. Biogas from landfills is commonly known as landfill gas and can have a methane content as low as 50%. State-of-the-art systems can produce biogas that is as high as 95% methane. The biogas can be burned directly or it can be upgraded or purified to remove most of the CO_2 and H_2S, as well as any other contaminants. The upgrading process typically leaves about 98% methane per unit volume of gas. The most common method of upgrading is water washing (also called scrubbing), where the gas under high pressure flows through water to remove the contaminants.

Steps in Anaerobic Digestion

The process of anaerobic digestion can occur naturally, or it can be speeded up by controlling temperature (optimally at 100° F or 38° C) and the environment in which the

Biogas Safety

Biogases are flammable, explosive, and/or toxic. Confined spaces with poor ventilation are particularly dangerous and should be entered only with appropriate safety measures. Such safety measures generally include self-contained breathing equipment, rescue harness, and a standby rescue person. Any person who wants to work in a biogas facility needs to have specialized safety training before beginning work.

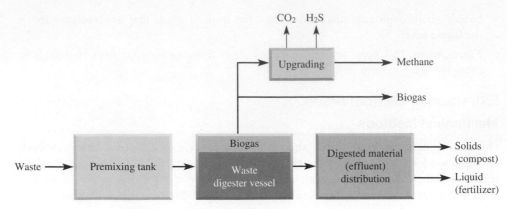

FIGURE 9-16 Functional Block Diagram of an Anaerobic Digester System

waste is placed. The organic waste is maintained in an airtight container. Digestion occurs in three steps:

1. The plant or animal waste decomposes, which breaks down the organic material into smaller units.

2. The broken-down material is converted to organic acids.

3. The acids are converted to methane gas.

Anaerobic digestion has been available commercially for many years as independent, for-sale units. In areas where unprocessed waste produces odor problems, anaerobic digestion can reduce the odor and the amount of liquid waste, and it can produce the side benefit of biogas fuel.

Types of Digesters

A functional block diagram of an anaerobic digester system for municipal sewage or livestock and poultry waste is shown in Figure 9-16. The process in such a system consists of premixing, digestion, extraction of biogas and methane, and distribution of digested waste material. The biogas can be burned for heat; the methane can be used to fuel an electrical generator. The solid effluent can be composted, and the liquid effluent can be used for fertilizer.

Batch Digester

In a batch digester, the waste material is loaded into the digester and allowed to be digested for a period of time, the length of which depends on temperature and other factors. When the digestion process is complete (typically in 10 to 20 days for a heated digester), the remaining material (effluent) is removed and the process is repeated, as illustrated in Figure 9-17.

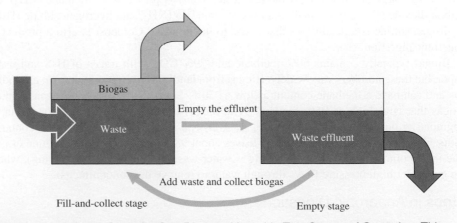

FIGURE 9-17 Batch Cycle for One Digester Vessel in Two Stages of Operation. This process results in an interrupted flow of biogas.

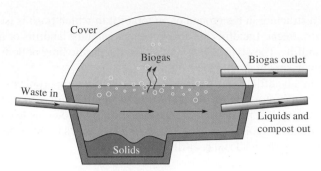

FIGURE 9-18   Continuous Anaerobic Digester

This type of digester is used frequently on farms to process manure and create biogas that can be used to offset energy needed on the farm for daily operations.

Continuous Digester

In this type of system, the waste material is continuously loaded into the digester and then moved through the digester tank, usually with paddle blades, at a rate that allows maximum digestion. As the new waste material comes in, the digested material is emptied. This process results in a continuous flow of biogas, as illustrated in Figure 9-18. A continuous flow system used to produce heat and electrical power is depicted in Figure 9-19. Additional digester tanks can be added in parallel to increase biogas production.

Landfills

Municipal landfills are a major source of methane gas through anaerobic digestion. Trash disposal in the United States, as well as in other countries, is a major problem. Trash is disposed of in three basic ways: It is recycled, burned (incinerated), or buried (landfill). About 50% to 60% of the trash is buried in landfills. Some countries, such as the United Kingdom, bury about 90%.

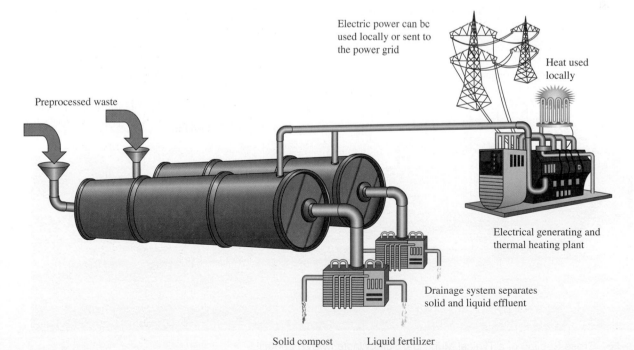

FIGURE 9-19   Anaerobic Digester System Used to Generate Electricity and Heat

A **landfill** is a structure in the ground or on top of it in which trash is isolated from the surrounding environment. Landfills are also known as sanitary landfills or municipal solid waste (MSW) landfills. They typically use a clay liner, a plastic liner or both to separate the trash from the environment around the landfill.

Storing trash in a landfill creates soil and water contamination problems. When rainwater, groundwater, trash moisture, or moisture from decomposition percolates through the layers of waste material, it becomes a contaminated liquid known as **leachate.** The term *leachant* is applied to any contaminated water, so it also applies to the water stored in hazardous waste sites like fly ash from coal-fired power plants.

Landfill Structure

A properly designed landfill isolates the contaminants from the surrounding environment by use of various material layers, leachate collection, and biogas collection. The cross section of a typical landfill is shown in Figure 9-20.

Bottom layer The purpose of the bottom layer of a landfill is to prevent contaminants from entering the groundwater and affecting the quality of the water supply in the area. The bottom layer typically consists of a puncture-resistant plastic liner and a layer of compacted clay. A fabric mat is used to help protect the plastic layer, as shown in Figure 9-20.

Leachate collection layer This is a layer of gravel with a system of pipes. Perforated pipes drain into a leachate pipe, where it is either pumped or gravity-fed (depending on the elevation of the leachate layer) into a surface collection pond. The drainage layer above the gravel is usually a sandy soil that allows the leachant from the trash cells to drain easily into the gravel layer. A compacted soil layer separates the drainage layer from the trash cells.

Cells Trash is compacted into cells that usually represent a day's collection. Using heavy equipment, the cell is compressed, covered with about six inches of soil, and then compacted again. The cells are arranged in layered rows, as shown in Figure 9-20. When a row is completed, another row layer is started on top of it.

Covering A plastic covering is placed on top when a section of the landfill is completed. Soil and then grass go on top of the covering for restoration of the landscape. The plastic

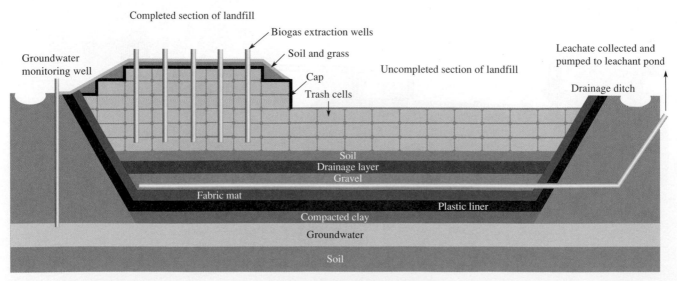

FIGURE 9-20 Cross Section of a Typical Municipal Solid Waste (MSW) Landfill

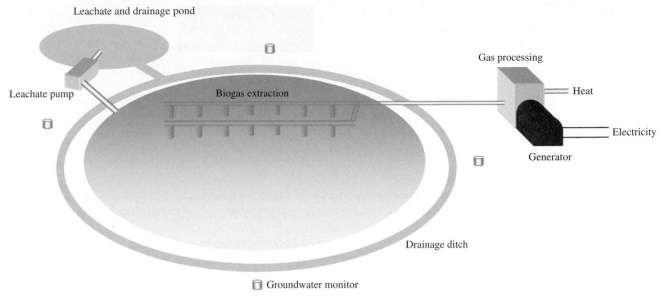

FIGURE 9-21 Completed MSW Landfill

covering prevents rain from entering the completed section and allows it to drain into ditches that surround the landfill.

Groundwater monitoring Wells are placed around a landfill in order to sample the groundwater. An increase in water temperature or an acidic pH can indicate leachate seepage.

Biogas collection As you know, biogas is a natural result of anaerobic digestion in rotting materials. Biogas contains methane, so it is very combustible and can be hazardous in landfills. The gas can also escape into the atmosphere if it is not controlled. Most landfills collect the biogas with a system of wells in the compacted trash cells area. The biogas is either burned off (flared) or used as renewable energy to produce heat and/or electricity, as shown in Figure 9-21.

SECTION 9-6 CHECKUP

1. Define the term *anaerobic digestion*.

2. Name two types of anaerobic digesters.

3. What does MSW mean?

4. What is leachate?

9-7 Biomass Combined Heat and Power

Combined heat and power (CHP) systems utilize combustion or gasification of biomass to produce electricity and heat directly. These systems are also known as cogenerators and typically use combustible garbage, wood chips, biogas, or other waste material as feedstock. Conventional power plants in general do not convert all of their thermal energy into electricity. Typically, more than half is lost as excess heat. CHP systems use heat that would otherwise be wasted and, as a result, can achieve efficiencies of up to 80%.

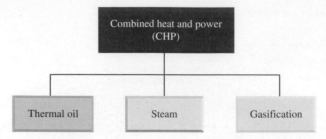

FIGURE 9-22   CHP Methods

The **combined heat and power (CHP)** is a process characterized by the production of both electrical power and heat. The heat is a by-product of the electrical generation. CHP is typically used in facilities such as hospitals, prisons, industrial plants, and small residential or business communities. Biomass CHP used three basic approaches: thermal oil, steam, and gasification, which are illustrated in Figure 9-22.

CHP Using Steam

The steam CHP process is based on the Rankine cycle, which was introduced in Section 1-2 for a nuclear reactor. An equivalent cycle is applied to CHP. The process of combusting biomass to produce heat and then converting the heat to electricity is the same basic four-step process of the Rankine cycle. Recall that the four steps are as follows: (1) pressurize the working fluid with a pump; (2) heat the working fluid to boiling or beyond boiling, thus converting it to a dry saturated vapor; (3) expand the dry vapor in a turbine, which extracts a large portion of the energy from the vapor, in turn cooling it and thus causing it to become a saturated vapor at reduced pressure; and (4) condense the wet vapor back to a low-pressure liquid by cooling it in a condenser at a constant temperature. The main difference in the standard process and the CHP system is the heat source and the extraction of lower-temperature heat to increase the overall usefulness of the plant. The lower-temperature heat is also referred to as **district heat,** which serves an area to fulfill a variety of its needs for lower-temperature heat, thus using heat that would otherwise be wasted. Recall that the Rankine cycle uses a closed loop; the fluid circulates continuously around a loop, carrying heat to the turbine and returning to the burner as it changes phase from a liquid to a gas and back again.

Variations of the Rankine cycle just described are designed to extract more energy than the standard cycle. One variation, called reheat, uses two turbines in series. After the vapor passes through the first turbine, it is reheated in the boiler before returning at lower pressure to the second turbine. The advantage of this method is that vapor does not condense in the first turbine, which can create problems with droplets striking the fast-moving rotor blades. The vapor is reheated and passes to a lower-pressure turbine. The process is more efficient than using a single turbine and tends to help reduce maintenance costs.

A portion of heat that cannot be used effectively by the electrical power side is used for district heat. This heat is extracted as lower-temperature heat from the flume and condenser, where steam is converted back to liquid water. Often, district heat is distributed as hot water for residential, public, and commercial buildings and has been widely adopted in Europe to reduce dependency on fossil fuels.

A diagram of a CHP plant that uses the Rankine cycle for electrical power and district heat is shown in Figure 9-23. Notice that district heating can be captured in two separate areas: in the hot flue gases, which would otherwise escape, and from the condenser, which would go to a cooling tower in traditional systems. Feedstock such as biogas, wood chips, combustible garbage, and so on, is fed into the combustion chamber (boiler), where it is burned (combusted). In a CHP plant, the working fluid is usually water. Water has a significant advantage over other fluids because of its thermodynamic properties, low cost, abundance, and unreactive chemistry.

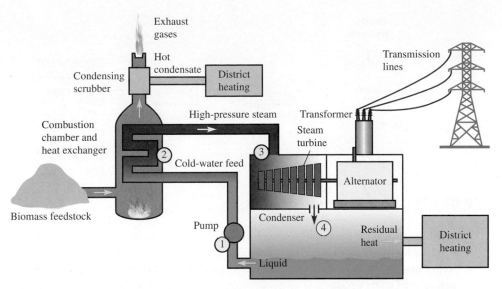

FIGURE 9-23 Simplified Diagram of a Steam CHP Plant. Biogas or methane from a landfill can also be used as fuel. The numbers indicate the four steps in the Rankine cycle.

CHP Using Thermal Oil

In some cases, a biomass plant uses an organic Rankine cycle (ORC) to generate electricity. The **organic Rankine cycle (ORC)** is a process identical to the basic Rankine cycle except it uses an organic fluid in place of water. The choice of fluid is based on the particular heat source, temperature, and flow. The working fluid has a boiling point lower than water, which enables it to extract energy from lower-temperature sources, such as waste heat from industrial processes. An ORC can operate with a lower-temperature heat source than is the case with a steam cycle, but it is less efficient. In cases where the heat source is lower quality, the ORC may be the best choice. The ORC cycle is described in more detail in Section 10-2 because it is also used with moderately hot geothermal sources.

CHP Using Gasification

Syngas is synthetic gas that contains hydrogen and carbon monoxide, as well as small amounts of other gases and water. Syngas is produced by a gasification process where preprocessed biomass rich in carbon is combined with a limited amount of oxidizer. The production of syngas is shown in the following formula:

> **Carbon fuel + oxygen → hydrogen + carbon monoxide + traces of carbon dioxide and water**

The product to the right of the arrow shows the constituents of the syngas.

Gasification differs from the other two methods previously discussed (the Rankine cycle and the organic Rankine cycle) because, with gasification, the biomass feedstock is not fully combusted. In the gasification process, the carbon fuel (biomass feedstock) is fed into a vessel called a gasifier, where steam and oxygen (air) are injected. After going through several stages of reaction of the biomass with the steam and oxygen, syngas is produced. The syngas can be used to fire a CHP system, as illustrated in Figure 9-24.

Gasification is generally more efficient than combustion and has almost no emissions. The gasification process consists of four stages:

1. Drying, in which the moisture content of the biomass feedstock is converted into steam

2. Pyrolysis, in which thermal decomposition of the biomass releases solids, liquids, and gases

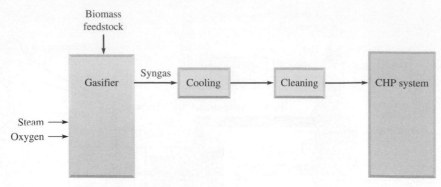

FIGURE 9-24   CHP System with Gasification

3. Oxidation, in which air is introduced and the reaction occurs between oxygen and solid carbonized fuel at a high temperature, thus producing carbon monoxide

4. Reduction, in which several complex reactions produce carbon monoxide and hydrogen as well as by-products

---

SECTION 9-7 CHECKUP

1. What does CHP stand for?

2. List three CHP methods.

3. How does reheat work?

4. What is the ORC?

CHAPTER SUMMARY

- Carbon is among the most abundant elements in the earth's crust, and it is the fourth most abundant element in the universe by mass after hydrogen, helium, and oxygen. Carbon is present in all known forms of life and is the chemical basis for all life on earth.

- The process by which carbon atoms are endlessly cycled around the biosphere is called the carbon cycle.

- The first step in the carbon cycle is photosynthesis, a process in plants, algae, and some species of bacteria. Photosynthesis uses energy from the sun to convert carbon dioxide and water into carbohydrates, which is stored in the plants, algae, or bacteria for food.

- Biomass feedstock is any raw material used to create an energy product, such as electricity, heat, or fuel.

- Ethanol is the most common biofuel around the world. It is an alcohol product made by fermentation of sugars in corn, wheat, rice, sugar beets, sugar cane, sorghum, potatoes, and other starchy food sources.

- The production of ethanol from cellulosic biomass is an alternative to starch-based sources.

- Biodiesel is a fuel derived from vegetable oils such as soybean oil or recycled cooking oils, animal fats, and recycled greases.

- Green diesel normally uses the same feedstocks as biodiesel, but it uses different processing methods.

- Algae is any of various aquatic, photosynthetic organisms that range in size from single-cell organisms to giant kelp. Algae convert carbon dioxide and sunlight into energy in the form of an oil.

- Anaerobic digestion is used as a source of renewable energy where a biogas consisting of methane and carbon dioxide is produced.

- The ideal Rankine cycle has four stages: compression, heating, expansion, condensation.

- Three methods for CHP are steam, thermal oil, and gasification.

KEY TERMS

algae   Any of various aquatic, photosynthetic organisms that range in size from single-cell organisms to giant kelp.

anaerobic digestion   A bacterial fermentation process in which microorganisms break down biodegradable material in the absence of oxygen.

biomass   Organic (carbon-containing) materials that have energy stored as carbohydrates, which can be used as feedstock for the production of renewable energy.

carbon cycle   The process by which carbon atoms are endlessly cycled around the biosphere.

carbon reservoir   A natural feature that stores carbon.

chemical conversion   A process that converts biomass into various forms of fuel without combustion.

combined heat and power (CHP)   A process characterized by the production of both electrical power and heat.

combustion   The burning of a substance; it is a rapid exothermic reaction that is the oxidation of a fuel, combining the fuel with an oxidant (usually oxygen).

ethanol   An alcohol made by fermentation of sugars in corn, wheat, rice, sugar beets, sugar cane, sorghum, potatoes, and other starchy food sources as well as cellulosic biomass.

fossil energy replacement (FER) ratio   The ratio of the energy delivered to the consumer to the fossil energy used at the production site.

organic Rankine cycle (ORC)   A cycle similar to the ordinary Rankine cycle except that thermal oil is used as the working fluid instead of water.

photosynthesis   A process in plants, algae, and some species of bacteria that uses energy from the sun to convert carbon dioxide and water into carbohydrates, which are stored in the plant, algae, or bacteria for food.

petroleum replacement ratio (PPR)   The ratio of the energy delivered to the consumer in the form of biofuel compared to the petroleum energy used in the process.

respiration   The process in which oxygen is used to break down organic compounds into carbon dioxide (CO_2) and water (H_2O).

thermal conversion   Any of several processes that use heat to extract energy from biomass without combustion by changing its chemical form with chemical reactions and interaction with oxygen.

CHAPTER TRUE/FALSE QUIZ

Determine whether each statement is true or false. Answers are at the end of the chapter.

1. All forms of life contain carbon.

2. Carbon is the most abundant element known.

3. In the photosynthesis process, plants use carbon dioxide and water to produce oxygen, and carbohydrates (glucose).

4. Biomass is nonorganic material used as feedstock for the production of renewable energy.

5. Burning wood is an example of combustion.

6. An example of co-firing is when biomass is used to replace a portion of the coal in a coal-fired power plant.

7. Ethanol, methane, and petroleum are all examples of biofuels.

8. Corn is the most commonly used feedstock for ethanol production.

9. A higher PRR means that less petroleum energy was used in producing the biofuel.

10. E10 is a blend of gasoline and ethanol.

11. In the biodiesel process, certain feedstocks must be pretreated by a process called transesterification.

12. Algae use photosynthesis to convert carbon dioxide and sunlight into energy in the form of an oil.

13. Raceway ponds can be used for algae production.

14. Oil can be extracted from algae only by mechanical crushing.

15. A photobioreactor is commonly used for algae production.

16. Anaerobic digestion is a process that uses microscopic viruses to produce methane gas.

17. Landfills are poor sources of biogas.

18. The CHP process is characterized by the production of both electrical power and heat.

19. Steam CHP is based on the ORC process.

20. Gasification is a noncombustion process for producing syngas.

CHAPTER MULTIPLE-CHOICE QUIZ

Complete each statement by selecting the one correct answer. Answers are at the end of the chapter.

1. The process by which carbon atoms are endlessly cycled around the biosphere is called
   a. the carbon reservoir
   b. the atomic cycle
   c. the carbon cycle
   d. photosynthesis

2. In the human body, carbon is
   a. the most abundant element
   b. the second most abundant element
   c. the third most abundant element
   d. not present

3. Plants process carbon dioxide and give off
   a. nitrogen
   b. carbon
   c. oxygen
   d. chlorophyll

4. Biomass is
   a. organic material
   b. used as feedstock
   c. used as a renewable energy source
   d. all of these

5. An example of a biomass material is
   a. scrap metal
   b. wood

   c. sand
   d. plastic bottles

6. E10 is
   a. gasoline with 10% ethanol
   b. diesel with 10% ethanol
   c. ethanol with 10% gasoline
   d. ethanol with 10% methane

7. The acronym FER stands for
   a. fuel-to-energy ratio
   b. fermentation energy ratio
   c. fossil energy ratio
   d. fossil energy replacement

8. Starch-based biomass includes
   a. sugar cane
   b. corn
   c. sugar beets
   d. all of the above

9. Biomass handling includes
   a. grinding
   b. chipping
   c. shredding
   d. all of these

10. One biomass source for producing biodiesel fuel is
   a. animal waste
   b. wood
   c. vegetable oil
   d. switchgrass

11. A closed-loop system for cultivating algae is a
   a. raceway pond
   b. fermenter
   c. photobioreactor
   d. digester

12. A method that is not used for separating algae from the water mix in which they are grown is
   a. fermentation
   b. filtration

   c. flotation
   d. centrifugation

13. A direct product of anaerobic digestion is
   a. ethanol
   b. biodiesel
   c. syngas
   d. biogas

14. Biogas typically contains about
   a. 60% CO_2 and 29% methane
   b. 60% methane and 29% CO_2
   c. 50% ethanol and 40% methane
   d. 98% methane

15. The acronym MSW stands for
   a. municipal solid waste
   b. maserated solid waste
   c. managed solid waste
   d. municipal steam-treated waste

16. In landfills, the contaminated liquid that percolates through the trash cells is known as
   a. residue
   b. leachate
   c. fermentate
   d. contaminate

17. In renewable energy, the acronym CHP stands for
   a. combined heat and power
   b. catalytic hydrogenated process
   c. centrifugal hydration process
   d. contaminate hydrolysis plant

18. As a working fluid, the organic Rankine cycle uses
   a. water
   b. syngas
   c. leachate
   d. thermal oil

CHAPTER QUESTIONS AND PROBLEMS

1. Describe the photosynthesis process in plant leaves.

2. Complete the following formula for the photosynthesis process.

      Sun energy + carbon dioxide + water →

3. Define respiration and state the process as a formula.

4. What is biomass feedstock?

5. List seven sources of biomass.

6. Explain the difference between combustion and thermal conversion.

7. What do the designations E10 and E85 refer to?

8. Refer to the graph in Figure 9-4. For every 1,000 BTUs of total fossil energy, approximately how many BTUs are in the resulting corn ethanol?

9. Refer to the graph in Figure 9-5. For every 1,000 BTU of petroleum energy, approximately how many BTUs are in the resulting corn ethanol?

10. Name the two categories of biomass sources for ethanol production.

11. Explain how biodiesel and green diesel differ.

12. What are algae and how are they a source of renewable energy?

13. What are the basic stages in cultivating and processing algae?

14. Name five methods for separating algae from the watery mix.

15. What is a raceway pond? Explain how it works.

16. What is a photobioreactor? Explain how it works.

17. What are the three methods used in biomass CHP systems?

18. List the four stages of an ideal Rankine cycle.

19. How does a gasification CHP differ from a steam or thermal oil CHP?

FOR DISCUSSION

Determine the biomass materials that are readily available in your area. Discuss how they are or could be utilized for energy production.

ANSWERS TO CHECKUPS

Section 9-1 Checkup

1. The process by which carbon atoms are endlessly cycled around the biosphere (earth and atmosphere)
2. Respiration, decomposition, burning of plants, burning of fossil fuels
3. Photosynthesis and respiration
4. Photosynthesis occurs in plants.
5. No. Animals and plants respire, but only animals breathe.

Section 9-2 Checkup

1. Organic materials used as feedstock for the production of renewable energy
2. Forests, agriculture, trash, sewage, animals, and industry
3. Ethanol
4. Land is taken out of food production.
5. Methane

Section 9-3 Checkup

1. Starch and cellulose
2. Corn
3. Cellulose ethanol does not use the edible portion of food crops.
4. Dry milling and wet milling
5. Fossil energy replacement ratio

Section 9-4 Checkup

1. Transesterification
2. Biodiesel has different chemical properties from petroleum diesel; green diesel is composed of long-chain hydrocarbons like petroleum diesel. Green diesel normally uses the same feedstocks as biodiesel, but it uses different processing methods.
3. Glycerin
4. Glycerin is used in certain soaps, preservatives, lotions, and certain pharmaceutical products

Section 9-5 Checkup

1. Algae can produce far more oil per acre than other biomass feedstocks.
2. Mechanical and chemical
3. Raceway pond and photobioreactor (PBR)
4. Filtration, centrifugation, flotation, and flocculation

Section 9-6 Checkup

1. Anaerobic digestion is a bacterial fermentation process in which microorganisms break down biodegradable material in the absence of oxygen.
2. Batch type and continuous type
3. Municipal solid waste
4. Leachate is a contaminated liquid resulting from water percolating through a landfill.

Section 9-7 Checkup

1. Combined heat and power
2. Steam, thermal oil, and gasification
3. Two turbines are in series. After the vapor passes through the first turbine, it is reheated in the boiler before returning at lower pressure to the second turbine.
4. Organic Rankine cycle

ANSWERS TO TRUE/FALSE QUIZ

1. T   2. F   3. F   4. F   5. T   6. T   7. F   8. T   9. T   17. F   18. T   19. F   20. T
10. T   11. F   12. T   13. T   14. F   15. T   16. F

ANSWERS TO MULTIPLE-CHOICE QUESTIONS

1. c   2. b   3. c   4. d   5. b   6. a   7. d   8. d   9. d   17. a   18. d
10. c   11. c   12. a   13. d   14. b   15. a   16. b

Geothermal Power Generation

CHAPTER OBJECTIVES

- Identify the five levels of the earth.
- Explain the operation of a binary-cycle steam electrical generating plant.
- Identify the five types of geothermal resources and their temperatures.
- Explain the operation of a geothermal heat pump.
- Identify the regions in the United States that are producing electricity with geothermal energy.
- List seven countries that produce electricity from geothermal energy.
- Explain the operation of a dry-steam electrical generating plant.
- List four agricultural applications that use hot steam or hot water from geothermal energy.
- Explain the operation of a flash-steam electrical generating plant.

KEY TERMS

Key terms are shown in bold and color. Definitions for key terms are provided at the end of the chapter and in the end-of-book glossary. Bold terms in black are defined in the end-of-book glossary only.

- enthalpy
- dry-steam plant
- flash-steam plant
- double-flash steam plant
- binary-cycle plant
- enhanced geothermal system (EGS)
- geothermal heat pump
- coefficient of performance (COP)

INTRODUCTION

As you learned in Chapter 1, geothermal energy is a clean source of renewable energy with many desirable attributes. When a geothermal plant is in operation, it emits only water vapor and a small amount of trapped gas, including some carbon dioxide in quantities much smaller than that released from fossil-fueled plants.

In locations where it is used for electrical production, a geothermal electrical generating plant can provide continuous electrical power, unlike other renewable energy systems that depend on sunlight or wind. All geothermal energy within the reach of humans is in the relatively thin earth's crust; holes drilled more than about 4 km become very expensive and generally not economical, although new drilling technology may extend the depth. For economically viable electricity generation, the geothermal resource needs to have enough energy that is reasonably accessible.

In almost all areas, geothermal heat pumps can be employed effectively. Other applications for geothermal energy include providing hot water and steam for heating systems and for agriculture and aquaculture.

10-1   Types of Geothermal Resources

Geothermal resources are found all over the world. Chapter 1 discussed locations of geothermal resources throughout the world and their environmental effects. In this section, the geothermal gradient is introduced and different types of geothermal resources are defined.

A high-quality geothermal source has three useful features: ample heat, water, and permeability and porosity of the ground near the heat source. The term **permeability** is a measure of the ability of a material to pass a fluid. When ground is permeable, water can move through it easily. Locations with these attributes are called high-enthalpy resources. The word **enthalpy** is a thermodynamic term that is the amount of energy in a system capable of doing mechanical work; it is a function of temperature, pressure, and volume. Enthalpy (abbreviated h) is the sum of the internal energy and the product of pressure and volume. Ideally a source should have high enthalpy and/or an adequate reservoir recharge of fluids.

Geothermal Gradient

The geothermal gradient refers to the increasing temperature at increasing depths within the earth, which occurs worldwide. For points not near a tectonic plate, the temperature increases with depth about 25° C/km, but it can be much higher when near hot igneous bodies. In some cases, the geothermal gradient is much less than average, particularly in subduction zones where cold, water-filled sediments are driven underground. In Iceland, the gradient can be as high as 200° C/km. A high temperature gradient is one indicator of useful geothermal energy. The best sources also have available fluids in the underground formation and porous rocks (rocks with open spaces) with high permeability (ability to allow fluids to move). At the present time, the only commercially viable sources are in a hydrothermal (hot, wet) environment with high permeability.

The geothermal gradient is of value for geothermal heat pumps (GHPs) because they can use the source at shallow depths. This is a relatively low-tech application for geothermal energy that takes advantage of the relatively constant temperature of the ground at a depth of a few meters (up to 100 meters), even in areas not suited for high-temperature applications. Geothermal heat pumps move heat in cold weather from the ground and put it back during hot weather, so they are useful in both heating and cooling applications. GHPs are discussed in Section 10-4.

Geothermal resources are commonly classified by temperature and amount of fluid (water) present that can carry energy to the earth's surface. High-temperature resources are greater than 150° C (302° F). Sources with fluids above 200° C are capable of driving steam turbines directly. Temperatures between 90° C and 150° C (194° F and 302° F) are considered to be in the moderate range but are still useful for electrical energy production in some cases. Moderate temperature resources are useful for space heating, drying, refrigeration cycles, and industrial processing.

Low-temperature resources have a temperature less than 90° C (194° F). The lower temperature resources are used for heating buildings, fish farming, bathing and even snow melting (in places like Iceland with a very large geothermal resource.) Where possible, the waste heat from a high-temperature application such as electrical generation may be used in another application such as space heating to optimize the resource.

High-Temperature Resources

High-temperature resources are efficient for producing electricity, which is their predominant application. High-temperature resources include molten magma, hot dry rock, geopressurized resources, and hot-water reservoirs. Each will be discussed in the following subsections.

Molten Magma

The hottest geothermal resource is approximately 650° C to 1,300° C and is found in the molten magma pools in and under volcanoes. Large pools of magma reside in chambers below the surface of the earth; occasionally, the magma can make its way to the surface and erupt as a volcano when pressure in the chamber builds. These lava pools contain significant energy but are difficult to tap. Experimental holes have been drilled in magma chambers in Hawaii and Iceland; however, there is no commercial use of this resource at this time. Any attempt to use this resource will require developing materials that can withstand the enormous temperatures and corrosive environment that magma is found in; it is at best a high-risk investment at this time.

Hot Dry Rock

Often the rock above magma chambers is very hot, but it often lacks water content and frequently lacks permeable rock formations. Underground hot dry rock formations are one of the most common forms of geothermal resource. While they can serve as a consistent source of heat, the lack of water and permeable rock are significant drawbacks. To be effective, pressurized cold water needs to be injected down a nearby injection well, and advanced methods such as hydraulic fracturing ("fracking") can be employed to increase permeability. Some projects in Europe, Japan, and the United States have been developed in hot dry rock formations. In these systems, permeability is enhanced artificially and water is piped to the region of hot dry rocks. When the water comes in contact with the hot rocks, it boils immediately, creating superheated steam at a temperature above 93° C (200° F). (Superheated means it is hotter than the normal atmospheric boiling point.) The steam can be sent directly to a steam generator to convert the energy to electricity.

Geopressurized Reservoirs

Geopressurized reservoirs consist of high-pressure hot brine (salt water) in deep underground reservoirs made of permeable rock that traps the water under an impermeable layer of cap rock. The water also contains dissolved natural gas that can be practical to develop with the geothermal resource. Usable geopressurized reservoirs are typically 3 km to 4 km deep (10,000 to 13,000 feet). A layer of impermeable cap rock (such as shale) covers the resource (see Figure 1-3 for an illustration). It is possible to capture the natural gas from this resource and use it for co-generation, but the water itself is unusable for heating or bathing because of the dissolved materials. Currently geothermal energy and natural gas are being co-produced from deep sedimentary rocks in the Gulf Coast of the United States by GreenTech Geothermal, LLC. Producing the two forms of energy together has some cost benefits. The large reservoirs of natural gas in the United States and elsewhere have spurred the development of related work in conversion of natural gas into diesel, jet fuel, and other liquid hydrocarbons. The process is called **gas-to-liquid (GTL),** and plans for developing manufacturing plants around the Gulf of Mexico to produce a variety of petrochemicals from the natural gas in these hot underground reservoirs are underway.

Hot-Water Reservoirs

Hot-water reservoirs are also called natural steam reservoirs; they have geothermal heated underground water and steam. Hot springs are often located along fault lines in the earth. These reservoirs are the most common usable form of geothermal energy for electricity production. The steam and hot water are used to drive steam turbines, which in turn drive the generators. The water typically reaches a temperature that is hot enough at pressure that is low enough to create steam.

Intermediate- and Low-Temperature Resources

Intermediate-temperature sources are too cool to power a steam generator directly, but they can still be useful for electricity production using an organic Rankine cycle (ORC) plant to take advantage of the lower boiling point for the working fluid. The systems are called binary systems and are discussed in Section 10-2. Several of these plants have been constructed in the United States (Mammoth Lakes) and in Europe. In addition, intermediate-temperature

Old Faithful

(*Source:* Courtesy National Renewable Energy Laboratory.)

One of the best known geysers in the world is located in Yellowstone National Park in Wyoming. This geyser is a cone geyser, and it erupts to a height of 90 to 185 feet. The eruptions occur between 45 and 125 minutes apart. The duration of each eruption is 1.5 to 5 minutes, and each eruption discharges between 3,000 and 8,000 gallons of hot water.

sources have provided hot water for heating offices and homes. These are district heating sources (which were described in Section 9-7). A limited number of communities in the western United States use geothermal district heat, but many more could benefit from this resource.

Geysers and Hot Springs

Hot springs are naturally occurring water resources in which geothermally heated water rises to the surface. The water is heated underground by hot rocks or magma. Relatively few have been developed for geothermal power; however, an exception in the United States is at Chena Hot Springs resort in Alaska (see Section 10-3). One of the best known hot springs used for geothermal power is the Blue Lagoon near Reykjavik, Iceland. In general, geysers and hot springs represent a very small fraction of the geothermal resource.

Geysers are natural hot springs that reach temperatures high enough to boil and intermittently eject a column of water and steam into the air. They are found in regions where magma is deep underground, heating layers of rock. Surface water entering the location of the hot rocks boils and builds pressure. When pressure builds sufficiently, the water is ejected violently from a surface vent.

One of the most active geyser regions in the world is in Yellowstone National Park in the United States. Yellowstone is located over a large, hot magma region; the iconic geyser in Yellowstone is Old Faithful, a major tourist attraction. Of course, no one is suggesting using Old Faithful for geothermal power; however, other geysers in volcanic regions of the world, such as South America, Japan, and the Philippines, are tapped for geothermal energy. In Iceland and New Zealand, they have been developed for providing heating resources.

Normal Ground Temperature Sources

Lower-temperature resources include the normal temperature a short distance from the surface in both saturated and dry rock. The distance and temperature vary from one location to the next, but most places in the world have usable temperatures 1 or 2 m underground for geothermal heat pumps because the earth's temperature remains rather constant at approximately 10° C at relatively shallow depths. This temperature is sufficient to keep a heat pump working efficiently when the outdoor temperature drops below freezing (when conventional heat pumps are not efficient). To take advantage of this heat, underground piping is installed for recirculating fluid, which adds to the initial cost; however, the long-term benefit of reduced electric usage pays for the initial higher cost.

SECTION 10-1 CHECKUP

1. What is the meaning of the term *enthalpy*?

2. What does the term *geothermal gradient* refer to?

3. Identify four forms of high-temperature geothermal resources.

4. What is a hot spring and geyser?

5. Identify an application for lower-temperature geothermal heat.

10-2   Geothermal Electrical Power

Capturing geothermal energy to produce electrical power is accomplished through dry steam flash and through binary-cycle steam power plants. The most common way of capturing the energy from geothermal sources is to tap into naturally occurring hydrothermal convection systems. The geothermal systems produce hot water or steam when cooler water makes its way deep below the surface of the earth, where it is continually heated, and then rises to the surface. When the steam is brought to the surface, it is used to drive turbines that turn electric generators.

Geothermal energy for generating electrical power comes from high and intermediate-temperature sources. The energy from these sources is used to turn turbines that rotate electrical generators to produce electrical power. Very hot water and steam naturally rise to the surface, where they are captured. The steam is directed across the blades of a turbine to drive it and spin the electric generator. In other applications where the steam does not make it all the way to the earth's surface, holes are drilled into the rock to depths that allow water to be injected into them, and the steam that is produced is captured. When the source is moderately hot but not hot enough for creating steam, the hot water can be sent through a heat exchanger to heat a secondary fluid.

Geothermal power plants can be based on three systems, all of which use hot water or steam from the source and then return the water to prolong the life of the source. In the simplest system, steam goes directly through the turbine, then into a condenser where the steam is condensed into water. In a second system, very hot water under high pressure is brought to the surface and depressurized (or flashed) into steam, then used to drive the steam turbine. In the third system, called a binary system, the hot water cannot be used directly, so it is passed through a heat exchanger, where it heats a second working fluid used in a closed loop with the turbine. These three systems are discussed in the following paragraphs.

Dry-Steam Plants

Dry steam (also called superheated steam) is steam so hot that it contains no liquid water in suspension. When a geothermal reservoir is located in a place that has steam at the surface, the steam can be piped directly to a steam turbine that drives an electric generator. This type of plant is called a **dry-steam plant**. The largest dry steam power production facility in the world is at the Geysers in Northern California. The only other large dry-steam plant was started in 1911 (following a demonstration that it could generate electricity in 1904) and it is located in Larderello, in Southern Tuscany, Italy. In other parts of the world, dry steam is rare.

The steam at the Geysers has a temperature greater than 235° C (455° F), so it has sufficient heat energy to turn the turbine and generator. The steam enters the turbine while it is very hot, and the energy in the steam is released. As the steam passes over the turbine blades, it loses energy. When the steam exits the turbine, it has less pressure and it is cooler. Some of it may start to condense in the turbine, creating water droplets. The water vapor completes the condensation process in a unit called the condenser, which is much cooler than the remaining steam. The cooler condenser extracts most of the remaining water from the air. This water can be reinjected down a second well to replenish the geothermal reservoir. At some dry-steam plants, adequate amounts of fresh water are introduced to restore the reservoir when some of the steam escapes to the atmosphere after it passes through the steam turbine. Figure 10-1 shows a basic diagram of a dry-steam plant.

In Figure 10-1 on page 312, you can see that dry steam is tapped by a well called the **production well.** Steam is under high pressure, and rocks and debris frequently come up from the well with the steam. A rock catcher is inserted between the steam pipe and the turbine to avoid damaging the turbine. In a dry-steam system, the steam is piped directly to the turbine. The turbine and generator are the basic energy converters in the system. Basically the turbine converts heat energy in the steam to mechanical work turning the generator. The generator converts the mechanical work to an electrical output.

Figure 10-2 on page 312 shows a cutaway view of a steam turbine. Steam is piped into the turbine, where it expands in nozzles and discharges at a high velocity and strikes moving blades. The input energy in the steam is proportional to the steam velocity and hence the available energy. As the steam strikes the turbine blades, it gives up its energy and expands. The steam is directed in an increasingly larger radius path around the turbine to the next stage. Succeeding stages are larger and consequently have higher tangential velocity due to the increased distance. The steam, now with lower energy and less pressure, is directed to progressively larger stages, where it gives up more of its energy at each pass. Finally it is exhausted from the turbine and is condensed back to a liquid.

The Geysers

(*Source:* David Buchla.)

The Geysers plant in Northern California was featured in an article in *Fortune* magazine in 1969 as "a development whose significance goes far beyond its size." The article went on to note that Pacific Gas and Electric (PG&E) in 1960 was the first US utility to generate geothermal power.

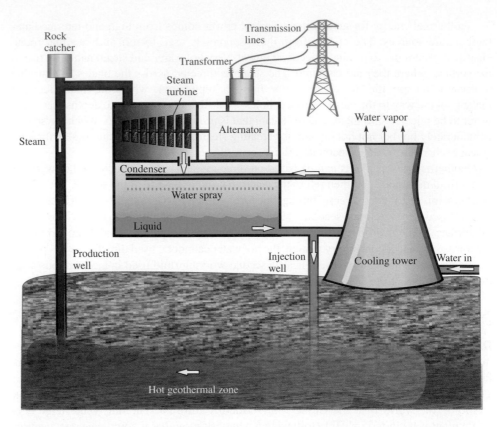

FIGURE 10-1   Dry-Steam Power Plant

As shown in Figure 10-1, the steam flows into a steam surface condenser after it passes through the turbine. Cool water from a cooling tower is sprayed over the remaining steam in the condenser, causing the steam to condense into water. The water is pumped out of the bottom of the condenser into an air-cooled tower, where it is cooled further. A second bypass condenser (not shown in Figure 10-1) is connected in a separate path for the hot input steam, which allows the turbine to be offline for maintenance or any other reason. If steam were to bypass the turbine and allowed to condense in downstream piping, pressure spikes and high temperatures could cause damage. The bypass condenser is a separate condenser that is ready to go online automatically should the turbine be offline.

For the cooling tower to provide a continuous supply of cool water, air is drawn through the bottom of the cooling tower, and air and water vapor is vented out the top to the atmosphere. Water that pools in the bottom of the cooling tower is pumped back down a second well called an **injection well.** This water replenishes the geothermal resource deep below the surface. Because some of the water makes its way back into the atmosphere, additional

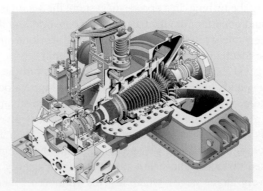

FIGURE 10-2   Cutaway View of a Steam Turbine   (*Source:* Courtesy of Siemens AG.)

FIGURE 10-3 Dry-Steam Geothermal Power Plant at the Geysers in California. The cooling towers in the background are emitting plumes of condensed water vapor. (*Source:* National Renewable Energy Laboratory.)

water is added to the injection well to keep producing sufficient steam. At the Geysers, this additional water comes from treated wastewater from the city of Santa Rosa. Figure 10-3 shows one of the plants at the Geysers, with water vapor escaping to the atmosphere.

Flash-Steam Plant

Flash steam is a mixture of pressurized hot water and steam that converts to steam when pressure is released. This combination of liquid and vapor in an underground reservoir is called a **liquid-dominated reservoir;** it is produced when large volumes of water from artesian wells come into contact with extremely hot rocks deep in the earth. The temperature of this resource for the flash-steam plant is greater than 182° C (360° F). In the **flash-steam plant**, the mixture of vapor and high-pressure hot water is pumped through a special control valve or orifice plate that expands it, reducing the pressure and causing some of the liquid to "flash" instantly into steam. The basic single-flash system is shown in Figure 10-4. Once the steam is separated

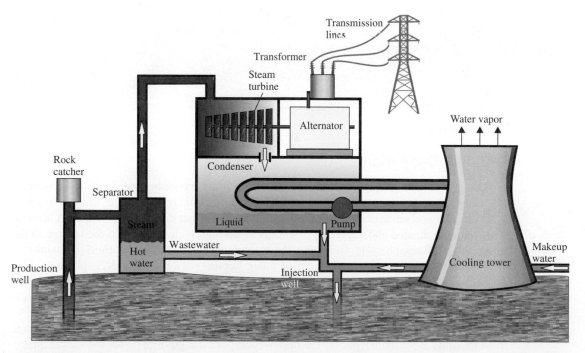

FIGURE 10-4 Single-Flash System. In a double-flash system, the hot water is under pressure and can create steam in a second separator.

from the remaining hot water and brine, the process is the same as in the dry-steam system. In some systems, waste hot water is sent to direct heat users (not shown). Some of the water is returned to the geothermal reservoir through the injection well, which includes additional makeup water to keep the resource in equilibrium and make up for lost water.

Double-Flash Steam Plants

When the source is very hot, the water that remains from the separator may be much hotter than the normal boiling temperature of 100° C (212° F) because it is still under pressure that is higher than atmospheric pressure. Additional steam can be created from the hot water by expanding it through a second separation system. This type of geothermal plant is called a **double-flash steam plant**. In a double-flash system, the second expansion creates lower-pressure steam, which is routed to a separate low-pressure turbine. About 20% more electrical power can be produced in a double-flash system but at greater cost due to the second turbine, which is a special low-pressure turbine. In addition, a double-flash system has extra piping and valves. The double-flash steam plant is used in several locations in California in spite of the additional cost. Figure 10-5 shows a diagram of a double-flash geothermal power plant.

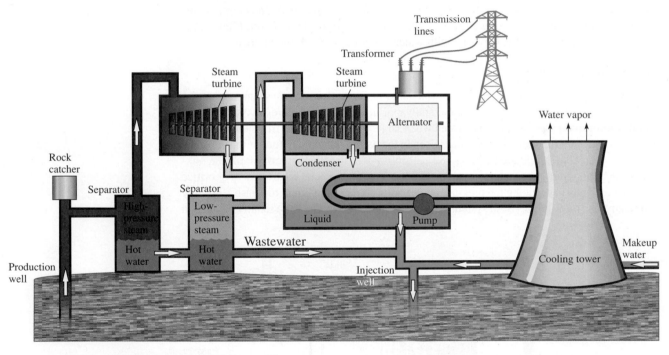

FIGURE 10-5   Double-Flash Geothermal Power Plant

Binary-Cycle Plants

The geothermal **binary-cycle plant** is different from dry-steam and flash-steam systems because the hot brine water and steam from the geothermal reservoir goes directly to a heat exchanger and never comes in contact with the turbine that drives the generator. A secondary fluid is heated and vaporizes. It circulates through a closed loop to the turbine in the traditional Rankine cycle. Generally, the secondary fluid is an organic fluid (contains carbon) that has a lower boiling point than water. For this reason, binary-cycle plants are also known as organic Rankine cycle (ORC) plants. ORC plants were discussed with respect to biofuel power production in Section 9-7. Figure 10-6 shows a diagram of a binary-cycle plant for geothermal electricity production.

Binary-cycle plants have two advantages for geothermal applications. The major advantage is the same as for biofuel plants: The binary-cycle plant can produce power with lower-quality heat sources. The fluid in the secondary loop of the heat exchanger is an organic fluid such as a thermal oil or isobutane that vaporizes at lower temperature. The fluid is selected based on the

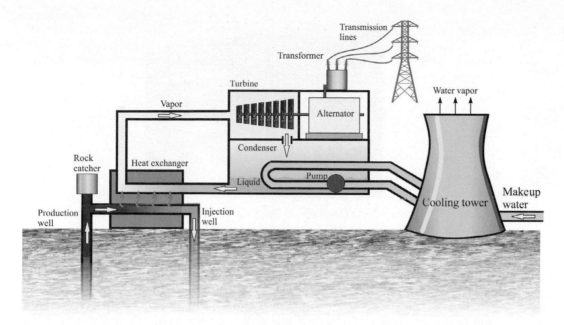

FIGURE 10-6 Binary-Cycle Plant. This type of plant is also known as an organic Rankine cycle (ORC) plant. The secondary fluid is shown in green in this diagram to distinguish it from the underground source.

temperature of the source. Because the vaporization temperature in the secondary loop is lower than that of steam, it can produce vapor even when the temperature of the brine solution is in the moderate range (typically 110° C to 176° C) or even lower. Lower-temperature operation allows ORC systems to operate in cases where a standard steam plant cannot.

The closed secondary loop is a second advantage to binary-cycle systems for geothermal applications. Geothermal sources frequently contain dissolved materials that may be corrosive and also may contain small debris carried up from the geothermal reservoir. Corrosive steam and debris is particularly hazardous to turbines because they spin at very high rates, so the binary-cycle plant avoids these problems by recycling the underground water and brine in a separate loop, where it gives most of its heat energy to a secondary fluid. The turbine is exposed only to this clean material, which is recycled in a separate closed loop.

Because the two loops in a binary-cycle system are completely enclosed, there is less waste and lower water usage than in other geothermal systems. By its nature, the binary-cycle plant does not release any pollutants and does not have any emissions except for a small amount of water vapor at the cooling tower. The main use of water is to cool the working fluid; it is important to cool it as much as possible for an efficient operation. The vapor from the cooling towers is supplied by a separate source and does not include anything from the underground source or the secondary loop. As a result, a binary plant pollutes the least of any of the geothermal types.

The binary-cycle electrical generating plant was first introduced in Russia in the late 1960s; the first operating plants in the United States started in the early 1980s. Figure 10-7 shows a binary-cycle geothermal plant in Raft River, Idaho. Today more systems are being brought online in locations where intermediate temperature geothermal resources are available (primarily in Alaska and the western United States.) Some of the sites include Chena Hot Springs, Alaska; San Emido, Nevada; Raft River, Idaho; and Neal Hot Springs in eastern Oregon. In some cases, deeper wells are being tested in locations in which geothermal electrical power was not thought possible.

Enhanced Geothermal Systems

Some deep underground heat sources may have the heat but they lack the natural permeability of the rock or needed water to exploit the resource. An **enhanced geothermal system (EGS)**

(*Source:* Photo courtesy of Alterra Power Corp, Soda Lake Geothermal.)

Northern Nevada played an important role in the westward migration in the nineteenth century, when pioneers traveled in wagon trains across the barren deserts of Nevada. Brady Hot Springs was a stopping point for vital water, but it was called the Spring of False Hope by the pioneers. Oxen that were pulling the wagons could smell the water and rushed forward to drink it, but they were unable to do so because it was scalding hot. At nearby Soda Lake, pipes bringing hot water from deep underground are close to the original Emigrant Trail. An aging sign marks the historical path, with the modern Soda Lake geothermal plant in the background.

FIGURE 10-7   Binary-Cycle Geothermal Electrical Generating Plant in Raft River, Idaho   (*Source:* Courtesy of U.S. Geothermal Inc.)

is one in which a geothermal site that is deficient in water or permeability is made productive by artificial means. Typically, an engineered underground reservoir is created where there is a heat source such as hot rock that lacks either the permeability or water (or both) to use for geothermal electrical production. Studies have indicated that EGS has the potential to make a significant contribution to the world energy mix, and potential EGS sites exist in many areas. In a typical EGS case, water is injected under controlled conditions to increase the permeability by fracturing rock and creating water storage. This technology had been implemented in demonstration plants and has the potential of opening a huge, clean energy source if it can be done cost effectively. Although EGS is normally associated with adding a reservoir to hot dry rocks, it can also be implemented in hot springs by increasing the permeability. At Brady Hot Springs in Nevada, the first application of EGS in the United States, the operator (Ormat) has developed fracture networks to increase the permeability of the rock. Several other sites are currently applying EGS technology to enhance productivity.

One aspect to EGS systems is the investigation of improved drilling techniques to make otherwise unproductive areas economically viable. Drilling cost represents one of the principal development costs of any geothermal resource. Modern drilling technology allows wells to be drilled over 4 km deep. Also, drilling technology includes horizontal drilling, which will enable reaching resources previously unavailable (such as under a city). Slanted drilling has been done for many years, but the advent of new tools and methods have potentially huge impacts on geothermal energy.

One new technology is the vaporization of hard rock (like granites and basalts) that can lead to much faster and cheaper drilling (over conventional drilling technology). Another technique is **hydrospallation drilling,** in which a stream of superheated high-pressure water is directed at the rock surface. The method is currently under development, but initial efforts show that hydrospallation drilling is effective and very fast. It eliminates the need for hoisting the drill out of the well, thus saving time and money. Another drilling advancement that falls in the domain of EGS is the development of materials for drill bits. These new materials can extend the life of drill bits, thus reducing the time required to replace bits. Improvements in conventional methods, such as developing better percussive drilling technology and use of small diameter core holes, also lower cost.

SECTION 10-2 CHECKUP

1. What is a dry steam and what is a dry-steam plant?

2. What is a double-flash system? How does it differ from a single-flash system?

3. What are two important advantages to a binary-cycle geothermal plant?

4. What is an enhanced geothermal system?

10-3   Low-Temperature Applications for Geothermal Heat

Lower-quality geothermal sources are also used around the world for various heating applications. Hot water from geothermal springs can be used to heat buildings, heat greenhouses and other agricultural applications, and provide process heat for industry (such as drying). Lower-temperature sources are also used for resorts and spas.

Archeological evidence has shown that humans have used geothermal heat from hot springs, for heating, cooking, and bathing, for more than 10,000 years. Currently, over 72 countries are using 16,000 megawatts of geothermal energy for heating applications and greenhouse production, aquaculture, and even snow melting. In industry, hot water is used to pasteurize milk, wash wool, dry wood and crops, sustain fish farms, and more. This section discusses lower-temperature resource applications; Section 10-4 will describe geothermal heat pumps.

Space Heating

Geothermal steam from a geothermal reservoir or hot spring can be used for space heating. The hot water or steam is routed through piping and heat exchangers and returned to the reservoir. The secondary loop in the heat exchanger has clean water running through it; the clean water is pumped through radiators or fan coil systems mounted in ductwork throughout the building or home. The air blows over the coils and moves heat into the conditioned space.

Geothermal heating is widely used in Iceland, which has over 250 separate thermal areas and more than 600 hot springs. Approximately 30% of Iceland's electricity comes from geothermal sources; in addition geothermal is supplies over 85% of the space heating requirements. After geothermal-generated steam has been used to generate electricity, the low temperature heat can provide heat for buildings. The lower-temperature steam passes through a secondary heat exchanger, and most of the heat is transferred to water. The hot water is pumped to storage tanks and then circulated through buildings. In some applications, the water is stored several hundred feet above the buildings, and gravity is used to circulate the water through the heating system of the building. The hot water is approximately 35° C (95° F), and it is pumped through heavily insulated pipes as much as 25 km from its source. The water loses less than 1° C while it is flowing through the piping. Combining electrical production and heating from the same geothermal energy source makes better use of the resource. Iceland's goal is to provide 100% of its electrical generation, heating, and cooling energy from renewable sources including geothermal and hydropower.

Cooling

Steam can be used as the heating source for absorption-type refrigeration systems. Geothermal heat is used in the absorption cooling process to provide air conditioning and refrigeration. Absorption units are designed in two basic configurations. For air-conditioning systems, lithium bromide/water ($LiBr/H_2O$) is used as the refrigerant. For refrigerators and applications below freezing temperatures, ammonia/water (NH_3/H_2O) is used as the refrigerant. If the system is used to provide refrigeration, for example, for meat and dairy processing, the steam can be used to provide the heat source for the ammonia used in the refrigeration system. The source of heat is the most expensive part of this type of refrigeration or cooling system, so getting the heat source from geothermal steam or hot water cuts the operating expense of this type of system.

In a $LiBr/H_2O$ absorption cooling system, a generator heats the refrigerant and the more volatile water boils off, creating a vapor at high pressure. The high-pressure vapor moves to a condenser unit, where it is condensed to a liquid at high pressure. Heat from compressing is removed, cooling the high-pressure liquid to near ambient temperature. The high-pressure water then passes through an expansion valve, where it immediately cools to chilled temperatures. The chilled water is used as a refrigerant; at this point, it picks up heat from the region to be cooled. Finally, the water is absorbed by the LiBr, and the process is repeated.

FIGURE 10-8 Using Geothermal Heat for Melting Snow   (*Source:* National Renewable Energy Laboratory.)

Melting Snow

Locations that have ample geothermal energy and are located where winters are cold enough for snow can benefit from snow-melting systems. When geothermal energy is used for melting snow in parking lots, on sidewalks, and in other public spaces, large pumps move hot water from the source through piping that is installed beneath the surface to be cleared. If the geothermal water is too hard or contaminated for piping, it is first pumped through the primary loop of a heat exchanger, and clean water is taken for snow removal from a secondary loop.

One problem with a snow-melting system is that, if the temperature is cold enough to for ice and snow, it is also cold enough to freeze the water in the piping. Thus, it is important to ensure that continuous flow occurs or that the systems are pumped down so that water does not remain in the piping. Figure 10-8 shows an example of geothermal heating used to melt snow and ice. These systems are used extensively in Iceland; in Klamath Falls, Oregon; and other regions that experience extreme winters and have geothermal energy available.

Snow-melting systems need a large reservoir of hot water. Such a reservoir can exist naturally, or large insulated tanks can be used to store the hot water before it is circulated through the piping. The system also needs one pump if the water is circulated directly through the piping and two pumps if a heat exchanger is used. In cases where hot water is used for heating a building, the water that exits the building is still over 10° C (50° F), which means enough heat energy remains to melt snow and ice. In these systems, the water that comes from the heating system is pumped outside through the underground piping to melt snow and ice.

Geothermal snow and ice removal is also being tested on highway bridges and over-passes that freeze more quickly, as well as on road surfaces that are on the ground. The road surface that runs on the ground receives enough heat from the earth to prevent the formation of ice until the ambient temperature remains below freezing for a prolonged period. Bridges and overpasses have cold air circulating above and below their surfaces, so they tend to freeze before road surfaces, thus causing more accidents in these areas. In locations where geothermal hot water is available, systems are being tested that pump hot water directly under the bridge or overpass to prevent the surface from freezing as quickly. The cost effectiveness needs to be evaluated to determine if it is useful to continue to pump water when temperatures drop well below freezing. Other locations that can benefit from melting snow include airports and helicopter ports on rooftops.

Greenhouses

For years, agricultural experiments have been determining the benefits of increasing temperatures to extend growing seasons. Other research is being conducted to determine the benefits of increasing temperatures in products to enhance growth and speed up the growing cycle. One application of using geothermal hot water or steam technology is to heat greenhouses. Figure 10-9 shows a greenhouse heated by geothermal energy. One of the

FIGURE 10-9   Greenhouse Heated by Geothermal Hot Water   (*Source:* National Renewable Energy Laboratory.)

largest expenses in the operation of a greenhouse is heating. The greenhouse must have large areas of glass or plastic that lets light reach the plants. These materials are very poor insulators; when the sun is not producing heat, the amount of heat lost through the materials is very large. Typically furnaces or boilers are used to heat the greenhouse, and in most cases the greenhouse is shut down during the coldest months of the year. In locations where geothermal energy is available, the cost of heating the greenhouse is reduced dramatically, and the months that the greenhouse can be operated are extended.

The main expense in the geothermal greenhouse is for the pumps, heat exchangers, and storage tanks. Generally, the cost of electrical power to run the pumps is the largest ongoing expense, but it is far less than the expense of gas or coal that the geothermal energy is replacing. In some locations, warm water is pumped through tubing belowground and used to warm ground where vegetables are planted. The warm ground allows seeds to be planted earlier and helps plants grow faster.

Fish Farms

Geothermal hot-water and steam systems are used in other agricultural businesses, such as raising fish and alligators. These businesses are called **aquaculture,** which is farming of aquatic plants and animals, including fish, shrimp, oysters, and seaweed. Aquaculture is centuries old; there is evidence the Chinese were using it in 2500 B.C. In states like California, Idaho, Oregon, and Washington, warm water is used in ponds for growing warm-water species of fish like tilapia. The warm waters help the fish to grow more quickly. Other uses of geothermal energy in agriculture include heat for growing food, and heat energy is also used in processing and drying fruits and vegetables. Heat from hot water and steam is also used for drying lumber and other products.

Food Drying

Some food crops such as onions, garlic, carrots, and celery are often used in other products like soups. To preserve the foods, they are dried in commercial operations, usually with natural gas dryers to extract the water. Moderate-temperature geothermal heat can be used to dry the air, which is then used for drying the food.

Resorts

Chena Hot Springs Resort, located on a hot springs north of Fairbanks, Alaska, has become a model for using geothermal energy in a variety of ways. It originally used the hot springs for bathing in its legendary healing mineral waters, where people relax in hot pools of geothermal water year-round. In 2006, Chena Hot Springs and United Technologies Corporation designed and built a 400 kW binary geothermal plant using two 200 kW organic Rankine

Alligator Farm

(*Source:* National Renewable Energy Laboratory.)

An alligator farm called Colorado Gators uses a hot spring in the San Luis Valley in Colorado to raise alligators and other reptiles. Even though the air temperatures in this area drop below freezing, the alligators thrive in the warm water. The farm is a sanctuary for exotic reptiles that got too big for their owners. These reptiles include pythons, tortoises, and iguanas that have become part of educational programs and rescue efforts. In addition, the geothermal hot springs allow the owners to raise and sell warm-water species of fish like tilapia. Thus, nearby Denver can obtain fresh fish that is native to Africa!

FIGURE 10-10   Geothermal Power Station at Chena Hot Springs, Alaska   (*Source: Courtesy of Chena Hot Springs.*)

cycle (ORC) modules that operates from a moderate temperature (74° C or 165° F) geothermal resource. A third unit is now in place. Figure 10-10 shows the power plant at Chena Hot Springs.

The fluid used in the secondary cycle is R134a, widely used for heating, ventilation, and air-conditioning (HVAC) systems. Cooling for the Rankine cycle is provided by cold water from a well in summer or air cooling in the cold Alaskan winters. The resort is completely independent of the grid, having years of experience running geothermal power with an average availability of 95% (backup is provided by diesel generators).

Hot water from the spring is also used to provide heat for a large greenhouse at Chena Hot Springs, where food is grown for guests. The resort wants to become an independent self-sustaining community. The temperature in the greenhouse maintains a reasonable growing environment all winter long despite extremely cold outside temperatures. The greenhouse supplements the limited amount of light on long dark winter nights with LED crop lighting. In 2004, the outside temperature dropped to −49° C (−56° F), yet the greenhouse was able to maintain a temperature of 25.5° C (78° F). Figure 10-11 shows the greenhouse in winter, when it uses a light-emitting diode (LED) lighting system. This lighting system produces red and violet light, which allows for an increase in production while decreasing energy use.

Chena Hot Springs uses the geothermal resource for several other applications. All of the rooms are heated with geothermal heat. The resort also uses the geothermal technology

FIGURE 10-11   Greenhouse at Chena Hot Springs, Alaska. The grow lights are red and violet light-emitting diodes (LEDs).   (*Source:* Courtesy of Chena Hot Springs.)

to produce hydrogen from water, with the idea of using excess generating capacity in off-peak hours to provide a reserve of hydrogen as fuel. The hydrogen can be used in fuel cells to produce electricity, or to power vehicles or other applications such as laundry services that need a heating source. The advantage of creating hydrogen is that it can be stored and used when other renewable energy resources, such as solar or wind, are not productive. Another application is to power an absorption-type chiller that brings the temperature low enough to keep an ice museum operating during the summer.

SECTION 10-3 CHECKUP

1. Identify four uses for hot water or steam.
2. Explain how hot water is used for space heating in Iceland.
3. Identify five ways geothermal energy is used in Chena Hot Springs, Alaska.
4. Explain how hot water is used in aquaculture.

10-4 Geothermal Heat Pumps

A heat pump is a mechanical system that uses a compressor to change the state of refrigerant so that heat can be removed or added to a conditioned space. The heat pump can move (pump) more energy than it consumes because of the compression cycle of the refrigerant. The geothermal heat pump is also called the ground source heat pump; it has liquid glycol running through tubing that is beneath the earth's surface. The temperature at this level is near constant, which improves the efficiency of the heat pump. The heat pump moves heat energy from one source to another.

Conventional Heat Pumps

The natural flow of heat is from a warmer region to a cooler one. A **heat pump** is a device that can reverse this natural direction using a basic refrigeration cycle. The heat pump can operate as a cooling system in the summer or as a heating system in the winter. The refrigeration cycle works by expanding a refrigerant material, which absorbs the heat of vaporization, or compressing vapor into liquid and releasing the heat of vaporization. To move heat from one location to another, one coil that gives up (or absorbs) heat is located in the space to be heated (or cooled) and the other coil in a standard heat pump is located in the space from which heat is taken (or released). The heat pump cycle has four main stages:

1. The refrigerant vapor is compressed and flows into a condenser coil. When air passes over the condenser, the vapor cools.
2. The vapor changes to a liquid as it moves through the condenser and gives up heat of vaporization. This causes the condenser coil to become hot; the heat is then released to the surroundings.
3. The liquid passes through an expansion valve and flows through an evaporator coil, where it suddenly expands from a liquid to a vapor.
4. When the refrigerant flows through the evaporator and changes from a liquid to a vapor, it absorbs heat from the air that moves over the evaporator coil.

A standard heat pump acts like a normal air-conditioning system in the summer to provide cool air to the indoor conditioned space. In the winter, a reversing valve is energized to reverse the refrigerant flow so that the indoor coil becomes the condenser, which gives up heat, and outdoor coil becomes the evaporator. Because the outdoor coil is the evaporator in the heating cycle, it gets heat from the outdoor air. As the outdoor temperature gets colder during winter, the air-to-air heat pump becomes less efficient.

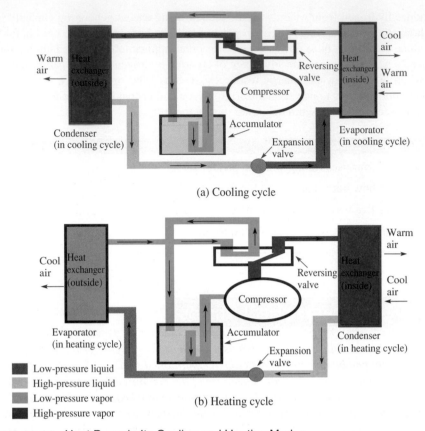

(a) Cooling cycle

Low-pressure liquid
High-pressure liquid
Low-pressure vapor
High-pressure vapor

(b) Heating cycle

FIGURE 10-12   Heat Pump in Its Cooling and Heating Modes

Figure 10-12 shows an example of a conventional heat pump cycle for an air-to-air system. Figure 10-12(a) shows the movement of refrigerant during the cooling cycle, in which the heat pump operates like an air-conditioning system. Air is moved over the indoor coil and heat is removed, which makes the conditioned space cooler. Figure 10-12(b) shows the heat pump when it is in the heating cycle. In the heating cycle, the reversing valve changes the direction of the refrigerant flow so that the high-pressure vapor moves through the indoor coil, where the fan moves air across it and moves heat from the coil to the indoor space like a regular furnace. The heat pump gets its name from the way it takes heat from outside air, even when it is cool during the winter, and pumps it into the indoor coil in the winter, where it is converted to temperatures of 27° C (80° F) by increasing refrigerant pressures. In the summer, it pumps heat from inside the conditioned space to the outdoor coil.

The standard heat pump can move heat from cold air in the winter because of the way the refrigeration cycle operates; it is possible to get the refrigerant to evaporate in the outside coil even at very low temperatures, but the heat pump will lose efficiency. When this occurs, the heat that causes the refrigerant to evaporate is absorbed by the refrigerant, and the compressor pumps the refrigerant in vapor form and compresses it as a hot gas at about 27° C (80° F) to 38° C (100° F). People inside the conditioned space may complain that this feels as if cool air is coming out of the air registers. Figure 10-13 shows a cutaway view of a heat pump; you can see the indoor coil, the outdoor coil, the reversing valve, and the compressor. The heat pump requires a two-circuit thermostat: One of the circuits is for the heating cycle; the other circuit is for the cooling cycle.

When the temperature is 10° C to 15° C (50° F to 59° F), the heat pump is very efficient. When the outdoor temperature drops below freezing, efficiency drops to a point where the amount of energy used to drive the compressor is equal to the heat energy that is put out. Thus, at temperatures below freezing, the heat pump has reached a balance point, and it is no longer efficient to run in the heat pump mode. At this point, the compressor is switched

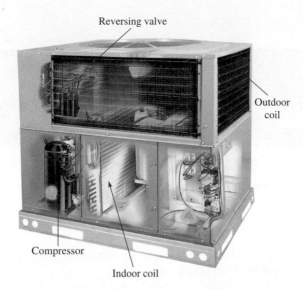

Reversing valve

Outdoor
coil

Compressor

Indoor coil

FIGURE 10-13 Cutaway View of the Basic Parts of a Heat Pump   (*Source:* Courtesy of Carrier Corporation.)

off and an electric resistance coil is used to provide the heat. When the outdoor temperature drops to approximately 7° C (45° F), the outdoor coil will be cold enough to create a layer of frost on its coils because it is evaporating refrigerant that flows through the lines. The frost must be removed periodically because it insulates the lines and causes the heat pump to be less efficient. During the defrost cycle, the heat pump reversing valve changes the direction of refrigerant flow so that the system operates as an air-conditioning system, and hot gas is pumped through the outdoor coils to melt the frost. During this time, the indoor coil operates as an evaporator, and a heating element must be used to warm the cool air that flows into the conditioned space. Operating an electric resistance coil during this time also reduces the heat pump efficiency.

Geothermal Heat Pumps

A **geothermal heat pump** uses a standard refrigeration cycle but receives heat from glycol that is circulated through a large amount of piping buried in the ground. Glycol picks up heat from the earth as it circulates through the piping. The piping remains at a fairly constant temperature that is equal to the ground temperature. For most locations, the temperature of the earth stays constant at approximately 10° C (50° F) all year long at only 1 or 2 m deep. Because of the constant temperature, the efficiency of the geothermal heat pump remains the same even with large variations in air temperature, including freezing temperatures. As you know, the conventional heat pump cannot operate efficiently in freezing temperatures because it uses the outside air as a heat source.

A geothermal heat pump has a large loop of underground tubing located below the frost line with a glycol solution running in a loop arrangement. Glycol is used as the solution because it has good heat transfer properties and has a much lower freezing point than water. Figure 10-14 shows a diagram for this type of system. The glycol solution (shown in green) is pumped through a long length of tubing buried in the ground and then is routed to a heat exchanger. Heat is picked up from the ground and transferred to the refrigerant as the compressor pumps it through the system. Because the tubing is underground, where the temperatures remain relatively constant year-round, the heat pump is more efficient throughout the heating season and does not need to be defrosted. In summer, the refrigeration system reverses the flow of refrigerant so that the 10° C (50° F) glycol becomes an efficient medium to help the heat pump condense its refrigerant when it is in the air-conditioning mode. Typically the conditioned space needs cooling when the outdoor temperature is above 27° C (80° F), and glycol is more efficient because it deposits heat in the cooler earth.

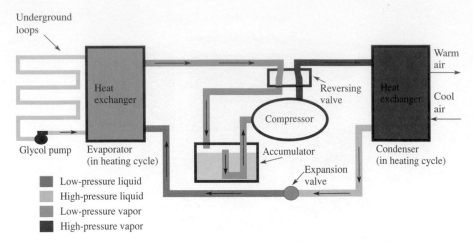

FIGURE 10-14 Geothermal Heat Pump in the Heating Cycle. The outside heat exchanger removes heat from the glycol.

Types of Geothermal Heat Pump Systems

Geothermal heat pump systems can have one of four basic arrangements of piping:

1. Closed-loop horizontal system
2. Closed-loop vertical system
3. Closed-loop system that uses a pond or lake
4. Open-loop system that uses two wells (in this type of system, one of the wells supplies water, and the second well returns it to a location below the earth's surface)

Closed-Loop Horizontal Piping Geothermal Heat Pump

The operation of the closed-loop horizontal system was explained in the discussion of the operation of a geothermal heat pump in the previous section. A closed-loop horizontal system requires a large surface area to accommodate the long sections of plastic tubing. Figure 10-15 shows a typical horizontal piping geothermal heat pump. The tubing is made of polyethylene (PE) pipe, which is used because it has ideal characteristics, including permanent flexibility and good thermal transfer properties so that it can transfer heat to the liquid flowing through it from the earth that surrounds it. The piping is buried in horizontal trenches below the frost line (which varies by location). The system requires approximately 2,500 square feet of surface area per ton of cooling to provide enough heat energy from the ground. A ton of cooling is 12,000 BTU/h, which is equivalent to 3.516 kW. The trenches must be placed far enough from each other to maximize the amount of heat transfer.

FIGURE 10-15 Typical Installation of Horizontal Piping for a Geothermal Heat Pump (*Source:* Courtesy of Daikin-McQuay International.)

FIGURE 10-16   Vertical Arrangement of Piping for a Geothermal Heat Pump (*Source:* Courtesy of Daikin-McQuay International.)

Closed-Loop Vertical System

In some locations, such as commercial buildings or schools, the large amount of land required for a horizontal system is not available, so the tubing is placed into the ground in a vertical arrangement. A closed-loop vertical heat pump system uses vertical tubes where space is limited or when ground is too difficult for trenching. The vertical system uses holes that are approximately 10 cm in diameter and are drilled about 6 m apart and 30 m to 120 m deep. Two pipes are placed into each hole and are connected at the bottom so that glycol can go down one pipe and return through the other. The top of one of the pipes is connected to a feeder manifold, and the second pipe is connected to a return manifold. The manifolds are connected to the first heat exchanger. Glycol (the main constituent of antifreeze for radiators) is circulated through the pipes. Figure 10-16 shows an example of a vertical geothermal heat pump.

Surface-Water or Lake-Loop System

Another type of closed-loop geothermal system is called a surface-water or lake-loop system. This system uses a lake or pond, into which loops of plastic tubing are laid. In some cases, the body of water is constructed on the building site to meet code requirements for runoff, or it might be placed on the property for aesthetic reasons. The size and the depth of the lake must be large enough to provide the heat energy needed for the building. Because the heat pump system gives up heat to the water in the summer, the size of the lake or pond must be large enough that the additional heat added to the water does not harm its ecosystem. Figure 10-17 shows a typical surface-water geothermal heat pump.

FIGURE 10-17   Surface-Water or Lake-Loop Arrangement of Piping for a Geothermal Heat Pump   (*Source:* Courtesy of Daikin-McQuay International.)

FIGURE 10-18   Open-Loop Well-Water Arrangement of Piping for a Geothermal Heat Pump   (*Source:* Courtesy of Daikin-McQuay International.)

Open-Loop Well-Water Geothermal Heat Pump System

The open-loop well-water geothermal heat pump system uses well water as the heat exchange fluid. The water is circulated up from a well, as shown in Figure 10-18. It is pumped through the heat exchanger for the geothermal heat pump, and then the water is returned through a recharge well. If water comes from an underground pool, two wells are drilled. One well brings water up to the surface, and the second well is a recharge well that returns the water after it has given up its heat. The amount of water available below the earth's surface must be sufficient to provide a constant temperature of at least 10° C (50° F). If the volume of water is insufficient, the temperature of the water will drop as it gives up its heat, causing the system to be inefficient. Any time a well is used for this purpose, it must be approved by appropriate governing bodies. Because it is wasteful of water, open-loop systems are not common and many local codes do not allow open-loop systems.

Measuring the Efficiency of a Heat Pump

The efficiency of a heat pump is determined by measuring the electrical energy consumed in the system by the compressor, fan motors, and pumps, and comparing this measurement to the energy that the system releases as heat into the living space during the same time. The efficiency is called the **coefficient of performance (COP)**. The energy that the heat pump converts to heat varies seasonally and will be larger when the outdoor temperature where the outdoor coil is located is warmer. When the temperature drops below freezing the COP may drop below 1, making resistance heating more efficient. In the air-to-air heat pump, the formula for COP is

$$COP = h_h/h_w$$
Equation 10-1

where

$\qquad$ COP = coefficient of performance (efficiency)

$\qquad$ h_h = heat delivered by the heat pump, in BTU

$\qquad$ h_w = equivalent electric energy supplied to the heat pump, in BTU

Because heat pump systems are rated in BTU and because electricity is measured in kilowatt-hours (kWh), you need to convert the units so they cancel (1 kWh = 3,413 BTU).

COP is a ratio, so it does not have units. If a heat pump delivers 2 units of heat for every unit of energy input, the COP is 2. Note that the COP is not a fixed number for a given heat pump; it varies according to the outside temperature. The COP of the heat pump for the heating mode becomes smaller as the outdoor temperature drops. When the COP for the heat pump drops below 1, it is more efficient to turn the compressor off and use a resistance electrical heater to provide heat directly to the heated space.

EXAMPLE 10-1

What is the COP of a heat pump that delivers 60,000 BTU and uses 7 kWh of electrical power?

Solution

$$\text{COP} = \frac{h_h}{h_w} = \frac{(60{,}000 \text{ BTU})\left(\dfrac{\text{kWh}}{3{,}413 \text{ BTU}}\right)}{7 \text{ kWh}} = \mathbf{2.51}$$

With geothermal heat pumps, the energy used must include any water pumps (or glycol pumps) that are used, as well as the compressor and any fans that move indoor and outdoor air. Typically the water from a ground source or from a lake or pond that is used with the geothermal heat pump stays at a constant temperature, and the COP remains fairly constant. The water from these sources can be as high as 50° F even when the outdoor temperature is below freezing, so the COP for the system may be as much as 5, indicating that the geothermal heat pump is very efficient.

EXAMPLE 10-2

What is the COP for a geothermal heat pump that has a water pump with a power rating of 2.0 kW, an indoor fan and outdoor fan with a power rating of 1.5 kW, and a compressor with a power rating of 4.0 kW? The heat pump produces 120,000 BTU/h.

Solution

The first step in this problem is to determine the electrical *energy* used from the power ratings given. In one hour, the energy delivered by the heat pump is 120,000 BTU. The total power rating of the heat pump, including pumps and fans, is 7.5 kW. The energy supplied to the heat pump in 1 h is 7.5 kWh.

$$\text{COP} = \frac{h_h}{h_w} = \frac{(120{,}000 \text{ BTU})\left(\dfrac{\text{kWh}}{3{,}413 \text{ BTU}}\right)}{7.5 \text{ kWh}} = \mathbf{4.69}$$

Two other performance criteria for heat pumps are in common use:

1. The **heating system performance factor (HSPF)** is a measure of a heat pump's heating efficiency over an entire heating season. It represents the total heating (in Btu) compared to the total electricity consumed (in watt-hours) during the same period.

2. The **seasonal energy efficiency ratio (SEER)** is a measure of an air conditioner's or heat pump's cooling efficiency over an entire cooling season. It represents the total cooling (in Btu) compared to the total electricity consumed (in watt-hours) during the same period.

Notice that, unlike COP, HSPF and SEER have units of Btu per Wh. To express these measures as pure ratios, the HSPF and SEER must be divided by 3.413 Btu/Wh.

SECTION 10-4 CHECKUP

1. How does a heat pump provide heat in the winter?

2. Identify the main parts of a geothermal heat pump.

3. What is the difference between the closed-loop horizontal heat pump and the closed-loop vertical heat pump?

4. Explain why the coefficient of performance of a geothermal heat pump is better than a standard heat pump in cold climates.

10-5   Environmental Impacts

The environmental impacts of geothermal energy were covered in Section 1-5, but the topic is expanded here to include both positive and negative impacts of air, water, and land use. Air and water pollution are very low, but in general still not zero. The water returned through reinjection wells has little or no pollution, but it can have an impact on water supplies. Some geothermal plants produce small amounts of solid material or sludge, which requires disposal.

Emissions from Geothermal Hot Springs

(*Source:* National Renewable Energy Laboratory.)

When steam rises from hot springs, small amounts of pollutants like hydrogen sulfide, sulfur dioxide, nitrous oxide and carbon dioxide are emitted. The amount of each pollutant is extremely low, so these pollutants are not considered as having a serious impact on the environment.

Emissions

When compared to conventional (fossil fuel) energy sources, the impact of geothermal energy on the environment is very small. Binary geothermal plants, which are the most widely used type of geothermal plants, produce almost no liquid or airborne emissions. In flash-steam plants, steam is the primary emission. Geothermal plants that use groundwater aquifers must recycle water to keep the reservoir replenished. In these plants, water pollution must be monitored continually. Some plants, like the Geysers in Northern California, use treated wastewater to replenish the groundwater aquifer. If injected water contains dissolved minerals or salts, they are usually put back into the reservoir, but the water is typically at levels that are well below the level where groundwater exists, so there is little or no possibility for groundwater pollution. If a geothermal plant produces solid materials, an approved disposal site is required; in some cases, the amounts of valuable solids, such as sulfur, zinc, and silica, can be processed and sold. In some geothermal plants, hydrogen sulfide may occur naturally in the steam and hot water, so a scrubber is used to remove these pollutants. Geothermal plants produce significantly less pollutants than those that are emitted from fossil fuel plants. Geothermal plants also emit less sulfur dioxide than fossil fuel plants, which means that geothermal plants do not produce as much acid rain from their emissions.

Air Pollution

One of the air pollution emissions from open-loop electrical generation plants like the Geysers in California is the release of small amounts of hydrogen sulfide gas (H_2S). Even at this very low release rate, the rotten-egg odor that is present near the plant is a rather small nuisance rather than a serious issue. In the event that larger amounts of hydrogen sulfide gas are present, technology is available to absorb or strip the pollutants before they are released into the atmosphere.

These plants also release small amounts of carbon dioxide, which varies widely from one type of plant to another and from one area to another. According to the US Environmental Protection Agency (EPA), the weighted average of CO_2 for all geothermal plants is 180 lbs/MWh. For coal-fired plants, it is about 2,249 lbs/MWh; for natural gas plants, it is 1,135 lbs/MWhr. On average, the release of CO_2 from geothermal plants is 8% of the average for coal-fired plants. The source of the CO_2 is dissolved gas that is underground, which is already in the CO_2 cycle and may make its way to the atmosphere over time via natural pathways. A small amount may come from marine sedimentary rocks that are high in carbonates. The positive side is that geothermal production does not create new CO_2, as does the burning of fossil fuels.

Other types of gas released into the air include sulfur dioxide (SO_2) and nitrous oxide (NO_X). These emissions are continually monitored, and they are lower for geothermal energy than for conventional fossil fuel. Some small amounts of sludge are created when the steam is condensed, and the sludge contains small amounts of hazardous material that needs to be disposed of. The only other air pollutants that may be released from a geothermal energy plant include small amounts of methane, ammonia, nitrogen, hydrogen, and some very small amounts of radon. The amount of these pollutants is extremely small and, for all intents and purposes, is not a problem.

Geothermal electrical energy production releases no nitrous oxide and very little CO_2, which is primarily from geothermal brine. Binary-cycle geothermal plants have zero air pollution because they are a closed-loop system. The US Department of Energy (DOE) is continuing research for reducing emissions from non-binary plants even more. One promising avenue of research is closed-loop EGS plants with emissions close to zero. Research into using supercritical CO_2 as an EGS fluid could result in zero CO_2 emissions and even allow carbon-negative geothermal plants.

Water Pollution

Water pollution at a geothermal energy site may occur above- or belowground. Both types of pollution are important, but pollution that occurs in the groundwater below the earth's surface can contaminate usable drinking water. One of the most common water pollutants from geothermal plants is saltwater (sodium chloride brines), which occurs naturally during the plant operation. The brine can be reinjected into the well, which may cause contamination of groundwater if leakage occurs beyond the well casing. In some cases the brine is stored aboveground in holding ponds that may run off into other surface-water sources, such as nearby rivers or streams, or it may leach into groundwater. When water is reinjected into wells, deep-well injection techniques must be used, and the well must be monitored to ensure the casing remains solid. Because of this type of contamination, drinking water sources near geothermal energy plants need to be tested continually for any sources of pollution. The wells used for reinjection have very thick and deep casing to protect against brine or other types of treated water contaminating the groundwater that is used for drinking and potable water. The water used in these plants is never released as surface water. Reinjection reduces surface-water pollution and increases the resilience of the geothermal reservoir. In some locations, certain poisons such as arsenic or boron have been found in geothermal waters, but these poisons are almost always naturally occurring and are not a product of using geothermal energy.

Overall, geothermal plants in operation around the United States and other countries have had a very positive record of controlling water pollution. On average, a geothermal plant that produces electricity uses less than 10 gallons to produce one megawatt-hour of electrical energy, whereas a coal-fired, natural gas-fired or nuclear-powered electrical generation plant may use in excess of 300 gallons of water to produce one megawatt-hour of electrical energy. In some locations, the consumption of the large amount of water is more of a concern than the small amount of water pollution caused by the plants. Figure 10-19 shows the Blue Lagoon in Iceland, which has become a sterling example of clean energy with clean water.

At the Geysers facility, 11 million gallons of treated wastewater from the city of Santa Rosa, California, are pumped daily for injection into the geothermal reservoir. Some plants use water-cooled condensers to turn the steam back to liquid, a process that uses a tremendous amount of water. When water is used to condense the steam, it may be

FIGURE 10-19 Geothermal Power Plant at the Blue Lagoon of Svartsengi. This geothermal power station is located in Keflavik, Iceland and is same plant as that shown on the front cover. About 30% of Iceland's electricity comes from geothermal sources—most of the geothermal resource is used for space heating.   (*Source:* David Buchla.)

diverted from agricultural use or human consumption. A newer technology uses an air-cooled condenser that does not use any water, and it has the potential to condense clean water from the steam, which can be used for human consumption. Unfortunately, air cooling is not as efficient as water cooling, but it is important in arid locations. In desert areas, where water is a precious resource, water requirements are an important issue for any type of power plant.

Land Use

Currently the land use for geothermal energy is predominately owned by the federal or state governments. Most of these sites were part of federal or state parks and provided sightseeing and recreational activities. In recent years, electrical power production and heat energy production from geothermal energy has expanded, and more power plants have been brought online. These plants have been designed to blend in with their surroundings and use a minimal amount of land resources. New technology is allowing exploration and development of geothermal energy resources where geothermal surface features, such as hot springs or eruptions, were not present. These areas were once thought to be of no use in the production of geothermal energy, but with new drilling and production systems, some of these areas are now under development.

Induced Seismicity

Induced seismicity is the creation of microquakes from human activity, as mentioned in Section 1-5. In the case of enhanced geothermal systems (EGSs), reservoirs are created by injecting water underground. This injected water may cause rocks to slip or move and produce a minor seismic event. Most of these tiny earthquakes cannot be felt at the surface, but they have led to some public concerns. The result is that the operators of geothermal plants must follow certain guidelines to avoid any potential larger event from becoming a hazard. In June 2012, the National Academy of Science released a report that found that induced seismicity posed little risk of causing significant movement of the earth.

Advantages and Disadvantages of Geothermal Energy

A concise way of describing the impact of geothermal energy on the environment is to provide a list of its advantages and disadvantages.

Advantages
- Geothermal energy is an excellent source for heating and creating electrical power with minimal pollution.
- Geothermal energy is sustainable, which means that it is a form of energy that can continue far into the future.
- Geothermal energy can be produced at lower cost after accounting for the initial start-up costs of drilling and infrastructure.
- Geothermal plants have much lower CO_2 emissions than comparable fossil-fueled plants. They emit no nitrous oxides.
- Useful minerals, such as zinc and silica, can be extracted as by-products.
- Geothermal plants can be online 90% to 100% of the time. Coal plants are typically online 75% of the time, and nuclear plants are online 65% of the time. Online production for wind energy varies from 38% to 80%, depending on location, and solar energy is generally available less than 50% of the time (assuming no energy storage).
- Flash- and dry-steam power plants emit far less carbon dioxide than fossil fuel plants do. They produce no nitrogen oxides and very little CO_2.
- Binary and hot dry rock plants have no gaseous emission at all.

Disadvantages

- Finding a suitable resource can take a long time because the site has to be tested for the available energy, before locating the power plant.

- Power plants that have extracted steam from a site for many years could suddenly stop producing steam if they are not operated and managed properly.

- When developing a piece of land for use as a geothermal power plant, developers have to be careful and know that harmful gases can escape from deep in the earth through holes drilled by builders. They must contain any leaked gases and dispose of them safely.

- Brine brought up from deep below the earth's surface can become an issue if the water is not injected back into the reserve after the heat is extracted.

- Extracting large amounts of water can cause land subsidence, and this can lead to an increase in seismic activity. To prevent this, the cooled water must be injected back into the reserve to keep the water pressure constant underground.

- Most geothermal sites need a large source of water to recharge the reservoir with water that previously escaped as steam and for cooling.

- There is a small potential of contamination of groundwater because water is continually pumped back deep into the earth.

- A very small potential exists for a hazard from induced seismicity.

SECTION 10-5 CHECKUP

1. Does geothermal electrical production create any air pollution? If so, what are the associated issues?

2. Explain how brine contamination may occur with groundwater at a geothermal plant.

3. What is induced seismicity? What are its causes?

CHAPTER SUMMARY

- Geothermal energy is available as steam; it can be used to turn steam turbines, which produce electricity.

- Molten magma is found in volcanoes and is the hottest form of geothermal energy, at approximately 2,000° C.

- Geopressurized reservoirs consist of high-pressure hot brine (salt)water in deep reservoirs. This resource is completely saturated with natural gas and is under extreme pressure. It exists at a depth of 3,000 to 6,000 m (10,000 to 20,000 ft).

- Geothermal power plants come in three basic designs, all of which pull hot water and steam from the ground: the dry-steam plant, flash-steam plant, and binary-cycle power plant.

- A dry-steam plant has steam that comes to the surface directly to a steam turbine. The steam turbine drives an electric generator.

- Flash steam is the most common geothermal reservoir; it is a combination of both steam and water.

- The combination of liquid and vapor is called a liquid-dominated reservoir. It is produced when large volumes of water from artesian wells come into contact with extremely hot rocks deep under the earth's surface.

- A double-flash steam plant is also called a dual flash steam plant. It can produce about 20% more electrical power than can a flash steam power plant because it throttles the low-pressure liquid to a second turbine called the low-pressure turbine.

- The geothermal binary-cycle power plant is different from dry-steam and flash-steam systems because the brine water or steam from the geothermal reservoir goes directly to a heat exchanger and never comes in contact with the turbine that drives the generator.

- In a steam heating system, the steam from the ground is routed through piping and heat exchangers and returned to the reservoir. The secondary loop in the heat exchanger has clean water running through it. The clean water is pumped through radiators or fan coil systems mounted in ductwork throughout the building or home.

- A heat pump is a device that utilizes a refrigeration cycle to move heat from a cooler region to a warmer one. The heat pump can be reversed for moving heat either into or from a conditioned space. Geothermal heat pumps move heat from the ground into the heated space.

- The balance point for a heat pump occurs when the outdoor temperature drops below a point where the amount of energy used to drive the compressor is equal to the heat energy that the system puts out. At this point, the heat pump can switch to resistance heating.

- Geothermal heat pump systems come in four basic types: closed-loop horizontal system, closed-loop vertical system, closed-loop system that uses a pond or lake, and open-loop system that uses two wells. In the open-loop system, one of the wells supplies water, and the second well returns it to a location below the surface.

KEY TERMS

binary-cycle plant   A geothermal power plant that uses the brine water or steam from the geothermal reservoir to heat and vaporize a secondary fluid with a lower boiling point to drive the turbine and electrical generator. Also known as *organic Rankine cycle* (ORC) *plant*.

coefficient of performance (COP)   A measure of efficiency for a heat pump; it is the ratio of the heat produced to the energy consumed, and it varies with the outside temperature.

double-flash steam plant   A geothermal plant with two pressure-reducing stages to create high-pressure and low-pressure steam. The high- and low-pressure steam is routed to two different turbines, which turn a generator.

dry-steam plant   A geothermal electrical plant that uses super-heated dry steam from a geothermal reservoir and routes it directly to a steam turbine and generator to produce electricity.

enhanced geothermal system (EGS)   A system in which a geothermal site that is deficient in water or permeability is made productive by artificial means.

enthalpy   The amount of energy in a system capable of doing mechanical work; it is a function of temperature, pressure, and volume.

flash-steam plant   A geothermal plant that creates steam from high-pressure hot water (brine) using a special control valve or orifice plate to reduce the pressure and cause some of the liquid to boil (flash) into steam. The steam is used to drive a steam turbine and generator to produce electricity.

geothermal heat pump   A heat pump that uses a standard refrigeration cycle; however, its heat source is glycol, which is circulated through a large amount of piping buried in the ground.

FORMULAS

Equation 10-1 $COP = h_h/h_w$

CHAPTER TRUE/FALSE QUIZ

Determine whether each statement is true or false. Answers are at the end of the chapter.

1. Geothermal energy can be used only as steam.

2. Geothermal energy and natural gas can be co-produced from the same source.

3. In a gas-to-liquid (GTL) system, the liquid is water.

4. A geyser is a natural hot spring that intermittently ejects a column of water.

5. The hottest geothermal resource used for steam is molten magma.

6. A dry-steam plant can operate with lower-temperature steam than a binary-cycle plant.

7. The best geothermal resources are concentrated in areas where tectonic plates interact.

8. A double-flash steam system uses a secondary fluid to generate steam.

9. The largest geothermal electrical generation facility in the United States is the Geysers in California.

10. In a dry-steam electrical generation plant, hot steam is brought up the injection well.

11. A heat pump can provide heat only for a building or residence.

12. The purpose of a cooling tower is to provide air conditioning for the workers in a geothermal plant.

13. With hydrospallation drilling, a stream of superheated high-pressure water is directed at the rock surface.

14. A heat pump uses a check valve to reverse the direction of refrigerant flow.

15. Geothermal hot water is used for agriculture and aquaculture applications.

16. Flash steam is a mixture of pressurized hot water and steam that converts to steam when pressure is released.

CHAPTER MULTIPLE-CHOICE QUIZ

Complete each statement by selecting the one correct answer. Answers are at the end of the chapter.

1. The geothermal gradient refers to the
   a. increasing density of the earth as a function of depth
   b. temperature increases in the earth as a function of depth
   c. variations in temperatures of hot-water pools
   d. differences between summer and winter temperatures

2. Geothermal resources are commonly classified by the
   a. depth and volume of magma present
   b. amount of hydrocarbons present
   c. porosity of rock at the source
   d. temperature and amount of fluid present

3. The purpose of an injection well is to
   a. replenish water in a geothermal source
   b. remove excess brine from the well
   c. lower instruments for testing the resource
   d. pressurize the well to bring up more heat

4. A disadvantage to geothermal heat pumps is that they
   a. cannot be used to cool
   b. are less efficient than resistive heat
   c. require a very hot geothermal source
   d. require underground plumbing and glycol

5. The hottest geothermal resource is
   a. hot dry rocks
   b. molten magma
   c. a hot-water reservoir
   d. a geopressurized reservoir

6. The most useful type of geothermal resource for generating electricity has
   a. geysers
   b. molten magma
   c. low permeability
   d. dry steam

7. Aquaculture includes
   a. growing trees
   b. drying vegetables
   c. producing fish
   d. all of these

8. The type of geothermal heat pump that requires the most surface area to be installed is the
   a. open-loop that uses two wells
   b. closed-loop vertical geothermal heat pump
   c. closed-loop horizontal geothermal heat pump

9. The type of geothermal plant that does not have steam piped directly to the turbine is the
   a. binary-cycle plant
   b. dry-steam plant
   c. flash-steam plant
   d. double-flash steam plant

10. A geothermal power plant that has a high-pressure and a low-pressure turbine is the
    a. binary-cycle power plant
    b. dry-steam plant
    c. flash-steam plant
    d. double-flash steam plant

11. The type of geothermal power plant that converts hot pressurized brine to steam in one step is the
    a. binary-cycle power plant
    b. dry-steam plant
    c. flash-steam plant
    d. double-flash steam plant

12. A geothermal heat pump
    a. operates only as a heater
    b. operates only as an air conditioner
    c. can act as a heater or an air conditioner

13. A geothermal heat pump
    a. has a fluid loop that extracts heat from the earth in winter
    b. uses an underground hot spring for hot water
    c. can provide heat only in the winter
    d. converts heat from the ground to electricity

CHAPTER QUESTIONS AND PROBLEMS

1. Explain how a geothermal heat pump can pump more heat energy into a building than it consumes.

2. Explain what the balance point is for an air-to-air heat pump.

3. Identify all of the ways that Chena Hot Springs Resort in Alaska uses geothermal energy.

4. Identify four ways that geothermal energy is used in agriculture or aquaculture.

5. What is a geothermal gradient?

6. Compare a geopressurized reservoir to a hot-water reservoir.

7. What is a hot springs?

8. What is a geyser?

9. Compare the temperature of a low-temperature resource to a high-temperature resource.

10. What is a hydrothermal resource?

11. List the main parts and explain the operation of a dry-steam power plant.

12. List the main parts and explain the operation of a flash-steam power plant.

13. List the main parts and explain the operation of the binary-cycle power plant.

14. Explain the differences among the open-loop geothermal heat pump that uses two wells, the closed-loop vertical geothermal heat pump, and the closed-loop horizontal geothermal heat pump.

15. Calculate the coefficient of performance for a heat pump that delivers 65,000 BTU and uses 6 kWh of electrical energy to do so.

FOR DISCUSSION

Assume that you are on a panel that is considering a nearby location for a geothermal energy system. What are important considerations for the decision?

ANSWERS TO CHECKUPS

Section 10-1 Checkup

1. Enthalpy is the amount of energy in a system capable of doing mechanical work; it is a function of temperature, pressure, and volume.

2. The geothermal gradient refers to the increasing temperature at increasing depths within the earth.

3. (1) molten magma, (2) hot dry rock, (3) geopressurized reservoirs, (4) hot-water reservoirs

4. Hot springs are naturally occurring, geothermally heated water sources; geysers are springs that intermittently eject jets of hot water and steam into the air.

5. Heating homes and businesses

Section 10-2 Checkup

1. Dry steam is superheated steam that has no water vapor in suspension. A dry-steam plant takes advantage of this resource by routing the steam directly to a turbine, which turns a generator.

2. A double-flash system has two stages of pressure reduction where brine is flashed into steam. A single-flash system has one stage of pressure reduction where brine is flashed into steam.

3. Binary-cycle geothermal plants can operate with lower-temperature resources and do not circulate the underground water through the turbine, thus avoiding problems with debris and corrosive substances.

4. A geothermal system that is deficient in water or permeability is made productive by artificial means.

Section 10-3 Checkup

1. (1) heating, (2) cooling, (3) snow melting, and (4) greenhouses

2. Hot water is circulated through insulated pipes from a central storage facility, where it is used to circulate through the heating system of buildings.

3. Chena Hot Springs uses geothermal energy to generate electricity, heat buildings, create hydrogen fuel, operate a greenhouse, and provide cooling.

4. Hot water can allow warm-water species to thrive in a cold environment.

Section 10-4 Checkup

1. In the heating cycle, the reversing valve changes the direction of the refrigerant flow so that the high-pressure vapor moves through the indoor coil, which acts like the condenser. The indoor fan moves air across the indoor coil and moves heat from the coil to the indoor space like a regular furnace.

2. Compressor, accumulator, reversing valve, expansion valve, underground tubing with glycol solution running through it, glycol pump, heat exchanger to exchange heat from glycol to refrigerant, heat exchanger to exchange heat between the refrigerant and air

3. The tubing for a closed-loop vertical-system heat pump is placed in the ground vertically and runs much deeper than the tubing for the closed-loop horizontal system heat pump. The vertical system is used where a larger area of land is not available for the horizontal placement of the tubing.

4. The COP of a geothermal heat pump tends to be more consistent because the ground temperature changes only slightly throughout the heating season. The standard heat pump gets heat from the air, which has wide variations in temperature during the heating season.

Section 10-5 Checkup

1. Air pollution is minor, but hydrogen sulfide (which produces a rotten-egg smell) may be dissolved in water and be released. CO_2 can be released also; currently the weighted average of all CO_2 releases is about 8% that of coal-fired plants per MWh produced.

2. When the brine from underground is returned to the ground, it needs to be reinjected at the level where it originated. Any problem with the well casing can contaminate drinking water at higher levels.

3. Induced seismicity is minor earthquakes caused by human activity; it may be caused by the effects of injected water in geological formations.

ANSWERS TO TRUE/FALSE QUIZ

1. F   2. T   3. F   4. T   5. F   6. F   7. T   8. F   9. T   10. F   11. T   12. F   13. T   14. F   15. T   16. T

ANSWERS TO MULTIPLE-CHOICE QUESTIONS

1. b   2. d   3. a   4. d   5. b   6. d   7. c   8. c   9. a   10. d   11. c   12. c   13. a

Energy from Water

CHAPTER OUTLINE

CHAPTER OBJECTIVES

- Identify the basic parts of a hydroelectric dam and describe the process of producing power.
- Explain how a pumped storage system stores and produces electrical power.
- Compare impulse and reaction turbines, and give examples of each type.
- Discuss the operation of Fourneyron, Francis, Kaplan, Pelton, Turgo, and crossflow turbines.
- Explain how a tidal stream generator produces electrical power and describe the effect of tides on power output.
- Identify the basic parts of a tidal barrage system and explain how it produces electrical power.
- Identify the basic parts of an oscillating water column system and explain how it produces electrical power.
- Explain each of the three basic wave machine types—the point absorber, the attenuator, and the termination—and give examples of each.
- Discuss some environmental concerns for implementing wave energy systems.

KEY TERMS

Key terms are shown in bold and color. Definitions for key terms are provided at the end of the chapter and in the end-of-book glossary. Bold terms in black are defined in the end-of-book glossary only.

- pumped storage system
- run-of-the-river (ROR)
- impulse turbine
- reaction turbine
- Francis turbine
- Kaplan turbine
- Pelton turbine
- tidal barrage system
- tidal stream generator (TSG)
- point absorber
- attenuator
- oscillating water column

INTRODUCTION

Hydroelectric energy is a useful renewable energy source; most hydropower comes from converting falling water to electricity. Some power comes from moving streams, rivers, and ocean tides. Some future possibilities for hydropower exploit temperature differences in the ocean. Nearly all energy from water is converted to electricity, but a tiny fraction is used in other applications.

Hydroelectric energy provides an important portion of the electrical energy for the residential, commercial, and industrial sectors. Worldwide, it accounts for over 6% of all energy and 17% of electricity production. The United States has over 80,000 dams of all sizes, and 2,000 of these produce electricity. About 48,000 dams over 15 m high exist in the rest of the world and produce electricity. Many early dams were not hydroelectric dams but were used to create head pressure for waterwheels that drove the machines for grinding grain or cutting timber. Later, many dams were constructed on rivers to control floods. Hydroelectricity was a side benefit of these flood-control projects. In this chapter, hydroelectric power, including tidal and wave power, are described. Basic power calculations are given.

The chapter closes with a discussion of wave energy systems. Currently, they are not cost-competitive with other energy sources, including offshore wind energy, but there is interest in developing this technology. Wave energy systems are broadly classified into three types of devices: point-source devices, attenuators, and terminators. Each is described in this chapter.

11-1   Energy in Moving Water

A hydropower resource can provide electric power by converting the energy of moving water into mechanical energy for spinning a turbine that produces electricity. Water is stored behind dams in large reservoirs. The energy in a reservoir is considered potential energy; the available energy is a function of the hydraulic head (average height above the discharge point) and the quantity of water. This potential energy is converted to kinetic energy as it falls.

The energy from moving water can be captured and converted to useful work directly or it can be converted into electricity. Moving water in a river or stream can turn a turbine that turns a generator. The water-powered paddlewheel was used for grinding grain for at least 2,000 years, but it was largely replaced by electrical power in the nineteenth century. Water can be stored behind a dam and released as needed to turn a turbine and generator to produce electrical power. Wave and tidal movement can also be captured to turn a generator.

Potential and Kinetic Energy in Water

Recall that potential and kinetic energy were defined in Section 2-1. These terms are common in hydrosystems, so they are reviewed here.

Recall that potential energy was defined as stored energy; it has the ability to do work because of position or configuration. Gravitational potential energy was given by the following equation:

$$W_{PE} = mgh$$ Equation 11-1

where

W_{PE} = potential energy, in J

m = mass, in kg

g = gravitational constant, in m/s^2

h = height in m

Kinetic energy was defined in the energy of motion. The amount of kinetic energy depends on the volume (mass) and the velocity and was given by the following equation:

$$W_{KE} = \frac{1}{2}mv^2$$ Equation 11-2

where

W_{KE} = kinetic energy, in J

m = mass, in kg

v = velocity, in m/s

Stored water is a form of potential energy. The potential energy in stored water is converted to kinetic energy by releasing it from a height. At the bottom, the fast moving water is used to spin a turbine, which is connected to a generator. The height the water drops is referred to as the head. **Head** is a height of water created by a vertical difference in elevation. It is usually expressed in a height unit (m or ft) from a reference. Each foot of vertical drop corresponds to 0.433 psi; in SI units, the height is measured in m and the pressure in Pascal (Pa). A 1 m drop corresponds to 9,794 Pa using SI units. In hydropower applications, the head is measured from the top of the water level to the inlet at the turbine. The distance is considered the gross head; the net head is the equivalent height after equivalent friction losses in piping are subtracted. To calculate the gross energy stored in a pond or reservoir, the volume and the average height are used. The reason for using the average height is to account for the difference in head between the top and the bottom. (Of course, not all of this energy can be recovered when converted to electricity.)

Waterwheel on a Grist Mill

(*Source:* Fotolia.)

This waterwheel, part of a mill in northwestern New Jersey, was built around 1810. It was used as a grist mill, and then for grinding plaster and talc and processing graphite. A small dam creates a reservoir, and the waterwheel is driven by a stream of water that is directed toward it.

FIGURE 11-1   Water Released After Passing Through Turbines in a Dam
(*Source:* Markus Haack/Fotolia.)

Runoff from rain begins as small streams that converge to form rivers and eventually flows to lakes, reservoirs, or directly to the oceans. Moving water is a form of kinetic energy due to its motion. For flowing water, the term *flow* is commonly used to describe *flow rate* (volume per time). Frequently, flowing water is captured in a reservoir, or it can be pumped up into the reservoir for storage and released later as needed. As mentioned, the stored energy in a reservoir or pond depends on the volume of water and the average depth from a reference point (typically a turbine). In North America, it is common to measure the volume in acre-feet because smaller units are impractical. An **acre-foot** is a volume of water that would cover an acre to a depth of 1 foot (1 acre-foot = 43,560 ft^3 = 1,233 m^3). For a typical family, this amount represents about a one-year supply of water.

Water has a density of 1,000 kg/m^3. Thus, the mass of 1 acre-foot of water is 1.233 × 10^6 kg. Although the total stored energy is relatively easy to calculate from the head and quantity of water, this does not take into account the efficiency. If water is stored behind a dam, it can be released so that it flows through a turbine that turns a generator to produce electrical power. A certain amount of energy in the water is lost in friction or escapes as kinetic energy. The water loses most of its kinetic energy as it passes through the turbine.

Figure 11-1 shows the water as it leaves the dam. This water may still have considerable kinetic energy. Hydroelectric is the most efficient method of large-scale power generation; it has an efficiency of 80% to 95% for large installations with high flow rates but less in installations with a low flow rate.

EXAMPLE 11-1

A small reservoir has 8,000 acre-feet of water at an average height of 24 feet.

(a) Calculate the total potential energy in the stored water.

(b) Assume the reservoir was drained in seven days. What is the average power in the water that was released? (Ignore any losses due to friction.)

Solution

(a) **Step 1.**   Convert the units. Find the mass of the water in kg and the average height in m (1 ft = 0.3048 m).

$$m = (8{,}000 \text{ acre-feet})\left(1{,}233\,\frac{\text{m}^3}{\text{acre-foot}}\right)\left(1{,}000\,\frac{\text{kg}}{\text{m}^3}\right) = \mathbf{9.86 \times 10^9\ kg}$$

$$h = (24 \text{ ft})\left(0.3048\,\frac{\text{m}}{\text{ft}}\right) = 7.32 \text{ m}$$

Step 2.   Find the energy in joules by substituting into Equation 11-1.

$$W_{PE} = mgh = (9.86 \times 10^9 \text{ kg})\left(9.8\,\frac{\text{m}}{\text{s}^2}\right)(7.32 \text{ m}) = \mathbf{707 \times 10^9\ J}$$

(b) Calculate the power. Recall from Chapter 2 that the equation for power is:

$$P = \frac{W}{t}$$

Substituting and converting units

$$P = \frac{W}{t} = \frac{707 \times 10^9 \text{ J}}{(7 \text{ da})\left(24\,\frac{\text{hr}}{\text{da}}\right)\left(3{,}600\,\frac{\text{s}}{\text{hr}}\right)} = 1.17 \times 10^6 \text{ W} = \mathbf{1.17\ MW}$$

Power in Moving Water

While most hydroelectric power is generated by large power projects located at dams, small streams and rivers can produce hydroelectric power directly. Typically, water is diverted into a pipeline that directs water to a turbine located in a powerhouse. In this case, no dam, with its significant impact on the environment, is required. Ideally the stream is fast moving; the available power is directly proportional to the velocity squared. A moving stream has kinetic energy, even if there is no head. This energy can be converted to electrical power.

The diversion system allows water to stand in a pool for excess dirt to settle and for a pipe called a **penstock** to fill and thus create a head. The inlet is screened to prevent any debris from entering the pipe and interfering with the turbine. Depending on the location of the inlet and the vertical drop of the inlet pipe, the water that reaches the turbine and generator has a combination of head (vertical drop) and flow from the stream motion. A basic concept drawing that illustrates a small-stream power system is shown in Figure 11-2.

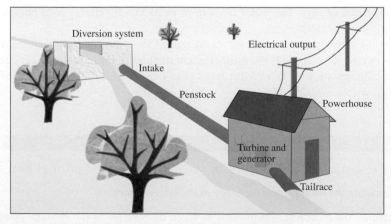

FIGURE 11-2   Basic Small-Stream Power-Generating System

Equivalent Head

The energy in water can take three different forms: (1) potential energy due to elevation, (2) potential energy due to pressure, and (3) kinetic energy due to motion. The total available energy is the sum of the three different forms.

Potential Energy Due to Elevation

You have already seen how a reservoir can store energy ($W_{PE} = mgh$). The h in this equation refers to height. (Recall that average height is used to find the total energy in a reservoir.) As you know, the head is the height, h, that creates pressure due to the column of water.

Potential Energy Due to Pressure

Water can have energy acquired from a pump or other means. Think of the pressure of water at your house; the water under pressure has the ability to do work and therefore contains energy. When water is under pressure, the pressure can be assigned an equivalent head that would be created by a column of water at a height that produces the same pressure. To identify equivalent head, the variable h_p is used to indicate it is the equivalent head due to pressure. The equation for equivalent head is:

$$h_p = \frac{p}{\gamma}$$

where

p is the pressure, in N/m^3 or lb/ft^3

γ is the specific weight of water, which is 9807 N/m^3 or 62.4 lb/ft^3

Kinetic Energy Due to Motion

As you know, moving water has kinetic energy, which is given by the following equation:

$$W_{KE} = \frac{1}{2}mv^2$$

To relate this to an equivalent head, the kinetic energy is assumed to be equivalent to the potential energy. The quantity h_{ke} is defined as the equivalent head due to kinetic energy as given in the following equation:

$$W_{KE} = \frac{1}{2}mv^2 = mgh_{ke}$$

$$h_{ke} = \frac{v^2}{2g}$$

The total equivalent head, h_{eq}, is the sum of h, h_p, and h_{ke}. That is:

$$h_{eq} = h + h_p + h_{ke}.$$

Substituting for h_p and h_{ke} gives the total equivalent head in known terms:

$$h_{eq} = h + \frac{p}{\gamma} + \frac{v^2}{2g} \qquad \text{Equation 11-3}$$

Equation 11-3 is useful if there is more than one form of energy in the water (elevation, pressure, and motion). The equation relates all the energy to an equivalent head. Recall that power is the rate energy is expended, as given in Equation 2-10 in Chapter 2. By substitution the power is,

$$P = \frac{W}{t} = \frac{mgh_{eq}}{t} = \left(\frac{mg}{t}\right)h_{eq}$$

TABLE 11-1

Some Useful Conversion Factors

To convert from	To	Multiply by
foot (ft)	meter (m)	0.3048
cubic feet (ft^3)	cubic meter (m^3)	0.02832
gallon (US)	cubic meters (m^3)	0.003785
gallon (US)	cubic feet (ft^3)	0.1336
cubic feet per second (ft^3/s)	cubic meters per second (m^3/s)	0.02832
pounds per square foot (lb/ft^2)	newtons per square meter (N/m^2)	47.88
foot-pounds per second (ft-lb/s)	kilowatt (kW)	0.001355

For moving water, the volumetric flow rate, Q_v, is an important parameter. Q_v is measured in m^3/s or ft^3/s. The product of specific weight, γ, and Q_v is the weight passing the turbine per time, which is:

$$\frac{mg}{t}$$

By substitution, the following general equation is found:

$$P = \gamma Q_v h_{eq} \qquad \text{Equation 11-4}$$

Equation 11-4 is a general equation for water power no matter what form it is in. If it is strictly a head due to gravity, it can be reduced to Equation 11-1. If it is strictly kinetic energy (no head), it reduces to Equation 11-2.

It is common in the United States to use English units, but in scientific work and other parts of the world, metric units are more common. Table 11-1 gives some conversion factors used in water power calculations for English and SI units.

Example 11-2 illustrates a case where both a head and kinetic energy are used to produce the power.

EXAMPLE 11-2

A small stream has a head of 8 m from the diverter to the turbine. The volumetric flow rate is determined to be 0.05 m^3/s.

(a) Calculate the total power available to the turbine (ignore pipe loss).

(b) Calculate the total power available to the turbine if the piping accounts for 20% loss.

Solution

(a) The specific weight, γ, for water (in metric units) is 9807 N/m^3. Substituting into Equation 11-4:

$$P = \gamma Q_v h_{eq} = \left(9{,}807\frac{\text{N}}{\text{m}^3}\right)\left(0.05\frac{\text{m}^3}{\text{s}}\right)(8\text{ m}) = \mathbf{3.92\,kW}$$

In English units, the γ for water is 62.4 lb/ft^3. The volumetric flow rate of 0.05 m^3/s is:

$$Q = \left(0.05\frac{\text{m}^3}{\text{s}}\right)\left(35.3\frac{\text{ft}^3}{\text{m}^3}\right) = 1.76\frac{\text{ft}^3}{\text{s}}$$

The equivalent head is:

$$h_{eq} = (8\text{ m})\left(3.28\frac{\text{ft}}{\text{m}}\right) = 26.2\text{ ft}$$

The power is:

$$P = \gamma Q h_{eq} = \frac{\left(62.4 \dfrac{lb}{ft^3}\right)\left(1.76 \dfrac{ft^3}{s}\right)(26.2 \text{ ft})}{737.6 \left(\dfrac{\text{ft-lb}}{s}\right)\Big/ kW} = \mathbf{3.90\ kW}$$

The difference is due to round-off error.

(b) If the piping accounts for a 20% loss, then the power available to the turbine is:

$$P = (0.80)(3.90\ kW) = \mathbf{3.12\ kW}$$

SECTION 11-1 CHECKUP

1. Give an example of potential energy in a hydroelectric application.

2. Give an example of kinetic energy in a hydroelectric application.

3. What two parameters are needed to calculate the amount of power that a stream can supply?

4. How does doubling the head in a small stream affect the power?

5. What two factors can change the amount of potential energy of water captured behind a dam?

11-2   Hydroelectric Dam Operation

Compared to thermal plants, hydroelectric plants are much more efficient in converting energy to electricity. Most hydroelectricity, by far, is generated in conventional hydroelectric dams. Another type of power dam is called run-of-the-river, which is discussed in this section. Microhydroelectric dams are also discussed in relation to small streams. Another type of dam is specifcally designed for pumped storage. This section explains the operation of these plants.

Conventional Hydroelectric Dam

The basic parts of a hydroelectric generating system are illustrated in Figure 11-3, which shows the dam with its reservoir and penstock (or tunnel), the turbine and generator, and

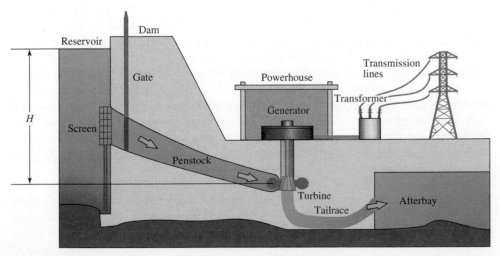

FIGURE 11-3   Basic Parts of a Hydroelectric Generating System

FIGURE 11-4 Generators for a Hydroelectric Dam. A hydroelectric dam can have multiple generators; each can have its own penstock and turbine.   (*Source:* Fotolia.)

the transmission substation. A large screen is located in front of the inlet gate to keep debris from damaging the turbine. The inlet gate controls the amount of water that flows through a penstock; large dams have multiple penstocks, gates, and generators. The penstock directs the water to the turbine blades to spin the shaft at a high speed. The shaft is connected to the armature of a generator. After giving up most of its energy, the water passes through a channel called the tailrace to a relatively shallow storage area called the afterbay.

Inlet gates in the penstock are usually completely open, but large drive motors can adjust the gates to control water flow. The inlet gates can also be completely closed to drain the penstock when it must be inspected or when maintenance is performed on a generator.

Figure 11-4 shows a series of generators mounted in a single dam. Each generator produces three-phase ac. Outside the generators, a substation uses transformers to increase the voltage for transmission. The generator outputs are connected in parallel using switching circuits within the substation. Because they are in parallel, a single generator (or more) can be taken offline for repairs or maintenance, or it can be removed from service if the electrical demand is down.

The operation of the hydroelectric dam in the production of electrical power is based on basic concepts. Water is stored behind the dam. As more water is stored, its level becomes higher, and its ability to produce electrical energy increases. Recall from Equation 11-4 that the higher the reservoir, the greater the head and quantity of stored energy. The dam has **spillways** that are built into it to allow water to be released and to avoid overfilling the reservoir during rainy periods. Water in the spillway does not go through the penstock and past the turbine, so its energy is lost. It is important that the water level is not allowed to increase to the point where water runs over the dam because this can damage the dam structure. During dry periods, the amount of water being released may exceed the incoming flow, so the level of the reservoir may drop. When this occurs, the available energy decreases because of the decreasing head.

When a generator is to be used, the inlet valve to the penstock is opened. Most dams have surge tanks connected to the penstock to reduce water hammering in the penstock. When water first enters the penstock, the surge tanks accept some of the water to reduce the quick increase of pressure, which could cause damage. The amount of electrical power the generators can produce depends on the current head in the reservoir and the quantity of water that is allowed to flow to the turbines.

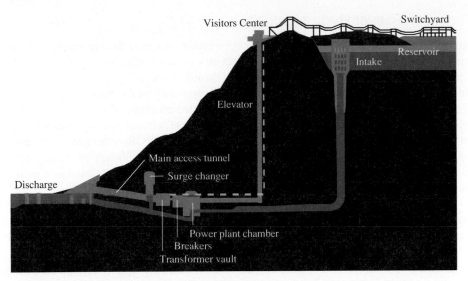

FIGURE 11-5 Raccoon Mountain Pumped Storage Hydroelectric Generating System   (*Source:* Tennessee Valley Authority.)

Pumped Storage System

A **pumped storage system** is a system of two dams, each with a reservoir. One is located at a much higher elevation than the other. Water is released from the higher reservoir to produce electrical power. The water is captured in the lower reservoir and pumped back during off peak hours to the higher reservoir, where it is used again. Figure 11-5 shows a basic diagram of the Raccoon Mountain facility, which belongs to the Tennessee Valley Authority (TVA) located on the Tennessee River. The dam is the largest rock filled hydroelectric dam in the TVA system; the dam is 70 m (230 ft) high and 1,800 m (5,800 ft) long. The system works with a hydroelectric generating station between the two reservoirs. (The lower reservoir is not shown in the figure.) During low-peak electrical hours, water is pumped from the lower reservoir into the higher reservoir. During the highest peak times, when electrical energy is needed on the grid, water is released from the upper reservoir, and it flows down through penstocks to turbines, as in a traditional hydroelectric dam, producing up to 1,600 MW for the grid. The water is released in the bottom of the lower reservoir, where it can be returned to the higher reservoir during times of low electrical demand.

The generator in a pumped storage facility can act as a motor, so technically it is called a motor/generator rather than just an ac generator or alternator. When water from the lower reservoir is to be pumped to the higher reservoir, power from the grid is applied to the motor/generator's field windings, which causes it to operate as a large motor. The turbine becomes a large pump. The result is that water from the lower reservoir is pumped back through the penstock up into the upper reservoir. The turbine acts as a normal turbine to turn the generator when water flows down, and it acts like a pump when water is moved up. The design of pumped storage turbines is complicated by this requirement.

The change from a generator to a motor is accomplished by switches. When operating as an ac generator, the output is connected to the output transformers and the grid. When the motor/generator operates as a motor, switches connect power from the grid through the transformers, which now operate as step-down transformers, and back to the motor/generator. The equipment used in the station is called the pumping generating unit because it operates as a motor and pump when water is pumped up to the higher reservoir, and it operates as a turbine and generator when water is released from the upper reservoir to the lower one.

The pump generating system is used to provide extra electrical power at the peak times when it is needed. Energy can be stored in various ways, but pumped water storage is one of the most effective. The ability to turn the stored energy back into electricity quickly during peak demand is a very useful tool for grid operators to meet demand.

Pumped Storage in an Abandoned Quarry

(*Source:* iwanstu/Fotolia.)

Thousands of abandoned rock quarries around the world can serve as reservoirs. If underground mining was also involved in the operation, the mine itself can serve as a lower reservoir. An abandoned quarry in Elmhurst, Illinois, is suitable for a pumped storage system and doesn't need a dam. The quarry has a deep underground mine that can serve as the lower reservoir, which eliminates the costs associated with obtaining land for a reservoir. The Elmhurst quarry is ideally located near a power substation and population centers, so it is being considered for service as a pumped storage system.

Run-of-the-River Hydroelectricity

Run-of-the-river (ROR) is a hydroelectric system that uses river flow to generate electricity. The system may include a small dam with storage for water, but many do not. ROR dams differ from ordinary dams because they do not flood massive amounts of land or interfere greatly with marine life; instead, they rely primarily on the river flow. When a dam is part of the system, the storage area behind it is called **pondage.** Run-of-the-river power plants are classified as with or without pondage. A plant without pondage has no storage and is therefore subject to seasonal river flows. This type of ROR system produces electrical power only when the river flow permits it. A system that has pondage can regulate water flow to some extent and can serve either as base-load source or can be used as a **peaking power plant,** which means it produces power when the demand is highest.

Run-of-the-river hydroelectric systems are typically selected for streams and rivers that do not have large swings in the volume of water during the year. Rivers with consistent flow can use an ROR system for base-load power. If flow is not consistent, the power level from the system is forced to drop, and the system is used only to supplement power from other sources. Up to 95% of the main stream of water in the river can be diverted to flow through the dam and then returned to the river, and this diversion has little effect on the main river flow. Figure 11-6 shows the Chief Joseph Dam, which is located on the Columbia River in Washington State. This concrete gravity dam is the largest run-of-the-river dam in the world. The dam has a small pondage lake; the power plant has a capacity of 2,620 MW and is the second largest hydroelectric power facility in the United States (behind the Grand Coulee Dam). The Columbia River is home to two additional very large ROR systems: the John Day Dam and the Dalles Dam.

Compared to conventional dams, with their large reservoirs, ROR dams are considered environmentally friendly because they have only a small effect on river flow, and they do not have large reservoirs with their accompanying environmental impact. Even with its smaller capacity, an ROR system with pondage can help with flood control to a limited extent.

Micro- and Small Hydroelectric Dams

Small hydroelectric dams generate between 100 kW and 10 MW. Small hydroelectric systems can be connected directly to the grid, or they can be used to provide electrical power to a building or business with grid power backup when the water level is not high enough

FIGURE 11-6 Chief Joseph Dam. The Chief Joseph Dam on the Columbia River in Washington State is the largest run-of-the-river dam in the world.   (*Source:* US Army Corps of Engineers.)

to produce all the power that is needed. Some small hydroelectric systems are installed in remote areas because they can provide electrical power for a small community, where the grid has not been extended into the area.

If a hydroelectric system generates less than 100 kW of electricity, it is called a microhydroelectric system. Many of the hydroelectric dams used by farmers, ranchers, homeowners, and small-business owners are basic microhydropower systems. In some cases, a 10 kW microhydropower system can provide enough power for a home or a small farm. If a property has water flowing through it, a microhydroelectric system can be installed on the stream. The water needs as little as 2 gallons per minute (GPM). Because most streams flow day and night, the microhydroelectric system can provide electrical power around the clock (unlike solar or wind energy systems).

The microhydroelectric system is considered to function as a run-of-the-river system (perhaps it would be more rightly named a run-of-the-stream system). The water is directed from the stream through a turbine and then sent back to the stream, which has little effect on the stream. There is no need for a dam and its reservoir. Generally, the microhydroelectric system is designed to prevent harm to fish or other wildlife in the system. Microhydroelectric systems are now being used in developing countries to provide small amounts of electrical power for refrigeration or pumping and purifying water.

Another type of very small hydroelectric system is called a picohydroelectric system. The picohydroelectric system has a generator that is smaller than 5 kW. The generator shaft is turned by a small turbine. If a water supply or stream has a drop of more than 1 m (3 ft), it has enough energy to operate a 5 kW generator. Picohydroelectric systems typically operate the same as larger run-of-the-river systems but without pondage, so they are environmentally friendly. Water is diverted using pipes or other devices to move the water from the main stream to the turbine and returned to the stream. The picohydroelectric system is very inexpensive, which means it can be installed on streams around the world, in developing countries, and especially in remote regions where other types of renewable energy systems would not function as well or cost too much to install.

SECTION 11-2 CHECKUP

1. How does a conventional hydroelectric dam produce electrical power?

2. How does a pumped storage system produce electrical power?

3. How does a run-of-the-river hydroelectric dam differ from a conventional hydroelectric dam?

4. Why does a pumped storage system have an upper and lower reservoir?

11-3 Water Turbines

As you know, a turbine is a rotary engine that extracts energy from a fluid and converts it to useful work. Almost all electrical power is produced with some type of turbine. Water turbines are different than the steam turbines used in most conventional power plants. This section includes a general discussion of water turbines and specific types used for converting hydropower to mechanical rotating motion. Special turbines for tides and wave generators are described in Sections 11-4 and 11-5, respectively.

The specific conditions of the head and flow largely determine the type of turbine that produces the most efficient operation in a given situation. Ideally a water turbine should transfer all of the kinetic energy in moving water into turning the shaft. While it is not possible for any mechanical device to be 100% efficient, well-designed water turbines can have very high efficiencies. The key to high efficiency is to minimize losses such as those resulting from turbulence, vibration, or heat. To minimize turbulence, the turbine blades must be smooth and the turbine must be well balanced with low-friction bearings.

FIGURE 11-7 Definition of Torque. A force, *F*, acts on a body at a perpendicular distance, *x*, from a pivot point that tends to produce rotation.

Fourneyron Turbine

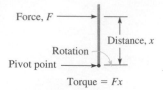

A Fourneyron turbine from the Adams Power Station at Niagara Falls, New York. The turbine was installed with the shaft oriented vertically.

(*Source:* Courtesy of John Ditchfield.)

Although small hydroelectric power plants existed earlier, the first successful, commercial hydroelectric plant in the United States was on the US side of Niagara Falls in New York State. Three Fourneyron turbines of 5,000 hp and 250 rpm were installed in 1893, making it one of the first power plants to use renewable energy. Today, Niagara Falls still generates power—currently it uses thirteen generators and develops over 2,500 MW of power.

An equation by the famous mathematician Leon Euler in 1754 is still used today by engineers to calculate the power from water turbines. The equation basically indicates that the power from a turbine is equal to the product of torque on the shaft and the rotational speed. **Torque** is the product of a force, *F*, and the perpendicular distance, *x*, from a fixed point that tends to produce rotation about the point; it is measured in newton-meters in the SI system and pound-feet in the English system. Torque is illustrated in Figure 11-7.

Impulse and Reaction Turbines

Water turbines are defined by the form of energy they convert to mechanical motion. Turbines are classified as either impulse or reaction turbines; most turbines are a mixture of the two classifications. An **impulse turbine** is a rotary engine that changes the direction of a high-velocity fluid, thus converting kinetic energy into mechanical rotating energy. In an impulse turbine, water is passed at high velocity (thus high kinetic energy) through a nozzle and focused on the turbine blades, causing them to rotate. Impulse turbines are like a pinwheel. As water hits the blade of the turbine, called a runner, the blade is pushed. Impulse turbines are used most often in applications with high-pressure heads and relatively low flow. The head is typically in excess of 300 m. Figure 11-8 shows a basic impulse turbine.

A **reaction turbine** is a rotary engine that develops torque by reacting to the pressure of a fluid moving through the turbine, thus primarily converting potential energy into mechanical rotating energy. In a reaction turbine, the rotating blades are completely encased in a pressure encasement and the fluid flows through a fixed guide mechanism onto rotating blades. Unlike an impulse turbine, the reaction turbine does not have nozzles. Water has both potential energy due to its pressure and kinetic energy due to its motion, but the primary mover is the pressure drop, which creates the reaction force. Due to the importance of water pressure in this process, the reaction turbine must remain submerged at all times. This stipulation often requires that the turbine be encased in water if the water levels at the site are not consistent. Most reaction water turbines are used with low or medium heads but with a larger volume of water.

Turbines can also subclassified as radial design or axial design. In a radial-flow turbine, the flow enters in the direction the turbine moves and exits along the axis of rotation. In an axial-flow turbine, the flow enters the turbine in the direction of the axis of the shaft and leaves the turbine in the same direction.

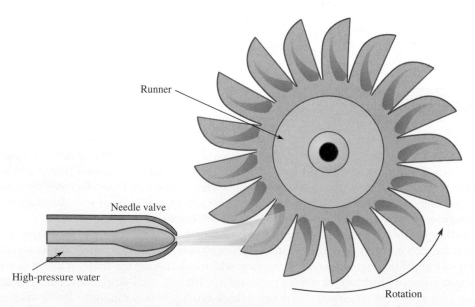

FIGURE 11-8 Impulse Turbine. Although one nozzle is shown here, several nozzles may be used to drive the turbine.

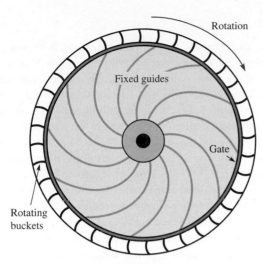

FIGURE 11-9 Top View of a Fourneyron Radial-Flow Water Turbine. Water flows from the inner ring and is turned by fixed guide vanes through gates to an outer rotating ring.

Types of Water Turbines

Impulse and reaction turbines are further divided into specific types of turbines. The type of turbine selected for a given application depends primarily on the flow and head. The following subsections discuss each type.

Fourneyron Turbine

A French engineer named Benoit Fourneyron developed the first reaction turbine in 1826, which is the predecessor to most turbines used in hydroelectric power plants. Fourneyron's turbine is a horizontal outward radial-flow waterwheel. His turbine was an advancement of a design created originally by his teacher Claude Burdin, which produced approximately 40 horsepower (hp). In 1837, Fourneyron improved the design and patented a 60 hp turbine that was approximately 30 cm in diameter. The turbine he designed was a reaction turbine that rotated at 2,300 revolutions per minute (rpm), and it could handle a wide range of heads: less than 1 m to 3 m and as high as 30 m at efficiencies up to 85%, which was much higher than the open waterwheel designs of his day. The Fourneyron turbine is shown in Figure 11-9. It used fixed guide vanes to direct water onto the moving runner blade with buckets that rotated and produced energy. The Fourneyron turbine operates completely submerged, which ensures that it has a smooth flow of water, thus creating a highly efficient design. The water flows from the inside through the fixed guide vanes and splashes against the moving outside wheel, which spins, thus producing the power. Today, the more efficient Francis and Kaplan turbines have largely replaced their predecessor as the reaction turbines of choice.

Francis Turbine

In 1849, James B. Francis improved a design on a radial inward-flow reaction turbine so that it produced power at over 90% efficiency. The **Francis turbine** is the most widely used water turbine in the world today. Unlike its predecessor, the Fourneyron turbine, water is directed from the outer circumference toward the center of the runner, which is the rotating center part of the turbine. Water from behind a dam flows through the guide vanes on the sides of the turbine, spins the runner, and exits through the bottom tailrace tube. Figure 11-10 shows the runner. The vanes are in a scroll case, which is a curved tube that diminishes in size, in a shape similar to a snail shell. The Francis turbine is used for medium head (30 m to 300 m) applications. Today, the Francis turbine is installed with a range of power outputs, from 1 kW to 820 MW.

FIGURE 11-10 Runner for a Francis Turbine (*Source:* Illustration courtesy of Hydro-Québec.)

FIGURE 11-11 Kaplan Turbine. The pitch angle of the blades can be adjusted to optimize power for different flows. (*Source:* Courtesy of GEA SRL/Orengine International Ltd.)

Kaplan Turbine

The **Kaplan turbine**, an evolution of the Francis turbine, is shown in Figure 11-11. It is a reaction water turbine that uses propellers with adjustable blades. The turbine is usually placed in a spiral-shape casing called a volute. It was developed by Austrian engineer Viktor Kaplan in 1913 and is generally considered for low-head, high-flow applications. By changing the pitch angle of the blades in tandem with the angle of the turbine, the Kaplan turbine can maintain a high efficiency even at very low flows.

As in all reaction turbines, a drop in pressure occurs as water passes through the Kaplan turbine. There is pressure on the upstream side of the turbine runner and suction on the downstream side. On the downstream, or outlet, side of the turbine is the draft tube, which connects the turbine to the tailrace. The tailrace is the channel leading away from the turbine. The draft tube has a unique curved hornlike shape that slows the water as it exits the turbine.

The Kaplan turbine has an efficiency curve that changes little with flow rate, so it is particularly suited to rivers where the amount of water flow varies greatly. The Kaplan turbine is often used for run-of-the-river power stations. Propeller turbines can reach very high rotation speeds that work well in very low head applications. Figure 11-12 shows a basic installation of a Kaplan turbine in a power station.

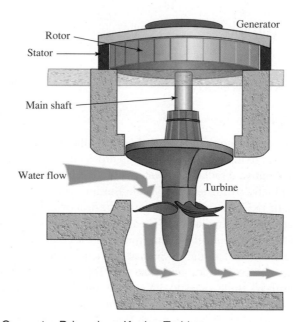

FIGURE 11-12   Generator Driven by a Kaplan Turbine

Pelton Turbine

The **Pelton turbine** (also called a Pelton wheel) is an impulse turbine whose runner has a number of cup-shape containers connected around its circumference. The runner is connected to a central hub that rotates and transmits energy to a shaft, as shown in Figure 11-13. Nozzles are positioned around the runner so that water flow is directed into the cups. When the water strikes the cups, it changes the potential energy of the water into kinetic energy by pushing the turbine wheel around. Pelton turbines have been modified over the years with different types of the cup-and-wheel apparatus. Each modification has made the Pelton turbine more efficient.

The turbine known today as the Pelton turbine was first designed by Samuel Knight in 1866, who worked as a millwright in California. He patterned his design after a high-pressure jet system used in mining gold, and his turbine was called the Knight wheel. The turbine uses a high head, over 30 m, where water drops into the cup-shape containers. The force of the water provides kinetic energy that moves the runner. As the runner rotates, the container carries the

FIGURE 11-13 Pelton Turbines at Walchensee Power Plant in Germany (*Source: Courtesy of Voith Hydro GmbH & Company.*)

water until it reaches the other side of the runner, where the container is inverted and the water is released out of it at a much lower velocity.

In 1879, the turbine that was originally designed by Samuel Knight was improved by Lester Pelton at Grass Valley, California. As Pelton watched the Knight wheel spin, he noticed that it had slipped out of its normal position and was offset, yet it seemed to spin faster. When it slipped, the cups became out of position so that water hit the cups near their edge instead of in the middle of the cup, which made the wheel more efficient. Pelton designed an improved waterwheel that had the cups aligned so that water would hit near their edges, and he included a double set of buckets that were side by side. This split the water jet in two. The water does a U-turn in Pelton's design, thereby extracting nearly all of its kinetic energy. In 1895, William Doble (who worked for Pelton) made a further refinement to Pelton's design by replacing the half-cylindrical buckets with buckets that were elliptical in shape. His buckets included a cut to force water more efficiently into each bucket. Doble's variation of the Pelton turbine can achieve up to 92% efficiency. Doble later took over the Pelton Company and did not change the name of the turbine because of the Pelton brand recognition. Pelton turbines are still preferred where there is a high head and a low-flow rate. Figure 11-13 shows Pelton turbines at Walchensee Power Plant in Germany. You can see the double cups in the new turbines.

Turgo Turbine

The **Turgo turbine** is a compact impulse turbine that was developed and installed in Scotland in 1919 by Gilbert Gilkes as an update to the Pelton turbine. The Turgo turbine uses half cups around the runner instead of full cups, like the Pelton turbine (see Figure 11-14 for a close-up of the cups). The half cup on the Turgo turbine allows water to enter and exit each cup faster and in larger amounts than with Pelton turbines, so it is better suited for higher flow rates. Water for the Turgo turbine flows through one or more nozzles directly onto the cups. Increasing the number of jets increases the speed of the runner by the square root of the number of jets. For example, if four jets are used, they will yield twice the specific speed as one jet on the same turbine. The Turgo turbine is usually used in applications where there is low to medium flow and with heads that are medium to high. It can be mounted horizontally or vertically.

FIGURE 11-14   Closeup of the Cups on a Turgo Turbine   (*Source:* Courtesy of Hartvigsen Hydro.)

Crossflow Water Turbines

The crossflow water turbine is a drum-shape impulse turbine. It is designed with long trough-shape blades (or horizontal slats) in a radial arrangement around a cylindrical runner, as shown in Figure 11-15. The turbine uses two spray nozzles to direct water flow at a 45° angle across the blades or one large wide rectangular water jet. Water striking the

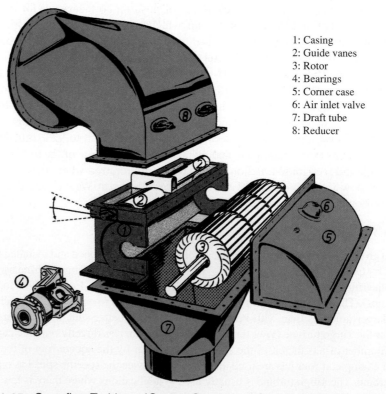

1: Casing
2: Guide vanes
3: Rotor
4: Bearings
5: Corner case
6: Air inlet valve
7: Draft tube
8: Reducer

FIGURE 11-15   Crossflow Turbine   (*Source:* Courtesy of Ossberger GmbH + Co.)

blades converts its kinetic energy to rotational energy. The flow of water is controlled to maintain the desired speed of the runner. In operation, water is sprayed onto the blades, and most of it then enters the turbine and passes through the blades again as it exits.

Crossflow turbines are used in smaller hydroelectric applications where the maximum power output typically does not exceed 2,000 kW and costs are critical. The head is usually between 10 m and 150 m, but the flow is relatively high. The average efficiency is about 80% for smaller units and 86% for larger ones. Because the water passes through the turbine across its blades, the water provides power once when it enters the blades and again as it leaves the blades on the opposite side, which improves its efficiency. When compared to the Pelton, a crossflow turbine of similar size can handle a greater amount of water flow.

The crossflow turbine is known by two other names: the Michell-Banki turbine and the Ossberger turbine. It was originally designed in 1903 by Anthony Michell. Donat Banki, a Hungarian professor, later improved the design and received a patent around 1925. In 1933, Fritz Ossberger, a German engineer, improved the patent, and the turbine remains much the same today.

Table 11-2 summarizes the types of water turbines discussed in this section.

TABLE 11-2

Reaction and Impulse Turbines

Reaction Turbines	Impulse Turbines
Fourneyron	Pelton
Francis	Turgo
Kaplan (propeller)	Crossflow (also called Michell-Banki or Ossberger turbine)

SECTION 11-3 CHECKUP

1. Explain how an impulse turbine operates.
2. Explain how a reaction turbine operates.
3. List three types of reaction turbines.
4. List three types of impulse turbines.

11-4 Tidal Power Generation

Tides are caused by gravitational interaction and rotation of the earth–moon system about a common center of mass located inside the earth. (There is also a minor effect from the sun's gravity.) As the earth rotates, the interaction causes two high tides and two low tides per day at any point on a coast. The change in water level between a high tide and low tide can be harnessed to produce electrical power, or the current can be used to turn turbines. Various methods of harnessing tidal energy are discussed in this section.

Causes of Tides

The gravitational pull of the moon causes the ocean to bulge in the part closest to the moon. The bulge in the ocean is known as a high tide. When the bulge reaches the coast, a high tide occurs. A second bulge can be found on the opposite side of the earth due to the earth's orbital motion. The earth–moon system actually rotates about a common center of mass that is located inside the earth and in the direction of the moon. This pivot point for rotation is the offset that causes the earth to experience a force that pushes water away from the center of mass on the opposite side, much like what happens if you swing a bucket of water in a circle. The bulge on the side toward the moon is responsible for one tide, but the effect of centrifugal force is responsible for the bulge on the opposite side. The actual time

between high tides is approximately 12 hours and 25 minutes, which is the time required for the alignment to occur again. Because of varying ocean depths, landmasses, and other factors at different locations on earth, the actual tides vary.

When the sun, moon, and earth all align, a stronger gravitational pull occurs on the oceans, and higher and lower tides called spring tides are produced. A **spring tide** has nothing to do with the spring season of the year; rather, it occurs at new moon and again at full moon (approximately twice a month). At first and third quarter, the net gravitational force of the sun and moon is not as pronounced, so lesser tides than normal, called **neap tides,** are produced (neap tides also occur twice each month). The currents associated with the tides depend on the particular location on earth; currents from tides can vary from 0 to over 2 m/s.

Tidal Barrage System

A **tidal barrage system** is designed to convert tidal power into electricity by trapping water behind a dam, called a **tidal barrage dam,** and generating power from the inflow and/or release of water. The tidal barrage dam is a large dam that stretches completely across an estuary, harbor, or river that connects to part of the ocean that has a tide. When the tide rises, sluice gates open to allow the tide to flow through tunnels and fill the area behind the dam. When the water is flowing into the area behind the dam as the tide is rising, it causes the turbine to rotate and produce electricity. When the tide reaches its highest level, the gates close, capturing the maximum amount of water behind the dam. When the tide drops, it creates a difference between the water behind the dam and the water in front of the dam that is at the level of the low tide. When the difference is large enough, the flow valves are opened so that the water flows under the dam through the tubes and past a generator turbine, causing it to turn and produce power. The barrage continues to fill and empty twice daily as the high tide comes in and then changes to a low tide. A conceptual plan for a tidal barrage system is shown in Figure 11-16.

Because the barrage dam stretches completely across the harbor or river, it must have a lock to accommodate shipping or boating. The cost for installing a tidal barrage dam is very high, and only a relatively few locations in the world have sufficiently high tides to make it economically viable. The world's largest tidal power system, the Sihwa Lake Tidal Power Station, is located in South Korea. This system, which opened in 2011, is a 254 MW tidal barrage dam west of Seoul, South Korea. The dam is 12.7 km (7.9 miles) long and uses ten 26 MW turbines positioned on the ocean floor. The area of the basin is huge, measuring 43 km^2 (17 mi^2). Because of water-use concerns, the turbines run only as sea water moves into the lake and not in the opposite direction.

A barrage dam even larger than the Sihwa project—the 18 km (11.1 mile) Severn Barrage Dam between Brean, England, and Lavernock Point, Wales—was proposed in 2010. Figure 11-17 shows an artist's conception of the project. It is a huge proposal, so it would

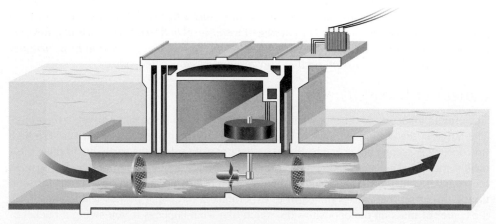

FIGURE 11-16   Plan View of a Tidal Barrage System

FIGURE 11-17 Proposed Severn Barrage Dam Project
(*Source:* Courtesy of Hafren Power, Ltd.)

require government support. As of September 2013, the plan has been shelved with Hafren required to improve the proposal before "serious consideration" is given to it. The plan had called for 1,026 bidirectional, slowly spinning turbines across the width of the estuary and includes locks for ship passage. While the massive Severn tidal project appears to be dead, there are a number tidal lagoons that are separated from the rest of the sea that may be exploited for energy in the future. The final chapter has not been written for the Severn and U.K is looking into smaller privately funded projects that will enable environmental impacts to be studied before attempting a project as massive as the Severn.

Power from a Barrage

The energy available from a barrage is calculated by starting with gravitational potential energy. Recall that gravitational potential energy is written as:

$$W_{PE} = mgh$$

The mass of water, m, is the volume times the density of water; that is:

$$m = \rho A h$$

The total available energy for water is one-half of the value because an average height is used to determine the available energy. Hence:

$$W_{PE} = \frac{1}{2}(\rho A h)gh = \frac{1}{2}\rho A g h^2 \qquad \text{Equation 11-5}$$

where:

W_{PE} = energy stored, in J

h = height of the vertical tide, in m

A = horizontal area of the barrage basin, in m^2

ρ = density of seawater = 1,025 kg/m^3 (Note that fresh water has a density of 1,000 kg/m^3.)

g = gravitational constant = 9.8 m/s^2

Equation 11-5 shows that the stored energy is proportional to the height squared. As the water level drops in the barrage, the power is dropping because the head is reduced. A problem with this method of generating power is that it depends on the height in the barrage, which drops dramatically when it is supplying energy (as opposed to a large reservoir, where the level is much less variable.)

EXAMPLE 11-3

(a) Determine the theoretical energy stored in a barrage if the height of the tide is 3 m in a barrage with an area of 300,000 m^2.

(b) The barrage drains in 10 hours; calculate the average power.

Solution

Substitute into Equation 11-5:

(a) $W_{PE} = \frac{1}{2}\rho A g h^2 = \frac{1}{2}\left(1{,}025\frac{kg}{m^3}\right)(300{,}000 \text{ m}^2)\left(9.8\frac{m}{s^2}\right)(3 \text{ m})^2 = \mathbf{1.36 \times 10^{10} \text{ J}}$

(b) $P = \dfrac{W_{PE}}{t} = \dfrac{1.36 \times 10^{10}\text{ J}}{(10 \text{ hr})\left(3{,}600\dfrac{s}{hr}\right)} = \mathbf{377 \text{ kW}}$

Tidal Stream Generator

A **tidal stream generator (TSG)** is an electrical generating system that is anchored on the bottom of a river or stream to capture energy from the natural flow of the stream over the turbine blades. The turbine is connected through a gearbox to a generator, which produces electricity. Conceptually, the system is like the wind turbines discussed in Chapters 7 and 8 except that it uses water instead of air as a fluid. Figure 11-18 shows a small three-blade tidal stream generator.

Streams that empty into the ocean have water levels that rise or lower with the tides in the area. When the tide rises, water can flow upstream in the lower portion of the river, and when the tide becomes lower, water flows back to the ocean. This return flow causes a very strong current that moves a large volume of water over the blades of the water turbine

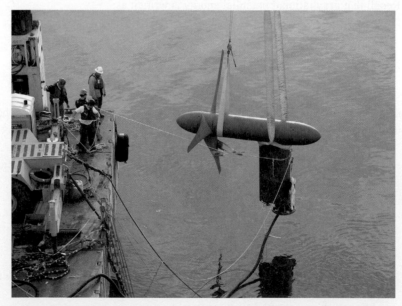

FIGURE 11-18 Three-Blade Tidal Stream Generator in the Process of Installation in the East River of New York. (*Source:* National Renewable Energy Laboratory.)

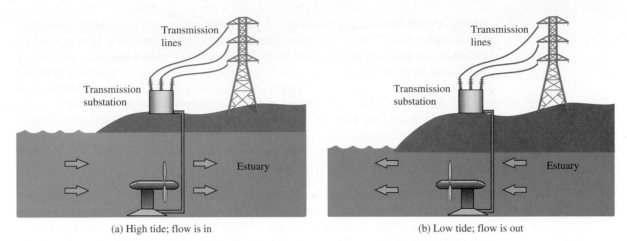

Transmission
lines

Transmission
substation

⇒   ⇒ Estuary

(a) High tide; flow is in

Transmission
lines

Transmission
substation

⇐   ⇐ Estuary

(b) Low tide; flow is out

FIGURE 11-19   Tidal Stream Generator. The movement of water in a tidal current generates power for both inflow and outflow.

generator, causing it to rotate. The turbine is designed so it can rotate to optimize power for incoming and outgoing tides. Figure 11-19(a) shows an example of a water turbine rotating when the tide is coming in; Figure 11-19(b) shows an example when the tide is going out. The main restriction to blade size is the depth of the water at low tide because the turbine needs to be anchored where it remains under water.

A number of tidal stream power systems have been installed and are operating throughout the world; others are in the planning stage. Marine Current Turbines Ltd, based in Bristol, England, has developed the world's largest tidal stream system, which is known as SeaGen. It is currently operating in Strangford Lough, which is a shallow bay on the east coast of Northern Ireland. SeaGen has a large steel tower embedded in the ocean floor that supports two 16 m diameter rotors connected on either side of a single monopile structure. This structure can raise and lower the turbines out of the water for maintenance. In operation, the turbines act like underwater "windmills" driven by the tidal current at a rotational speed of up to 14 rpm. They are connected through a gearbox to a generator. The system has been operational for a number of years and generates approximately 1.2 MW of electricity for the Northern Ireland grid. Figure 11-20 shows a view of SeaGen with the propellers in the raised position.

FIGURE 11-20   SeaGen Tidal Stream Energy System. SeaGen is a 1.2 MW tidal energy system located off the coast of Ireland.   (*Source:* Courtesy of Siemens AG.)

In the United States, a large tidal stream generation project is located in the East River in Manhattan, New York. The project is called Roosevelt Island Tidal Energy (RITE). Verdant Power designed and installed several water turbines, each with three fixed blades. The 5 m (16 ft) blades rotate at approximately 40 rpm in the current. Each turbine/generator produces 35 kW of electricity. The project is scheduled to install 100 to 300 turbines in an array in the final phase. The system is designed to provide over 1 MW when all the turbines are installed and running. The turbine has a peak efficiency of 38% to 44% when the current is 1 to 2 m/sec.

Crossflow Tidal Energy Turbine

The crossflow turbine was introduced primarily as a small system turbine in Section 11-3. It can also be used in tidal applications, where the crossflow turbine is less prone to cavitation (voids) in fast flows or water that is not as deep. Cavitation can create shock and vibration and loss of efficiency. The output of the generator is currently connected to the grid in Maine, producing a small amount of power (approximately 180 kW), but it is the first in North or South America producing commercial power.

The blades are turned by the tides flowing across the long side of the turbine, which drives a generator. The tidal energy turbine called TidGen is shown in Figure 11-21 prior to installation. Figure 11-22 shows an artist's drawing of the turbine mounted on the ocean floor. Eventually, the system is planned to add 20 units to bring the power up to nearly 4 MW. The first unit was installed in Cobscook Bay in Maine and became operational in September 2012. Cobscook Bay opens to the Bay of Fundy, which has tides that range over 15 m (50 ft) a day, said to be the world's largest tides.

Dynamic Tidal Power

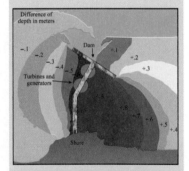

Dynamic tidal power (DTP) is a proposed technology for capturing energy in tidal flows using an off-shore dam shaped like the letter T. The dam does not enclose an area, but instead traps water due to a difference of tidal levels on either side of it. Tides not only have a height difference between high and low tides, but they can also have a height difference in places where the tide runs horizontally along the shore. Water moves through the turbines located in the face of the dam to generate power.

FIGURE 11-21   Tidal Energy Crossflow Turbine   (*Source:* Courtesy of Ocean Renewable Energy Company.)

Advantages and Disadvantages of Tidal Energy

Generating electrical power from tidal energy has advantages and disadvantages. One of the biggest advantages of tidal energy is that the tides are predictable and occur all along the ocean front; however, the current is not constant but fluctuates as tides move in and out. The location of turbines underwater is an advantage because it eliminates noise and a large visual presence. The main disadvantage of tidal energy systems is the initial expense of installation and the upkeep of a marine system located in saltwater. Other concerns include the effects of any system on fish, seals, marine life, and birds. Initial studies of environmental effects are positive. The SeaGen system, which has extensive operating experience, is located in an area designated as a special protection area for bird and seal populations, so the SeaGen system has been monitored carefully

FIGURE 11-22   Artist's Drawing of Tidal Energy Crossflow Turbine. The tidal energy crossflow turbine is shown mounted on the ocean floor.   (*Source:* Courtesy of Ocean Renewable Energy Company.)

after construction, as discussed in Section 1-6. The SeaGen system and the TidGen system are providing valuable baseline data for future projects in harnessing tidal energy.

A concern for barrage dams is the effect on fish migration patterns as well as fish mortality due to passing through turbines. Another concern is the changing patterns for sediments. When water containing sediments slows, the sediments that were suspended in the water tend to be deposited. Sedimentation tends to decrease the intertidal area, which can affect shellfish and can kill clams. It can also affect other marine life, including fish by reducing fish spawning areas, and bird habitat. The extent of effects on wildlife varies depending on the specific location. Depending on the source of the sediments, the net effect is cumulative changes in deposition and erosion patterns that can have negative environmental impacts. At Annapolis in the Bay of Fundy, most sediment is derived from upstream of the barrage dam, so there is a deficit of sediments below the dam, creating erosion along the shore of the basin region after the dam.

SECTION 11-4 CHECKUP

1. Explain what causes the tides to rise and fall.

2. How often do spring tides occur? What causes them?

3. What is a tidal barrage power system? How does it work?

4. Explain how to estimate the total stored potential energy in a tidal barrage.

5. What is the advantage of a crossflow turbine in a tidal application?

11-5   Wave Power Generation

Ocean waves produce a large amount of energy that can be harnessed to produce electrical power. A number of concepts and prototypes as well as a few actual systems have been developed that have delivered electrical power from wave energy. The costs so far have not proven competitive with other sources, however new devices are being developed that may prove to be competitive. Unlike solar and wind energy, wave energy tends to be constantly available, even more so than tides. This section explains several of the methods currently under development around the world to convert wave power into electrical power.

Ocean Waves

Waves are generally categorized as either longitudinal or transverse. In a longitudinal wave, the individual particles move back and forth in the direction that the wave moves. In a transverse wave, the individual particles move up and down in a direction that is perpendicular to the wave motion.

Water waves are created by the wind as it blows across a large stretch of water. The air transfers energy to the water and causes a wave that is a combination of longitudinal and transverse motion. The individual water particles move in an elliptical pattern as a result. As the wind speed increases, the waves become larger; their size also increases as they move onshore.

Large stretches of the ocean are open, so waves can build a considerable amount of energy. As they approach shore, they are affected by water depth and can become quite large. With water waves, the wave height is higher as the energy in the wave increases. **Wave height** is twice the amplitude and is measured between the lowest point, called the trough, and the top, called the crest. The **wavelength** is a common measurement for any wave and is the horizontal distance between two consecutive points on a wave (in water it is typically from a crest of one wave to the crest on the following wave.) These definitions are illustrated in Figure 11-23.

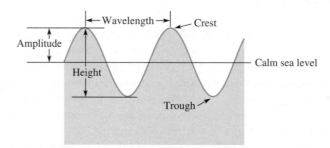

FIGURE 11-23 Wave Definitions for Water Waves

When the wave is at its trough, it compresses the water directly below it slightly, and this extra pressure helps push the water back and upward, adding to the force of the next crest. The higher the crest, the more surface the wind has to push against, and over time swells develop. If the wind speed increases, it pushes more water; this causes the crests to become higher, which increases the wave energy. As waves become higher, they naturally tend to get longer, an important design factor for certain wave energy devices.

The oceans and seas cover approximately 70% of the world's surface. A large amount of energy is available in the waves that roll across the oceans every day, twenty four hours a day. A number of machines can convert wave power into electrical power, and they are generally known as wave energy converters, which form the basis of a wave energy system that includes cabling and shore installations. Wave energy converters can be classified into three basic types: point absorbers, attenuators, and terminating devices. Each will be discussed in the following subsections.

Point Absorbers

A **point absorber** is a floating wave energy converter that is in a fixed position. It moves up and down from wave motion. The motion with respect to a fixed reference is captured and the energy is converted to electricity using electromechanical or hydraulic energy converters. Point absorbers work by creating relative motion between a floating part and a submerged assembly. The motion can be used to drive a hydraulic motor and generator to create electricity. Because they are axially symmetrical, point absorbers can work with waves moving in any direction. Several companies have point absorbers under various stages of development, but one that is ready for deployment is from Ocean Power Technologies.

FIGURE 11-24   An Ocean Power Technology Point Absorber   (*Source:* National Renewable Energy Laboratory.)

Figure 11-24 shows a 150 kW Ocean Power Technologies (OPT) power buoy installed off the coast of Scotland. This buoy has a floating hull that is constrained to vertical reciprocal motion along a spar connected to a massive underwater hull that tends to stay in place due to inertia. Mechanical energy is captured by coupling the two hulls with a structure that has a push rod that moves in response to wave motion. The push rod acts on a power take-off (PTO) housed in the spar. This reciprocating motion is converted to rotary motion that turns a turbine. An undersea substation pod collects the power and transmits it, via a marine power transmission cable, to the shore. Much of the development work on this system has been supported by the United States Department of Energy.

OPT has received the first license issued in the United States by the Federal Energy Regulatory Commission to begin construction of phase 1 of a wave park consisting of up to ten power buoys, each rated for 866 kW of peak power. The wave park will be constructed 2.5 miles offshore near Reedsport, Oregon. This first phase of the project will produce up to 1.5 MW. After this first phase is completed, OPT plans to construct more buoys that will eventually produce up to 50 MW at the Reedsport location.

Another point absorber was called a wavebob, which is shown in Figure 11-25. The wavebob system was developed between 1999 and 2013, when unfortunately the company that developed it had to close its doors due to lack of funding. The wavebob could not compete economically with other energy sources without government support; however, the system had several important innovations and was able to produce power that was environmentally friendly. It is possible that future innovators will find a method to apply the ideas this company pioneered. For this reason, some of the innovative features of the wavebob are briefly discussed here.

Like the OPT system, the **wavebob** system uses a point absorber floating device that is composed of two structures: an upper structure called the torus, which is a donut shape, and a lower tank that is connected by a shaft to a central PTO unit inside the torus. The PTO tends to move up and down in sync with the motion of the lower tank, while the floating torus moves with the water surface. Relative motion between the inner PTO and the outer torus moves a fluid into an accumulator that released the fluid under constant pressure to a hydraulically driven motor that drives a generator to produce electrical power.

The wavebob had innovative features, such as the ability to change the mass of the lower tank automatically as conditions change by adding or purging water, thus forming a variable

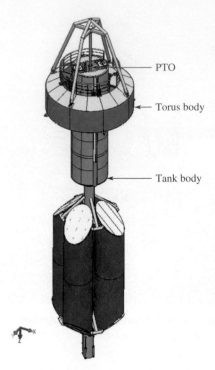

FIGURE 11-25   The Operating Parts of a Wavebob Power Generating System   (*Source:* Courtesy of Wavebob Ltd.)

mass ballast. In rough seas, a massive device could destroy itself; so in those conditions, the wavebob could purge the ballast, making itself lighter and able to withstand the rough conditions. This feature also allows the wavebob to tune its natural frequency to the waves and thus absorb the maximum amount of energy even as the seas change.

Figure 11-26 shows another smaller point absorber device called a floating wave energy device. This device has an upper part that moves up and down with respect to the lower part. As the upper part moves up and down, it pumps hydraulic fluid into an accumulator and then to a hydraulic motor that rotates a generator, just as in the wavebob. These units can be anchored in large arrays near a common cable to connect their output to the grid.

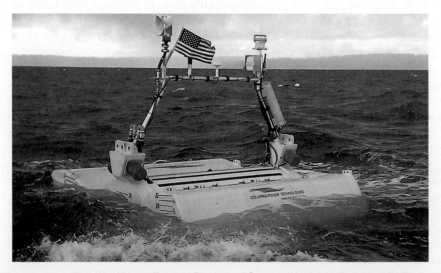

FIGURE 11-26   Floating Wave Energy Device   (*Source:* National Renewable Energy Laboratory.)

A point absorber that uses a different idea from offshore generation is from a US company called Atmocean. Working in conjunction with Sandia Labs, the Atmocean device uses the point absorber called Wave Energy Sequestration Technology (WEST) to drive pistons that pressurize hoses. The high-pressure seawater is brought to the shore, where it ultimately drives electrical generators. The offshore units are much lighter than other offshore systems, and the WEST device eliminates expensive undersea power cables and the inherent difficulties with electricity generated at sea. By using seawater as a fluid, an accidental break in a hose will have no environmental impact. The units can be deployed far from the shore to capture the energy from the largest waves.

The Atmocean technology comes with an expected positive benefit to the environment. The WEST system creates upwelling and thus enhances the production of phytoplankton, the base of the ocean food chain, which should help ultimately in increasing fish stocks. It also has a positive benefit in absorbing CO_2 from the increase in phytoplankton. Currently a number of units are being tested in sea trials, and the tests include data on various biological and environmental effects.

Attenuators

Attenuators are composed of large semisubmerged cylindrical sections joined together at hinge joints and placed parallel to the direction of the waves (perpendicular to the wave front). A passing wave causes each of the tubes to rise and fall independently, like the movement of a giant snake. The movement tugs at the universal joints linking the tubes. The joints use a hydraulic cylinder to act as a pumping system and push high-pressure oil through a series of hydraulic motors that drive electrical generators. The system is anchored to the seafloor, where electrical cables transfer electrical power to shore from the floating generator.

The **Pelamis wave energy system** is an attenuator system manufactured by Pelamis Wave Power Ltd. The latest version consists of five semisubmerged cylindrical sections; each section is 180 m long and 4 m in diameter, and weighs 1,350 tons (which is mostly sand ballast). Figure 11-27 shows a second-generation system installed in 2010 off the west coast of the Orkney Islands in the British Isles. Figure 11-28 shows a cutaway diagram of one of the cylindrical sections. The resistance in the hydraulic cylinders can be adjusted automatically to maximize power for a given wave height. Pelamis Wave Power

FIGURE 11-27 Second-Generation Pelamis Wave Energy System. This system is located at EMEC North Berth, 2 km west of the Orkney Islands in the British Isles. (*Source:* Courtesy of Pelamis Wave Power Ltd.)

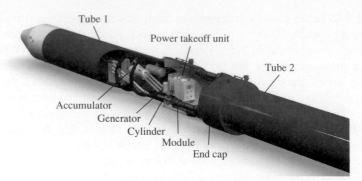

FIGURE 11-28   Cutaway Diagram of a Pelamis Wave Energy System   (*Source:* Courtesy of Pelamis Wave Power Ltd.)

systems have been actively tested since 2004, and they have been installed in wave farms located primarily around the British Isles. Some problems have occurred due to the constant attack from the sea, which has caused many delays. To date, the wave farms have delivered thousands of hours of power to the grid. The Orkney project is a 50 MW wave farm, and when this project is completed, it will have up to 66 Pelamis machines connected to the United Kingdom grid.

Terminating Devices

Terminating devices trap waves within the device and use the up-and-down motion of the water to extract energy. The **oscillating water column** is an example where captured water moves up and down in a tube, forcing air though an opening to turn a turbine. The turbine is connected to a generator that produces electrical power, similar in concept to a wind machine.

The tube for an oscillating water column is mounted at the edge of a cliff and extends below the water. Figure 11-29 shows the horizontal and the vertical sections of the tube, and the basic operation for this type of generator as waves flow in and out of the tube. Incoming waves push a large volume of air up and out of the chamber, as shown in Figure 11-29(a). As the wave recedes, as shown in Figure 11-29(b), the air rushes back into the tube and across the turbine. The turbine is a variable pitch device that turns in the same direction no matter which way the air moves.

The first commercial version of this type of system went into production in 2011 in Scotland. Previously, a demonstration project was constructed in 2000 in Italy; this early version has been connected to the grid ever since, collecting valuable operating information. Several projects are in the planning and development stages. Possible applications include using the power generated to desalinize seawater and to convert wave energy to hydrogen for fuel cells and other applications.

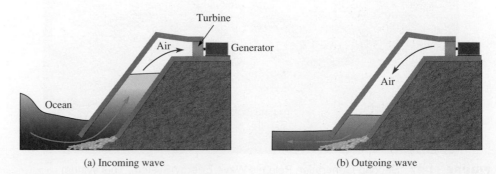

(a) Incoming wave   (b) Outgoing wave

FIGURE 11-29   Oscillating Water Column

Environmental Effects of Wave Power

Wave energy systems have not undergone extensive environmental studies, but several issues have been identified that may require further study, especially in environmentally sensitive areas. These issues include navigation hazard for small boats and shipping, as well as access to fishing grounds. Environmental disturbances to fish, marine life, and certain species of birds (such as the marbled murrelet, which dives for food) need to be assessed. There may also be a positive effect on birds by providing shelter and resting places. A very large floating system may affect migrating species like whales and may disrupt their feeding patterns. Shorelines and sediment deposition can be altered when energy is taken from waves. Pollution is possible if hydraulic fluid escapes; this scenario was made more real by a test wave power buoy that sank in 2012 off the Oregon coast. Recreational sites could also be disturbed.

Islay island, off the coast of Scotland is the site of one of the world's first oscillating water column generators, which was completed in the year 2000. The generator can generate 500 kilowatts of power using two wind turbines located at the top of an inclined concrete column. An inlet allows seawater to enter and leave the chamber, which creates a strong air flow at the top for the turbines. The power is distributed to 400 homes in the area and has a facility to recharge a local electric bus in overnight hours.

SECTION 11-5 CHECKUP

1. How does the measurement of water wave height differ from measuring sine waves?

2. With reference to wave energy, what is an attenuator?

3. What are the major differences between the OPT point absorber system and the WEST system?

4. How does an oscillating water column convert wave motion to electrical energy?

5. What are the concerns for wildlife in relation to wave energy systems?

CHAPTER SUMMARY

- Energy from falling water can be captured from hydroelectric dams, moving water in a stream, wave energy, and tidal energy.

- The amount of gravitational potential energy can be calculated from the formula $W_{PE} = mgh$.

- The amount of electrical power that can be produced by flowing water in a stream can be found if the volume of flowing water is known.

- Hydroelectric dams provide almost one-fifth of the world's electricity.

- Hydroelectric power is the cheapest electrical power.

- A conventional hydroelectric dam has spillways, a penstock with a control gate, a turbine, and a generator.

- Hydroelectric dams provide electrical power and also provide flood control for rivers.

- A run-of-the-river dam does not need a large reservoir like a conventional dam, so it is considered more environmentally friendly.

- A pumped storage system can provide peaking power.

- A turbine is a rotary engine that extracts energy from a moving fluid and converts it to useful work.

- An impulse turbine is a rotary engine that changes the direction of a high-velocity fluid, thus converting kinetic energy into mechanical rotating energy.

- A reaction turbine is a rotary engine that develops torque by reacting to the pressure of a fluid moving through the turbine, thus primarily converting potential energy into mechanical rotating energy.

- The tides in the oceans cause the level of the water to rise and fall. The tides are caused by gravitational interaction between the earth and the moon and the earth's orbital motion..

- When the sun, moon, and earth all align, they exert an exceptionally strong gravitational force, which causes very high and very low tides called spring tides.

- A tidal stream generator (TSG) uses a water turbine to turn a generator and produce electrical power when a stream of water caused by tides or a river flow past it.

- A tidal barrage power system consists of a large dam that stretches completely across an estuary, harbor, or river that connects to an ocean that has a tide.

- The top of a wave is called the crest, and the bottom of the wave is called the trough. The vertical distance between the crest and the trough is called the wave height. The wavelength is measured from the crest of one wave to the crest of the next wave.

- Wave energy converters can be classified as point absorbers, attenuators, or terminating devices.

- Point absorbers have a floating buoy that moves up and down with wave motion and drives a submerged assembly containing a hydraulic motor and generator to create electricity.
- The Pelamis wave energy system is an attenuating device that consists of five semisubmerged cylindrical sections. Relative

motion between the cylinders pumps hydraulic fluid to a hydraulic motor and generator to generate electricity.

- An oscillating water column consists of a tube or chamber mounted above the ocean that is driven by air pushed by rising and falling wave motion. The air motion drives a wind turbine.

KEY TERMS

attenuator   With respect to wave energy devices, a device that extracts energy from wave power by converting relative motion between large semisubmerged cylindrical sections to electricity.

Francis turbine   A reaction water turbine that directs water from the outer circumference toward the center of a runner. Water flows through a scroll case, which is a curved tube that diminishes in size similar to a snail shell.

impulse turbine   A rotary engine that changes the direction of a high-velocity fluid, thus converting kinetic energy into mechanical rotating energy.

Kaplan turbine   A reaction water turbine that uses propellers with adjustable blades. The turbine is usually placed in a spiral casing called a volute.

oscillating water column   A fixed device for producing electrical power from waves. It consists of a large tube that extends over a cliff and into the ocean. Wave action causes water to rise in the tube and displace air, which rotates a wind turbine.

Pelton turbine   An impulse turbine in which water moves under it (impulse) rather than water falling over it. It is among the most efficient types of water turbines. Also known as *Pelton wheel*.

point absorber   A floating wave energy converter that is in a fixed position. It bobs up and down from wave motion. The motion with

respect to a fixed reference is captured and the energy is converted to electricity.

pumped storage system   A system of two dams, each with a reservoir. One is located at a much higher elevation than the other. Water is released from the higher reservoir to produce electrical power. The water is captured in the lower reservoir and pumped back during off peak hours to the higher reservoir, where it is used again.

reaction turbine   A rotary engine that develops torque by reacting to the pressure of a fluid moving through the turbine, thus primarily converting potential energy into mechanical rotating energy.

run-of-the-river (ROR)   A hydroelectric system that uses river flow to generate electricity. The system may include a small dam with storage for water, but many do not.

tidal barrage system   A system designed to convert tidal power into electricity by trapping water behind a dam, called a tidal barrage dam, and generating power from the inflow and/or the release of water.

tidal stream generator (TSG)   An electrical generating system that uses a water turbine to turn a generator and produce electrical power when a stream of water caused by tides or a river flows past it.

FORMULAS

Equation 11-1	$W_{PE} = mgh$	Gravitational potential energy
Equation 11-2	$W_{KE} = \frac{1}{2}mv^2$	Amount of kinetic energy of motion
Equation 11-3	$h_{eq} = h + \frac{p}{\gamma} + \frac{v^2}{2g}$	Total equivalent head

Equation 11-4	$P = \gamma Q_v h_{eq}$	Amount of electrical power available in water
Equation 11-5	$W_{PE} = \frac{1}{2}(\rho Ah)gh = \frac{1}{2}\rho Agh^2$	Total available energy for water

CHAPTER TRUE/FALSE QUIZ

Determine whether each statement is true or false. Answers are at the end of the chapter.

1. Water stored behind a hydroelectric dam is an example of kinetic energy.

2. The penstock on a hydroelectric dam is a large tube that connects the reservoir to the turbine.

3. A run-of-the-river (ROR) hydroelectric dam is considered more environmentally friendly than a traditional hydroelectric dam.

4. A microhydroelectric dam produces more than 100 kW of electricity.

5. Water flowing in a stream is an example of kinetic energy.

6. A pumped storage system has two reservoirs and is used to provide electricity at peak times.

7. The Pelton turbine is an impulse turbine and has very high efficiencies.

8. The amount of electrical power that can be produced from falling water decreases as the height of the falling water increases.

9. The spillway on a hydroelectric dam allows excess water to be released from behind the dam without the water going through the penstock and turbine.

10. A high tide is caused by a bulge in the ocean that is produced by the gravitational pull of the moon and sun on the earth.

11. A spring tide occurs only in the spring season of the year, during the months of March, April, May, or June.

12. A tidal stream generator is anchored on the bottom of a river or stream and has a turbine that produces electrical power.

13. A tidal barrage power system consists of a large dam that stretches completely across an estuary, harbor, or river that connects to a body of water that has a tide.

14. The top of a wave is called the trough, and the bottom of the wave is called the crest.

15. A point absorber power generator is a fixed or terminating device.

16. Wave energy power systems have no effect on wildlife.

CHAPTER MULTIPLE-CHOICE QUIZ

Complete each statement by selecting the one correct answer. Answers are at the end of the chapter.

1. The Pelamis wave energy machine uses
   a. a point absorber system to produce electrical power
   b. hydraulic cylinders to pump oil through hydraulic motors to turn generators
   c. a barrage to produce electrical power
   d. an oscillating water column to produce electrical power

2. An oscillating water column uses
   a. a tube or chamber and waves that cause air to compress and flow past a turbine blade that rotates and turns a generator
   b. water from a reservoir to flow through a penstock and turn a turbine and generator
   c. hydraulic cylinders to pump oil through hydraulic motors to turn generators
   d. a barrage to produce electrical power

3. A buoy connected to a fixed assembly to generate power is an example of a(n)
   a. barrage
   b. attenuating system
   c. oscillating water column
   d. point absorber

4. A tidal stream generator uses
   a. an oscillating water column to produce electrical power
   b. a water turbine that is anchored on the bottom of a river or stream to turn a generator
   c. hydraulic cylinders to pump oil through hydraulic motors to turn generators
   d. water that flows through a penstock to produce electrical power

5. Water stored behind a hydroelectric dam is an example of
   a. kinetic energy
   b. potential energy
   c. electrical energy
   d. all of these

6. A pumped storage system has
   a. a single reservoir that allows water to flow through a penstock to turn a turbine
   b. two reservoirs; water is pumped from the lower one to the higher one during low usage times
   c. two reservoirs; power is generated when pumped from the lower one to the higher one
   d. a single reservoir that has its water pumped past a turbine, which turns a generator

7. A run-of-the-river (ROR) hydroelectric dam
   a. is a form of pumped storage system
   b. produces the same electrical power at all times of the year
   c. can have a small storage area called pondage to store water behind the dam
   d. all of these

8. A microhydroelectric system produces
   a. less than 100 kW of electrical power
   b. less electrical power than a picohydroelectric system
   c. more electrical power than a traditional hydroelectric dam
   d. more than 10 MW of electrical power

9. A traditional hydroelectric dam
   a. does not have a penstock
   b. uses the spillway to bring water to the turbines
   c. stores water in a reservoir to be used as needed for power production
   d. includes a pumped storage reservoir

10. The power from a hydroelectric dam is related to the
    a. volumetric flow rate
    b. area of the reservoir
    c. both (a) and (b)
    d. none of these

11. The unit for torque in the SI system is the
    a. kilogram-meter
    b. kilogram-meter/second
    c. newton
    d. newton-meter

12. An example of a reaction turbine is the
    a. Pelton
    b. Turgo
    c. crossflow
    d. Francis

13. A reaction turbine is one that develops torque by reacting to the
    a. velocity of a fluid
    b. nozzle size used in the turbine
    c. pressure of a fluid
    d. mass of a fluid

14. A good turbine for cases when there is a high head but relatively low volume would be a
    a. Pelton
    b. Kaplan
    c. crossflow
    d. Francis

15. The potential energy in a tidal barrage is given by the equation $W_{PE} = \dfrac{1}{2} \rho A g h^2$. In this equation, A stands for the area of the

    a. inlet pipe for the turbine(s)
    b. outlet pipe for the turbine(s)
    c. barrage reservoir
    d. dam

16. A crossflow tidal turbine can generate power from
    a. incoming tides only
    b. outgoing tides only
    c. either incoming or outgoing tides
    d. neap tides only

CHAPTER QUESTIONS AND PROBLEMS

1. What is the amount of power a hydroelectric dam can produce if the height of the water is 50 m, the flow rate is 3,000 m/s, and the coefficient of efficiency is 0.7?

2. List the basic parts of a traditional hydroelectric dam and explain how it produces electrical power.

3. Explain why a run-of -the-river (ROR) is considered more environmentally friendly than a traditional hydroelectric dam with a large reservoir.

4. List the basic parts of a tidal barrage power system and explain how it produces electrical power.

5. List the basic parts of a point absorber power generating system and explain how it produces electrical power.

6. What is the amount of power available in falling water in watts (W) where the water falls 5.5 m, the flow is 0.30 m³/s, the density of the water is 1,000 kg/m³, and the gravitational constant 9.8 m/s²?

7. Find the energy in joules of the mass of 175 million kg of water at an average height of 8 m.

8. List the basic parts of a tidal stream generator and explain how it produces electrical power.

9. Explain the difference between a microhydroelectric system and a picohydroelectric system.

10. List the basic parts of a pumped storage system and explain how it produces electrical power.

11. a. What is the energy available in the reservoir in Example 11-1 if the level has dropped so that the average height is now 5 m?
    b. Assume the reservoir in Example 11-1 can now be drained in five days. What is the average power in the water as it is released? Ignore friction losses.

FOR DISCUSSION

The building of the Hetch Hetchy dam in a beautiful valley in the Sierra Nevada Mountains had foes from the outset (including John Muir), but it was constructed to provide water to the city of San Francisco, which had been devastated by the 1907 earthquake.

Today, there is a move to remove the dam (although when put to a vote in December 2012, it was defeated by voters in San Francisco.) How do you feel about removing this dam? What problems need to be addressed if it were removed?

ANSWERS TO CHECKUPS

Section 11-1 Checkup

1. The water in a storage reservoir behind a hydroelectric dam represents potential energy.

2. The flow of water through a turbine represents kinetic energy.

3. The head and the kinetic energy of the stream

4. It is doubled.

5. The area of the reservoir and the depth of the water

Section 11-2 Checkup

1. Water is released from a reservoir through a penstock to a turbine. The turbine is connected to a generator that converts the kinetic energy in the water to electrical energy.

2. A pumped storage system converts the potential energy in an upper reservoir to electricity as it falls. It can pump the water back for reuse during peak demand, but it always loses some energy in the process of pumping.

3. A run-of-the-river system may have storage but it primarily uses the kinetic energy of the river to generate electricity. A conventional hydroelectric dam stores the energy in the lake until it is released.

4. Pumped storage transfers water back and forth between an upper and lower reservoir in order to meet peak demands. Pumping water always requires using more energy than can later be obtained.

Section 11-3 Checkup

1. An impulse turbine has water moving with high kinetic energy through a nozzle aimed at turbine blades to cause them to rotate.

2. A reaction turbine creates torque by reacting to the pressure of a fluid moving through the turbine, thus converting potential energy to kinetic energy.

3. Three reaction turbines are (1) Fourneyron, (2) Francis, and (3) Kaplan.

4. Three impulse turbines are (1) Pelton, (2) Turgo, and (3) crossflow.

Section 11-4 Checkup

1. The moon provides most of the gravitational pull that creates the tide closest to the moon. The tide on the opposite side is due to the orbital motion of the earth.

2. Spring tides are larger than average because of the alignment of the sun and moon with the earth and they occur approximately twice each month.

3. A tidal barrage system includes a dam that traps water from changing tides and generates power from the inflow and/or release of water.

4. The potential energy is the mass of the water behind the dam multiplied by its average height.

5. A crossflow turbine can generate power from incoming or outgoing tides.

Section 11-5 Checkup

1. Sine waves are normally measured from center to peak (amplitude); water waves are measured from trough to crest.

2. A device that extracts energy from wave power by converting relative motion between large semisubmerged cylindrical sections to electricity

3. The OPT system uses a submerged fixed assembly that captures wave energy from the relative motion between the two parts, and converts this motion to electricity at sea, where it is transmitted on undersea electrical cables. The WEST system transmits high pressure sea water to shore and generates electricity from shore based generators.

4. An oscillating water column captures the up-and-down motion of waves in a tube, which forces air in and out of the tube. The air motion is converted to energy by a wind turbine.

5. Wave machines can affect the migration patterns of whales and other migrating species, and they may affect birds that dive for fish. Pollution is possible if hydraulic oil escapes into the ocean.

ANSWERS TO TRUE/FALSE QUIZ

1. F 2. T 3. T 4. F 5. T 6. T 7. T 8. F 9. T   10. T 11. F 12. T 13. T 14. F 15. F 16. F

ANSWERS TO MULTIPLE-CHOICE QUESTIONS

1. b 2. a 3. d 4. b 5. b 6. b 7. c 8. a 9. c   10. a 11. d 12. d 13. c 14. a 15. c 16. c

Fuel Cells

CHAPTER OUTLINE

CHAPTER OBJECTIVES

- Identify the three basic parts of a hydrogen fuel cell.
- Explain the operation of a hydrogen fuel cell and how it produces electrical current.
- Describe the reforming process to produce hydrogen fuel.
- Write the chemical reaction for a hydrogen fuel cell.
- Explain why fuel cell stacks are required.
- Identify six types of fuel cells.
- Identify the major parts of a fuel cell electric vehicle.
- Explain how a fuel cell can be powered by gasoline, methane, or other type of hydrocarbon fuel.
- Identify the major parts of a stationary fuel cell.
- Describe how a stationary fuel cell provides electrical power and heat to power an absorption air-conditioning system.

KEY TERMS

Key terms are shown in bold and color. Definitions for key terms are provided at the end of the chapter and in the end-of-book glossary. Bold terms in black are defined in the end-of-book glossary only.

- fuel cell
- electrolysis
- membrane electrode assembly
- reformer
- fuel cell stack
- fuel processing unit
- heat recovery system
- proton exchange membrane fuel cell (PEMFC)
- direct-methanol fuel cell (DMFC)
- alkaline fuel cell (AFC)
- phosphoric acid fuel cell (PAFC)
- solid oxide fuel cell (SOFC)
- molten-carbonate fuel cell (MCFC)

INTRODUCTION

Fuel cells oxidize a fuel (usually hydrogen) and produce electricity, water, and heat as by-products. Fuel cells using hydrogen as a fuel need to obtain it either in gaseous form or from a compound such as methanol. Hydrogen is not available in its natural state but must be generated from a source such as a fossil fuel or water. Any electrical source can be used to produce hydrogen, including renewable sources such as wind, solar, hydropower, or biomass.

Once formed, hydrogen is an explosive gas that turns into a liquid only at cryogenic temperatures and tends to be corrosive. For these reasons, it is difficult to store but easier to transport in pipelines. Despite the storage difficulties, hydrogen offers some attractive advantages, including reducing dependence on oil for transportation and for distributed power applications (which is power generated close to the point of use and is therefore more efficient). For distributed power applications, fuel cells can provide process heat or heat that can be used for other purposes such as space heating. Fuel cells are also an excellent way to provide power in conjunction with intermittent sources such as solar or wind.

A single fuel cell produces only a small voltage, so fuel cells must be connected in series stacks to provide a usable voltage. The stacks can be connected in parallel to increase the current capability. Fuel cells are used to power vehicles, including buses, forklifts, and cars, as well as some stationary power systems. Fuel cell power systems can be connected to the grid, or they may provide enough electricity to power a building or an industry that is remote or is off the grid. In this chapter, fuel cell operation, types of cells and applications are discussed.

12-1   Basic Fuel Cell Operation

Fuel cells combine a fuel (usually hydrogen in some form) with an oxidizing agent (usually oxygen). In the hydrogen fuel cell, hydrogen and oxygen react to form water as by-product. Electrical current is produced when electrons are freed during the process, which is clean, quiet, and more efficient than burning fuels.

A **fuel cell** is a device that converts electrochemical energy into dc, much like a battery. One difference is that a battery stores its chemicals inside; a fuel cell has a constant flow of fuel into the system from an outside source. Early major users of fuel cells were NASA and the military because of their very specialized requirements and because the high cost of manufacture was not the main issue. Fuel cells were first used by NASA in 1962 in the *Gemini* space program, a manned space program. Fuel cells replaced battery power as a power source on the shorter flights of the *Mercury* space program, which preceded *Gemini*. Improved alkaline fuel cells were used for the longer flights to the moon on the *Apollo* missions, and later on the space shuttle. NASA went on to fund 200 research contracts for fuel cell technology. Today, renewable energy systems are able to take advantage of this research.

Today fuel cells are used to produce electrical power for newer spacecraft; remote undersea stations; and mobile vehicles such as automobiles, trucks, buses, forklifts, and tractors. Some larger fuel cells provide power to buildings or small cities as a stationary electrical plant. These units are highly reliable and can bring power closer to the end user, and thus save distribution costs.

Oxidation-Reduction Reactions

Like all physical systems, chemical reactions tend to go from a higher energy state to a lower state (energy decreases). Hydrogen gas, when in the presence of oxygen, can react explosively, producing light and heat after the addition of a small amount of activation energy (heat from a flame). The reaction for burning hydrogen is

$$2\,H_2(g) + O_2(g) \longleftrightarrow 2\,H_2O(l) + \text{energy}$$

This chemical equation shows that two molecules of hydrogen can react with one molecule of oxygen to produce two molecules of water plus a net change in energy (called free energy). The process involves breaking the chemical bonds in the gases (H_2 and O_2), which absorbs energy. New bonds are formed in the water molecule, which releases energy, and the system becomes stable at a lower energy. The free energy of the system has decreased and has appeared as heat and light from the reaction. Not all of the energy in the reaction is available for work—some goes into an increase in entropy, a quantity that measures randomness of disorder in a system. This energy is not available for work, so there is an automatic limit on the efficiency of a fuel cell.

The chemical reaction for the combustion of hydrogen is called an oxidation reaction (in this case, combining hydrogen with oxygen). The process involves transferring electrons from hydrogen to oxygen; hydrogen loses electrons and oxygen gains electrons. **Oxidation** is the process in which one reactant loses one or more electrons, and **reduction** is the process in which a reactant gains one or more electrons. The two processes always occur together, so the type of reaction is referred to as a redox (from *red*uction *ox*idation) reaction.

The reaction described for oxidation of hydrogen is one of the most common reactions used for fuel cells. In a fuel cell, the free energy is turned into electricity and heat. The reactions vary from one cell to another, but all fuel cells use a redox reaction in which electrons are separated from a fuel (oxidation), travel in an external circuit, and combine (reduction) with oxygen to complete the process. Like fuel cells, batteries also use redox reactions in which electrons travel in the external circuit to complete the reaction. The main difference is that, in a fuel cell, a constant supply of fuel is supplied, whereas in a battery, the reactants are stored within the battery.

You may have noticed that the arrow in the equation for oxidation of hydrogen is double-headed, implying that the reaction is reversible. To separate the resulting water molecules

Sir William Grove, a British scientist who is often referred to as the father of the fuel cell, performed research in 1839 on reversing the electrolysis of water. Groves found that it was possible to free electrons in the process, and the free electrons could power a small electrical load. In 1889, Charles Langer and Ludwig Mond completed further research in Britain and coined the term *fuel cell*. In 1932, Francis Bacon, at Cambridge University in England, developed what was perhaps the first successful fuel cell device, with a hydrogen-oxygen cell using alkaline electrolytes and nickel electrodes.

back into hydrogen gas and oxygen gas, energy must be added. **Electrolysis** is the breaking down of a substance that contains ions by applying an electric current between two electrodes and separating the substance into its components. The electrolysis of liquid water breaks it down into hydrogen and oxygen gases, which is one way to obtain pure hydrogen fuel for use in a fuel cell.

Although hydrogen is the most common element in the universe, it is not available in its elemental form on earth and must be generated from some other source. For this reason, it is not considered to be a primary source. It is important to understand that the energy to produce hydrogen is never fully recovered. If an electric power plant that burns a fossil fuel is used to create the hydrogen used in a fuel cell, the net effect is more steps in the process, and each step loses a little of the available energy.

Characteristics of Fuel Cells

Figure 12-1 shows a simplified diagram of a typical fuel cell, which has three basic elements: (1) an **anode** that separates the hydrogen fuel into positive ions and electrons; (2) a **cathode** that forms oxygen ions from oxygen molecules and electrons, and combines negative oxygen ions and the positive hydrogen ions to form water; and 3) an **electrolyte**, which is a conductive substance that passes hydrogen ions from the anode to the cathode. The type of material for the electrolyte differs in various types of fuel cells, but it must meet certain requirements, such as mechanical strength and resistance to impurities, as well as being a good conductor.

Within the cell, hydrogen ions migrate across a barrier and combine at the cathode to complete the reaction. The anode and cathode provide the electrical connection to the external electrical circuit. The electrolyte material is sandwiched between the anode and cathode. Together, the components in a fuel cell, which includes the anode and cathode electrodes, the electrolyte, and the catalyst, form an assembly called the **membrane electrode assembly**, where the chemical reactions occur.

The anode is made of porous material such as carbon and is coated by platinum. It must be capable of allowing hydrogen ions to pass while discriminating against electrons and hydrogen molecules. The catalyst in the cathode breaks apart oxygen molecules, forming two oxygen ions. The cathode is made of porous material that is coated by a material such as platinum or nickel and allows the hydrogen ions to combine with the oxygen ions to form water molecules (H_2O), which are eliminated from the cell as a by-product. Both the anode and cathode are constructed with a catalyst. (Recall that a catalyst was introduced in Chapter 9 as a chemical that helps speed up the reaction but emerges from the reaction unchanged.) This means that both the anode and cathode must be able to facilitate the reactions within the fuel cell; ideally they should have high electrical conductivity as well. The best materials for the electrodes depend on the type of fuel cell.

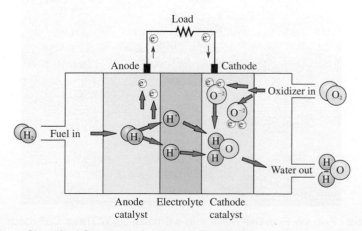

FIGURE 12-1 Simplified Structure of a Typical Fuel Cell

Water is removed from the fuel cell at the bottom right side in Figure 12-1. Some fuel cells also create extreme heat during the process, so the heat can be used to heat buildings or to create steam, which can be used to power a steam turbine to make electrical power. By controlling the means by which such a reaction occurs and directing the reaction through a heat exchanger, it is possible to harvest the heat energy.

Some fuel cells must operate on pure hydrogen and pure oxygen, while others have been designed to operate on hydrocarbon fuels such as methane, butane, natural gas, coal, or gasoline. These fuel cells use a **reformer**, which separates the hydrogen from the hydrocarbon fuel. Hydrocarbons are especially useful for fuel cells in vehicles because transporting pure hydrogen is expensive and dangerous.

Fuel Cell Systems

The fuel cell itself is but one part of the overall fuel cell system. Fuel cell systems are used for applications such as stationary power units and for transportation, that is, electric vehicles. A fuel cell system has three basic parts: the fuel cell stack; the fuel processing unit; and a heat recovery system that processes the excess heat that is a by-product of the fuel cell operation. Systems that have ac output for the grid have a standard electrical inverter to change the dc output to ac. Some systems also have a water containment subsystem.

Fuel Cell Stack

The **fuel cell stack** consists of a group of fuel cells that are connected and bound together to provide increased electrical power. The output voltage of a single fuel cell is very small (about 0.7 V), so fuel cells are connected (stacked) in series to increase the voltage and in parallel to increase the amount of current they can provide. The stack is typically rated by the total wattage it can provide. The exact number of cells and how they are connected depends on the current, voltage, and power requirement of the load. The system includes an inverter if the load requires ac. Figure 12-2 shows a 5 kW fuel cell stack and a physically smaller 25 W fuel cell stack.

Fuel Processing Unit

The **fuel processing unit** is the portion of a fuel cell system that converts the input fuel into a form usable by the fuel cell. If hydrogen is fed to the system, a processor may not be required, or it may be needed only to filter impurities out of the hydrogen gas. In solar energy systems, the system may be used to prepare pure hydrogen and oxygen from water; in which case, the fuel processor is not needed. Larger fuel cell systems frequently use methane or other hydrocarbon to produce electrical power, so the fuel processor is required to extract the hydrogen from the hydrocarbon.

FIGURE 12-2   Five kW Fuel Cell Stack with a Smaller 25 W Three-Cell Stack (*Source:* National Renewable Energy Laboratory.)

Heat Recovery System

Fuel cells typically operate at high temperatures (up to 600° C to 700° C), so they produce heat as a by-product. The **heat recovery system** collects excess heat for another use, which increases the overall energy efficiency of the fuel cell system. The excess heat can be used for hot water and/or steam generation. Steam can be used to drive a turbine and generator to produce electricity. Hot water can be used to provide heat for buildings or power an absorption air-conditioning system.

EXAMPLE 12-1

How many 25 watt fuel cell stacks are needed to produce 5 kW?

Solution

The total number of 25 watt fuel cell stacks needed to produce 5 kW: 5 kW/0.025 kW = **200 fuel cell stacks**.

Regenerative Fuel Cells

Regenerative fuel cells produce electricity from hydrogen and oxygen, and they generate heat and water as by-products, just like other fuel cells. The regenerative fuel cell systems also receive extra hydrogen and oxygen from water that is processed from electrical power produced by a solar panel or wind turbine. Electricity from a solar cell or wind generator is used to recycle water from the fuel cell into hydrogen and oxygen through electrolysis. This process is unique because all the energy is produced through renewable energy systems. It has been used by NASA on space flights that get electrical power from solar panels as well as fuel cells. In this case, the hydrogen and oxygen can be stored for use in the fuel cell when solar energy is not available.

Obtaining Hydrogen

Hydrogen is the basic fuel for most fuel cells. As mentioned previously, hydrogen is not available naturally and must be produced from another source, so it is not consdered to be a primary source. Electrolysis of water has been mentioned as a useful method for obtaining hydrogen from any source of electricity, including renewable sources. Electrolysis can produce very pure hydrogen, but it is simply the reverse fuel cell reaction, so it requires a source of electric power and always requires more energy to produce the fuel than can be obtained from the fuel cell. If the source of electricity for producing the fuel is a renewable source, such as solar or wind, the hydrogen may serve as an energy storage mechanism, available whenever needed and without depleting nonrenewable sources.

Another method used to produce hydrogen is a reaction called steam reformation in which methane (CH_4) and steam are passed over a catalyst to produce carbon monoxide (CO) and hydrogen (H_2). The thermal or catalytic process of breaking down a large molecule into smaller ones is called **reforming** (such as what happens when gasoline is separated from oil). The carbon monoxide can be burned to create additional hydrogen and carbon dioxide. The chemical reactions involved in steam reformation are:

$$CH_4(g) + H_2O(g) \rightarrow CO(g) + 3H_2(g)$$
$$CO(g) + H_2O(g) \rightarrow CO_2(g) + H_2(g)$$

A problem with steam reformation is that the resulting hydrogen is not as pure as with electrolysis, so is not suitable for all types of fuel cells. Also, using a fossil fuel such as methanol, natural gas, propane, gasoline, coal gas, or even coal powder to obtain hydrogen is not carbon-neutral except in the case of biofuels, so this may not be the most desirable way to obtain hydrogen.

No matter how it is obtained, the hydrogen normally is stored as a cryogenic liquid. Hydrogen liquifies, boils, and condenses at −252.5° C (−422° F) at atmospheric pressure.

Reforming Natural Gas to Produce Hydrogen

(*Source:* National Renewable Energy Laboratory.)

Natural gas is a hydrocarbon that contains hydrogen. The steam reformer is used to externally separate hydrogen from natural gas. This operation allows some types of fuel cells to operate on natural gas, which is easier to store and transport. Natural gas is also available through pipe lines to many industrial sites where it is used for heating; it can be used to supply the reformer to provide hydrogen for fuel cells.

If hydrogen is stored under pressure, its temperature can be higher before it changes from a liquid to a vapor. Generally, hydrogen is transported in vehicles as a liquid at cryogenic temperatures, which is one of the problems associated with its use. Converting gaseous hydrogen to a liquid requires a signifcant amount of energy. Hydrogen is highly flammable and any accident involving the exposure of liquid hydrogen to the environment means immediate evaporation into a gaseous state, which can be explosive. Thus, stringent safety measures need to be in place for transporting liquid hydrogen and detecting any leaks. Hydrogen gas must be under very high pressure, but gaseous transport by pipeline offers advantages over other forms of moving it. No matter how it is stored, hydrogen can enter the microstructure of many metals, so special tanks and pipelines need to be constructed. Newer hydrogen storage tanks are reinforced with composite carbon fibers that allow hydrogen to be stored at high pressure. To reduce costs, most hydrogen is prepared close to the place where it will be used.

Coal is one of the least expensive fuels and it is available in most of the world. Coal can be gasified so that hydrogen is separated from it and used to power fuel cells, making it a cleaner way to use coal than burning it. Certain types of fuel cells have been using hydrocarbon fuels to produce hydrogen for many years. These fuel cells have had problems because they create carbon deposits on the anode, which causes the fuel cell to shut down in less than a half hour.

A problem with preparing hydrogen from fossil fuels is that the reformer creates a small amount of pollutants, and the hydrogen is not as pure and is not suitable for some types of fuel cells. Still, the level of pollution from reforming is much smaller than when the fuel is burned. Some reformers are designed to operate directly with a fuel cell. In these systems, the hydrocarbon fuel is processed through the reformer immediately prior to being used by the fuel cell. The reformer makes it easy to use the hydrocarbon fuel, but it does increase the cost and the weight of the fuel cell system.

Researchers at Georgia Tech have devised a vapor-deposition technique for growing nanostructures from barium oxide nanoparticles on the fuel cell's anode. The nanostructures start a water-based chemical reaction that oxidizes carbon deposits as they form. This system has been so successful that it allows the fuel cell to run on coal gas–powering the fuel cell for up to 100 hours, with no signs of carbon deposits.

Hydrogen Production Using Electricity from Wind or Solar

Hydrogen for fuel cells can be produced easily by using electricity to separate hydrogen from water or other materials. The biggest drawback to this process is that, in many cases, the electricity used for this process was produced by burning hydrocarbon fuels such as coal or natural gas. It is possible to use electricity produced by wind turbines or solar panels to make hydrogen. This process is considered a renewable form of energy because the electrical power to isolate the hydrogen comes from a renewable energy system, and it may help by providing a way to store the energy from solar or wind for use as needed.

Carbon Nanotube as a Catalyst in Fuel Cells

Nanotechnology is being used in several ways with fuel cells. One way is to reduce the amount of expensive platinum that is used as the catalyst in certain types of fuel cells. Nanotechnology allows platinum nanoparticles to be produced. The nanoparticles are used to provide a thinner coating on the anode, which reduces the amount of platinum in the fuel cell and reduces the cost of production. Nanotechnology is also used to create lighter, more efficient fuel cell membranes. Researchers at the University of Dayton in Ohio found that an array of carbon nanotubes could be used as a catalyst in some fuel cells. This new material makes the fuel cell more durable, and the material can be made less expensively than more traditional materials.

Advantages and Disadvantages of Fuel Cells

Fuel cells are very reliable. Their advantages for producing electricity, particularly in remote locations, include no moving parts, quiet operation, and heat as a by-product. They also produce clean water as a by-product. A disadvantage of some fuel cells is that they use expensive platinum catalysts. Platinum is expensive because it is in high demand for many other applications. Fuel cells that use hydrogen gas as a fuel also have an issue with storing the hydrogen gas safely. As we noted, hydrogen gas must be stored under high pressure.

SECTION 12-1 CHECKUP

1. List the three basic parts of a fuel cell. Explain what each part does.

2. What is the difference between oxidation and reduction?

3. Explain how a fuel cell produces electrical current.

4. Explain what a reformer does.

5. What are the benefits of carbon nanotubes for certain fuel cells?

12-2 Types of Fuel Cells

As you know, most fuel cells use hydrogen as a basic fuel—either directly or by reforming a fossil fuel. There are five important types of fuel cells, and each has a different chemistry. Some types of fuel cells work well for use in stationary power plants. Others may be useful for small portable applications, others for vehicle applications; they are particularly suited for applications in which pollution free vehicles are needed. In this section, the five most important types of fuel cells are discussed.

Various types of fuel cells differ primarily by the electrolyte and catalytic electrode used. Selection for a particular application depends on criteria such as cost, efficiency, weight, fuel storage, and so forth. The primary types of fuel cells discussed in this section are the proton exchange membrane fuel cell (PEMFC), direct-methanol fuel cell (DMFC), alkaline fuel cell (AFC), phosphoric acid fuel cell (PAFC), solid oxide fuel cell (SOFC), and molten-carbonate fuel cell (MCFC).

Proton Exchange Membrane Fuel Cell

The **proton exchange membrane fuel cell (PEMFC)** is also called a polymer electrolyte membrane fuel cell. A polymer is composed of molecules with high molecular weight that can be formed by the addition of many smaller molecules. The PEMFC uses hydrogen fuel and oxygen (obtained from air) as an input and produces pure water and electricity. It has a high power density and high operating temperature of 50° C to 100° C (122° F to 212° F) and the electrolyte is a polymer membrane (hence the alternate name). It has a wide range of efficiency (25% to 58%). It warms up and begins generating electricity quickly, which makes it very useful in transportation vehicle applications. General Electric developed the first proton exchange membrane fuel cells for the *Gemini* space missions in the early 1960s. PEMFCs were replaced by alkaline fuel cells in the *Apollo* program and in the space shuttle program.

Prior to the PEMFC, fuel cells were used only in special applications where remote electrical power was needed. These early fuel cells were solid oxide types that required very expensive materials and could be used only for stationary applications due to their size. Their operation also required large budgets. When the PEMFC was first designed, it used sulfonated polystyrene membranes for electrolytes, but later they were replaced by solid organic polymer polyperfluorosulphonic acid, which has proved superior in performance and durability.

Figure 12-3 shows a diagram of a PEMFC. The electrolyte in the PEMFCs is a solid polymer. The electrodes use porous carbon that contains a platinum catalyst. The fuel cell uses either pure hydrogen from storage tanks or from reforming a hydrocarbon fuel with an onboard reformer. In Figure 12-3, hydrogen gas is input to the cell, where it reacts with the anode, which in turn releases electrons as current. The cell needs oxygen, but it can be

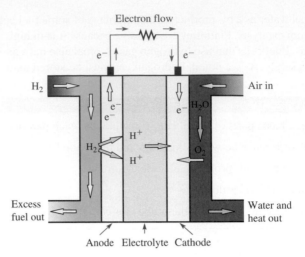

FIGURE 12-3 Operation of the Proton Exchange Membrane Fuel Cell (PEMFC)

extracted from air and it does not require corrosive fluids. Oxygen from the air reacts with the remaining hydrogen ions to produce water (water and heat are by-products).

The PEMFC is light weight compared to other fuel cells and produces sufficient power to be useful for providing power for vehicles. The PEMFC is also used for stationary applications. One disadvantage is that, if onboard hydrogen is used, it must be stored as a compressed gas in pressurized tanks. In this form, it is not possible to store enough hydrogen to travel over 300 miles in a vehicle, so this technology cannot compete with traditional gasoline- or diesel-fueled vehicles at this time. If onboard reformers are used, higher-density liquid fuels, such as methanol, ethanol, natural gas, liquefied petroleum gas, and gasoline, can be used to provide the hydrogen. The onboard reformer is expensive and has additional costs, including maintenance; however, it is used in a number of applications.

Direct-Methanol Fuel Cell

The **direct-methanol fuel cell (DMFC)** is a subcategory of a proton exchange membrane fuel cell. This fuel cell was invented by college researchers in cooperation with the Jet Propulsion Laboratory (JPL) in 1990. Figure 12-4 shows a diagram of the DMFC. It uses a polymer electrolyte similar to the one in the PEMFC. Methanol is used as the fuel, and

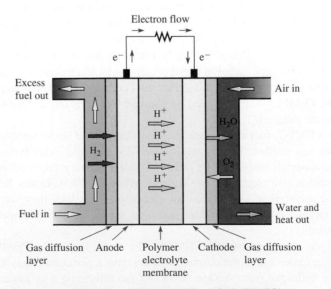

FIGURE 12-4 Operation of the Direct Methanol Fuel Cell (DMFC)

hydrogen is separated by a steam reformer. Methanol and water are supplied to the fuel cell, as shown at the bottom left of the figure, and they react with the anode. Hydrogen is extracted from the methanol, and electrons from the hydrogen are free to flow as current out from the anode and return back through the cathode. Oxygen is supplied through the air and is ionized at the cathode. Oxygen ions combine with hydrogen ions to make water. The operating temperature range of this fuel cell is about 50° C to 120° C (122° F to 248° F). Typical units have a power rating between 25 W and 5,000 W.

The direct methanol fuel cell can be used in vehicles because it uses methanol in liquid form to provide the hydrogen source. Lower operating temperatures mean that this fuel cell does not need a large, heavy heat shield. A disadvantage is that the efficiency is low, so the DMFC is more suited for portable applications, where energy and power density are more important than efficiency. Another disadvantage is that carbon dioxide is a by-product and it is released into the atmosphere, just as it would be from the combustion of methanol.

Alkaline Fuel Cell

The **alkaline fuel cell (AFC)** is a very efficient fuel cell that requires pure hydrogen fuel and pure oxygen. It uses an aqueous (water-based) electrolyte solution of potassium hydroxide (KOH) in a porous stabilized matrix. AFCs have been used in the US space program since the 1960s to produce electricity and fresh water for spacecraft. The *Apollo* missions and the space shuttle used alkaline fuel cells to produce electrical power and fresh drinking water. The AFC is very susceptible to contamination, so it requires pure hydrogen and oxygen, which makes this fuel cell expensive to operate and limits its application.

Figure 12-5 shows a diagram of the AFC. Pure hydrogen is supplied to the fuel cell at the anode, and pure oxygen is supplied at the cathode. The anode and cathode are made of lower-cost, nonprecious metals such as nickel. The electrolyte is a solution of potassium hydroxide (KOH) in water, which ionizes to form potassium ions (K^+) and hydroxyl (OH^-) ions. In the AFC, hydrogen gas is oxidized to hydrogen ions and combines with the hydroxide ions, which produces water (H_2O) and releases two electrons. The electrons flow through the external circuit and return to the cathode, where they reduce oxygen to form more hydroxide ions and water. The AFC operates at temperatures of 100° C to 250° C (212° F to 482° F). More recent AFCs operate at temperatures of 23° C to 70° C (74° F to 158° F). This fuel cell has an operating efficiency of 60% to 70%. Excess heat is removed from the fuel cell as a by-product, and it is hot enough to provide steam to power a steam turbine. The heat can also be used to heat buildings.

The AFC fuel cell works well in spacecraft or in undersea locations where the atmosphere can be controlled. A disadvantage is the requirement for pure oxygen and pure hydrogen, and both gases must be supplied continuously. The fuel cell is also poisoned easily by

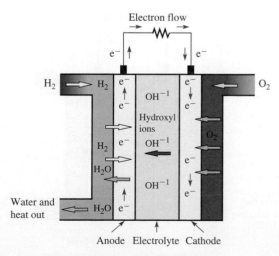

FIGURE 12-5 Operation of the Alkaline Fuel Cell (AFC)

carbon dioxide (CO_2), which affects the cell's lifetime. AFCs do not currently have lifetimes beyond about 8,000 operating hours, so they tend to be less cost-effective than other types.

Phosphoric Acid Fuel Cell

The **phosphoric acid fuel cell (PAFC)** is equivalent in structure to the proton exchange membrane fuel cell (PEMFC), but it has liquid phosphoric acid as the electrolyte. The electrolyte is contained in a Teflon-bonded, silicon carbide matrix. It uses an external reformer to separate hydrogen from a hydrocarbon fuel. PAFCs are typically used in small, stationary, power generation systems, but research is being conducted for application in larger vehicles such as buses.

The PAFC operates at around 150° C to 200° C (300° F to 400° F). This operating temperature is hot enough to provide external heat as well as electricity. If gasoline or diesel is used as a basic fuel, sulfur must be removed from the fuel prior to use or it will damage the electrode catalyst. PAFCs are more tolerant of impurities in fossil fuels that have been reformed into hydrogen than are PEMFCs. The electrical efficiency for PAFCs is 40% to 50%, and when energy produced by the waste heat is considered, the efficiency rises to about 80%. Today, PAFCs are used in commercial electrical production.

The PAFC can tolerate a concentration of carbon monoxide (CO) of about 1.5%, which is a larger concentration than can be tolerated by other types of fuel cells. Another advantage to the PAFC is that the phosphoric acid electrolyte can operate above the boiling point of water. A disadvantage of the PAFC is that, when compared to other fuel cells of similar weight and volume, it produces less power. Also, PAFCs are expensive because of the platinum catalyst and the need for corrosion-resistant materials (because of the acid).

Solid Oxide Fuel Cells

The **solid oxide fuel cell (SOFC)** is named for the solid oxide electrolyte used. The solid oxide electrolyte consists of yttria-stabilized zirconia, a zirconium oxide–based ceramic. Figure 12-6 shows a diagram of a standard solid oxide fuel cell. Hydrogen fuel (H_2) is applied to the anode, which is oxidized to hydrogen ions. The electrons are given off to an external circuit as electrical current. (Recall that oxidation is a loss of electrons.) The oxygen from air is reduced to oxygen ions that diffuse through the solid electrolyte to the porous anode. (The acronym AN OX and a RED CAT reminds you that oxidation occurs at the anode and reduction occurs at the cathode.) The electrolyte is a hard, nonporous solid oxide that allows only the oxygen ions (rather than hydrogen ions) to pass and blocks electrons. A traditional SOFC must be operated at high temperatures (800° C to 1,000° C) to achieve reasonable conduction of the oxygen ions.

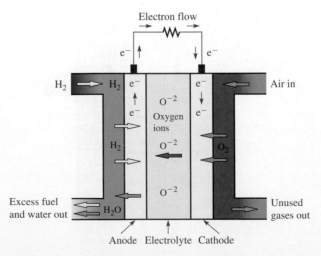

FIGURE 12-6 Operation of a Solid Oxide Fuel Cell (SOFC)

In the traditional SOFC, air is taken in so that oxygen (O_2) can diffuse through the cathode. The unused gas is routed out of the cell. The fuel cell produces water, which must be taken out of the cell. The solid oxide fuel cell takes several minutes to come up to temperature before it can produce power continuously to the external circuit, so this type of fuel cell is not useful for applications where it needs to be turned on and off frequently. It also produces a large amount of heat in the electrochemical reaction, and this heat must be removed from the fuel cell to ensure that it operates safely. The high operating temperature allows the SOFC to separate (reform) hydrogen from other fuels internally so that expensive reforming equipment is not needed.

SOFCs were first designed in the 1930s and have been improved over the years. Two major improvements have been announced by Department of Energy's Pacific Northwest National Laboratory (PNNL). A small SOFC with up to 57% efficiency is the first breakthrough. It uses methane (CH_4) fuel that is reformed and cleaned. This solid oxide fuel cell can generate between 1 and 100 kW of power, making it useful in smaller applications. The second major breakthrough for SOFCs is an electrolyte made of scandium in cubic zirconia, which can pass oxygen faster and at temperatures far lower than older SOFCs. The anode has traditionally been made of nickel; however, some new composite materials made of oxides of strontium titanate and ceria can replace the traditional nickel. Research is continuing on these fuel cells because of these recent promising developments.

An SOFC can use hydrogen and carbon monoxide as fuel. It can also run on other common hydrocarbon fuels such as diesel, natural gas, gasoline, or grain alcohol. Some versions are being tested to operate with coal. An advantage of the SOFC is the solid electrolyte, which does not require a pump to circulate the hot electrolyte, as in some other fuel cells, and fuel is reformed internally. In addition, SOFCs do not require the expensive platinum catalyst that is needed in other fuel cells. The SOFC tolerates sulfur in fuels better than other fuel cells, has a resistance to carbon buildup, and is not poisoned by carbon monoxide (CO). Another advantage of SOFCs is that the hard, nonporous ceramic electrolyte can be shaped like cylinders rather than the platelike configuration of other fuel cells. Figure 12-7 shows a diagram of a solid oxide fuel cell in a tube structure.

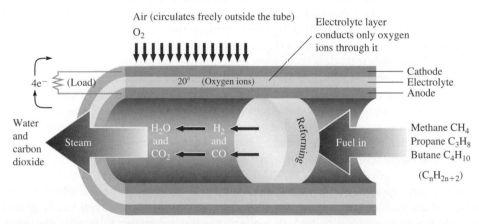

FIGURE 12-7 Tube Structure of a Solid Oxide Fuel Cell (*Source:* Courtesy of Acumentrics, Inc.)

A disadvantage of the SOFC is that the ceramic parts cannot withstand heating and cooling cycles very well. Another disadvantage is that the SOFC requires heat shielding, which makes it too heavy to use in some lighter transportation systems such as cars, although it is being investigated for use in trucks. The new developments cited earlier may alleviate the need for the heavy heat shielding in the near future.

Molten-Carbonate Fuel Cell

The **molten-carbonate fuel cell (MCFC)** is similar to the solid oxide fuel cell (SOFC), but it uses carbonate ions as the charge carrier in a high-temperature liquid solution of lithium,

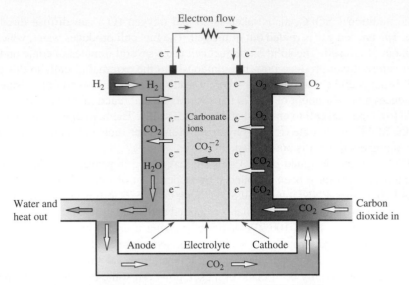

FIGURE 12-8   Operation of a Molten-Carbonate Fuel Cell (MCFC)

potassium, or sodium carbonate as an electrolyte. The MCFC operates at a temperature of 600° C to 700° C (1,112° F to 1,292° F), so it can generate steam that can be used to generate more power. It is best suited for large stationary power generators. Because the MCFC has an operating temperature that is a bit lower than solid oxide fuel cells, it can use less exotic materials, which makes it a little less expensive to manufacture and operate.

The MCFC was first designed in the early 1930s along with the SOFC. Research on both devices continued to provide improvements until the 1950s, when researchers found that fused molten-carbonate salts could be used as an electrolyte in the fuel cell. Because MCFCs operate at extremely high temperatures, nonprecious metals such as nickel can be used as catalysts in the anode and cathode, rather than the platinum that is used in some other fuel cells.

Figure 12-8 shows a diagram of an MCFC. Hydrogen is ionized at the anode by the catalyst. After the carbonate ions (CO_3^{-2}) pass through the electrolyte, they combine with the hydrogen ions (H^+) to form water (H_2O) and carbon dioxide (CO_2), which is returned to the input. The carbon dioxide and electrons from the external circuit combine with oxygen to supply more carbonate ions, and the process continues. The net reaction is hydrogen plus oxygen, which produces water as in other fuel cells. High-temperature MCFCs can extract hydrogen from a variety of fuels, such as natural gas, diesel, or coal, and they are not subject to the contamination issues of other types of fuel cells. Because the MCFC operates at very high temperatures, these fuels can be converted to hydrogen within the fuel cell itself without a separate reformer, which also reduces cost.

An important advantage of an MCFC is that it can be used as a large stationary power plant that can supply power to the main load or at peak times as needed. In addition to its ability to use various fuels, these plants are clean and quiet. The fuel cell can be placed close to the point where it is used, so it allows electric companies to provide electricity without a large investment in transmission lines. One other advantage of MCFCs is that the waste heat can be captured and used, thus raising the overall efficiency.

One major disadvantage with molten-carbonate technology is that it is more difficult working with a very hot liquid electrolyte rather than a solid electrolyte. Another disadvantage is that the chemical reactions at the anode use carbonate ions from the electrolyte, making it necessary to inject carbon dioxide at the cathode and thus requiring a supply of carbon dioxide.

Summary of Fuel Cell Technologies

Table 12-1 compares specific parameters—the electrolyte, operating temperature, and electrical efficiency—for the fuel cells discussed in this section.

TABLE 12-1

Comparison of Fuel Cell Types

Fuel Cell Type	Electrolyte	Operating Temperature	Electrical Efficiency
Proton exchange membrane fuel cell (PEMFC)	Solid organic polymer polyperfluorosulphonic acid	50° C to 100° C	25% to 58%
Direct-methanol fuel cell (DMFC)	Polymer similar to PEMFC	50° C to 120° C	25% to 40%
Phosphoric acid fuel cell (PAFC)	Liquid phosphoric acid	150° C to 200° C	40% to 50%
Alkaline fuel cell (AFC)	Potassium hydroxide	100° C to 250° C	60% to 70%
Solid oxide fuel cell (SOFC)	Yttria-stabilized zirconia	600° C to 1,000° C	35% to 43%
Molten-carbonate fuel cell (MCFC)	Liquid solution of lithium sodium or potassium carbonates	600° C to 700° C	45% to 47%

(*Source:* US Department of Energy.)

SECTION 12-2 CHECKUP

1. What attributes are necessary for the anode of a fuel cell?
2. Why isn't hydrogen considered a primary fuel source?
3. What type of reaction always takes place at the anode of a fuel cell?
4. How does a PEMFC produce electrical power?
5. What is an advantage to the high temperature operation of an SOFC?
6. How does an MCFC differ from an SOFC?

12-3 Vehicle Applications

Fuel cells are used to provide power for automobiles, trucks, and buses. Hydrogen fuel or a compound fuel such as gasoline can be used to generate electrical power to operate electric drive motors. Hydrogen fuel cell cars have been tested since 1991. At this time, fuel cell vehicles are still not competitive with gasoline vehicles on a per mile basis, but the cost gap is closing.

Fuel Cell Electric Vehicles

A **fuel cell electric vehicle (FCEV)** is a hydrogen-powered vehicle that uses electricity from a fuel cell to drive its electric motor. The hydrogen fuel is generally obtained from natural gas or other fossil fuel. Thirteen car manufacturers have worked on developing FCEVs. Development has been slow. The cost per vehicle was originally thought to be too expensive for these vehicles to become viable compared to gasoline and hybrid electric cars. The current cost to build a production model, fuel cell vehicle is still much more than a comparable gasoline car. Various government programs have enhanced the development of hydrogen for automotive applications, the infrastructure for fuel stations, and advanced technology fuel cells to make fuel cell vehicles practical and more cost-effective. Of all the fuel cells, the PEMFC and the PAFC have shown the most promise for use in vehicles.

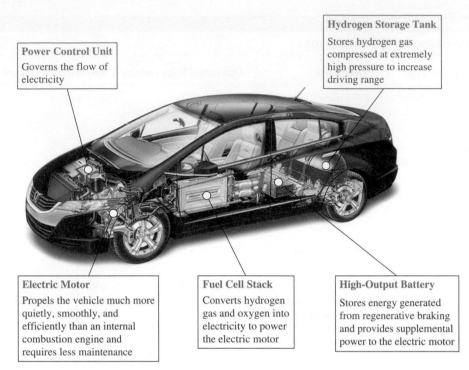

FIGURE 12-9 Basic Parts of a Fuel Cell–Powered Vehicle   (*Source:* Courtesy of American Honda Motor Co., Inc.)

Labels on figure:

Power Control Unit
Governs the flow of electricity

Hydrogen Storage Tank
Stores hydrogen gas compressed at extremely high pressure to increase driving range

Electric Motor
Propels the vehicle much more quietly, smoothly, and efficiently than an internal combustion engine and requires less maintenance

Fuel Cell Stack
Converts hydrogen gas and oxygen into electricity to power the electric motor

High-Output Battery
Stores energy generated from regenerative braking and provides supplemental power to the electric motor

---

Self-Contained Automotive Fuel Cell

(*Source:* Courtesy of ClearEdge Power Corporation.)

United Technologies Company produces a stand-alone self-contained hydrogen fuel cell that can be used to provide electrical power for vehicles. This fuel cell uses proton exchange membrane (PEM) fuel cell auxiliary power units (APUs) for use in vehicles. The APU unit provides energy for all the car's onboard electrical needs, including climate control, even when the engine is off. The unit produces 5 kW, and it runs on pure hydrogen. Research vehicles have logged more than 3,000 hours of operation.

A typical fuel cell–powered automobile is shown in Figure 12-9. Many of the systems used in a fuel cell–powered vehicle have similar components compared to a typical gasoline-powered automobile: a transmission, battery, radiator, and cooling system. In some ways, the fuel cell–powered car is simpler because it does not have a gas engine, with all of a gas engine's requirements. It has instead a fuel cell stack, electric motor, and power control unit. The hydrogen fuel is stored in the hydrogen tanks at approximately 1,500 to 5,000 psi. When the tanks are filled, they can provide enough hydrogen fuel to power the automobile for 175 to 190 miles. Hydrogen gas under pressure is provided to the fuel cell, and the fuel cell creates dc, which is sent to a battery or ultracapacitor. The dc is converted to ac if the motor is an ac synchronous drive motor. The vehicle shown in Figure 12-9 has regenerative braking, which captures a portion of the vehicle's kinetic energy (which would otherwise be wasted in braking) and uses it to charge the battery or ultracapacitor.

Some automobiles use a dc permanent magnet motor or a brushless dc motor (brushless dc motors were discussed in Section 5-4. Oxygen is supplied to the fuel cell when fresh air is pumped into the cell. The battery charge level and the ultracapacitor charge level and power to the motor are controlled by the power control unit. The fuel cell produces heat when it is operating, so it must have water flowing through it for cooling. A radiator is used to remove heat from the cooling water. The ac synchronous motor powers the drive wheels though a transmission, so it can accelerate and decelerate the automobile smoothly. As long as hydrogen flows through the fuel cell, it produces electrical power.

As mentioned previously, the two most common type of fuel cells used in automobiles is the PEMFC and the PAFC. The low operating temperature for these cells means that they don't take long to warm up and begin generating electricity. Figure 12-10 shows a fuel cell in an automobile.

Phosphoric acid fuel cells have recently been used because they can reach temperatures up to 220° C (428° F), which ensures the water will vaporize and can be expelled easily. The higher temperature allows for greater efficiencies. They can also use reformed hydrogen from gasoline more efficiently than the PEM fuel cells. During the reforming process,

FIGURE 12-10   Fuel Cell–Powered Vehicle   (*Source:* National Renewable Energy Laboratory.)

gasoline is heated to convert it from a liquid to a vapor. The vaporized gasoline is processed into hydrogen in a **partial oxidation (POX) reactor,** which is a method to produce hydrogen and carbon monoxide from a hydrocarbon using a catalyst at high temperature to complete the reaction. When gasoline is the hydrocarbon, sulfur in the gasoline is converted to hydrogen sulfide, and it is filtered from the vapor. The carbon monoxide is separated in the process by adding water vapor in the form of steam, creating carbon dioxide (CO_2) that is eliminated from the reformer. If carbon monoxide were allowed to come into contact with the fuel cell parts, contamination would result, so this process ensures all carbon monoxide is removed. The process also separates hydrogen from water; the hydrogen is added to the fuel supply. The fuel must pass through a heat exchanger to cool it to approximately 80° C, which is the operating temperature of the fuel cell.

One of the concerns of using fuel cells in automobiles is that they are difficult to start and are slow to reach the operating temperature when the ambient temperature is below −10° C (14° F). Also, if the water (which is a by-product) is not eliminated, it can freeze at temperatures below 0° C and damage the system.

Buses

A number of cities have been testing fuel cell–powered buses for some time. The bus frame is larger, so it can handle more and larger hydrogen tanks than a car frame can. Intercity buses do not need to have the same driving range as cars, so fuel cells can be recharged, which helps reduce pollution within the city. Figure 12-11 shows a bus used in Burbank, California.

Hydrogen Refueling Stations

(*Source:* National Renewable Energy Laboratory.)

Hydrogen fuel is available at a limited number of refueling stations, where the hydrogen is stored and dispensed under pressure. The fuel pump and hoses look similar to those used with gasoline pumps. As more vehicles that use hydrogen are operated on the nation's highways, more refueling stations will become available. The production of hydrogen will also need to be expanded.

FIGURE 12-11   Large Transit Bus Powered by a Fuel Cell   (*Source:* National Renewable Energy Laboratory.)

FIGURE 12-12   Fuel Cell–Powered Vehicle Refueled with Hydrogen That Is Under Pressure (*Source:* National Renewable Energy Laboratory.)

Refueling Stations for Hydrogen Fuel Cell Vehicles

One kg of hydrogen is approximately equivalent to 1 gal of gasoline. Hydrogen refueling stations will need to be located approximately 200 km (125 mi) apart so that vehicles can locate a refueling station before they run out of fuel. Hydrogen fuel must be stored at high pressures (1,500 to 5,000 psi), depending on the temperature of the fuel. Currently, hydrogen storage tanks and refueling pumps are being added to traditional gasoline or diesel refueling stations. Figure 12-12 shows a typical refueling hose and fitting for a hydrogen refueling station. The hoses and fittings used to dispense the fuel must also be rated for the higher pressures, as well as the fitting on the automobile fuel connection equipment. Fire protection equipment must be provided specifically for hydrogen because it burns invisibly. Currently hydrogen fuel cell vehicles can obtain up to the equivalent of 60 miles per gallon (mpg). The tanks that store the hydrogen onboard the vehicle must contain the gas at high pressures, so the tank size is limited to maintain vehicle weight and space. Thus, the amount of hydrogen stored in a vehicle allows it to travel perhaps 300 km (186 mi), but newer vehicles may have a better range.

Fuel Cell–Powered Working Vehicles

Fuel cells have several applications in working vehicles such as forklifts, mining locomotives, tractors, and so on. Typically, a fuel cell–powered working vehicle can operate for an entire 8-hr shift on one tank of fuel (methanol, natural gas, or hydrogen) and can be refilled rapidly (in as little as 90 s in some cases). Standard battery-powered working vehicles need to be out of service for several minutes for batteries to be changed or for hours if the batteries are recharged in the vehicle.

One early application of fuel cells in a working vehicle was an experimental tractor built by Allis-Chalmers in 1951. The tractor used 1,008 fuel cells in combination to power a 20 hp dc motor. The old tractor currently resides in the Smithsonian in Washington, DC. In 2011, New Holland demonstrated a new prototype for fuel cell farm tractor (see Figure 12-13) that was put to use in the summer of 2012 in a farm in Italy. It has fewer components than a traditional tractor and replaces hydraulic implements with electrical driven implements, so it should have improved reliability over its traditional cousins. Many farms have renewable energy (solar or wind) available, so the potential for farmers to make their own fuel from electrolysis of water using renewable energy is an attractive idea that can reduce or eliminate fuel costs for farmers.

Another working vehicle that is a good fit for fuel cells is the forklift; many are currently in service. Figure 12-14 shows a fuel cell–powered fork lift. One of its main features is that it does not create any emissions like a gas-powered forklift, so it can operate safely indoors or in enclosed spaces (like warehouses or boxcars). Typically, fuel cell forklifts are powered by PEMFCs, and some newer models are powered by direct-methanol fuel. At an installation with many forklifts, the turnaround time for refueling is particularly

FIGURE 12-13 New Holland NH2 Fuel Cell–Powered Tractor   (*Source:* National Renewable Energy Laboratory.)

important, and the fuel cell forklifts can be turned around faster than battery-powered ones. An example is Walmart, which has over 500 PEMFC forklifts for use in its warehouses. The PEMFC forklifts have meant increased productivity compared to the battery-powered forklifts Walmart once used. These PEMFC forklifts have also eliminated problems with storing, charging, and disposing of lead and acid.

Mining locomotives require a zero emission vehicle. In the past, mining locomotives needed to be battery-powered to avoid leaving poisonous fumes in mines. Five prototype mining locomotives have been constructed for use in a South African platinum mine. The locomotives have an integrated fuel reformer to extract hydrogen from natural gas fuel and then use it in PEMFC stacks.

Other Applications

Small boats can use fuel cells to avoid problems with pollution, particularly with outboard engines. The average outboard motor produces approximately 140 times the hydrocarbons produced by a modern car, and the exhaust on some outboards is pumped directly into the water and causes water pollution. Fuel cell–powered boats eliminate nearly all this pollution. Fuel cells typically used for small boats have a net output that is approximately 6.5 kW.

FIGURE 12-14 Forklift Powered by a Hydrogen Fuel Cell   (*Source:* National Renewable Energy Laboratory.)

Fuel cells are being investigated for electrical systems on some aircraft. The electricity production is efficient. A side benefit is that the water by-product can be used for the aircraft's water and waste system. The result is lower weight, greater efficiency, and lower emissions. Airbus has tested fuel cells to power an Airbus A320 hydraulic system and found that the systems worked perfectly with the fuel cell. Airbus is also looking to replacing the auxiliary power unit (APU) with a fuel cell. The APU is a generator used for powering air conditioning and starting the engines when the plane is on the ground.

SECTION 12-3 CHECKUP

1. Identify two types of fuel cells that are typically used for hydrogen fuel cell electric vehicles.

2. What is the purpose of a battery or ultracapacitor in a fuel cell–powered automobile?

3. What are the main advantages of fuel cell power in working vehicles?

4. How are fuel cells being tested for use by aircraft?

12-4   Stationary Fuel Cell Applications

Larger stationary units can provide electrical power to buildings, and they can be interconnected to provide stand-alone power to small cities or supplemental and backup power. They are particularly useful where access to the grid is an issue. To replace power from the grid, the dc output from the fuel cell is converted to ac.

Fuel cells are an efficient way to generate baseload (continuous) power for smaller utility plants, although the capital cost for the fuel cell unit is generally higher than that of a natural gas turbine. Fuel cells can be used in combined heat and power (CHP) applications to boost overall efficiency by using the excess heat. Fuels can include biogas from wastewater treatment and food processing, or natural gas, which is reformed to provide hydrogen for the fuel cells.

One of the largest installations of a fixed fuel cell power plant is under construction in Bridgeport, Connecticut, where a 14.9 MW electrical plant, consisting of five fuel cell power generation units and an organic Rankine cycle (ORC) turbine, are being built. (The ORC is discussed in Section 9-7.) By power plant standards, 14.9 MW is small, but this plant can still power about 15,000 homes. An advantage to a plant like this is that it is scalable, meaning power generation can be added incrementally as demand grows.

Another large installation located at Samsung/GS Power in Anyang, South Korea, is shown in Figure 12-15. This group of fuel cells uses 12 United Technology Company PureCell 400 stationary fuel cells, which provide a total of 4.8 MW of electrical power.

FIGURE 12-15   Fuel Cell Utility Installation at Samsung/GS Power, in Anyang, South Korea.   (*Source:* Courtesy of ClearEdge Power Corporation.)

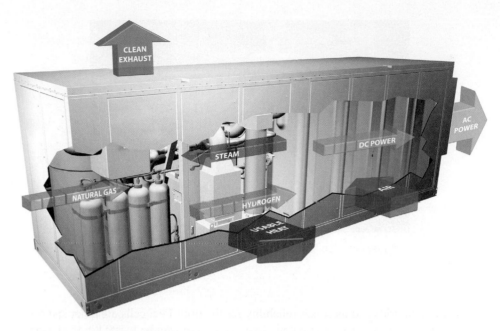

FIGURE 12-16 Internal Parts of the PureCell 400 Stationary Fuel Cell Power System (*Source:* Courtesy of ClearEdge Power Corporation.)

The power station satisfies 5% of the town of Anyang's power needs while also providing nearly 20 million BTU/hr of usable thermal energy.

Figure 12-16 shows the sections of one of the PureCell 400 system's units. The fuel cell uses phosphoric acid fuel cell technology and has three modules: the power module, the cooling module, and the remote monitoring system module. The fuel cell is powered by natural gas, which is reformed to provide hydrogen, and can produce 400 kW of electrical power plus enough excess heat to provide up to 1.7 million BTU/hr of heat. As in the case of all fuel cells, the output power of the stack is dc, so it must be sent through an inverter to provide ac for the grid.

The solid oxide fuel cell is another type of fuel cell that is useful for distributed power applications, where the electricity is generated close to the point of use. These fuel cells are scalable, modular, and efficient. Because of the high-temperature operation, the SOFC is suited to a combined cycle operation, where the SOFC is combined with a micro gas turbine to produce electricity from both the fuel cell and the turbine. In Tokyo, Mitsubishi has achieved over 3,000 hours on a 200 kW SOFC unit with a micro gas turbine that has a combined efficiency of over 52%.

As in other high-temperature fuel cells, the excess heat can be used for process heat or building heat, or it can be used to produce steam to power a steam turbine for additional electrical power. The solid oxide fuel cell produces enough primary and secondary power through the steam turbine generator to provide electricity for a large building or small village. The SOFC is compact in size; is a clean source of electrical power; and operates very quietly, which makes it attractive for urban settings.

SOFC fuel cells are of great interest for power applications for the reasons mentioned. Research is being conducted on advanced SOFCs at the National Energy Technology Laboratory (NETL), which evaluates new SOFC fuel cells. One technology of interest is planar SOFCs with thin ceramic electrolytes that can operate at lower temperature and with less expensive interconnects. Figure 12-17 illustrates a test platform for fuel cells that can obtain experimental data on new fuel cells or on related issues such as the fuels themselves.

Distributed Power

Fuel cells are also useful for distributed generation, which means the power is generated near the point of use. This reduces (and in some cases eliminates) the need for transmission

Portable PEM Fuel Cells for Military Use

(*Source:* National Renewable Energy Laboratory.)

Portable PEM fuel cells are being used by military units to provide electrical power at remote sites. These remote sites need electrical power for refrigerating food, providing fresh water, and for surgical and hospital facilities. Normally a diesel generator would be used for power, but it requires large quantities of fuel over time. Fuel cells can provide the electricity and potable water at the same time. The self-contained models are small enough to be moved with the forces and be set up quickly.

FIGURE 12-17   A test platform for testing fuel cells at NETL.   (*Source:* US Department of Energy.)

and distribution grids and increases reliability for the user. Fuel cells are quiet and do not rely on combustion, so there are no emissions. Thus, they can be located in the basement of a hospital, for example, without disturbing patients. Power generated off the grid can provide power to a large store, factory, or community. In distributed applications, high-temperature fuel cells can use the excess heat to produce more power with an ORC turbine. The ORC can power additional electrical load for the business or industry. The excess heat can also be used directly to provide process heat for drying, heat treating, space heating, chemical processing, or air conditioning using an absorption chiller. (An absorption chiller can provide air conditioning without electricity.)

Stationary fuel cells offer an excellent way to back up the grid by providing power during outages for businesses and other commercial users. Supermarkets are particularly vulnerable to long power outages because of refrigeration and freezer requirements, so a backup power system is important. Several large stores have installed stationary fuel cell power systems made by United Technology Company (UTC Power) to provide electrical backup power. The system is CHP and includes stored natural gas, a reformer, fuel cells, and an inverter, and it is very efficient because it also provides heat, which can be used for space heat or hot water. A system like this can provide continuous electrical power in remote locations, or it can be used for backup power during a power outage.

Fuel cell–powered stationary power plants also contribute points toward Leadership in Environmental and Engineering Design (LEED™) certification. LEED™ is an internationally recognized green building certification system. It was developed by the US Green Building Council (USGBC) in March 2000. LEED provides building owners and operators with a framework for identifying and implementing practical and measurable green building design, construction, operations, and maintenance solutions.

SECTION 12-4 CHECKUP

1. Identify the basic parts of a stationary fuel cell system for power generation.

2. What is the advantage of combining an ORC turbine with fuel cells at a power station?

3. What is meant by distributed power generation?

4. What advantage do fuel cells have for distributed power generation?

5. Why is it important for supermarkets to have backup power?

CHAPTER SUMMARY

- A fuel cell is a device that converts electrochemical energy into dc voltage. Fuel cells combine a fuel (usually hydrogen or a hydrocarbon) with an oxidizing agent (usually oxygen).

- The hydrogen fuel cell has three basic parts: (1) an anode that separates the hydrogen ions and electrons, (2) a cathode that helps combine oxygen and returning hydrogen ions to form water, and (3) the electrolyte.

- All electrochemical reactions in a fuel cell use an oxidation-reduction chemical reaction. Oxidation involves the loss of electrons; reduction involves the gain of electrons.

- Some fuel cells must operate on pure hydrogen and pure oxygen, while others have been designed to operate on hydrocarbon fuels such as methane, butane, natural gas, coal, or gasoline. A reformer can separate the hydrogen from the hydrocarbon fuel.

- The membrane electrode assembly includes the anode and cathode electrodes, the electrolyte, and the catalyst. The type of fuel cell is usually defined by the electrolyte that it uses.

- Fuel cells have several advantages, such as no moving parts, quiet operation, production of electricity, and production of heat as a by-product. They also produce clean water as a by-product, which can be used as a potable water source.

- Six common fuel cells include the proton exchange membrane fuel cell (PEMFC), which is also known as the polymer electrolyte membrane fuel cell); the direct-methanol fuel cell (DMFC); the phosphoric acid fuel cell (PAFC); the alkaline fuel cell (AFC); the solid oxide fuel cell (SOFC); and the molten-carbonate fuel cell (MCFC).

- The electrolyte in the PEMFC is a solid polymer. It has porous carbon electrodes that contain a platinum catalyst.

- The DMFC is a subcategory of the PEMFC.

- The AFC uses an electrolyte that is an aqueous (water-based) solution of potassium hydroxide (KOH) retained in a porous stabilized matrix.

- The PAFC is typically used in small, stationary, power generation systems. Research is being conducted to learn more about its use in large buses.

- The SOFC is a high-temperature fuel cell that is compatible with common fuels and uses an inexpensive nickel oxide catalyst.

- The MCFC is similar to the SOFC in that it operates at high temperatures of 600° C to 700° C (1,112° F to 1,292° F). It can generate steam, which can be used to generate more power. It is best suited for large stationary power generators.

- Hydrogen fuel cell electric vehicles (FCEVs) have been tested since the early 1990s. The FCEV uses a fuel cell to produce electrical power for an electric drive motor.

- Fuel cells are applied to various vehicles and are especially useful for powering indoor working vehicles.

- The PEMFC is used in vehicles because it is the most compact of all fuel cells.

- A partial oxidation (POX) reactor processes vaporized gasoline into hydrogen and carbon dioxide.

- One kg of hydrogen is approximately equal to 1 gal of gasoline.

- Fuel cells can be used to provide electricity for the grid, or they can provide distributed power to a large store, factory, or community.

KEY TERMS

alkaline fuel cell (AFC)   A very efficient fuel cell that requires pure hydrogen fuel and pure oxygen. It uses an aqueous (water-based) electrolyte solution of potassium hydroxide (KOH) in a porous, stabilized matrix.

direct-methanol fuel cell (DMFC)   A fuel cell that is a sub-category of the proton exchange membrane fuel cell (PEMFC). Hydrogen is separated from methanol fuel by a steam reformer and supplied, with water and air, to the fuel cell. The electrolyte is a polymer similar to the PEMFC.

electrolysis   The breaking down of a substance that contains ions by applying an electric current between two electrodes and separating the substance into its components. The electrolysis of liquid water breaks it down into hydrogen and oxygen gases.

fuel cell   A device that converts electrochemical energy into dc directly by using a constant flow of fuel (usually hydrogen) from an outside source.

fuel cell stack   A group of fuel cells that are connected and bound together to provide increased electrical power.

fuel processing unit   A portion of a fuel cell system that converts the input fuel into a form usable by the fuel cell.

heat recovery system   A part of the fuel cell that processes the excess heat for another use such as heating water or creating steam.

membrane electrode assembly   The components in a fuel cell that include the anode and cathode electrodes, the electrolyte, and the catalyst.

molten-carbonate fuel cell (MCFC)   A fuel cell similar to the solid oxide fuel cell (SOFC) but with carbonate ions as the charge carrier in a high-temperature solution of liquid lithium, potassium, or sodium carbonate as an electrolyte.

phosphoric acid fuel cell (PAFC)   A fuel cell that is equivalent in structure to the proton exchange membrane fuel cell (PEMFC), but it has liquid phosphoric acid as the electrolyte. The electrolyte is contained in a Teflon-bonded, silicon carbide matrix.

proton exchange membrane fuel cell (PEMFC)   A type of fuel cell that uses hydrogen fuel and oxygen (obtained from air) to produce pure water and electricity. It uses porous carbon electrodes that contain a platinum catalyst.

reformer   A device that extracts hydrogen from another fuel such as methane.

solid oxide fuel cell (SOFC)   A fuel cell named for the solid oxide electrolyte that it uses. The fuel is hydrogen, which is supplied as a gas along with oxygen from the air. The electrolyte is a hard, non-porous, solid oxide or ceramic compound made of yttria-stabilized zirconia.

CHAPTER TRUE/FALSE QUIZ

Determine whether each statement is true or false. Answers are at the end of the chapter.

1. A fuel cell produces ac that is converted to dc by an inverter.

2. The fuel cell has three basic parts: the anode, the cathode, and the electrolyte.

3. The anode is the terminal at which a reduction reaction occurs.

4. Both batteries and fuel cells have an oxidation-reduction chemical reaction driving the process.

5. A fuel cell power system has four parts: the fuel cell stack, the fuel cell processing unit, the electrical inverter, and the heat recovery system.

6. During the reforming process, methanol gas is cooled to a point where hydrogen is separated from it.

7. The proton exchange membrane fuel cell is also known as a polymer electrolyte membrane fuel cell.

8. The direct-methanol fuel cell (DMFC) is a subcategory of the proton exchange membrane fuel cell (PEMFC).

9. The phosphoric acid fuel cell (PAFC) is typically used in small, stationary, power generation systems, and research is being conducted to learn about its use in larger buses.

10. An individual fuel cell produces 12 V dc.

11. Vaporized gasoline can be processed into hydrogen in a partial oxidation (POX) reactor.

12. Hydrogen fuel is stored as a liquid and is not under pressure when used in a fuel cell power electric vehicle.

13. Fuel cells cannot be used with forklifts.

14. A car with a fuel cell does not require an electric motor.

15. Regenerative braking can be used to save energy in a fuel cell–powered vehicle.

16. Distributed power is generated near the point of use.

17. All fuel cells have hydrogen ions as the moving charge carrier in the electrolyte.

CHAPTER MULTIPLE-CHOICE QUIZ

Complete each statement by selecting the one correct answer. Answers are at the end of the chapter.

1. An advantage of the molten-carbonate fuel cell is that it
   a. sequesters carbon dioxide
   b. runs at low temperatures
   c. can extract hydrogen from various fuels
   d. has a liquid electrolyte

2. All fuel cells can provide
   a. dc electrical power
   b. ac electrical power
   c. process heat
   d. all of the above

3. When hydrogen is supplied to the anode of a fuel cell, it separates into
   a. water and carbon dioxide
   b. carbon dioxide and hydrogen ions
   c. hydrogen ions and electrons
   d. oxygen and water

4. The reforming process
   a. separates hydrogen from hydrocarbon fuels
   b. separates water from hydrocarbon fuels
   c. combines water and hydrogen into a new fuel
   d. combines oxygen molecules with water to make a new fuel

5. The electrolyte in a proton-exchange membrane fuel cell passes
   a. electrons but not hydrogen ions
   b. both electrons and hydrogen ions
   c. hydrogen ions but not electrons
   d. hydrogen molecules and electrons

6. The fuel cell that operates at the highest temperature is the
   a. proton exchange membrane fuel cell (PEMFC)
   b. direct-methanol fuel cell (DMFC)
   c. phosphoric acid fuel cell (PAFC)
   d. solid oxide fuel cell (SOFC)

7. The two fuel cells that are most likely to be used in automotive applications are the
   a. PEMFC and the PAFC
   b. PEMFC and the SOFC
   c. PAFC and the SOFC
   d. PAFC and the MCFC

8. An advantage of the solid oxide fuel cell is that it
   a. runs cooler, so can be used in cars
   b. has no requirement for a platinum catalyst
   c. can use water as a fuel
   d. produces ac without the need for an inverter

9. All fuel cell–powered vehicles have
   a. an inverter
   b. an internal combustion engine
   c. an electric motor or motors
   d. all of these

10. A fuel cell powered vehicle does not have a
    a. radiator
    b. battery
    c. transmission
    d. carburetor

11. An advantage to fuel cell–powered vehicles over gasoline-powered vehicles is that they
   a. have a much greater range
   b. emit only water vapor for exhaust
   c. do not have a cooling system
   d. feature all of these

12. Solid oxide fuel cells
   a. send dc to the grid
   b. have a molten electrolyte
   c. can convert common hydrocarbons to fuel
   d. feature all of these

13. In a solid oxide fuel cell, platinum is
   a. used in the anode
   b. used in the cathode
   c. used in both the anode and the cathode
   d. not used

14. An advantage to an alkaline fuel cell is that it
   a. can use any hydrocarbon for fuel
   b. is not susceptible to contamination
   c. is less costly than other types
   d. can provide fresh drinking water as a by-product

CHAPTER QUESTIONS AND PROBLEMS

1. State the purpose of steam reformation and describe the chemical reactions for it.

2. List at least three issues associated with transporting hydrogen fuel.

3. How can renewable sources and water be used to form molecular hydrogen?

4. Describe the operation of a direct-methanol fuel cell (DMFC).

5. What is the main difference between a phosphoric acid fuel cell (PAFC) and a proton exchange membrane fuel cell (PEMFC)?

6. Explain how a fuel cell can use a hydrocarbon such as methanol or natural gas for its fuel.

7. What are the disadvantages to the alkaline fuel cell (AFC) that make it more expensive to operate than other fuel cells?

8. Why aren't solid oxide fuel cells (SOFCs) used in cars?

9. What have been hindrances for the adoption of more fuel cell–powered vehicles?

10. What are three applications for fuel cell vehicles?

11. List the main components of a fuel cell–powered electric vehicle and give the purpose of each.

12. Identify the main parts of a stationary fuel cell power plant.

13. Indicate at least three applications for process heat.

14. Explain why an ORC is used in conjunction with some electrical fuel cell plants.

15. What are the advantages to using distributed power?

FOR DISCUSSION

The partial oxidation (POX) reactor can convert gasoline to hydrogen for a fuel cell. Discuss the pros and cons of converting gasoline to hydrogen for a fuel cell–powered vehicle versus burning the gasoline directly in a gasoline-powered vehicle.

ANSWERS TO CHECKUPS

Section 12-1 Checkup

1. The basic parts of a fuel cell are the anode, the cathode, and an electrolyte. The anode separates hydrogen into ions and electrons, the cathode forms oxygen ions that combine with hydrogen ions to form water, and the electrode is a conductor that passes hydrogen ions to the cathode.

2. Oxidation is a loss of electrons; reduction is a gain in electrons.

3. A fuel cell, like a battery, completes an oxidation-reduction reaction, which is a transfer of electrons between reactants. The electrons are transferred in the external circuit, forming current.

4. A reformer converts a hydrocarbon fuel to hydrogen fuel.

5. Carbon nanotubes can reduce the amount of platinum required for the fuel cell and can serve as a catalyst in some fuel cells.

Section 12-2 Checkup

1. The anode needs to be a porous conductor that facilitates the oxidation of hydrogen.

2. It does not occur naturally on earth but is the product of a chemical reaction.

3. Oxidation

4. The PEMFC uses pure hydrogen and air (oxygen) as inputs and oxidizes hydrogen at the anode, and uses a platinum catalyst to facilitate the reaction. At the cathode, oxygen accepts electrons to become oxygen ions. Electrons travel in the external circuit, while the hydrogen ions are attracted to the cathode. At the cathode, oxygen ions combine with the hydrogen ions to form water.

5. The SOFC runs hot enough to reform fuel internally. Also it is not as sensitive to sulfur and carbon monoxide buildup as other fuel cells.

6. At the anode of an MCFC, hydrogen is ionized by a catalyst and electrons move through the external circuit to the cathode. Carbon dioxide and oxygen from the air are ionized at the cathode to form carbonate ions, which travel through the solid electrolyte to the anode. At the anode, the hydrogen ions combine with the carbonate ions to form water and carbon dioxide, which is returned to the cathode.

Section 12-3 Checkup

1. The PEMFC and the PAFC

2. The battery or ultracapacitor stores energy from regenerative braking and provides supplemental power to the electric motor.

3. Pollution-free operation in restricted environments (indoors or underground mining operations) and elimination of problems associated with lead-acid batteries.

4. To operate certain onboard electrical equipment and the auxiliary power unit (APU) on the ground

Section 12-4 Checkup

1. The power module, the cooling module, and the remote monitoring system module

2. The ORC can use heat from the fuel cells that is otherwise lost to generate additional electricity.

3. Electricity that is generated close to the point of use

4. Fuel cells are efficient, quiet, non-polluting, and do not require a lot of space.

5. The most important reason is to maintain refrigerators and freezers to prevent food spoilage.

ANSWERS TO TRUE/FALSE QUIZ

1. F **2.** T **3.** F **4.** T **5.** T **6.** F **7.** T **8.** T **9.** T   **10.** F **11.** T **12.** F **13.** F **14.** F **15.** T **16.** T **17.** F

ANSWERS TO MULTIPLE-CHOICE QUESTIONS

1. c **2.** a **3.** c **4.** a **5.** c **6.** d **7.** a **8.** b **9.** c   **10.** d **11.** b **12.** c **13.** d **14.** d

Generators

CHAPTER OBJECTIVES

- Calculate the reluctance of a basic magnetic circuit consisting of a toroid with an air gap.
- Compare Ohm's law for electrical circuits with the corresponding equation for magnetic circuits.
- Define electromagnetic properties including permeability, reluctance, and magnetomotive force.
- Identify the basic parts of a dc generator and explain their functions.
- Explain the difference between separately excited and self-excited dc generators and discuss the advantages of each type.
- From a wiring diagram, identify series, shunt, and compound dc generators.
- Explain why a dc generator needs brushes and commutator segments.
- Identify the basic parts of a three-phase ac induction generator.
- Explain the operation of a rotating armature (stationary field) ac generator.
- Explain the operation of the rotating stator (stationary armature) ac generator.
- Explain the operation of the doubly fed ac induction generator.

KEY TERMS

Key terms are shown in bold and color. Definitions for key terms are provided at the end of the chapter and in the end-of-book glossary. Bold terms in black are defined in the end-of-book glossary only.

- permeability
- reluctance ($\mathcal{R}$)
- magnetomotive force (mmf)
- ampere-turn (At)
- magnetic field intensity (H)
- shunt generator
- compound generator
- self-excited shunt generator
- synchronous generator
- exciter
- induction generator
- squirrel-cage rotor
- doubly fed induction generator (DFIG)

INTRODUCTION

A generator was defined in Section 2-7 as a device that converts mechanical energy to electrical energy. The electrical energy can be in the form of either alternating current or direct current. As you know, an alternator is an ac generator that converts mechanical energy to ac. It is common practice to refer to large ac generators as generators and smaller ones as alternators; however, either term is correct. The ac or dc generator can be used as a stand-alone generator that provides electricity to a home or small business from a renewable energy source, or the generator can be used to produce large amounts of electricity for the grid. AC voltage from the generator can be produced at any frequency, and it can be converted to 50 Hz or 60 Hz ac for the grid if necessary. Other types of ac generators can be controlled to produce the correct grid frequency and phase.

This chapter explains the operation of different types of generators that you may encounter with renewable energy systems. You will learn about their advantages and disadvantages. You will also learn why the frequency and voltage is controlled on some generators and not on others.

13-1   Magnetism and Electromagnetism

Magnetism and electromagnetism were introduced in Section 2-7. These topics are reviewed here with an expanded discussion, including basic equations and definitions of magnetic terms. A magnetic circuit has some similarities to electrical circuits. In an electrical circuit, voltage causes current that is opposed by resistance. In a magnetic circuit, current causes flux that is opposed by reluctance. One significant difference between electrical and magnetic circuits is the concept of magnetic hysteresis, which is a property of magnetic materials whereby a change in magnetism lags the application of the magnetic field intensity.

Review of Magnetic Fields, Flux, and Flux Density

Any discussion of generators involves magnetism, so magnetic concepts are reviewed in this section. Recall from Section 2-7 that a permanent magnet, such as a bar magnet, has a magnetic field surrounding it, which consists of lines of force, or flux lines. Flux lines radiate from the north pole (N) to the south pole (S) and back to the north pole through the magnetic material.

The group of force lines going from the north pole to the south pole of a magnet is called the magnetic flux, symbolized by ϕ (the lowercase Greek letter phi). The number of lines of force in a magnetic field determines the value of the flux. The more lines of force, the greater the flux and the stronger the magnetic field. The weber was defined previously as 10^8 lines of force. The weber is a very large unit; thus, in most practical situations, the microweber (μWb) is used. One microweber equals 100 lines of magnetic flux.

The magnetic flux density was introduced briefly in Chapter 2 and is reviewed here. Magnetic flux density is the amount of flux per unit area in the magnetic field. Its symbol is B, and the SI unit is the tesla (T). One tesla equals one weber per square meter (Wb/m^2). Flux density can also be expressed in units of the Gauss (G) (10^4 G = 1 T). The following formula, which was given in Section 2-7 as Equation 2-19, expresses the flux density:

$$B = \phi/A \qquad \text{Equation 13-1}$$

where

B = flux density, in tesla (T)

ϕ = flux, in Webers (Wb)

A = cross-sectional area of the magnetic field, in m^2

The Earth is a Magnet

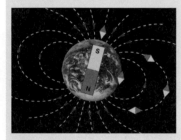

(*Source:* Courtesy of NASA)

The earth is sometimes referred to as a magnet in space. Basically the earth is a spherically shaped magnet. The magnetic field around a spherical magnet is essentially the same as if a bar magnet were located inside the earth, as shown in the diagram. The earth's North Pole might be better named as the earth's North-Seeking Pole because it attracts the north end of a compass. When a compass lines up in the earth's magnetic field, it is the north end of the compass magnet that seeks the north pole.

EXAMPLE 13-1

Find the flux density in a magnetic field in which the flux in 0.1 m^2 is 600 μWb.

Solution

$$B = \frac{\phi}{A} = \frac{600\ \mu\text{Wb}}{0.1\ \text{m}^2} = \mathbf{6{,}000 \times 10^{-6}\ T}$$

How Materials Become Magnetized

Ferromagnetic materials such as iron, nickel, and cobalt become magnetized when placed in the magnetic field of a magnet. You may have seen a permanent magnet pick up paper clips, nails, iron filings, and so on. In these cases, the object becomes magnetized (that is, it actually becomes a magnet itself) under the influence of the permanent magnetic field

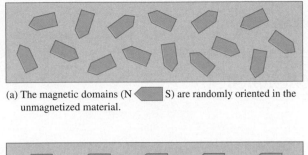

(a) The magnetic domains (N ◀▬ S) are randomly oriented in the unmagnetized material.

(b) The magnetic domains become aligned when the material is magnetized.

FIGURE 13-1 Ferromagnetic Domains

and becomes attracted to the magnet. When removed from the magnetic field, the object tends to lose its magnetism. Ferromagnetic materials have minute magnetic domains created within their atomic structure by the orbital motion and spin of electrons. These domains can be viewed as very small bar magnets with north and south poles. When the material is not exposed to an external magnetic field, the magnetic domains are randomly oriented, as shown in Figure 13-1(a). When the material is placed in a magnetic field, the domains align themselves, as shown in Figure 13-1(b). Thus, the object itself becomes a magnet.

Recall that a magnetic field is produced whenever there is current in a wire. When there is current in the wire, the magnetic flux lines form invisible concentric circles that may be visualized with iron filings, as you saw in Figure 2-24. The field is stronger when it is close to the conductor and becomes weaker with increasing distance from the conductor. As the current is increased, the number of flux lines also increases.

Left-Hand Rule

The direction of the lines of force surrounding the conductor is shown in Figure 13-2. The situation can be visualized with the **left-hand rule.** Imagine that you are grasping the conductor with your left hand, with your thumb pointing in the direction of electron flow. Your fingers indicate the direction of the magnetic lines of force. If current is reversed, the field also reverses.

If the wire is formed into a coil, as in Figure 13-3(a), the circular patterns around each loop reinforce each other and a stronger magnetic field is produced. The more loops of wire in the coil, the stronger the magnetic field. The coil has a magnetic field that resembles a

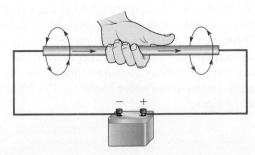

FIGURE 13-2 Left-Hand Rule. Electron flow is from left to right in the conductor.

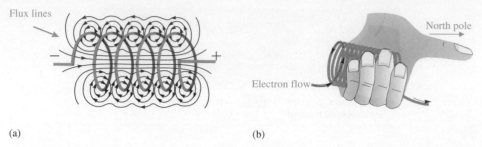

Flux lines

North pole

Electron flow

(a)   (b)

FIGURE 13-3 (a) Magnetic Field Around a Coil. (b) The Left-Hand Rule for a Coil.

bar magnet. The coil's magnetic field can be strengthened even more by wrapping the coil about a magnetic core; this is what is done in motors and generators. In Figure 13-3(b), the left hand rule for coils is illustrated; if the fingers curl in the direction of electron flow, the thumb points to the north pole of the magnet.

Electromagnetic Properties

Recall that electromagnetic coils are used in transformers, relays, solenoids, motors, and generators. These devices are all electromagnets with a magnetic core. The main reason that an electromagnet is used instead of a permanent magnet is that the magnetic field of the electromagnet can be energized and de-energized by interrupting the current flow through the wire. Materials used as magnetic cores have properties that affect the field that is formed. Three important properties related to electromagnetism—permeability, reluctance, and magnetomotive force—are discussed in the following subsections.

Permeability (μ)

Permeability is a measure of the ease with which a magnetic field can be established in a given material. The higher the permeability, the more easily a magnetic field can be established. The symbol of permeability is μ, and its value varies depending on the type of material. The permeability of a vacuum (μ_0) is $4\pi \times 10^{-7}$ Wb/At · m (weber/ampere-turn · meter) and is used as a reference. Ferromagnetic materials typically have permeabilities that are hundreds of times larger than that of a vacuum, indicating that a magnetic field can be set up with relative ease in these materials. Ferromagnetic materials include iron, steel, nickel, and cobalt and the alloys of these elements.

The **relative permeability (μ_r)** of a material is the ratio of its permeability (μ) to the permeability of a vacuum (μ_0). Because μ_r is a ratio, it has no units.

$$\mu_r = \frac{\mu}{\mu_0}$$ Equation 13-2

Reluctance ($\mathcal{R}$)

The opposition to the establishment of a magnetic field in a material is called **reluctance** ($\mathcal{R}$). Reluctance is a concept used in magnetic circuits and is analogous to resistance in electric circuits. Unlike resistance, which dissipates energy, reluctance stores energy. The value of reluctance is directly proportional to the length (l) of the magnetic path, and is inversely proportional to the permeability (μ) and to the cross-sectional area (A) of the material. It is expressed in ampere-turns/weber (At/Wb). The following equation defines reluctance in mathematical terms:

$$\mathcal{R} = \frac{l}{\mu A}$$ Equation 13-3

where

$\mathcal{R}$ = reluctance, in At/Wb

l = length, in m

μ = permeability of the material, in Wb/At $\cdot$ m

A = area, in m^2

EXAMPLE 13-2

What is the reluctance of a solid 6.0 mm diameter toroid (a toroid has a doughnut shape) that has a mean length of 12 cm and a relative permeability of 3,200?

Solution

Start by converting the dimensions to m and find the area:

$$l = 0.12 \text{ m}$$

The radius is 3 mm = 0.003 m

$$A = \pi r^2 = \pi (0.003 \text{ m})^2 = 2.83 \times 10^{-5} \text{ m}^2$$

The permeability of the core is found from Equation 13-1:

$$\mu = \mu_0\mu_r = (4\pi \times 10^{-7} \text{ Wb/At} \cdot \text{m})(3{,}200) = 4.02 \times 10^{-3} \text{ Wb/At} \cdot \text{m}$$

$$\mathcal{R}_{\text{CORE}} = \frac{l}{\mu A} = \frac{0.120 \text{ m}}{\left(4.02 \times 10^{-3}\dfrac{\text{Wb}}{\text{At} \cdot \text{m}}\right) \times (2.83 \times 10^{-5} \text{ m}^2)} = \mathbf{1.05 \times 10^6 \text{ At/Wb}}$$

A typical magnetic circuit has one or more air gaps, so the total reluctance of the magnetic circuit increases when there is an air gap. The total reluctance in a magnetic circuit is the sum of the reluctance of each element, just as the total resistance in an electrical circuit is the sum of the resistances. An air gap increases the reluctance dramatically over solid magnetic materials such as iron. Example 13-3 illustrates this concept.

EXAMPLE 13-3

Assume the toroid in Example 13-2 has a 2 mm air gap cut in it.

(a) What is the reluctance of the air gap?

(b) What is the reluctance of the total toroid and air gap? (Assume the toroid has the same mean length as in Example 13-2.)

Solution

Assume the magnetic field lines in the gap have the same area as the toroid (2.83×10^{-5} m^2)

(a) $\mathcal{R}_{\text{AIR}} = \dfrac{l}{\mu_0 A} = \dfrac{0.002 \text{ m}}{(4\pi \times 10^{-7} \text{ Wb/At} \cdot \text{m}) \times (2.83 \times 10^{-5} \text{ m}^2)} = \mathbf{5.62 \times 10^7 \text{ At/Wb}}$

(b) $\mathcal{R}_T = \mathcal{R}_{\text{CORE}} + \mathcal{R}_{\text{AIR}} = 1.05 \times 10^6 \text{ At/Wb} + 5.62 \times 10^7 \text{ At/Wb} = \mathbf{5.72 \times 10^7 \text{ At/Wb}}$

As you can see in Example 13-3, the air gap is responsible for most of the reluctance. For this reason, the air gap is kept as small as possible in generators and motors to minimize the reluctance. Rotors should have minimal clearance with the fixed magnetic path.

EXAMPLE 13-4

Mild steel has a relative permeability of 800. Calculate the reluctance of a mild steel core that is 10 cm long and has a cross section of 1.0 cm by 1.2 cm.

Solution

First, determine the permeability of mild steel:

$$\mu = \mu_0\mu_r = (4\pi \times 10^{-7}\,\text{Wb/At} \cdot \text{m})(800) = 1.00 \times 10^{-3}\,\text{Wb/}\Lambda\text{t} \cdot \text{m}$$

Next, convert the length to meters and the area to square meters:

$$l = 10\,\text{cm} = 0.10\,\text{m}$$

$$A = 0.010\,\text{m} \times 0.012\,\text{m} = 1.2 \times 10^{-4}\,\text{m}^2$$

Substituting values into Equation 13-4:

$$\mathcal{R} = \frac{l}{\mu A} = \frac{0.10\,\text{m}}{(1.00 \times 10^{-3}\,\text{Wb/At} \cdot \text{m})(1.2 \times 10^{-4}\,\text{m}^2)} = \mathbf{8.33 \times 10^5\,\text{At/Wb}}$$

Magnetomotive Force

As you have learned, current in a conductor produces a magnetic field. The cause of a magnetic field is called the **magnetomotive force (mmf)**. Magnetomotive force is something of a misnomer because, in physics, mmf is not really a force; rather, it is a direct result of the movement of charge (current). The unit of mmf, the **ampere-turn (At)**, is established on the basis of the current in a coil of wire. The formula for mmf is:

$$F_m = NI \qquad\qquad \text{Equation 13-4}$$

where

F_m is the magnetomotive force, in ampere-turns

N is the number of turns of wire

I is the current, in amperes

Figure 13-4 is the basic electromagnetic circuit originally shown in Figure 2-25(a) and is repeated here for reference. The figure shows that a number of turns of wire carrying a current around a magnetic material creates a force that sets up flux lines through the magnetic path. The amount of flux depends on the magnitude of the mmf and on the reluctance of the material, as expressed by the following equation:

$$\phi = \frac{F_m}{\mathcal{R}} \qquad\qquad \text{Equation 13-5}$$

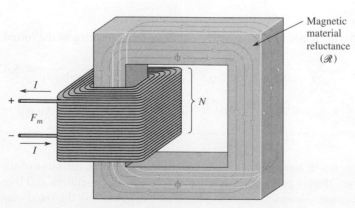

FIGURE 13-4 Basic Electromagnetic Circuit

Equation 13-5 is known as the Ohm's law for electromagnetic circuits because the flux (ϕ) is analogous to current, the mmf (F_m) is analogous to voltage, and the reluctance ($\mathcal{R}$) is analogous to resistance. Like other phenomena in science, the flux is an effect, the mmf is a cause, and the reluctance is an opposition.

One important difference between an electric circuit and a magnetic circuit is that, in magnetic circuits, Equation 13-5 is valid only up to a certain point before the magnetic material saturates (flux becomes a maximum). Another difference that was already noted is that flux does occur in permanent magnets with no source of mmf. In a permanent magnet, the flux is due to internal electron motion rather than to an external current. No equivalent effect occurs in electric circuits.

EXAMPLE 13-5

How much flux is established in the magnetic path of Figure 13-5 if the reluctance of the material is 2.8×10^5 At/Wb?

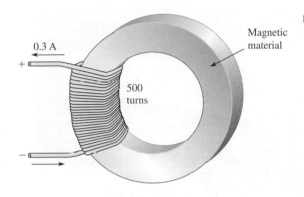

FIGURE 13-5

Magnetic material

0.3 A

500 turns

Solution

$$\phi = \frac{F_m}{\mathcal{R}} = \frac{NI}{\mathcal{R}} = \frac{(500)(0.3 \text{ A})}{2.8 \times 10^5 \text{ At/Wb}} = 5.36 \times 10^{-4} \text{ Wb} = \textbf{536 } \boldsymbol{\mu}\textbf{Wb}$$

EXAMPLE 13-6

There is 0.1 ampere of current through a coil with 400 turns.

(a) What is the mmf?

(b) What is the reluctance of the circuit if the flux is 250 μWb?

Solution

(a) $N = 400$ and $I = 0.1$ A:

$$F_m = NI = (400 \text{ t})(0.1 \text{ A}) = \textbf{40 At}$$

(b) $\mathcal{R} = \dfrac{F_m}{\phi} = \dfrac{40 \text{ At}}{250 \text{ } \mu\text{Wb}} = \textbf{1.60} \times \textbf{10}^\textbf{5} \textbf{ At/Wb}$

In many magnetic circuits, the core is not continuous. For example, an air gap cut into the core increases the reluctance of the magnetic circuit. This means that more current is required to establish the same flux as before because an air gap represents a significant opposition to establishing flux.

Electromagnet

An electromagnet is based on the properties that you have just learned, and it is created when electric current is in a conductor. If electrical current is stopped, the magnetic field immediately collapses and ceases to exit. When current is increased in the conductor, the number of flux lines increases and the magnetic strength gets stronger. The magnetic flux lines of a permanent magnet cannot be turned off because the current is the motion of electrons in domains that are aligned. Generators (introduced in the next section) and motors are common in renewable energy systems and are based on electromagnetic principles. Figure 13-6 shows a motor where you can see the rotor (a rotating electromagnet) inside the stator (a fixed electromagnet). The case of the motor forms part of the magnetic path for the fixed magnetic field. The ability of the electromagnet to change its magnetic field by changing the amount of current in its wires is perhaps the most important concept for both motors and generators.

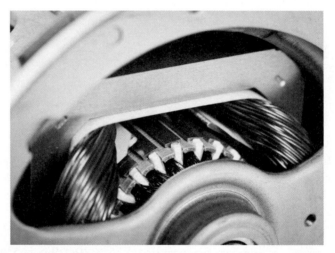

FIGURE 13-6   Typical Motor with Electromagnets   (*Source:* Dmitry Naumov/Fotolia.)

Magnetic Field Intensity

The **magnetic field intensity** (H) (also called magnetizing force) in a material is defined as the magnetomotive force (F_m) per unit length (l) of the magnetic material, as expressed by the following equation:

$$H = \frac{F_m}{l}$$

Equation 13-6

where

$$F_m = NI$$

The unit of magnetic field intensity (H) is ampere-turns per meter (At/m). Note that the magnetic field intensity (H) depends on the number of turns (N) of the coil of wire, the current (I) through the coil, and the length (l) of the material. It does not depend on the type of material.

Because $\phi = F_m/\mathcal{R}$ as F_m increases, the flux increases. Also, the magnetic field intensity (H) increases. Recall that the flux density (B) is the flux per unit cross-sectional area. $B = \phi/A$, so B is also proportional to H. The curve showing how these two quantities (B and H) are related is called the *B-H* curve, or the hysteresis curve. The parameters that influence both B and H are illustrated in Figure 13-7.

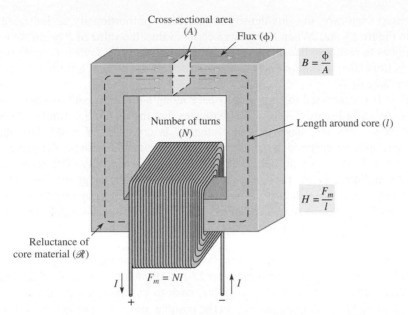

FIGURE 13-7 Parameters That Determine the Magnetic Field Intensity (H) and the Flux Density (B)

Hysteresis Curve and Retentivity

Hysteresis is a characteristic of a magnetic material whereby a change in magnetization lags the application of the magnetic field intensity. The magnetic field intensity (H) can be readily increased or decreased by varying the current through the coil of wire, and it can be reversed by reversing the voltage polarity across the coil.

Figure 13-8 illustrates the development of the hysteresis curve. Let's start by assuming a magnetic core is unmagnetized so that $B = 0$. As the magnetic field intensity (H)

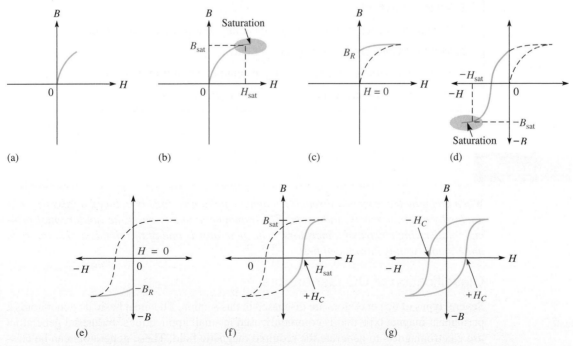

FIGURE 13-8 Development of a Magnetic Hysteresis (B-H) Curve

is increased from zero, the flux density (B) increases proportionally, as indicated by the curve in Figure 13-8(a). When H reaches a certain value, the value of B begins to level off. H continues to increase, but B reaches a **saturation** value (B_{sat}) when H reaches a value (H_{sat}), as illustrated in Figure 13-8(b). Once saturation is reached, a further increase in H does not increase B.

Now, if H is decreased to zero, B falls back along a different path to a residual value (B_R), as shown in Figure 13-8(c). This indicates that the material continues to be magnetized even when the magnetic field intensity is removed ($H = 0$). The ability of a material, once magnetized, to maintain a magnetized state without the presence of a magnetizing force (magnetic field intensity) is called **retentivity.** The retentivity of a material is indicated by the ratio of B_R to B_{sat}. Reversal of the magnetic field intensity is represented by negative values of H on the curve and is achieved by reversing the current in the coil of wire. An increase in H in the negative direction causes saturation to occur at a value ($-H_{sat}$) where the flux density is at its maximum negative value, as indicated in Figure 13-8(d).

When the magnetic field intensity is removed ($H = 0$), the flux density goes to its negative residual value ($-B_R$), as shown in Figure 13-8(e). From the $-B_R$ value, the flux density follows the curve indicated in Figure 13-8(f) back to its maximum positive value when the magnetic field intensity equals H_{sat} in the positive direction. The complete B-H curve is shown in Figure 13-8(g) and is called the hysteresis curve. The magnetic field intensity required to make the flux density zero is called the coercive force (H_C).

In motors, generators, and transformers, the current in the coils is constantly changing. Assume that the current in a coil goes to zero and the magnetic core material has residual magnetism. It requires additional current in the opposite direction of the initial current to demagnetize the material. Demagnetizing the core uses energy that is lost, and this lost energy is known as hysteresis loss. This type of loss is undesirable for motors, generators, and transformers, so magnetic materials with low hysteresis (small area bounded by the hysteresis curve) are selected for these devices.

SECTION 13-1 CHECKUP

1. What is a weber?
2. What is the difference between permeability and relative permeability?
3. What is Ohm's law for a magnetic circuit?
4. Why is it desirable to keep the air gap small in a motor or generator?
5. What causes hysteresis loss in motors and generators?

13-2 DC Generators

When a conductor is moved through a magnetic field, a voltage is induced across the conductor by electromagnetic induction. Electromagnetic induction is the fundamental principle underlying electrical generators. A dc generator is one of the simplest generators to understand, so it is introduced first.

Classification of DC Generators

Several types of dc generators are discussed in this section. The most basic dc generator is a permanent magnet type that is commonly used in small applications. Industrial generators use electromagnets to generate the required magnetic field. These generators can be classified according to how the magnetic field is generated. Figure 13-9 shows an overview of this classification method; each type of generator is discussed in this section.

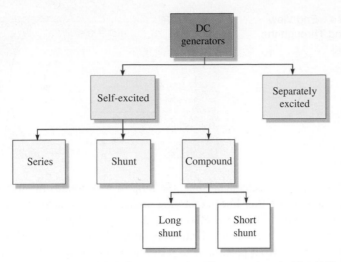

FIGURE 13-9 Classification of DC Generators by Electromagnetic Field Windings

Simple DC Generator

Figure 13-10 shows a greatly simplified view of a dc generator. It consists of a single loop of wire that rotates in a magnetic field. Keep in mind that practical dc have many turns wrapped on an iron core. For simplicity, a permanent magnet is shown for the field. Notice that each end of the loop is connected to a split-ring arrangement. Recall from Section 2-7 that this conductive metal ring is called a **commutator**; it is a rotary electrical switch that periodically changes the direction of current from the rotor to the external circuit. As the wire loop rotates in the magnetic field, the split commutator ring also rotates. Each half of the split ring rubs against the fixed contacts, called **brushes,** which are carbon blocks that press against the rotor of a motor or generator to make an electrical connection between the fixed part and the rotating part. The brushes connect the rotating wire to the external circuit.

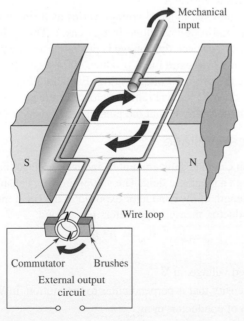

FIGURE 13-10 Basic DC Generator

FIGURE 13-11 End View of Loop Cutting Through the Magnetic Field

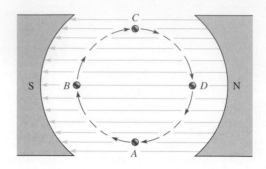

Driven by an external mechanical force, the loop of wire rotates through the magnetic field and cuts through the flux lines at varying angles, as illustrated in Figure 13-11. At position A in its rotation, the loop is effectively moving parallel with the magnetic field. Therefore, at this instant, the rate at which it is cutting through the magnetic flux lines is zero. As the loop moves from position A to position B, it cuts through the flux lines at an increasing rate. At position B, it is effectively moving perpendicular to the magnetic field and is thus cutting through a maximum number of lines. As the loop rotates from position B to position C, the rate at which it cuts the flux lines decreases to minimum (zero) at C. From position C to position D, the rate at which the loop cuts the flux lines increases to a maximum at D and then to a minimum again at A.

As you have learned, when a wire cuts magnetic field lines, a voltage is induced. The amount of induced voltage is proportional to the number of loops (turns) in the wire and the rate at which it is moving with respect to the magnetic field. The angle at which the wire moves with respect to the magnetic flux lines determines the amount of induced voltage because the rate at which the wire cuts through the flux lines depends on the angle of motion.

Figure 13-12 illustrates how a voltage is induced in the external circuit as a single loop of wire rotates in the magnetic field. Assume that the loop starts in the instantaneous vertical position (position A), so the induced voltage is zero. As the loop rotates, the induced voltage builds up to a maximum at position B, as shown in the figure. Then, as the loop continues from B to C, the voltage decreases to zero at C where it is back to zero.

During the second half of the revolution, the brushes switch to opposite commutator sections, so the polarity of the voltage remains the same across the output. Thus, as the loop rotates from position C to D and then back to A, the voltage increases from zero at C to a maximum at D, and back to zero at A.

Figure 13-13 shows how the induced voltage varies as a single wire loop in a dc generator goes through several rotations (three in this case). This voltage is dc because the polarity does not change. However, the voltage is pulsating between zero and its maximum value. Figure 13-14 shows how extra loops produce extra segments of voltage that smooth out the ripple and raise the average voltage rating. Each additional loop added to the generator ensures that the dc voltage level remains closer to the peak voltage. Capacitors can also be used as filters to help smooth out the pulsing dc ripple.

Calculating Induced Voltage

The following equation calculates the amount of induced voltage that a wire loop can produce as it moves through a magnetic field. The values are flux density, measured in webers per square meter; the length of the conductor exposed to the field, measured in meters; and the velocity of the conductor, measured in meters per second.

$$v_{ind} = B_\perp lv$$ Equation 13-7

where

v_{ind} = induced voltage, in V

$B_\perp$ = flux density that is perpendicular to the motion, in Wb/m^2

l = length of conductor, in m

v = velocity of the conductor, in m/s

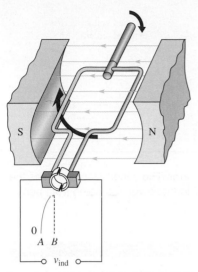

Loop moves from *A* to *B*. At *B* the maximum flux lines are cut and the voltage is at the peak.

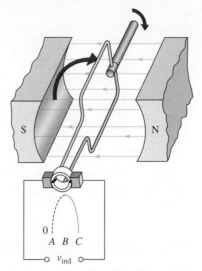

Loop moves from *B* to *C*. At *C* the minimum flux lines are cut and the voltage is zero.

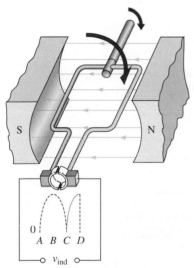

Loop moves from *C* to *D*. At *D* the maximum flux lines are cut and the voltage is at the peak.

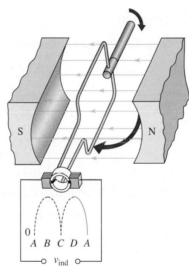

Loop moves from *D* to *A*. At *A* the minimum flux lines are cut and the voltage is zero.

FIGURE 13-12   Basic Operation of a DC Generator

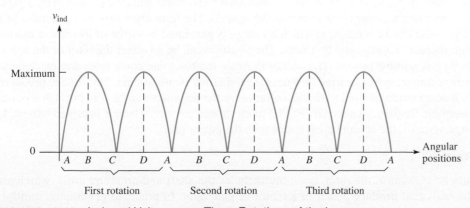

FIGURE 13-13   Induced Voltage over Three Rotations of the Loop

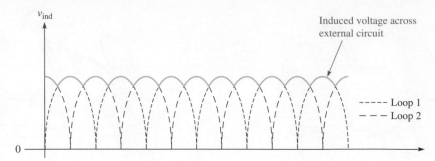

FIGURE 13-14   Induced Voltage for a Two-Loop Generator. The second loop doubles the amount of segments, which causes less variation in the induced voltage going from peak to zero.

EXAMPLE 13-7

What is the induced voltage in a single-loop conductor that is 35 cm long and passes through a magnetic field that has 82,400 lines/sq cm and moves at a rate of 135 cm per second?

Solution

Step 1:   Convert units to mks form:

$$B_\perp = \left(\frac{82,400 \text{ lines}}{\text{cm}^2}\right)\left(\frac{10^4 \text{ cm}^2}{\text{m}^2}\right)\left(\frac{\text{Wb}}{10^8 \text{ lines}}\right) = 8.24 \text{ Wb/m}^2$$

$$l = \left(\frac{35 \text{ cm}}{100 \text{ cm/m}}\right) = 0.35 \text{ m}$$

$$v = \left(\frac{135 \text{ cm/s}}{100 \text{ cm/m}}\right) = 1.35 \text{ m/s}$$

Step 2:   Substitute into Equation 13-7 for induced voltage

$$v_{ind} = B_\perp lv = (8.24 \text{ Wb/m}^2)(0.35 \text{ m})(1.35 \text{ m/s}) = \textbf{3.89 V}$$

Industrial DC Generator

Figure 13-15 shows a small industrial dc generator that was taken out of service and disassembled. The dc generator is divided into two parts: the moving portion called the rotor (which includes the fan, armature, commutator, and shaft [shown in the figure]) and a stationary part called the stator (which includes the end plates and case [shown in the figure] and the stator windings [not shown in the figure]). The term **armature** in a generator or a motor refers to the winding in which a voltage is generated by virtue of its relative motion with respect to a magnetic flux field. The armature can be on either the rotor or the stator; the key is *relative* motion. The energy to rotate the rotor can come from any mechanical energy source, such as a wind turbine or a geothermal steam turbine. The voltage produced by a dc generator is used to charge batteries, or the dc power can be sent through a power electronic frequency converter (PEFC) or electronic inverter, where it is turned into ac. In large systems, the ac can be interfaced with the power grid.

Rotor

In a dc generator, the rotor assembly includes the shaft and armature coils, which are the coils that produce power in a motor or generator. In practical generators, multiple coils are pressed into slots in a ferromagnetic-core rotor assembly. The armature coils

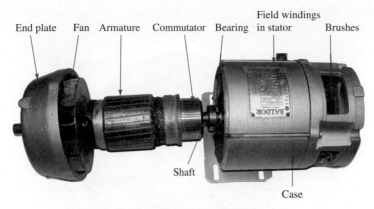

End plate   Fan   Armature   Commutator   Bearing

Field windings
in stator   Brushes

Shaft

Case

FIGURE 13-15 Industrial DC Generator. The brushes are shown in Figure 13-16, and the field windings are shown in Figure 13-17. (*Source:* Tom Kissell.)

are connected to the commutator, which is a rotating switch that connects coils through brushes to the output terminals. The entire assembly rotates in the magnetic field. The fan blades on the end of the shaft ensure that sufficient air moves through the generator to keep it cool when the armature is rotating. The shaft is supported by a bearing on each end.

Brushes and Commutator

Most generators have a number of rotor coils, and the commutator constantly switches different coils to the output. **Commutation** is the provision of a unidirectional current from the generator as different coils are switched to the output. Thus, one of the brushes always has positive voltage potential, and the other always has negative potential. In effect, the commutator and brushes serve as a mechanical rectifier.

The commutator is divided into segments, with each pair of segments connected to the end of a coil. The voltage from several coils is combined because the brushes can contact more than one of the commutator segments at once. The loops are sequentially connected to the output when they are near maximum voltage, so the pulsating output voltage is much smoother than is the case with only one coil or loop, as shown previously. Figure 13-16(a)

(a) The commutator with a carbon brush to show the relative size of each

(b) The brush rides directly on the commutator

FIGURE 13-16 Commutator and Brush (*Source:* Tom Kissell.)

FIGURE 13-17   Stator Assembly   (*Source:* Tom Kissell.)

shows a close-up view of how the ends of each coil on the armature are soldered to a commutator segment. A single carbon brush is also shown in this image to show the relative size. Two carbon brushes—one on each side of the commutator—are used to make contact between the commutator and the output terminals. All of the output current must pass through these brushes. Figure 13-16(b) shows a brush mounted in place against the commutator. A very strong spring holds the brushes against the commutator to maintain a low-resistance connection. Carbon brushes wear down eventually because of the contact with the commutator and because of electrical arcing. When the brush is worn down, the spring can be pulled back and a new brush can be placed into the holder.

Stator

Recall that that the stationary part of a generator is called the stator. Figure 13-17 shows the stator with the rotor removed. The rotor fits inside the stator, which is the stationary part of the generator that has the field windings mounted onto a steel core. Notice that the core is shaped to follow the rotor's curvature and thus keep the air gap to a minimum (review Example 13-3 to see why this is important). The core, case, rotor, and air gap are all part of the magnetic circuit for the stator. The windings in the stator are connected to an external voltage source, which causes current in the stator coils, thus creating the strong magnetic field that is directed across the air gap and into the rotor. As the armature rotates from an external energy source, the armature coils cross the magnetic field lines from the stator and voltage is induced across the armature coils.

Separately Excited and Self-Excited Generators

DC generators are classified by how they obtain current for the field electromagnets (called excitation current) and how they are wired. We will look first at excitation methods.

As you have seen, nearly all dc generators use an electromagnet to create the required magnetic field (very small generators can use a permanent magnet). An electromagnet requires current to produce the field (except for a small residual field). Current can be provided for the field in one of two ways: (1) use a separate source for field current or (2) use a portion of the generator's own output to supply current for the field. Both methods have advantages and disadvantages, which will be discussed later in this section.

Series, Shunt, and Compound DC Generators

The second factor in classifying dc generators is how the field and armature coils are wired for a self-excited generator. They can be wired in series, shunt, or compound generators. The term **shunt** means "parallel," and in the case of generators, this means that the field coils are in parallel with the armature coils. Keep in mind that schematics generally show a single field coil, but it is split to put part of it on each pole of the magnetic assembly.

Series DC Generators

In a series-wound dc generator, the field windings are in series with the armature windings and the load. Because of the series connection, the field current, armature current, and load current are all identical. If the load increases (less resistance), current increases; hence the field current and magnetic field increase. The result is increased output voltage for the increased load current, which tends to raise the output current even more. This is classic positive feedback, an undesirable situation for most applications. For this reason, the series-wound dc generator has only limited applications.

Shunt Generators

A **shunt generator** is one where the field winding and armature winding are in parallel. Unlike the series-wound generator, current in the field windings is not part of the load current. However, there is still a loading effect on a dc shunt generator. When the output current increases, more voltage is dropped across the resistance of the armature coil, causing the output to decrease. Another effect that reduces the output is called armature reaction. Armature reaction is an effect caused by armature current whereby current in the armature generates an mmf that tends to distort and decrease the magnetic flux from the field coils. The decrease from these two effects means that a self-excited generator supplies less current to the field windings, which reduces the magnetic field and reinforces the tendency for the voltage to drop under load.

Compound Generators

A **compound generator** has two field windings: one in series and one in parallel. The load current and the series field winding are the same, so an increase in load current tends to increase the magnetic field from the series winding and hence the output tries to increase. The increase in load current in the shunt winding tends to compensate for this by decreasing the output. The series and shunt windings combine and tend to keep the magnetic field strength relatively constant, so the output voltage tends to be independent of the load in the typical case; this is called a flat compounded generator. The compensation can keep the output to a 1% to 2% variation over the full range of loads. Compound generators have the most consistent output for varying loads of any of the three types.

Separately Excited DC Generator

Figure 13-18 shows a diagram of a **separately excited shunt generator.** Notice that the armature is represented as a circle and a brush is shown on each side of the armature. The

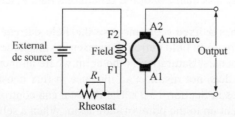

FIGURE 13-18   Separately Excited DC Generator. The separately excited dc generator has the field voltage connected to an external dc source.

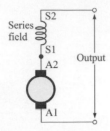

FIGURE 13-19 Self-Excited Series DC Generator. The self-excited series dc generator has the field winding in series with the armature and load (not shown).

two wires connected to the brushes are identified as A1 and A2 to indicate that they are armature wires and this is where electrical power is removed for the load. Recall that separate excitation means that there is an external supply for the field current. In Figure 13-18, the field voltage comes from a battery and is connected to terminals labeled F1 and F2. A rheostat (a type of variable resistor that controls current) is shown to adjust the field current, but this is usually some form of electronic control. This armature supplies current directly to the load and the field does not share load current. With no load, the output voltage is the same as the armature voltage. When a load is connected, the output drops because of an *IR* drop in the armature and because armature reaction, as described for shunt generators, which tends to shift the magnetic field due to armature current.

A separately excited generator is controlled by varying the source current, which can be useful in renewable energy systems such as wind turbines in which the shaft speed may vary. Controlling the excitation can help maintain a constant dc output. The disadvantage of this method is that a separate source of dc must be available, which adds to the cost.

Self-Excited Series DC Generator

A self-excited generator is one in which some of the output is used to provide current for the field. The self-excited series dc generator has the field winding in series with the armature; thus, the armature current, field current, and load current are the same. Figure 13-20 shows the arrangement. When the generator is started, a small residual magnetism produces a field and the output is small. As current is supplied to the load, the increased current causes the output voltage to increase. Thus, the voltage varies with load current, an undesirable characteristic for generators. Although a series-wound dc generator has very limited application by itself, series field coils are very useful in compound generators, as you will see.

The generator starts from residual magnetism in the stator's poles. This is a case where a small amount of hysteresis is an advantage because the core "remembers" the last state it was in and begins to output a small voltage. If the generator has not been used for a long time, it may be necessary to apply an external source to the field windings to start it; this is called **flashing the fields**.

Most dc generators are self-excited because the method is simpler; they use the residual magnetism in their fields to produce the required field at start-up. Once voltage is produced, the self-excited generator can tap some portion of the output current to control the generator's output. The voltage regulation tends to be less efficient than that found in separately excited generators.

Self-Excited DC Shunt Generator

The majority of dc generators are self-excited shunt generators. A **self-excited shunt generator** is a generator that supplies voltage from its armature to create the field current. The field and armature windings are parallel. Figure 13-20 shows the configuration. In the schematic shown, a rheostat (R_1) is shown in series with the shunt field to control the current in the field. This connection allows a small fraction of armature current to be routed back through the field winding to create the magnetic field. The amount of power the field needs is usually less than 2% of the generator's output, and the rheostat is set for the required amount.

The field winding has a fixed resistance, so the field current is proportional to the applied voltage. An increase in field current increases the magnetic field up to the point at which core saturation occurs. Saturation is the point where an increase in magnetic field intensity (*H*) in a core does not result in an increase in flux density (review the discussion of hysteresis curves in Section 13-1). The rheostat can control the output voltage by adjusting the field current up to the point of saturation. When a self-excited shunt generator is turned at a constant speed, output voltage remains fairly constant. Because the field

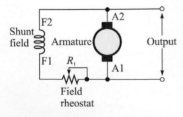

FIGURE 13-20 Self-Excited Shunt Generator. The self-excited shunt dc generator has the field winding and field rheostat in parallel with the armature and load (not shown).

is connected in parallel with the armature, the voltage applied to the field also remains fairly constant. This equilibrium continues as long as the load is constant and it remains at constant speed. If the voltage is used for charging a bank of batteries, a small change in voltage is not a problem.

Long-Shunt and Short-Shunt Compound Generators

As you have learned, a flat compounded generator uses both a series field and a shunt field to maintain nearly constant output voltage over a range of loads. The series field consists of a few turns of very large wire and it is connected in series with the armature, whereas the shunt field consists of many turns of smaller-diameter wire. The shunt field coil can be connected in one of two ways. The first is to connect field coils, armature, and rheostat, as shown in Figure 13-21 with the connections labeled. In this configuration, the combination of the shunt field coil and its series rheostat across are both in parallel with the combination of the armature coil and the series field coil. This configuration is called a long-shunt generator.

The second way to connect the coils is to connect the circuit shown in Figure 13-22. In this configuration, the combination of the shunt field with its series rheostat is connected across the armature only. This arrangement is called a short-shunt generator. In both the long-shunt and short-shunt configurations, the coils can be wound on the same pole pieces.

Controlling the Output Voltage

The output voltage of a dc generator can be controlled in one of two ways. A prime mover is the source of kinetic energy that is converted to electrical energy by the generator; it can be a steam turbine, a hydroelectric generator, a wind turbine, or another source. If the prime mover rotates with constant speed, then field current control is the method that is used to vary the output. The first method is to maintain a constant rotational speed and vary the field current to vary the output voltage. The output voltage follows the field current before saturation.

The second method is to vary the rotational speed of the generator. The output voltage is proportional to the rotational speed. Frequently, a renewable energy system does not have a consistent rotational speed (some wind turbines, for example) but needs to have a constant output voltage. In this case, the field current can compensate for speed changes. The field current can be controlled automatically by an electrical circuit called a voltage regulator. A **voltage regulator** is an electrical circuit that automatically maintains a constant output voltage level over variations in load. If the voltage regulator uses solid-state devices such as insulated gate bipolar transistors (IGBTs) or operational amplifiers (op-amps), the voltage can be controlled automatically, even as the load changes or the speed changes. If the voltage regulator is an older style, you may need to adjust the field current manually to set the voltage to the level that the application requires.

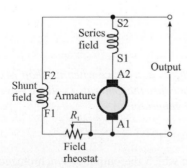

FIGURE 13-21 Long-Shunt Compound Generator

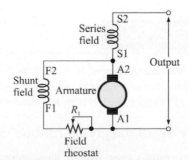

FIGURE 13-22 Short-Shunt Compound Generator

Polarity in DC Generators

The polarity of a dc generator can be reversed in one of two ways. First, the direction of rotation can be changed so the polarity of the armature voltage is reversed. If the generator is driven by a wind turbine or steam turbine, this may not be an option. Second, the polarity of the field winding with respect to the armature winding is reversed. In some cases, you will have trouble getting the polarity the way you want it and you may need to flash the field with an external voltage to change the polarity. (Flashing the field restores a small amount of residual magnetism in the core by briefly connecting a low voltage external source to the field connections when the field has been isolated.) Another way to get the polarity correct to the load is to reverse the wires from the load that are connected to the + and − terminals on the generator. The net result is the same, but reversing the load does not change the polarity of the generator.

EXAMPLE 13-8

You need to adjust the output voltage of a dc generator.

(a) Explain two ways you can change the output voltage of the generator.

(b) Describe two ways you can change the polarity of the voltage that is sent to the load if you could not reverse the direction of rotation.

Solution

(a) One way is to change the speed of the armature if possible. The second way is to adjust the voltage regulator.

(b) One way to change the polarity is to reverse the polarity of the field winding. The second way is to change the polarity of the armature terminals where the load is connected.

SECTION 13-2 CHECKUP

1. How is induced voltage created?
2. Explain the parts of a simple dc generator.
3. What is the function of the armature in a dc generator?
4. What is the function of the field windings in a dc generator?
5. What type of dc generator tends to have a constant voltage regardless of the load?

13-3 AC Synchronous Generators

A synchronous generator is a machine that produces ac voltage when its shaft is rotated. A synchronous generator is called synchronous because the generated voltage waveform it produces is synchronized with the rotation of the generator.

Overview of the Synchronous Generator

A **synchronous generator** is an ac generator in which the output is synchronized to the position of the rotor. The frequency of the voltage produced by the synchronous generator depends only on the speed at which its shaft is turned and the number of poles it has. This makes the synchronous generator very efficient for producing electrical power for utility companies because it produces power at line frequency on a continual basis when its rotor is rotated at a constant rate.

Large synchronous generators require an excitation voltage for the field. This voltage comes from a separate power source such as a smaller auxiliary dc generator called an **exciter** to supply field current. Usually, the exciter is mounted on the main shaft. Different types of exciters include separate exciters that are dc generators, static exciters (with no rotating parts), and shaft-driven dc exciters. Current from the exciter is usually controlled by an automatic or manual regulator.

You will encounter two types of synchronous generators when you are working with renewable energy systems. In one type, the armature is the rotor, and current from the armature is generated in the rotor; this is called a rotating armature ac generator. In this case, slip rings and brushes are used to pass current from the rotor through insulated porcelain bushings to the electrical terminals on the frame of the generator. The other type has the field on the rotor and the armature on the stator. In this case, slip rings and brushes may not be necessary because power is produced in the stationary stator, and rotor current can be supplied from a separate rotating exciter that is mounted on the same shaft. This is called a rotating-field ac generator. In either case, the rotor shaft is connected to a prime mover that causes it to spin.

Recall that the armature is the coil that generates power from a generator. In large generators, the field rotates and the armature windings are on the stator. Three-phase is standard for utilities because it can be transmitted at lower cost, and a three-phase generator is significantly smaller than a single-phase generator of the same rating. The electrical frequency of the three-phase output voltage depends on the mechanical speed of the rotor and the number of poles in the generator as mentioned previously.

Rotating Armature AC Generator

The rotating armature generator is also called the stationary field generator. In a small rotating-armature generator, the magnetic field can be supplied by permanent magnets surrounding the rotor or by electromagnets.

Because the armature is in the rotating assembly, slip rings and brushes are used to take current from the rotor and pass it to the output, as discussed in Section 2 7. In addition to hundreds of windings, the practical rotating-armature generator usually has many pole-pairs in the stator that alternate as north and south poles around the periphery. Opposite poles are positioned next to each other so that the rotor generates a complete sine wave as it passes each pair of poles. When the prime mover turns the rotor, the armature windings cut the magnetic flux lines from the field and generate a sinusoidal wave.

The poles for the field winding are part of the magnetic path; the path includes the rotor, air gap, stator poles, and case, but not the bottom plate. The bottom plate is made of a nonmagnetic material to eliminate induced current. The field windings are wound on the poles. A single-phase generator has two slip rings that are connected to the coil on the rotor. Rotating armature ac generators are typically used for low-power applications, usually less than 5 kVA, because the current through the slip rings and brushes is low. Most rotating-armature generators produce only single-phase.

Rotating-Field AC Generator

All large ac synchronous generators are rotating-field generators, which are universally used by utility companies. The rotating-field ac generator is also called the stationary armature generator. Because the armature windings are on the stator, larger amounts of power can easily be generated and moved to the load or to the grid (there are no moving contacts between the armature and the output terminals). In very small, rotating-field ac generators, permanent magnets may be used for the rotor field; however, most rotating-field generators use an electromagnet for the rotor (this is known as a wound rotor). A **wound rotor** is a rotor core assembly that has a winding made up of individually insulated wires. In a generator, dc is supplied to the rotor to provide the magnetic flux for the rotating field.

Slip Rings and Carbon Brushes

(*Source:* Tom Kissell.)

Slip rings are smooth rings that are mounted on a rotor and connected to one end of the rotor coil. Brushes ride directly on the slip rings and make electrical contact with the external terminals. Brushes wear down eventually, so they must be inspected periodically and replaced as needed.

Because dc is provided, the electromagnet has fixed polarity (like a bar magnet). As the rotating magnetic field sweeps by the stator windings, the magnetic field from the rotor cuts through armature windings in the stator and power is generated.

Utility companies are particularly concerned about the efficiency of their generators. As generators are made larger, their efficiency improves. A large machine actually weighs less per kW produced than a small machine, and with the increase in efficiency, larger is better from a utility company perspective. The only drawback is that large generators require some form of cooling. Three basic cooling systems—air cooling, compressed hydrogen cooling, and water/oil cooling systems—are in use. The required cooling system depends on the specific type of generator and power output. For example, large, low-speed, multi-pole generators are easier to cool than high-speed generators.

In a rotating-field ac generator, the current for the field winding is usually produced from an exciter. As mentioned previously, several different types of exciters are available, but it is common for the exciter armature and the main field rotor to move together on a common shaft. The exciter can be a dc generator or it can be an alternator that uses diodes to convert its output to dc. The net result is the required dc that is used to create the rotor field.

Figure 13-23 shows a diagram of a large rotating-field ac generator. It has a dc exciter generator on the left end of the rotor, and the rotating field in the main generator. The armature for the main generator is made of coils of wire pressed onto the stator poles. These coils are connected as three separate windings located 120° apart to produce three-phase (3φ) voltage. The current in the rotating field is controlled by the exciter, which in turn controls the output. The output voltage from the ac exciter is three-phase ac that is routed through a rotating six-diode bridge rectifier, where it is turned into dc. Because the exciter armature and diodes are mounted on the same shaft as the field windings of the main generator, the two wires that provide positive and negative dc to the main generator field can be connected directly without the need for slip rings and brushes. This means that the generator can run long periods between maintenance times. The field current can be controlled directly with a regulator, or the output voltage of the exciter can be controlled to increase or decrease the main generator output.

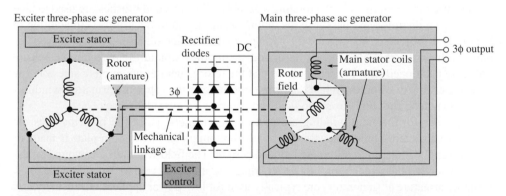

FIGURE 13-23   Rotating-Field Synchronous Generator with an Exciter that Supplies Current for the Rotor Field

Figure 13-24 shows a large three-phase synchronous generator that can produce up to 75 MVA of power. This is an example of a rotating-field generator that uses an exciter to provide field current. The rotor on the synchronous generator may be made as a salient pole or a nonsalient pole. The term **salient** means projecting beyond a surface, level, or line. The salient poles consist of wires wrapped tightly around magnetic pole pieces that project from the rotor. This design is limited to lower-speed generators, so it is useful for some smaller wind turbines and some-low speed hydroelectric turbines. Generators with nonsalient poles are used for higher speeds and are useful in fossil-fuel and nuclear plants, where they typically spin at 3,600 rpm to take advantage of the high-pressure steam. The higher rotational speeds produce stronger centrifugal forces, which would pull salient pole

FIGURE 13-24 Synchronous Generator for a Utility   (*Source:* Courtesy of GE Energy.)

rotors apart. Nonsalient poles are also called turbine poles. They can be made as a long steel cylinder. The rotor is made by pressing windings into slots of a pole piece, and this design can withstand the higher speeds produced by steam turbines.

Three-Phase Synchronous Generators

Figure 13-25(a) shows a simplified three-phase generator. The rotating field is shown as a permanent magnet. When the prime mover rotates the field past the three stator windings, a three-phase sine wave is produced. Figure 13-25(b) shows the three-phase output from the generator. A sine wave is produced as the pole of the field winding is rotated past the armature winding in the stator. If the north pole generates the positive half of the sign wave, the south pole generates the negative half cycle. Because the armature windings are mounted 120° apart in the stator, the sine waves are separated by 120°. Most generators produce three-phase because it is more efficient. If the final output needs to be dc, three-phase is easy to convert to dc using diodes. The synchronous generator does not have slip, so the output frequency is constant when the speed is held constant.

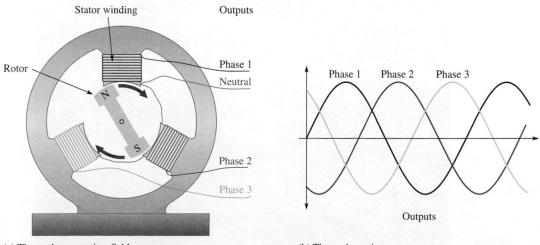

(a) Three-phase rotating-field generator

(b) Three-phase sine wave

FIGURE 13-25 Three-Phase Rotating-Field Generator and Three-Phase Sine Wave. Figure 13-15(b) shows the output for the generator. For simplicity, the rotor in Figure 13-25(a) is shown as a permanent magnet.

Synchronous Generators Used in Wind Turbines

The synchronous generator is generally used in wind turbines when the generator is connected directly to the grid and does not use an inverter. A primary advantage of synchronous generators for wind turbines is that they can receive voltage from the grid and act as an electric motor if the blades are not turning. If the wind speed is low, the generator can act as a motor to begin turning the blades. The voltage from the grid helps the motor come up to near-synchronous speed and starts the blades turning fast enough so that the wind can take over. If the motor were not used to turn the blades during start-up, the wind turbine would not be able to start to harvest energy until wind speeds are higher. As the wind begins to pick up and the blades begin to harvest energy, the voltage from the grid is automatically disconnected from the synchronous machine, and the wind turbine blades begin to turn the shaft fast enough that it is generating electricity. This transition occurs above a wind speed of approximately 6 mph.

Another advantage of using the synchronous generator in a wind turbine is that, when dc is provided to its rotating-field coil, a very strong magnetic field is created, and the synchronous generator has almost no slip. Thus, if the generator is connected correctly to the grid, it's shaft runs at near its design speed at all times, which ensures that it produces voltage with a frequency near its rated 60 Hz. Figure 13-26 shows a synchronous generator used in a wind turbine.

Calculating the Speed of a Synchronous Generator

The frequency of a synchronous generator is determined by the number of poles in the armature and the speed of the turning rotor. The equation for frequency from a synchronous generator is:

$$f = \frac{N_p \times \text{rpm}}{120}$$

Equation 13-8

where

$\quad\quad f = $ frequency, in Hz

$\quad N_p = $ number of poles

$\quad \text{rpm} = $ rotational speed, in revolutions per minute

FIGURE 13-26   Generator for a Wind Turbine   (*Source:* National Renewable Energy Laboratory.)

EXAMPLE 13-9

What is the frequency of an induced voltage for a generator that has four poles and turns at 1,800 rpm?

Solution

$$f = \frac{N_p \times \text{rpm}}{120} = \frac{4 \times 1,800}{120} = \textbf{60 Hz}$$

EXAMPLE 13-10

What is the frequency of an induced voltage for a generator that has four poles and turns at 1,500 rpm?

Solution

$$f = \frac{N_p \times \text{rpm}}{120} = \frac{4 \times 1,500}{120} = \textbf{50 Hz}$$

EXAMPLE 13-11

How fast does a twenty-four-pole generator need to rotate to produce 60 Hz?

Solution

$$rpm = \frac{120 \times f}{N_p} = \frac{120 \times 60 \text{ Hz}}{24} = \textbf{300 rpm}$$

Table 13-1 summarizes the number of poles and rpm required to produce 50 Hz or 60 Hz, which are the two most common frequencies for electrical grids throughout the world. The number of poles for a generator is always in pairs, so the number of poles is always an even number. The higher the number of pole-pairs, the lower the rotational speed for the generator to produce a given frequency.

Figure 13-27 shows a second set of field poles on a generator. When a second set of poles is added, the output voltage of the generator has an extra sine wave for each revolution of the

TABLE 13-1

Number of Poles Needed to Generate 50 Hz or 60 Hz

Number of Poles	RPM for 50 Hz	RPM for 60 Hz
2	3,000	3,600
4	1,500	1,800
6	1,000	1,200
8	750	900
10	600	720
12	500	600
14	429	514
16	375	450
18	333	400
20	300	360
40	150	180

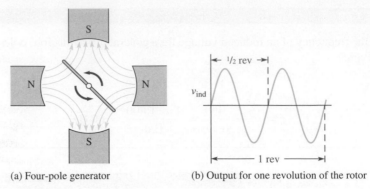

(a) Four-pole generator   (b) Output for one revolution of the rotor

FIGURE 13-27 Output Sine Wave for a Generator with Four Field Poles. Notice that there are two cycles at the output for every revolution of the rotor.

generator. If additional poles are added, more sine waves are produced during each revolution of the rotor.

SECTION 13-3 CHECKUP

1. What are the main parts of a synchronous generator?
2. How does a rotating-armature generator produce voltage?
3. How does a rotating-field generator produce voltage?
4. How does a permanent magnet synchronous generator produce voltage?

13-4 AC Induction Generators and Permanent Magnet Generators

*An induction generator is a type of **asynchronous generator**, meaning the waveform that is generated is not synchronized to the rotational speed. Induction generators are widely used in wind turbines and some smaller hydroelectric installations due to their simplicity. Another type of asynchronous generator is the permanent magnet generator. Permanent magnet generators are simple and reliable, and can work at low speeds, so they are ideal for use with wind turbines. With the advent of very strong permanent magnets, many wind turbine manufacturers are choosing low-speed permanent magnet generators for large turbines. Both induction and permanent magnet generators are covered in this section.*

Induction Machines

An **induction generator** is an asynchronous electrical machine that can function as a motor or as a generator. In the case of an asynchronous motor, the rotor spins less than the synchronous speed of the field; as a generator, it spins faster than the synchronous speed. An induction generator always starts as an induction motor, the most common type of motor in the world. As you know, a motor converts electrical energy to mechanical energy, and a generator does the opposite. In an induction motor, the rotor constantly tries to keep up with a rotating field in the stator (the synchronous speed), which is created by the applied ac. The rotor slips and does not turn as fast as this rotating magnetic field; if the rotor could catch up, no torque would be generated because there would be no relative motion between the rotor and the field. In a motor, ac that is applied to the stator is converted to mechanical rotational power that is taken from the rotor (hence, it is the armature). Slip is the difference in speed between the rotor speed and synchronous speed of the rotating stator field; as the load increases in a motor, slip increases and the motor slows. Maximum torque depends on the motor, but it is around 80% of synchronous speed.

FIGURE 13-28 Squirrel-Cage Rotor. In a motor, this type of rotor is the armature; in a generator, it generates the field.   (*Source:* Tom Kissell.)

Induction machines use two basic types of rotors. A common type of rotor is the **squirrel-cage rotor**, which was named many years ago because of its resemblance to an exercise wheel for a pet squirrel. Figure 13-28 shows a squirrel-cage rotor, which has aluminum bars connected on each end for conduction. The bars are embedded in an iron core, which produces a low-reluctance magnetic path. When ac power is applied to the stator, a voltage is induced in each rotor bar. The voltage is given by Equation 13-7 (repeated here for reference):

$$v_{ind} = B_\perp lv$$

The rotor has very little resistance to current because the conductive bars are short-circuited by the end rings; thus, a high current is developed. This current creates magnetic poles in the rotor that are attracted to the rotating magnetic field in the stator by the induced magnetic field in the rotor. Thus, the rotor is dragged along by the moving stator field at a speed that is nearly that of the rotating field. The rotor tries to keep up with the rotating field but cannot (if it ever did, the relative motion would be zero, and no voltage would be induced in it). This is the normal operation of any ac induction motor, similar to that found in the motor of a household appliance such as a refrigerator or a washing machine.

Another type of rotor is the wound rotor, which is more common in a three-phase machine. A three-phase machine usually has three windings on the rotor connected in a wye configuration and mounted on an iron core. The windings are connected through slip rings and brushes. The wound rotor has the option of connecting external resistors into the windings to limit current during start-up; this setup is useful in larger induction machines. For normal running, the resistors are shorted out by shorting the brushes together. The wound rotor has the advantage of being able to vary speed by varying the resistance. It also has higher starting torque than a squirrel-cage rotor due to the number of windings. Wound rotors are used in doubly fed generators in wind machines (covered later in this section).

Singly Fed Induction Generators

If instead of taking mechanical power from the rotor, the rotor is driven by a prime mover such as the wind, it can be moved faster than the synchronous speed and starts producing electrical power instead of consuming it. Thus, the basic induction motor becomes an induction generator. Electrical power is now taken from the stator, which now becomes the armature. Because of this dual nature, an induction machine is sometimes referred to as a motor/generator. All generators require a magnetic field either from some form of excitation current or from a permanent magnet. In the case of the induction generator, after passing synchronous speed, the magnetic field is induced into the rotor from ac that is applied to the stator. A prime mover such as the wind turns the rotor faster than the synchronous speed, and power is generated, which returns power to the grid from the stator windings.

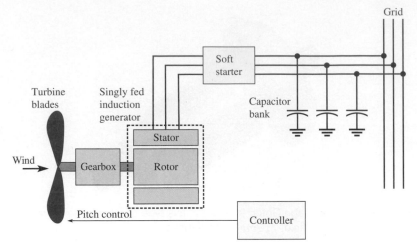

FIGURE 13-29   Basic Singly Fed Induction Generator (SFIG) Used in a Wind Turbine

The simplest induction generators are referred to as singly fed induction generators (SFIGs), which use a squirrel cage described previously. Because squirrel-cage induction machines look inductive, power factor correction capacitors are added to generators. In addition, a soft-starter unit is often used to reduce inrush current during start-up. Figure 13-29 shows a basic SFIG system used in a wind turbine.

Figure 13-30 shows a cutaway view of a small SFIG for a wind turbine. Notice that the rotor is mounted in close proximity to the stator to reduce the air gap. The main parts are the stator, which houses the armature windings; the squirrel-cage rotor, which provides the rotating field; and the end plates, which house the bearings that support the ends of the rotor shaft. The electrical terminals for the generator are located in the terminal box of the generator so connections can be made easily.

As a generator, the rotor turns faster than the synchronous speed, which is called negative slip. The synchronous speed is inversely proportional to the number of poles; if a generator has more poles, it will have a lower synchronous frequency. For a four-pole generator with a 50 Hz output, the synchronous speed is 1,500 rpm; for a 60 Hz output, it is

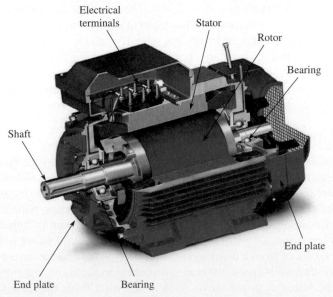

FIGURE 13-30   Cutaway View of an Induction Generator   (*Source:* Courtesy of Aerostar Inc.)

FIGURE 13-31   Forty kW Induction Generator for a Wind Turbine   (*Source:* Tom Kissell.)

above 1,800 rpm. If the number of poles doubles, these synchronous speeds are halved. To convert a slow-moving wind turbine to higher speed generally requires adding a gearbox to the system or adding many poles to the generator. To produce power, the wind speeds need to be above the transition speed; otherwise, the motor/generator acts as a motor.

When induction generators are used in larger wind turbines, they are designed as three-phase ac machines. The ac voltage is typically increased to 12,470 V or more and connected to the grid. Figure 13-31 shows a 40 kW (medium-size) three-phase induction generator for a wind turbine. The generator is 0.7 m long.

The stator coils are the armature coils on an induction generator, and the ends of these coils are connected to terminals that are accessible in a terminal box. In a true induction machine, the rotor creates the magnetic field only through induction as it turns past the stationary coils, so no slip rings or brushes are required. This is a great advantage in cases where minimum maintenance is important. Another advantage to induction generators is safety. If the grid goes down, the generator loses its field and stops so that it cannot send power to the grid. The drawback to induction generators is they are less efficient than are synchronous generators.

Doubly Fed (Double-Excited) Induction Generators

As in the case of singly fed machines, doubly fed machines can operate either as a motor or a generator. As a motor, they are useful for driving variable-speed devices such as certain tools or pumps. The **doubly fed induction generator (DFIG)** is particularly useful for wind turbines and is used in many larger turbines. A doubly fed induction generator has a wound rotor that is connected to a different source of ac than the stator. It typically has a three-phase wound rotor connected through brushes and slip rings to a secondary ac source that can be controlled for frequency, phase, and voltage. If the secondary field is 0, then the DFIG acts like an asynchronous generator, and the output frequency depends strictly on the rotor's rotational speed and the number of poles. When the rotor has a secondary frequency included, the rotational speed of the magnetic field is a combination of the rotor speed and the ac fed to the rotor. The magnetic field's rotational speed can either be increased or decreased by changing the phasing of the ac to the rotor. If the magnetic field due to the rotor's applied ac rotates in the same direction as the rotor is moving, the frequency induced in the stator is higher; conversely, if it rotates in the opposite direction as the rotor is moving, the frequency induced in the stator is lower. The bottom line is that the magnetic field's net rotational speed can be tightly controlled to generate an exact match to the stator's frequency and thus synchronize to the utility frequency despite variations in the rotor speed. This is a significant advantage to wind turbines because the rotor can vary to follow changing winds without affecting the output frequency for the grid.

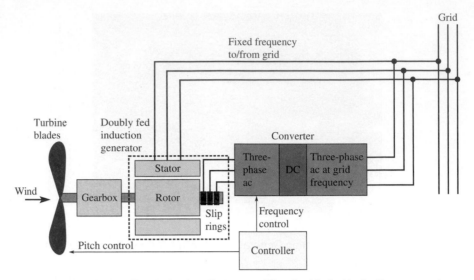

FIGURE 13-32 Doubly Fed Induction Generator. The doubly fed induction generator can maintain an exact match of the grid frequency to return electricity to the grid.

The secondary frequency is provided directly to the rotor through slip rings and brushes, without the losses experienced when the rotor receives its voltage through induction. The controller determines the optimum characteristics of the ac for the rotor and also controls the blade pitch, which determines the speed of the rotor and optimizes power for the given conditions. The stator is connected directly to the utility line, so it is either 50 Hz or 60 Hz, depending on what is used in that particular location. A block diagram of a typical DFIG system is shown in Figure 13-32.

The DFIG is more complicated than a singly fed generator and thus costs more at the outset, but it has higher overall efficiency than an SFIG and can harvest energy at various wind speeds. It is also very useful when the amount of energy surpasses the machine's rating intermittently. Other generators normally have to be taken offline (or be operated with reduced load) if they are exposed to conditions that exceed the design rating. When a DFIG is used, the generator can accept the extra input energy; the generator is allowed to speed up for a short period of time and it continues to produce the grid frequency. This continuous operation improves the overall efficiency of the generator.

One of the important control factors that make the doubly fed induction generator widely used in wind turbines and microhydraulic systems is that an ac-dc-ac converter is used to control the frequency of the voltage fed to the rotor. The slip rings and brushes in this system carry only the current for the field; some power is produced via the rotor, but it is only about 20% of the total. Because the rotor current is small compared to the total current, the brushes can be smaller and thus have less wear.

If the doubly fed induction generator is used with a wind turbine, it can produce power with a constant utility frequency in wind speeds from 6 mph to 50 mph. This allows the wind turbine to accept gusting winds and allows the blades to harvest the extra energy when the wind speeds are very high, which in turn improves the wind turbine's efficiency. If the wind turbine is very large (2 MW or larger), the control system can incorporate individual wind turbine blade adjustments and nacelle directional yaw adjustments to harvest the maximum amount of wind available. The doubly fed induction generator is also used in microhydro-generation and other renewable energy systems where the generator speed may be variable.

Variable-Speed Induction Generators

Another option for handling a variety of wind speeds is a generator configuration called a variable-speed induction generator (VSIG), which uses a large number of poles and requires no gearbox, with its associated losses and maintenance issues. Mechanical loading in the

Vibration Testing

(*Source:* Courtesy of ABB.)

One of the most critical tests of any generator is balancing the generator to ensure that rotor vibration is within acceptable limits. Testing needs to include tests of the maximum over speed. Designers need to specify the shaft stiffness and balancing requirements and test technicians verify that the generator operates within these specifications.

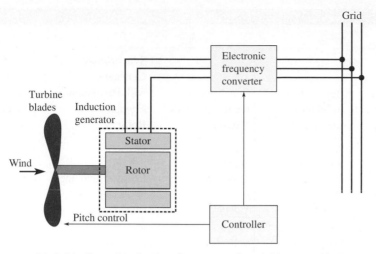

FIGURE 13-33   Variable-Speed Induction Generator. A variable-speed induction generator uses a full-scale electronic frequency converter to match the grid frequency.

drive train is thus reduced because the generator is completely decoupled from the grid. The configuration uses electrical or permanent magnet excitation and allows the generator to optimize power at an uncontrolled frequency. A full-scale frequency converter performs reactive power compensation and conversion to grid-quality ac. For wind turbines, the system can take advantage of an input in wind speeds ranging from a few mph during start-up to over 40 mph winds. The drawback is that the power electronics converter and the large multipole generator are expensive. Figure 13-33 shows a block diagram of the system.

Permanent Magnet Generators

The permanent magnet (PM) generator relies on a very strong permanent magnet to provide the original rotating magnetic field for the generator. The rotor in this generator is a very strong permanent magnet that continually puts out a very strong magnetic field. The stator is made of coils of wire that are mounted in the stationary part of the generator. When the rotor begins to turn, the magnetic field cuts across the stator coils, and a voltage is generated in the stator coil. Because the stator coils are in the stationary part of the generator, the permanent electrical connections at the terminal end of each coil allow voltage to be taken off the stator coils without brushes or slip rings. The main advantage of the PM generator is its simple design (it does not require an exciter). It requires very little maintenance because it can be completely sealed and it does not use brushes. Figure 13-34 shows the internal parts of a permanent magnet generator used in a wind turbine.

Early PM generators were small generators used primarily for charging batteries, or they were used with a small inverter to provide a low-power ac. Today, PM generators are used in

Permanent Magnet Low-Speed Generator

(*Source:* Courtesy of The Switch.)

This permanent magnet low-speed generator can produce power up to 4.25 MW and can withstand over-power for limited periods of time. Permanent magnet generators use expensive rare earth neodymium magnets and can have either inner or outer rotors. The variable-speed generator is frequently used in wind generators and hydroelectric systems.

FIGURE 13-34   Permanent Magnet Generator   (*Source:* Courtesy of Northern Power Systems Inc.)

TABLE 13-2

Comparison of Various Generators

	DC Generator	AC Induction (Asynchronous) Generator	Doubly Fed Induction Generator	Synchronous Generator with Exciter	Synchronous Generator with Permanent Magnet Rotor	Permanent Magnet Generator
Advantages	No frequency control needed; output is DC	Simple design; few parts; inexpensive	Provides controlled frequency of output voltage	Provides controlled frequency of output voltage	Provides controlled frequency but does not need exciter, slip rings, or brushes	Simple design; few parts; does not require dc voltage for excitation
Disadvantages	Brushes and commutator wear out periodically	Frequency not controlled	Needs voltage from grid to excite field	Needs exciter voltage and brushes and slip rings	Difficult to disassemble because of permanent magnet	Frequency not controlled
Frequency control	No frequency; output is dc voltage	Not controlled	Controlled by providing fixed frequency to field	Controlled when synchronized	Controlled when synchronized	Not controlled
Speed of rotor	Variable	Variable	Variable	Constant	Constant	Variable
Brushes and slip rings	No	No, uses induction	Yes	Yes, unless permanent magnets used for field	No, uses permanent magnet	No, uses permanent magnet
Brushes and commutator	Yes	No	No	No	No	No
External voltage or exciter needed	No	No	No	Yes	No	No

wind turbines of various sizes, including very large ones such as offshore turbines. Some wind turbines are low-speed direct drive turbines that rotate between 11 and 17 rpm and produce up to 4.25 MW. Other wind turbines that use PM generators can use one or two stage gearboxes that turn the generator at 150 to 400 rpm and produce 3.2 MW, or they can use two or three stage gearboxes that turn between 1,000 to 1,500 rpm and produce up to 1.6 MW. When the PM generator is used in a wind turbine, the blade speed is not controlled, and it produces a single- or three-phase ac power at a variable voltage and frequency. Because the PM generator's output voltage and frequency varies with its rotational speed, a full power converter (FPC) is used to provide three-phase electricity at fixed frequency and voltage levels to match the grid. The FPC uses transformers and robust line conditioners to ensure that the electricity meets the strict network requirements for harmonics, flicker, and fault ride through.

Table 13-2 compares the generators discussed in this chapter on the basis of their advantages, disadvantages, and some basic characteristics.

SECTION 13-4 CHECKUP

1. What are the main parts of an asynchronous induction generator?

2. How does a doubly fed generator produce electrical power at a fixed frequency?

3. What is the difference between a squirrel-cage rotor and a wound rotor?

4. Name some of the applications of a variable-speed induction generator.

CHAPTER SUMMARY

- Magnetic fields are described by flux lines that radiate from the north pole to the south pole of a magnet.

- The stronger the magnetic field, the more flux lines.

- When unlike poles of two permanent magnets are placed close together, they attract each other. When two like poles are brought close together, they repel each other.

- For a coil, the left hand can be used to find the North Pole with the fingers grasped in the direction of electron flow.

- The mks unit of magnetic flux is the weber (Wb).

- The magnetic flux density is the amount of flux per unit area in the magnetic field. Its symbol is B, and its unit is the tesla (T). The centimeter-gram-second (CGS) unit is the Gauss.

- An electromagnet is produced when there is current in a coil of wire.

- The ease with which a magnetic field can be established in a given material is measured by the permeability (μ) of that material.

- The opposition to the establishment of a magnetic field in a material is called reluctance ($\mathcal{R}$).

- Magnetomotive force (mmf) causes magnetic flux; the mks unit is the ampere-turn (At).

- Relative motion between a conductor and a perpendicular magnetic field produces an induced voltage.

- A commutator is a rotating switch that has independent conductors and is designed to switch the current in a rotor coil as the structure rotates.

- If the generator uses an external voltage source to create the initial magnetic field, it is called a separately excited generator.

- If a self-excited generator has not been used for a long time, it may be necessary to provide an external source to the field windings to start it. This step is called flashing the fields.

- When a generator supplies voltage from its armature to create the field current, it is called a self-excited generator.

- A compound generator uses both a series field and a shunt field.

- A synchronous generator produces a sinusoidal waveform whose sine wave has a peak that corresponds to a physical position of the rotor.

- An exciter generator supplies dc to the rotor for a synchronous generator.

- Salient poles are protruding magnetic field poles on the rotor of a motor or generator. They are only used in lower-speed motors and generators.

- In some rotating-field synchronous generators, the exciter generator produces ac voltage, and diodes are used to convert this ac voltage to dc voltage for the field in the main generator.

- The frequency of a synchronous generator is determined by the number of poles in the armature and the speed at which the rotor is turned.

- The permanent magnet generator relies on a very strong permanent magnet (PM) to provide the original rotating magnetic field for the generator.

- The variable-speed generator system uses a permanent magnet generator that is combined with an electronic inverter.

KEY TERMS

ampere-turn (At)   The SI unit of magnetomotive force (mmf).

compound generator   A generator that has two field windings: one in series and one in parallel. Typically, the output voltage tends to be independent of the load.

doubly fed induction generator (DFIG)   An ac induction generator that has a wound rotor connected to a different source of ac than the stator. AC from the grid or from an inverter is supplied to the fields in the stator. Also known as *double-excited induction generator*.

exciter   A smaller auxiliary generator that supplies field current for a larger generator.

induction generator   An asynchronous electrical machine that can function as a motor or as a generator.

magnetic field intensity (H)   The magnetomotive force per unit length of the magnetic material. The unit is the ampere-turns per meter. Also known as *magnetizing force*.

magnetomotive force (mmf)   The cause of a magnetic field; it is measured in ampere-turns.

permeability (μ)   A measure of the ease with which a magnetic field can be established in a given material.

reluctance ($\mathcal{R}$)   The opposition to the establishment of a magnetic field in a material.

self-excited shunt generator   A generator that supplies voltage from its armature to create the field current. The field and armature windings are parallel.

shunt generator   A generator that has its field winding and armature connected in parallel. The armature supplies both the load current and the field current.

squirrel-cage rotor   A rotor for a motor or generator that is an aluminum "cage" with conducting bars that are shorted together at each end by a ring. The cage surrounds an iron core that provides a path for the magnetic field.

synchronous generator   An ac generator in which the output frequency is synchronized to the position of the rotor.

FORMULAS

Equation 13-1	$B = \dfrac{\phi}{A}$	Flux density		Equation 13-5	$\phi = \dfrac{F_m}{\mathscr{R}}$	Flux
Equation 13-2	$\mu_r = \dfrac{\mu}{\mu_0}$	Relative permeability		Equation 13-6	$H = \dfrac{F_m}{l}$	Magnetic field intensity
Equation 13-3	$\mathscr{R} = \dfrac{l}{\mu A}$	Reluctance		Equation 13-7	$v_{ind} = B_\perp lv$	Induced voltage
Equation 13-4	$F_m = NI$	Magnetomotive force		Equation 13-8	$f = \dfrac{N_p \times \text{rpm}}{120}$	Frequency

CHAPTER TRUE/FALSE QUIZ

Determine whether each statement is true or false. Answers are at the end of the chapter.

1. The tesla (T) and the gauss (G) are both units for magnetic flux density.

2. The unit for measuring magnetomotive force (mmf) is the volt.

3. Ohm's law for a magnetic circuit gives the relationship among flux density, magnetomotive force, and reluctance.

4. Saturation occurs in a magnetic core when an increase in magnetic field intensity does not have a corresponding increase in flux density.

5. To induce a voltage across a coil, a magnet can be moved in and out of the coil.

6. The ease with which a magnetic field can be established in a given material is measured by the permeability of that material.

7. Commutation is the process of providing a unidirectional current from the generator.

8. The term *shunt* means that the windings of a generator are connected in series with each other.

9. The speed of a dc generator can be controlled with a rheostat in the field windings.

10. The opposition to the establishment of a magnetic field in a material is called reluctance.

11. A self-excited dc generator normally has enough residual magnetism in the field coil to start the generator so that it produces an output when it is first turned on.

12. The DFIG requires that the rotor maintain constant speed in order to synchronize the output ac to the grid.

13. A compound dc generator has two shunt windings.

14. A long shunt refers to the length of wire in the shunt.

15. Current from a dc generator passes through slip rings.

16. A synchronous generator produces a sinusoidal waveform, and the peak of the sine wave corresponds to a physical position of the rotor.

17. Induced voltage can occur in a wire when the wire is moved at right angles to a magnetic field.

18. To increase flux in a generator, increase the air gap.

19. An exciter is a small motor used to start a large generator.

20. Synchronous generators can have either a rotating armature or a rotating field.

CHAPTER MULTIPLE-CHOICE QUIZ

Complete each statement by selecting the one correct answer. Answers are at the end of the chapter.

1. A component not found on a large alternator is
   a. an exciter
   b. a commutator
   c. slip rings
   d. diodes

2. To transfer voltage from its rotating part to the output, a dc generator uses
   a. brushes and commutator segments
   b. brushes and a stator
   c. brushes and slip rings
   d. all of these

3. A three-phase alternator has
   a. three individual stator windings, and each produces a sine wave that is 120° out of phase with the others
   b. three separate rotors that spin simultaneously and produce sine waves that are 120° out of phase with each other
   c. three separate shafts for the wind turbine gearbox to turn
   d. all of these

4. The frequency that the alternator produces can be increased by
   a. increasing the speed of the rotor rpm
   b. increasing the amount of exciter current
   c. adding more commutating segments

5. The strength of an electromagnet coil can be increased by
  a. increasing the current that flows through the coil
  b. increasing the number of coils or the amount of wire in the coils
  c. providing an iron core for the coil
  d. doing all of these

6. The south poles of two bar magnets brought in close proximity to each other produce
  a. a force of attraction
  b. a force of repulsion
  c. an upward force
  d. no force

7. A magnetic field is always associated with
  a. iron
  b. magnetic domains
  c. moving charge
  d. salient poles

8. By convention, magnetic field lines are drawn from
  a. the north pole to the south pole
  b. the south pole to the north pole
  c. inside to outside the magnet
  d. front to back

9. Reluctance in a magnetic circuit is analogous to
  a. voltage in an electric circuit
  b. current in an electric circuit
  c. power in an electric circuit
  d. resistance in an electric circuit

10. The SI unit of magnetic flux is the
  a. tesla
  b. weber
  c. volt
  d. ampere-turn

11. The SI unit of magnetomotive force is the
  a. tesla
  b. weber
  c. ampere-turn
  d. electron-volt

12. When there is constant current in a coil of wire,
  a. the coil's resistance increases
  b. a magnetic field is created
  c. a force is created across the coil
  d. the magnetic field is cancelled

13. If the number of turns in the coil of a generator armature is increased, the voltage induced across the coil
  a. remains unchanged
  b. decreases
  c. increases

14. A doubly fed induction generator is mainly used to produce
  a. a variable frequency as its shaft changes speed
  b. a constant frequency as its shaft changes speed
  c. constant dc that does not change as rotor speed changes
  d. none of these

15. An induction generator
  a. has more component parts than a dc generator does
  b. must have brushes and a commutator or slip rings to generate power
  c. does not need brushes, a commutator, or slip rings to generate power

16. The main advantage of a permanent magnet generator is that
  a. it can have a smaller exciter than similar generators
  b. it is simpler and requires little maintenance
  c. it can use permanent magnets on the rotor and the stator
  d. all of these

CHAPTER QUESTIONS AND PROBLEMS

1. Identify the basic parts of an ac alternator.

2. What is the instantaneous voltage of a single-loop conductor that is 30 cm long, passes through a magnetic field that has 82,000 lines/sq cm, and moves at a rate of 130 cm/s.

3. Explain the operation of the rotary-armature type of ac generator.

4. Explain how voltage is taken from the rotor of a dc generator.

5. Explain two ways you can get more voltage from an ac generator.

6. Name two ways to change the polarity of a dc generator.

7. What is the frequency of an induced voltage for a generator that has 10 poles and turns 600 rpm?

8. Name two ways in which exciter voltage can be supplied to the rotor of an alternator.

9. Explain how a doubly fed alternator operates.

10. Explain the difference between commutator segments and slip rings.

11. Explain the difference between a wound-rotor armature and a squirrel-cage rotor.

12. Explain why laminated steel pieces are typically used in a squirrel-cage rotor.

13. Assume that you monitor the output from a voltage regulator applied to the field windings on an induction generator, and you observe the voltage increase by 20%.
  a. Explain why the magnetic flux density may not increase by 20%.
  b. Why does the voltage regulator change the field voltage?

14. Why are almost all large, synchronous generators rotating-field generators?

15. Synchronous generators are widely used for wind turbines. Explain what advantage they have over an induction generator in this application.

16. What is the speed of a 24-pole synchronous generator if the output frequency is 50 Hz?

17. Why do large synchronous generators use an exciter?

18. When does an induction motor/generator act like a motor and when does it act like a generator?

FOR DISCUSSION

Wind machines are used by farmers to protect sensitive crops from frost damage. Could they be turned into a wind generator for generating power on a windy day? If so, what kind of generator would be best? What applications for the power would you suggest?

ANSWERS TO CHECKUPS

Section 13-1 Checkup

1. The SI unit for magnetic flux

2. Permeability is a measure of the ease of establishing a magnetic field and is measured in Wb/At · m. Relative permeability is a ratio of permeability in a magnetic substance to the permeability of a vacuum. Because it is a ratio, relative permeability is dimensionless.

3. Ohm's law for a magnetic circuit states that the flux is the magnetomotive force divided by the reluctance. In equation form: $\phi = \dfrac{F_m}{\mathcal{R}}$.

4. The air gap represents a high reluctance path to the magnetic circuit, thus reducing the flux.

5. Hysteresis loss is energy expended to remove residual magnetism in a magnetic core.

Section 13-2 Checkup

1. By relative motion of a conductor to a magnetic field such that magnetic field lines are cut by the motion

2. A dc generator includes a stator that provides a magnetic field; a rotor coil, which acts as the armature; a commutator to keep a unipolar output; brushes to take power from the armature; bearings to allow the rotor to spin; a cooling fan and a case, which forms part of the return path for the magnetic circuit

3. In a dc generator, the armature is the rotor and generates the voltage across the rotor coils as they cut magnetic field lines. The output of the armature is conducted through a commutator and brushes to the final output.

4. The field windings carry current and generate a magnetic field for the rotor.

5. A compound generator

Section 13-3 Checkup

1. The main parts of a synchronous generator are the rotor (which generally is the field winding or, in small generators, a rotating permanent magnet, a method of supplying the field coils with electricity) and the stator (which is usually the armature where power is removed). Very large synchronous generators have an exciter that provides field current. If the exciter produces ac, there are diodes to rectify it for the required dc for the main generator's field. Other generators have bearings, a cooling fan, and a case.

2. In a rotating armature generator, the armature is a coil that cuts lines in the magnetic field provided by the stator and transmits the output through slip rings and brushes to the outside.

3. In a rotating-field generator, the field is provided by either a coil or a permanent magnet and provides a rotating field that generates voltage in fixed stator coils as it spins. The output is taken from the stator windings.

4. The permanent magnet provides the rotating magnetic field that induces voltage in the stator coils as it passes them.

Section 13-4 Checkup

1. The main parts of an asynchronous induction generator are a rotor (consisting of electrical loops, either wired or in the form of a squirrel cage, that is spun by a prime mover to something above the synchronous speed); a stator (which is a rotating electrical field provided by ac); and the usual bearings, fan, and case.

2. A doubly fed generator is a rotating-field generator that is connected to two ac sources: one on the wound rotor and one on the stator. The rotor's ac is supplied through slip rings and brushes from a controlled ac source that combines with the rotor speed to produce a net rotation of the field that exactly matches the stator frequency. The stator is connected to the ac utility line.

3. A squirrel-cage rotor is an aluminum cage with conducting bars that are shorted together by a ring at each end. The cage surrounds an iron core that provides a path for the magnetic field. A wound rotor has wire windings instead of an aluminum cage.

4. A variable-speed induction generator can be used in a wind turbine that converts the output to dc and back to grid-quality ac using an inverter. It is also useful in applications that are independent of the grid.

ANSWERS TO TRUE/FALSE QUIZ

1. T   2. F   3. T   4. T   5. T   6. T   7. T   8. F   9. F   17. T   18. F   19. F   20. T
10. T   11. T   12. F   13. F   14. F   15. F   16. T

ANSWERS TO MULTIPLE-CHOICE QUESTIONS

1. b   2. a   3. a   4. a   5. d   6. b   7. c   8. a   9. d   10. b   11. c   12. b   13. c   14. b   15. c   16. b

The Electrical Power Grid

CHAPTER OBJECTIVES

- State the advantages for three-phase ac for grid transmission and large industrial users of electricity.
- Show, with a diagram, the difference between a delta winding and a wye winding on a three-phase transformer.
- Show how a tapped secondary on a delta three-phase transformer can be used to supply split, single-phase power.
- Identify the three main parts of the electrical grid and how it distributes power.
- Compare the smart grid with the original grid, and describe advantages to the utility and to the customer.
- Discuss the advantages of microgrids and where they are used.
- Describe a substation and the important functions it performs.
- Discuss at least four power quality issues for connecting a renewable source to the grid.

KEY TERMS

Key terms are shown in bold and color. Definitions for key terms are provided at the end of the chapter and in the end-of-book glossary. Bold terms in black are defined in the end-of-book glossary only.

- personal protective equipment (PPE)
- arc flash
- electrical grid
- transmission system
- substation
- distribution system
- smart grid
- ground fault interrupt (GFI) circuit breaker
- microgrid
- power quality
- harmonic
- uninterruptible power supply (UPS)
- low-voltage ride through (LVRT)
- flicker

INTRODUCTION

Power is distributed in an interconnected network called the grid, which includes power generation, transmission, and distribution. The grid structure is rooted in history, with the original power companies operating independently of each other. Later, the benefits of connecting systems together (load sharing and backup power) led to the development of the massive systems that we collectively call the grid. New smaller distribution systems, called microgrids, are being developed to meet some needs. A microgrid is a smaller power grid that can operate either by itself or connected to a larger utility grid.

This chapter focuses on grids and microgrids. To understand grids, you need to understand three-phase power and the three-phase transformers that convert three-phase voltages to voltages used for transmission and distribution, so the chapter begins with these topics. After an overview of the grid system, information about the new smart grid and its features are explained. Basic connection issues for renewable energy systems, including power quality issues, are discussed.

14-1 Three-Phase AC

Single-phase ac voltage was introduced in Chapter 2 and is generated in some smaller renew-able energy systems. Power companies almost universally generate three-phase (3ϕ) ac, and they supply it to industries as three-phase or convert it to single-phase (1ϕ) for residential and small commercial operations. Larger industrial users use the three-phase ac directly for powering motors and other industrial components. Three-phase was introduced briefly in Chapter 2 and is covered in more detail in this section because of its importance in the power grid.

Review of Single-Phase AC

As you know, ac is a sinusoidal waveform, which means it has the shape of the mathematical definition of a sine wave. It is common practice to refer to electrical waveforms with this shape as sine waves. An ac voltage can be specified as peak voltage, peak-to-peak voltage, or root-mean-square (rms) voltage. The frequency of the wave is defined as the number of cycles that occur in 1 s and is given the units of Hertz (Hz). For example, if 1,000 cycles occur in 1 s, the frequency is 1 kHz. The reciprocal of frequency is called the period and has units of time. Thus, a 1 kHz frequency has a period of 1 ms.

The frequency of electrical energy is determined by the generator or inverter that produces it. Generally the frequency distributed by power companies is 50 Hz or 60 Hz, depending on the country (and, in some cases, by the region within the country). In some specialized applications, other frequencies are used. On military ships and planes, 400 Hz is common for radio and electronic systems. In some countries, low-frequency ac (16⅔ Hz and 25 Hz) is used for electric trains in order to reduce reactive losses in overhead lines and improve commutation in the large series motors used in locomotives. Lower frequencies were originally supplied by large rotary frequency converters, but today they can be supplied with static frequency converters.

In residential and smaller commercial uses, the voltage supplied by the utility company is **single-phase voltage** (1ϕ), which means it has only one continuously varying sinusoidal waveform. The minimum requirement to deliver single-phase ac is only two wires, but it is standard in North America for the utility company to supply two "hot" wires of opposite polarity and a **neutral** wire, which is the current carrying return wire. This configuration of two hot wires and a neutral is referred to as **split-phase.** In North America, the voltage from each hot wire to neutral is 120 V, and between the two hot wires it is 240 V. The 240 V is used on larger appliances, such as air conditioners and clothes dryers, and the lower voltage is used on smaller appliances and lighting. In much of the rest of the world, the standard varies from 220 V to 240 V at 50 Hz. Some of these countries also have a lower voltage available.

Single-phase voltage is used in residential and in commercial applications because it is less expensive to distribute than three-phase power. The loads for residential and small commercial establishments are typically small (lights, appliances, etc.) and they are designed for low-voltage, single-phase.

Definition of Three-Phase Voltage

Recall that the term *three-phase* refers to three ac voltages that have the same magnitude but are separated by 120°. Figure 14-1(a) illustrates three-phase voltage. The first phase is called phase A, the second phase is called phase B, and the third phase is called phase C. Phase B is 120° shifted from phase A, and phase C is shifted another 120°. Figure 14-1(b) represents the same waveforms but drawn as a phasor diagram. A **phasor** is a rotating vector, which is a quantity with magnitude and direction. Phasors allow basic trigonometry relationships to be applied to sinusoidal waves.

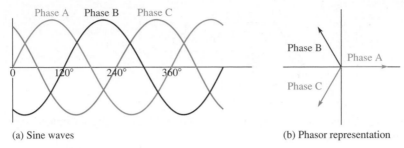

(a) Sine waves   (b) Phasor representation

FIGURE 14-1   Three-Phase Voltage

Advantages of Three-Phase Systems

In general, three-phase voltage is more economical to produce and transmit than is single-phase, so most generating and transmission systems are three-phase. Three-phase voltage is stepped up to very high voltages to be transmitted over long distances. The higher transmission voltages cause current to be lower. For this reason, conductors can be smaller and weigh less, resulting in a net savings to the utility company. In a few cases, very high voltage dc has been used for very long transmission lines, but the conversion process from ac to dc and back is expensive and rarely done.

For industrial applications, three-phase voltage is applied directly to motors and some other circuits. When three-phase voltage is used to power motors, it creates a rotating magnetic field that allows the motor to produce high torque to move loads on its shaft, without the need for separate starting windings. Because the three-phase voltage has a phase shift of 120° between each pair in its three phases, it can create a rotating magnetic field when it is applied to the stator of three-phase motors. The rotating magnetic field causes the rotor to spin. Three-phase ac motors have other advantages: They are simpler to maintain, and they use smaller wiring than comparable single-phase motors and do not need special starting circuits, so they are widely used in industries that require a lot of motors.

Three-Phase in Power Systems

In almost all commercial power generators, three-phase ac is generated at the source. The generator can be turned by a wind turbine, steam turbine, or other prime mover. (Recall that, with geothermal generators, the steam comes directly from heated groundwater.) Because maintaining the exact frequency is critical, wind turbines may use a power electronic frequency converter, as described in Section 7-5 or a doubly fed induction generator as described in Section 13-4. In solar energy systems that use photovoltaic panels, dc is produced by the system. In this case, the dc is sent through an inverter that converts it to ac. In large systems, the ac is a three-phase voltage that can be connected to the grid for transmission to end users.

Single-phase voltage is derived from three-phase voltage. Converting three-phase to single-phase can be done in different ways. A common conversion method is from a delta (Δ) connected transformer. The **delta** configuration is a three-phase wiring method named for the Greek letter delta and takes its triangular shape in diagrams. (Three-phase transformers are covered in detail in Section 14-4 but are briefly mentioned here to show one method for converting three-phase to single-phase.)

In North America, the center-tap from one winding of a delta connection is neutral (N). Across each winding in the secondary is 240 V, so the voltage from the center-tap to the coil ends is 120 V (see Figure 14-2). The corners of the delta are the connection points for the three-phase output and are identified as L_1, L_2, L_3. In this case, only L_1, L_2, and N are taken for the output to a single-phase customer. Industrial locations can have all four wires (L_1, L_2, L_3, and N) brought out in order to have 120 V, 240 V, and 3ϕ 240 V available for large three-phase motors (this configuration is called a **four-wire delta**). It is important

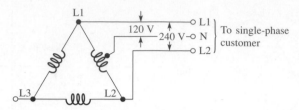

FIGURE 14-2 Conversion of Three-Phase to Single-Phase Using a Center-Tapped Secondary on a Delta-Wired Transformer

that L_3 is *not* used with the N. This combination would produce 208 volts, which is not used in residential applications because it could load the transformer unevenly and possibly cause overheating. The L_3 wire is marked with an orange color so that it can be identified and used only for three-phase. Although 208 V is used in some industrial applications, it is developed from a wye connected transformer instead of a delta connected transformer. The **wye** configuration is a three-phase wiring method that is named for the letter Y because its diagram is shaped like the letter Y.

True Power, Apparent Power, and Reactive Power

Power is a fundamental concept in electrical work and many renewable energy systems, so it was introduced in Section 2-4. It was discussed further in Section 6-4 with respect to inverters. In this section, power concepts are reviewed and then applied to a three-phase circuit. As you learned in Section 2-4, the power in a purely resistive circuit can be calculated by multiplying voltage by current. Power in purely resistive circuits is sometimes referred to as **true power (P_{true}),** and it is measured in watts. True power is the actual power dissipated in a circuit, usually in the form of heat. Most loads are *not* purely resistive. If the load has inductance or capacitance, the current and voltage shift in phase with respect to each other, depending on the amount of resistance and reactance present. **Reactive power (P_r)** is the rate at which energy is stored and alternately returned to the source by a capacitor or an inductor. It is measured in units of **volt-amps reactive (VAR)**. Reactive power is "borrowed" power; it shuttles back and forth between the source and the load, but does no useful work. In a reactive component, you can calculate the reactive power by finding the product of voltage across the component and current in the component (ignoring any resistive loss).

Most practical loads are not purely resistive or purely reactive but are a combination of both. In this case, the product of rms voltage and rms current without regarding phase differences is called the **apparent power (P_a).** Apparent power is measured in units of volt-amps (VA) to distinguish it from true power and reactive power. Apparent power is always equal to or larger than true power. Apparent power is the power that *appears* to be transferred between the source and the load, and it consists of two parts: One part is the true power that is actually dissipated, the other is the reactive part that tends to shift the phase. Motors act like an inductive load. You may see on a motor nameplate the units for power as VA because it is specified as apparent power.

Power Factor

Power factor was introduced in Section 6-4 as it applies to inverters in single-phase circuits. It was defined in terms of the phase angle between current and voltage ($PF = \cos\theta$, where θ = the phase angle between voltage and current). An equivalent definition that is useful is to define it in terms of the ratio of true power to apparent power. For an induction motor, the power factor changes with the load, so it is normally specified under full load. The equation for power factor written as a ratio of two powers is:

$$PF = \frac{P_{\text{true}}}{P_a}$$

Equation 14-1

where

PF = power factor

P_{true} = true power, in W

P_a = apparent power, in VA

EXAMPLE 14-1

Determine the power factor for a circuit if the true power is 1.56 kW and the apparent power is 1.84 kVA.

Solution

$$PF = \frac{P_{true}}{P_a} = \frac{1.56 \text{ kW}}{1.84 \text{ kVA}} = \mathbf{0.847}$$

Expressed as a percentage, the PF is 84.7%.

Three-Phase Power

The total power delivered to a three-phase load is more complicated because of the phase shifts in the circuit. The methods for finding power are described in more detail in the next section, but you can find total power delivered to a balanced load if you know the voltage between two lines and the current in one line. The product of line voltage and line current is the apparent power, P_a. The total power is given by

$$P_{total} = \sqrt{3} \, (PF)P_a \qquad \qquad \text{Equation 14-2}$$

where

P_{total} has the units of watts or kilowatts (W or kW)

EXAMPLE 14-2

A three-phase motor is running on 480 V and has a line current of 51.5 A.

(a) Determine the apparent power.

(b) If the power factor is 0.95, determine the total power supplied from the source.

Solution

(a) $P_a = VI = (480 \text{ V})(51.5 \text{ A}) = \mathbf{24.7 \text{ kVA}}$

(b) $P_{total} = \sqrt{3} \, (0.95)(24.7 \text{ kW}) = \mathbf{40.7 \text{ kW}}$

SECTION 14-1 CHECKUP

1. What is split-phase?
2. How do you label the three wires that supply 120/240 V single-phase in North America?
3. What are the major advantages of three-phase voltage to utilities?
4. Why is a three-phase motor more efficient than a single-phase motor?
5. What is the difference between true power and apparent power?
6. What is power factor? What units are used to represent it?

14-2   Three-Phase Transformers

The most common use for a transformer is to step up or step down voltage, as you saw in the discussion in Section 2-8 about basic single-phase power transformers. In this section, single-phase transformers are reviewed with additional information; however, the main focus of the coverage here is the three-phase transformer. It has many similarities with the single-phase transformer studied earlier. Three-phase transformers are used in large utility-scale renewable energy systems.

Single-Phase Lines to a Residence

(*Source:* Tom Kissell.)

Electrical power from a single-phase transformer is connected to homes through a weather head. Single-phase voltage is supplied by two hot wires and neutral (in the United States) that is provided by a very strong steel wire that is wrapped around the insulated wires to provide support. The steel wire is connected securely to the mast on the roof and at the transformer to minimize the droop when they are run overhead. The wires at the bottom of the mast in this photograph are telephone and cable television lines.

Review of Single-Phase Transformers

Recall that transformers were defined as a passive component in which two or more magnetically coupled coils are wound in a single core. The primary application is to convert ac from one voltage to another in both electronic and power circuits (although there are a few other applications). Tapped, single-phase transformers are widely used in residential areas to step down the grid voltage to the requirements for household use. (In North America, this is 120 V/240 V.) Important equations relating the input and output voltage in a transformer were given in Equation 2-22 as $n = N_{sec}/N_{pri} = V_{sec}/V_{pri}$, where n is the turns ratio, N_{sec} is the number of turns in the secondary winding, and N_{pri} is the number of turns in the primary. This definition is one given by the Institute for Electrical and Electronics Engineers (IEEE) for electronics power transformers, but other definitions are based on a different category of transformer (such as an instrument transformer) that use turn ratio (singular); these transformers use the ratio of the higher-voltage winding to that of the lower-voltage winding. If you are working with either turn ratio or turns ratio, it is important to understand the definition that is being used.

Another useful relationship for single-phase transformers is introduced here. No device is 100% efficient, and transformers are no exception. Power transformers, which are designed for a single frequency, are very efficient and, in general, you can assume they are 100% efficient (or ideal) for basic calculations. Given this assumption, the power delivered to the primary is equal to the power delivered to the load, and no power is lost internally in the transformer; that is:

$$P_{pri} = P_{sec}$$ Equation 14-3

where

P_{pri} = power delivered to the primary, in W

P_{sec} = power delivered to the load, in W

This simplified equation does not consider transformer heating; however, it is useful for many basic transformer calculations. By substitution, it is evident that, in an ideal transformer, $V_{pri} I_{pri} = V_{sec} I_{sec}$. Rearranging:

$$\frac{V_{sec}}{V_{pri}} = \frac{I_{pri}}{I_{sec}}$$

This is just another way to write the turns ratio as defined previously. Notice that for a step-down transformer, voltage is stepped *down*, but current is stepped *up* by the same amount. Likewise in a step-up transformer, voltage is stepped *up*, but current is stepped *down* by the same amount. Keep this in mind when you are selecting a fuse for a transformer circuit or calculating power delivered to a load.

Because ac circuits have phase shifts, the power rating on a transformer is not generally shown in watts but instead in volt-amperes (VA). The **volt-ampere (VA) rating** is the maximum allowed apparent power dissipated for a device and is found by multiplying the voltage (volts) by the current (amperes). The reason VA is used rather than watts is that the product of voltage and current can be higher if there is a reactive load, and the VA rating rather than wattage rating is what determines the safe operating limit.

EXAMPLE 14-3

Calculate the primary current for a step-down transformer that steps down from 120 V to 12 V for a 10 Ω load. Choose a fuse for the primary side with this load.

Solution
The turns ratio is:

$$n = \frac{V_{sec}}{V_{pri}} = \frac{12 \text{ V}}{120 \text{ V}} = 0.1$$

The secondary current is 12 V/10 Ω = 1.20 A. By substitution:

$$n = \frac{I_{prl}}{I_{sec}} = \frac{I_{prl}}{1.2 \text{ A}} = 0.1$$

$$I_{pri} = (1.2 \text{ A})(0.1) - 120 \text{ mA}$$

Choose a fuse larger than 120 mA (⅛ A) to avoid blowing the fuse due to a slight variation in current. A ¼ A fuse is a reasonable choice.

Selecting a Single-Phase Transformer

If you are connecting a piece of equipment to a new transformer or if you are selecting a transformer for a renewable energy system, you need to size the transformer correctly so it does not overheat. A transformer that is too small will overheat. One that is too large will cost more than necessary.

Start by determining the required load voltage and load current. For most devices, you can obtain this information from a nameplate (data plate) on the device. Next calculate the VA (or kVA) by multiplying the load voltage times the load current (or read it from the nameplate). In most cases, the nameplate shows the required frequency as well, which should be the same as the frequency of the source voltage. Verify that the load requires single-phase voltage. If a nameplate is not available, you can generally get the specifications from a user manual or the manufacturer.

At this point, you can select a transformer with a standard kVA capacity that is equal to or greater than the load requirement. Transformer specifications are normally listed by kVA rating and either turns ratio or the input/output voltage ratings. Depending on the transformer, you may need to select the correct voltages on the primary and secondary side by connecting the source and load to one of several taps on the transformer.

Overview of Three-Phase Transformer Operation

Three-phase transformers can be integrated as one unit or they can consist of three individual single-phase transformers that are wired together to operate in a three-phase system. Figure 14-3 shows a cutaway view of a three-phase transformer wound on a single iron core. When the three windings are on a single core, the transformer can be manufactured at a lower cost as a single unit.

Figure 14-4(a) shows the transformers in the yard of a coal-fired power station that is in the process of converting to co-generation with biofuel. Notice that this plant uses groups of three single-phase transformers to step up the voltage for transmission. This simplifies replacement if one transformer goes out. Large transformers like this can get hot under load. Three separate transformers can be cooled easier than a single one can, and they can better handle overloads when they occur.

Figure 14-4 (b) shows a utility pole with three single-phase transformers to transform three-phase at the distribution end. The main advantage to using three individual transformers in this case is that one or two of the transformers can be larger than the other(s) if the load is unbalanced. Also, the weight is distributed to the three individual transformers,

FIGURE 14-3 Cutaway View of a Three-Phase Transformer (*Source:* Courtesy of Acme Electric.)

(a) Single-phase transformers used to step up a three-phase voltage for transmission

(b) Three single-phase transformers used to step down three-phase voltage for distribution

FIGURE 14-4   Single-Phase Transformers Used in Three-Phase Systems (*Source:* Part (a), National Renewable Energy Laboratory; part (b), Tom Kissell.)

which, in some cases, makes it easier to handle. If a failure occurs on one, the replacement is simplified because only the one with the failure needs to be replaced. Otherwise, the three individual single-phase transformers operate exactly like the integrated three-phase transformers. The choice about which to use depends on the application.

Transformers can be designed for outdoor use by making the enclosure waterproof so that rain or snow does not get into the electrical connections. Some transformers are specifically designed for indoor use and their enclosure is not watertight. Some transformers in larger buildings are located in a special room called a transformer vault, which allows the transformers to be inspected from time to time.

Windings on a Three-Phase Transformer

Figure 14-5 shows a three-phase transformer that has three separate primary and three secondary windings wound on a common core. The primary winding of each transformer

FIGURE 14-5   Three-Phase Transformer. The windings are on a common core and are separated by insulators.   (*Source:* David Buchla.)

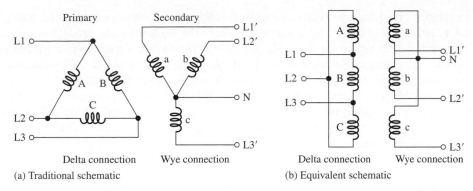

FIGURE 14-6 Delta Wye Connected (Δ-Y) Transformer. The circuits in parts (a) and (b) are identical.

phase is mounted close to its secondary winding for effective coupling of the primary voltage to the secondary winding. The three primary windings are identical to each other, and the three secondary windings are identical to each other. The leads for all six of the windings are available externally.

Delta and Wye Connections

The three primary and three secondary windings are available externally for connecting in either of two common configurations: a delta (Δ) or a wye (Y). Figure 14-6(a) shows the primary winding connected in a delta and the secondary connected in a wye configuration. Figure 14-6(b) shows the same schematic drawn so that is similar to what you see as a physical installation. Some three-phase transformers may use other terminal identification numbers or letters, but the connections are equivalent. One common naming convention is to label the high-voltage windings with the letter H and the low-voltage windings with the letter X (neutral is X0).

When all leads are brought out, a three-phase transformer can be connected in any of four combinations: Δ-Y, Y-Δ, Δ-Δ, or Y-Y, although only the first three are generally used. The first three combinations have certain advantages and disadvantages. For example, the wye connection has a central connection point from which neutral can be brought out. A variation on a standard delta connection is making a center-tap from one of the windings available on a delta secondary. This configuration was shown in Figure 14-2 to illustrate a method for splitting a phase, which is the norm in residential homes in North America. In this case, the center-tap is the neutral winding. When connecting a delta, it is important to check for proper polarity before applying power because an incorrect configuration can lead to extremely high current.

In Section 14-1, you saw how a delta connected secondary can be used to provide single-phase voltage. Single-phase voltage can also be obtained from a wye connected transformer. The wye connected transformer can provide 208 V, which is used in some commercial and industrial locations but rarely in residential locations, where loads require 240 V. Instead of a center-tap on one of the coils, the center-tap for the neutral connection is exactly in the midpoint where the three windings are joined; that midpoint is called the wye point. Any one of the three lines can be used for a single-phase supply when combined with neutral. In industrial and commercial applications, wye connected transformers with 480 V line to line provide 277 V line to neutral, and this voltage is used to power fluorescent lighting.

Important points about the configurations of three-phase transformers are discussed in the next subsections.

Delta-Wye

The delta-wye connected transformer has a delta connection on the primary side and a wye connection on the secondary side. In Europe, this type of connection is called a delta-star transformer. The delta-wye is one of the most widely used connections. It can be configured as either a step-up or step-down transformer. The neutral can be brought out from the secondary side where the three windings are joined so that a single-phase voltage is available for lighting or electrical outlets in an office. The secondary side of the delta-wye transformer must be grounded at the wye point according to the National Electrical Code (in the United States) as protection against shorts. When the delta-wye transformer is used as a step-up transformer, the wires on the secondary side that provide high voltage are smaller because of the lower current. This feature makes this type of transformer very useful for renewable energy systems because it can step up the voltage for transmission over long distances.

Wye-Delta

The wye-delta three-phase connected transformer is most commonly used in a step-down transformer application. The wye connection on the high-voltage primary side provides a place for a ground to be connected at the wye point. The wye connection on the high-voltage side reduces insulation costs. It has the advantage of being stable when driving unbalanced loads.

Delta-Delta

The delta-delta connection is useful when three single-phase transformers are connected as a three-phase transformer. An advantage of this configuration is that if one of the transformers fails, it can be replaced while the other two continue to operate and supply power at a reduced level (this is called an open delta). The delta-delta connection can be used as a step-up or a step-down transformer. This type of connection is required by National Electrical Code (NEC) to be grounded on its secondary windings or to have ground fault protection in each phase in the secondary winding.

Wye-Wye

The wye-wye connection is rarely used. It has problems with unbalanced loads, which can cause overheating.

Tapped Transformers

In large distribution transformers, it is common for the primary or secondary to have tapped terminals. Taps near the end of a winding enable fine adjustment of the voltage at the site so that it matches the load requirements more closely. The reason it is necessary to make adjustments at the site is that voltages from utilities are not the same in all locations due to different drops along the distribution path. Taps are used to adjust consistently high or low voltages, not for normal fluctuations.

Taps are also used as a center-tap. Center-tapped transformers with the tap on the secondary are widely used in electronic applications. They are also used in three-phase systems, as was previously illustrated with the three-phase to the single, split-phase transformer.

Phase and Line Voltages and Currents

Voltages and currents are typically described in three-phase systems as either phase voltages and phase currents or line voltages and line currents. The phase voltage or current refers to the components in the wye or delta. Thus, a phase voltage is across one of the coils, and a phase current is the current in the coil. The line voltage or current refers to the connections to L_1, L_2, and L_3; thus, a line voltage is between two of the external lines. Notice that the line voltage is equal to the phase voltage for a delta connection because both are measured across the same two points. Likewise, the line current is equal to the phase current for a wye connection because the current is in the same wire. Figure 14-7 illustrates representative line and phase voltages and currents for the delta and wye configurations.

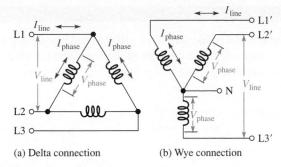

(a) Delta connection   (b) Wye connection

FIGURE 14-7 Representative Phase and Line Voltages and Currents. In the delta connection, $V_{line} = V_{phase}$ and $I_{line} > I_{phase}$. In the wye connection, $I_{line} = I_{phase}$ and $V_{line} > V_{phase}$.

In a delta connected winding that has a balanced load, the line current splits into two phase currents. The phase currents are not one-half of the line current because of the phase angle differences. Rather, they are related by a factor of $\sqrt{3}$. The equation for phase current in a delta connection with a balanced load is:

$$I_{phase} = \frac{I_{line}}{\sqrt{3}}$$

Equation 14-4

Likewise, for a balanced wye connection, the phase voltage is related to the line voltage by this same factor of $\sqrt{3}$. Thus, the equation for phase voltage in a wye connection with a balanced load is:

$$V_{phase} = \frac{V_{line}}{\sqrt{3}}$$

Equation 14-5

EXAMPLE 14-4

Draw a phasor diagram for a wye connected secondary with a 120 V phase voltage. Use it to show that the line voltage is 208 V by finding the difference in voltage between L_1 and L_3. This configuration is used in some industrial installations that require 208 V.

Solution

The phase voltages (L_1, L_2, and L_3) are shifted by 120° from each other as shown in Figure 14-8(a). Figure 14-8(b) shows the phasor diagram for L_1 (blue) and L_3 (green). To find the difference between L_1 and L_3, reverse the direction of L_3 as shown in Figure 14-8(b), creating the phasor $-L_3$. Add the inverted phasor ($-L_3$) to L_1 to find the difference ($L_1 - L_3$). You can find the length of $L_1 - L_3$ graphically or with basic trigonometry. From this, you can show that the difference is $\sqrt{3} \times 120 \text{ V} = 208 \text{ V}$.

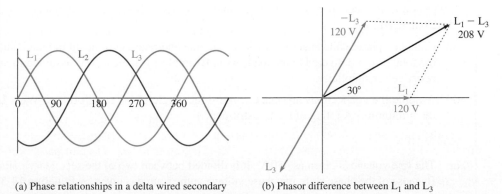

(a) Phase relationships in a delta wired secondary   (b) Phasor difference between L_1 and L_3

FIGURE 14-8 Graphical Solution to Example 14-4

EXAMPLE 14-5

A delta connected secondary has 480 V to a 100 Ω balanced load, as shown in Figure 14-9. Calculate the line and phase currents in the load, and the total power delivered to the load. The load is typically a motor in a three-phase system, but to simplify the problem, resistors are shown. Assume the PF is 1.

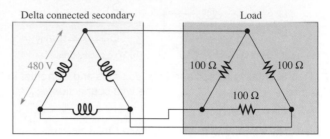

Delta connected secondary   Load

480 V   100 Ω   100 Ω

100 Ω

FIGURE 14-9   Delta Connected Secondary and Load

Solution

The line voltage and phase voltages across the secondary and the load are the same and are equal to 480 V because of the parallel connections. The phase current in the load can be found by Ohm's law:

$$I_{phase} = \frac{V_{phase}}{R_{load}} = \frac{480\ V}{100\ \Omega} = \textbf{4.80 A}$$

As given in Equation 14-4, the line current is larger then the phase current in a delta by $\sqrt{3}$; thus:

$$I_{line} = \sqrt{3}(4.80\ A) = \textbf{8.31 A}$$

The power in each resistor is the phase voltage multiplied by the phase current. There are three resistors, so the total power is:

$$P_{total} = 3 \times V_{phase}\, I_{phase} = 3(480\ V)(4.80\ A) = \textbf{6.91 kW}$$

Notice that this same result can be obtained by applying Equation 14-2, using line current and line voltage to calculate the apparent power.

A wye connected secondary and load has similar calculations. Because of phase relationships, the phase voltages rather than the phase currents include a factor of $\sqrt{3}$ in the calculation. Example 14-6 illustrates a wye connected secondary and load.

EXAMPLE 14-6

A wye connected secondary has 480 V, as shown in Figure 14-10, and is connected to a 75 Ω balanced load.

(a) Calculate phase voltage across the load, the line and phase currents in the load, and the total power delivered to the load. As in Example 14-5, the load is assumed to be resistive for simplicity.

(b) Assume the connections are made using three single-phase transformers. What is the minimum VA rating of the transformers?

Solution

(a) The line voltage is given as 480 V. It is divided between two of the secondary coils. Because of the phase shift, the phase voltage at the source is given by Equation 14-5:

$$V_{phase} = \frac{V_{line}}{\sqrt{3}} = \frac{480\ V}{\sqrt{3}} = \textbf{277 V}\ \text{(this is the voltage from any line to N)}$$

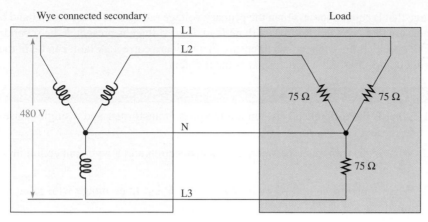

FIGURE 14-10   Wye Connected Secondary and Load

The phase voltage is identical in the load because of the parallel connection.

The phase current in the load can be found by Ohm's law:

$$I_{phase} = \frac{V_{phase}}{R_{load}} = \frac{277 \text{ V}}{75 \text{ }\Omega} = \textbf{3.70 A}$$

The line current is the same in the wye connection because of the series connection.

$$I_{line} = \textbf{3.70 A}$$

The power in each resistor is the phase voltage multiplied by the phase current. There are three resistors, so the total power is:

$$P_{total} = 3 \times V_{phase} I_{phase} = 3(277 \text{ V})(3.70 \text{ A}) = \textbf{3.07 kW}$$

(b) Each of the three transformers must be rated for one-third of the total, as long as the load is resistive (PF = 1, so kW = VA). Therefore:

$$\frac{3.07 \text{ kW}}{3} = \textbf{1.03 kVA} \text{ (note that the transformer is rated in VA, not W)}$$

Troubleshooting a Transformer

Troubleshooting high-power transformers with voltages above 200 V requires special training and special personal protective equipment. **Personal protective equipment (PPE)** consists of safety equipment such as safety glasses, safety shields, insulated gloves, hard hats, and suits to protect against arc flash for working on power systems with voltages above 200 V. Larger voltages create an electrical hazard called **arc flash** (or arc blast), which is a high-current, high-energy arc between two points that has the potential to cause injury or death. If you work for a company that does work on systems with voltage above 200 V, you will be required to be trained in the procedures for working safely using the appropriate safety gear.

If the transformer you are testing is disconnected from power and there are no external paths for current (such as a load), you can test each of the windings with an ohmmeter for continuity. Each of the coils should have some amount of resistance that indicates the amount of resistance in the wire in each coil. A measure of infinite resistance (∞) indicates that the winding has an open; a measure of 0 ohms indicates that one of the windings is shorted. Because transformers are basically coils of wire, the ohmmeter should reveal a faulty transformer. If the transformer passes the continuity test, you can apply power to its primary winding. If the transformer windings are not shorted or open, you can apply an ac voltage that is equal to or less than the primary voltage rating of the transformer and be able to measure the secondary voltage with no load. If no voltage is present at the secondary, be sure to check all the connections to ensure that they are correct. A fault can be in the load, making it appear as if the transformer is the problem.

SECTION 14-2 CHECKUP

1. What is the difference between a three-phase transformer and a single-phase transformer?

2. What is the difference between a delta connection and a wye connection in a transformer?

3. What is the difference between phase voltage and line voltage with reference to a transformer?

4. What type of winding has phase voltage equal to line voltage?

5. What type of winding has phase current equal to line current?

14-3   Grid Overview

The electrical grid in North America is a combination of various elements that provide a means of taking electrical power from where it is produced to where it used. This section introduces the grid in general and key aspects of its operation.

The **electrical grid** is defined as an interconnected network of power stations, transmission lines, and distribution stations that transmit power from suppliers to consumers. It provides a point of connection for renewable energy systems that produce electrical power. The long-distance movement of electricity is referred to as transmission; the **transmission system** moves bulk electricity from the transmission substations located at power plants and renewable energy systems to distribution substations. A **substation** is a network of switching equipment, control equipment, and transformers that are used to convert voltage to a different level (discussed in more detail in Section 14-5). Substations form the basic interface to the transmission system. After the distribution substation, the final delivery system to the consumer is called the **distribution system**.

Control of the Grid

Electrical power from both renewable energy sources and other generating plants (coal, nuclear, hydropower) is put onto the grid. The grid control system provides a means of turning on and off sections of the grid to isolate one or more sections when there is a problem, and to move electrical power from one section of the grid to another when the demand changes. In the United States, the three major grid sectors are isolated so that a problem in one sector can be contained and not affect the other two. Control centers in sectors and subsectors monitor the grid continually to ensure that sufficient electricity is available and that blackouts are limited to a small area if they do occur. In recent years, switching equipment on the grid has been upgraded with sophisticated computer and network controls that allow more efficient operation of the grid. This technology is being placed in the substations at industrial users and commercial users to allow the grid controllers to determine which loads can be disconnected if there is a problem anywhere on the grid. These more sophisticated control systems allow load sharing, which lets the grid control companies move electricity from area to area as the load shifts throughout the day. All of these controls are being integrated into something that is now being called the **smart grid**. While there is some disagreement as to the definition of the smart grid, it generally refers to the electrical distribution technology that uses computer-based remote control and automation to improve the delivery efficiency of electricity and provide information to the customer to optimize power use. In Section 14-4, you will learn more about the smart grid.

Multiple Three-Phase High-Voltage Transmission Lines for Moving Extra Power

(*Source:* Tom Kissell.)

When larger renewable energy systems are added to the grid, high-voltage transmission lines may need to be doubled to move the extra power. Because the transmission company owns the adjacent land, it is more cost-effective to place the additional high-voltage transmission lines in parallel with an existing set.

Transmission and Distribution Systems

The present grid has roots in the early days of electrical generation. One of the biggest obstacles facing the early producers of electrical power was how to get the power to consumers efficiently. Multiple power plants needed to be connected together so that their total output would be available as the demand for the electrical power changed over time from distant locations. Another issue was how to maintain continuous power to consumers when power plants needed to be disconnected from the grid for maintenance or other conditions. Cooperation between producers was needed to control the movement of electrical power through the transmission lines and to provide maintenance for the growing complexity of the generation and distribution system.

Figure 14-11 shows the North American Electric Reliability Corporation's (NERC's) interconnections. You can see that the grid in the map consists of four major interconnections: Quebec, Eastern, ERCOT (Electric Reliability Council of Texas), and Western. Each of the subsections of each interconnection is listed in the legend at the bottom of the map. The Quebec Interconnection, the Eastern Interconnection, and the Western Interconnection all include large sections of Canada.

In nearly all cases, the voltage from various sources is stepped up with a three-phase transformer for transmission, and then it is stepped down in substations located near the customer.

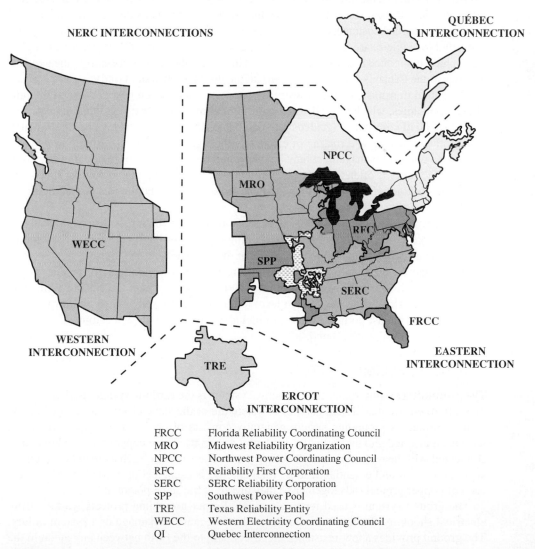

FRCC	Florida Reliability Coordinating Council
MRO	Midwest Reliability Organization
NPCC	Northwest Power Coordinating Council
RFC	Reliability First Corporation
SERC	SERC Reliability Corporation
SPP	Southwest Power Pool
TRE	Texas Reliability Entity
WECC	Western Electricity Coordinating Council
QI	Quebec Interconnection

FIGURE 14-11 North American Electrical Interconnections for the Electrical Grid (*Source:* Reproduced through permission from NERC.)

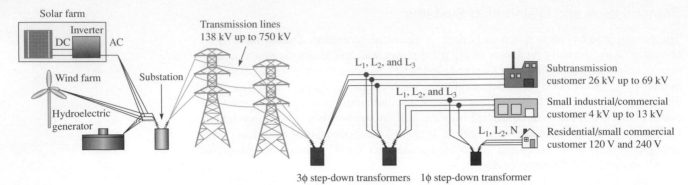

FIGURE 14-12 Basic Electric Power Generation, Transmission, and Distribution Systems. Distribution substations reduce the voltage and send it to the end users.

The voltages used for the transmission system are separated into different voltage levels for long-distance transmission. In North America, one of the highest voltages used is 765,000 V. Other lines transmit voltage at 500,000 V, 345,000 V, and 230,000 V. Typically, higher voltages are used for longer distances because they are more efficient. For large commercial and industrial customers, the three-phase voltage is usually supplied at high voltages, and the customer provides the transformers to reduce the voltage to the final value. For residential and smaller commercial customers, the three-phase is stepped down by the power company and converted to single-phase. In some areas of the world (some parts of Europe, for example), three-phase is supplied to houses. Figure 14-12 illustrates the basic process from the generating sites to the transmission system (green), to the distribution system (red).

The actual transmission lines, towers, transformers, and switch gear are owned by individual companies, which are called **transmission companies (transcos)** that manage this hardware. **Grid companies (gridcos)** are companies that manage the grid function, which is the interconnecting and routing of electricity through the electrical system. Some parts of the grid and grid management may be controlled by some local or regional utility companies. Electricity is often treated as a commodity, and it is bought and sold on short- and long-term contracts. Some of the ownership of the equipment and the sale of electricity are regulated by government agencies.

The final step in getting electricity to the customer is distribution. A distribution substation is a large set of transformers and switching equipment that is located at the edge of a city or near an industrial site that steps down the voltage from the transmission site to a lower voltage for distribution. Feeder lines transmit this lower voltage to the customers' sites, either on overhead lines or buried lines. Service systems near the customers include the transformers that drop the voltage level to the final delivered voltage and provide the service drop. The last portion of the grid includes any wiring at the customers' sites and grounding to a local earth ground.

Grounding System

The **grounding system** for electrical energy systems is the earthing system used as a safety measure to ensure that certain parts of a system are at the same electrical potential as the earth's surface. A grounding system generally consists of one or more copper rods that have been inserted into the earth to a depth of 2.4 m (8 ft). These copper rods are bonded or connected with heavy copper conductors to parts of the system, such as metal cabinets, to keep them at ground potential. If more than one rod is used, a copper conductor connects each of copper ground rods together so they are all at the same potential.

The ground system is used for two reasons. Proper grounding protects workers from electrical shock in case the frame becomes "hot" because of damage or a system failure. The ground provides a low-resistance electrical path to the earth between any metal in the system frame and the ground rods, thus causing high current and a circuit breaker to trip. If the system were not grounded, and the circuit breaker did not trip, anyone who came into contact with any metal parts could receive severe electrical shock or be killed by electrocution. If you come upon a circuit breaker that trips open, you must test the system to see if a short circuit has occurred before restoring it. When you close the circuit breaker and it immediately opens again, you should suspect a short circuit somewhere in the system, and you must determine the cause before trying again.

The second reason a system is grounded is to help protect the system from damage in the event of a lightning strike or other fault. If a renewable energy system is struck by lightning, the large amount of electrical energy in the lightning bolt must be dissipated quickly into the earth, where it will not cause damage. If the electrical system or its structures are not grounded properly, the location where the lightning strike enters the system can be severely damaged. The ground system must be checked periodically to ensure that it is connected correctly and that all connections provide a low-resistance path to the earth. In some installations, additional lightning arrestors may need to be added to protect the system from lightning strikes. Providing a grounding system is required by the electrical code, and it must be maintained and inspected annually to ensure that it is operational.

Ground Fault Interrupt Circuit Breakers

An important safety feature (such as protecting personnel from serious electrical shock while they work near wet locations) is the **ground fault interrupt (GFI) circuit breaker**. This special circuit breaker can trip if the hot and neutral current differs by more than a few milliamps, which can occur when a person is receiving an electrical shock. The circuit breaker consists of a small current transformer that encloses parts of the hot and neutral lines. When the lines have the same current, the fluxes from each cancel in the core. If they differ, a sensitive circuit detects the difference and trips a breaker. This feature is important in certain renewable energy systems where personnel are working in wet conditions, such as a rooftop. If a person is receiving an electrical shock, the current in the person returns to ground, rather than through neutral, and causes the hot and neutral lines to differ, thus tripping the GFI breaker. While a shock may still be the result, it will be very brief and less likely to be lethal.

Islanding

Islanding was defined in Section 6-6 as a condition when a grid-tie source (called a distributed generation [DG] resource), continues to operate and provide power to a certain location and remains connected to the electrical grid after the grid no longer supplies power. Islanding is a consideration for renewable systems: Local suppliers can provide more reliable power and often more efficiently because these small islands may reduce the transmission and infrastructure costs. In addition to increased reliability, efficiency is enhanced when the consumer is near the source. A safety hazard for power company employees associated with islanding is particularly important in renewable energy systems. If a distributed source such as a homeowner's solar electric system is connected to the grid and the grid goes down, the solar electric system may continue to energize power lines and pose a serious danger to power company workers. It is important that inverters and other equipment connected to the grid have anti-islanding circuitry to remove this hazard when the grid goes down.

Delivery to the Customer

Electrical power is routed to the customer either on overhead or underground wiring. Figure 14-13 shows an overhead connection with all of the switch gear and metering. When electrical power is transmitted overhead, it can be overloaded slightly without damaging the wiring, but it is exposed to storm damage.

Power can be installed underground by a special machine (see Figure 14-14) that installs a conduit underground without digging a trench. It installs sections of pipe as it drills horizontally approximately 3 ft below the surface. When the drill and pipe reach the destination, the pipe is withdrawn one segment at a time, and it pulls the plastic electrical conduit into the hole it has created. The conduit has a pull wire installed inside it to enable pulling the electrical cable through the conduit. Underground wiring is used in congested areas and

The National Electric Code and Safety Considerations

Renewable energy systems that generate electricity abound with potential safety hazards. When wiring is initially installed in a system, it important that it is installed according the electrical codes that apply to that area. The standards for electrical wiring are set in the United States by the National Electrical Code (NEC), which is published by the National Fire Protection Association (NFPA), a nonprofit corporation. The code is revised every three years. The NEC publishes a guide, *Essentials of Distributed Generation Systems,* that is covers solar photovoltaic, wind turbine, fuel cell, microturbine, and engine-generator system technologies.

Flashover

A serious problem can occur when high-voltage lines come into contact with each other, or when one of the lines makes contact with a lower-voltage feeder line or grounded metal parts. Either situation can occur during a storm (a tree falling across lines, for example). The low-voltage systems could be subject to a high voltage, which can cause a condition called flashover. Flashover is a serious hazard and can cause a fire or other serious result. A properly grounded neutral can force most of the current through the neutral line to ground, which can trip breakers and help avoid more serious consequences.

FIGURE 14-13
Overhead Connections
to a Commercial
Building   (*Source:*
Tom Kissell.)

FIGURE 14-14   Underground Power
Installation. A specialized machine
buries conduit underground for
running electrical power below the
surface.   (*Source:* Tom Kissell.)

where unsightly wiring is not wanted, but it costs much more to install. Frequently, spare wiring is installed in case a fault develops.

SECTION 14-3 CHECKUP

1. What are three segments of the grid?

2. What is the interface to the transmission system?

3. What is the distribution system?

4. Why is the voltage level raised for long-distance transmission on the grid?

5. What does GFI stand for?

14-4   Smart Grid

The smart grid was defined previously as the electrical distribution technology that uses computer-based remote control and automation to improve the delivery efficiency of electricity. In this section, issues with the smart grid are discussed and microgrids are introduced.

In the standard grid, information and energy in the grid travel in one direction: from the point where it is generated to the point where it is consumed. In the United States, the smart grid was initially called the Modern Grid Initiative was started (in the United States) at the National Energy Technology Laboratory (NETL) in January 2005. The mission of the initiative was to accelerate grid modernization. Later, this initiative was renamed the Smart Grid Implementation Strategy (SGIS). The smart grid will allow information to move to and from consumers when it is fully implemented. The smart grid concept has been embraced throughout the world, with huge turnouts to events like the International Smart Grid Expo.

One aspect of the smart grid is centralized collection of information. For example, the electric meter in a residence or industrial location will have the ability to report details of how much energy is being consumed at all times. In some cases, it will be able to send specific information from certain compatible appliances. (Appliance manufacturers are currently developing appliances that can report their energy use.) Data that is collected will enable improved predictions about energy use patterns. Power companies could offer electricity at different prices during off-peak times so that consumers could operate certain loads when the cost of electricity is less. For example, more power is usually available late at night and early in the morning, when demand is less (called off-peak times). It may be possible to shift certain usage (irrigation or water heating) to these off-peak times to save money. Another way the smart grid helps customers is to apply credit for any renewable electricity produced and provided to the grid.

Microgrids

A **microgrid** is a subset of the smart grid that has the ability to operate independently of the main grid (the macrogrid). A microgrid can be a small electric utility that can be connected to the larger grid, or it can stand alone as an independent system when it is advantageous to do so. Microgrids can manage and distribute power efficiently from multiple power sources, including renewable energy sources. The two-way communication from source to user enables utilities and consumers to respond quickly to changing conditions. Because of their smaller scale, microgrids are easier to control (using the latest controllers), experience lower line losses, place less demand on infrastructure, and can choose from the optimal sources. Over 100 smart microgrids are already at work, and the number is expected to grow rapidly. Microgrids are generally set up to achieve specific local goals, such as reliability, connection to renewable energy sources, and control of a smaller regional area. Smart microgrids generate, distribute, and regulate electricity for customers such as college campuses, hospitals, data centers, military bases, large industrial users, or municipalities that want more control over their electrical power. They are used all over the world and are particularly useful in developing countries that do not have a massive grid structure in place.

Microgrids now have significant power-generating capacity, which is expected to expand and may eventually include most power. It is envisioned that many more microgrids will be implemented at locations where they can provide an efficient operation. They may ultimately be a model for how the smart grid is implemented. Microgrids allow a smaller area to be connected and tested, and do not require full implementation over the entire grid. An example of a very small version of the microgrid occurred during Hurricane Sandy, which hit the East Coast of the United States in 2012. One building in Manhattan was able to run a CHP system to provide power to the building for five days after the hurricane while surrounding areas were blacked out.

Interoperability and Support of Standards

Designers of the smart grid must consider the wide variety of standards that must be integrated into each segment of the smart grid. These standards are published by organizations like International Standards Organization (ISO), Open Geospatial Consortium (OGC), International Electrotechnical Committee (IEC), and Common Information Model (CIM). These key organizations are working to build a truly smart grid. When a system meets these international standards, it can be integrated much more easily.

When fully implemented, the smart grid can result in innumerable benefits for utilities and customers. It has the ability to boost reliability, reduce response times to outages, increase operational efficiency, and improve the safety and security of the transmission lines. The smart grid can also help utilities to continue to meet the world's energy needs and advance the grid to new renewable energy locations.

Smart Grid Implementation Issues

The smart grid requires much more than just installing new smart meters for customers or monitoring their large appliances. The smart grid will need to address six issues, which are listed in Table 14-1. These issues are discussed in detail in the paragraphs that follow.

Distribution Control

In addition to its other benefits, the smart grid will help with power distribution, automatic switching control to isolate problem areas, and provision of bidirectional information, which will help pinpoint outage areas precisely so that repairs can be made quickly. Smart control will also allow faster response to problems by switching to backup systems as the need arises; a secondary benefit is less stress on the current infrastructure.

The smart grid, with residential and commercial controllers, will allow end users to enter into contracts with energy producers and agree to have some of their largest loads turned off by the smart grid for short periods of time when the grid is becoming overloaded.

Identifying the Extent of Widespread Outages

(*Source:* Warren Rosenberg/Fotolia)

When an ice storm or a hurricane causes outages, it is important to identify the extent of the outage. If work crews are sent to the correct locations, they can make repairs to get the largest sections of the grid back into service and thus provide power to the largest number of customers. New smart grid technology will be able to identify the extent of an outage and key locations to get the most customers back online quickly.

TABLE 14-1

Smart Grid Implementation Issues

1. Control of electrical distribution by automating substations and switch gear with digital protective relays, weather prediction system, software, and integration tools.

2. Sensing and measurement technologies that consist of wide-area monitoring systems, dynamic line-rating technologies, fiber-optic temperature monitoring systems, and special protection systems.

3. Improved interfaces and decision support technologies to enhance a person's ability to interface and work with the grid when using appliances and other electrical equipment.

4. Advanced components for energy technologies, such as fuel cells, microgrid technologies, ultracapacitors, sodium sulfur (NaS) and lithium (Li) ion batteries.

5. Integrated communications consisting of technologies such as broadband over power lines, fiber optics to the home, and hybrid fiber coax (HFC) architecture.

6. Cybersecurity standards to make the system resilient to attack and provide for rapid restoration capabilities because the new grid will have network capability.

A **brownout** is reduced voltage from the grid when demand exceeds supply (the voltage drops 10% or more). If the brownout continues for a longer period of time, large electrical loads that have motors, such as air-conditioning compressors and pumps, may be severely damaged or overheated. One method of alleviating a brownout is allowing the control system to remove some larger loads until the voltage level comes back up. The smart grid will allow large industrial users to permit sections of their electrical equipment to be turned off for short periods of time in return for better pricing. Controllers may be able to avoid a brownout by routing electrical power from a nearby sector to the area that has the low-voltage problem.

Sensing and Measurement

Advanced sensing equipment allows the smart grid to determine the health and integrity of all switch gear and other equipment used to control the electrical power as it is moved over the grid. When a major power outage occurs at a utility with a smart grid infrastructure and technology-enabled workforce, the utility is automatically notified which customers are affected via downed smart meters and other sensors. These characteristics allow utilities to identify where the outage has occurred and where to send resources to get the service restored as quickly as possible.

The same sensors that monitor the grid will allow demand response (DR) control to reduce peak demand. **Peak demand** is the period of greatest electrical energy consumption during a specific period of time. The peak demand varies depending on the location, time of day, day of the week, and weather. Energy needs to be available to match the peak demand or a brownout will occur. If electrical power cannot be redirected where the demand exceeds the supply, the affected region will have to be disconnected to prevent it from taking down a larger sector of the grid. Sensors provide the information required by demand response control to reduce the high demand during peak times.

Interfaces and Decision Support

The grid sometimes has problems that require operators and managers to make decisions. Decision support and improved interfaces in the smart grid will enable more accurate and timely human decision making at all levels of the grid, including the consumer level. The interface and decision support will also enable advanced operator training.

Homeowners will be able to set up their smart grid interface to provide power to certain electrical loads (such as heating hot water or recharging an electrical car) at off-peak times. This feature will allow them to better match the times when they consume electricity for some loads and match it to the time when the cheapest electrical energy is available.

On the transmission side, the smart grid will enable control of grid sectors that are using the most power so that it can be redistributed as needed. It will also allow the interconnection system to identify problems, and to disconnect and isolate small sections with those problems faster, thus ensuring that they do not affect the entire grid.

Advanced Components

Some experts believe that as much as 30% of the power in the grid may be lost to inefficiency, either through heat loss due to older wiring, or because of poor equipment and connections. Advanced components in substations and at customer sites will improve efficiency. One component of great interest is advanced sodium-sulfur batteries, which allow for peak leveling. These batteries are long-lasting, compact, and efficient. They have been in use by some utilities to level peak loads. They have been widely applied in Japan for storing energy when demand is low and using the energy when demand is high. These batteries may be particularly valuable additions to wind farms when winds are intermittent.

Another battery of interest is the Li ion battery, which may help the smart grid become more efficient. In the case of electric trains, the process of stopping trains has required dissipating the energy of motion into braking resistors. In conventional systems, the energy is fed back; however, there is usually too much energy for the grid to absorb all at once, so most of it is wasted as heat. By absorbing the energy in Li ion batteries, the power can be recovered efficiently for aid in starting the train again. The Li ion batteries are located in a substation, and initial pilot projects have proven that the idea for electrical storage is a sound one.

Integrated Communications

The communication method employed by the smart grid is to identify each step in the transmission process and each end user with an Internet protocol (IP) address that allows selected equipment at that location to send and receive data. The data includes current energy demand, time of day, and temperature and other environmental conditions. The smart grid will use the transmission lines as the means for sending and receiving this data. Because each point in the grid will have its own specific address, it can be accessed much like a large computer network, where information can flow to and from each node in the network. The information is similar to a computer network, so it can be displayed on a small display monitor or on a computer screen in the home and at commercial or industrial establishments to enable consumers to understand their energy usage.

Cybersecurity

When information is put on the smart grid, it will be vulnerable to unauthorized access, just like computers on the Internet. The smart grid will need cybersecurity measures to protect it from being hacked or interrupted by malicious intruders. Grid transmission must be protected and secured. As on the Internet, security threats can change, and security must be ever-vigilant to respond to these threats. The requirements for optimum integrity and functionality of the grid is often incompatible with security procedures, so it is important for utility companies to implement security procedures that are similar to those used in information technology (IT): changing passwords, removing and disabling unused accounts and services, setting up firewalls, and maintaining event files in order to keep the grid safe.

Currently, computers control the grid at every level, including the generators, substations, and distribution systems. Most of the systems use common operating systems, well understood by hackers. Even though these systems are generally not connected to the Internet, they are still vulnerable to attack. Some systems use radio commands sent via low-power radio that can be intercepted. Another vulnerable area includes the business side of the utility, which a hacker could use to find usernames and passwords to simulate a legitimate employee who requires access to the control system. In one simulated attack, government officials were able to infiltrate a site and create havoc by rapidly cycling an electronic circuit breaker, causing a generator to lose sync and be destroyed.

The consequences of a coordinated attack on any nation's electrical grid could be devastating, Nearly all industrial production in the affected area would grind to a halt. A lengthy outage would strain resources, from gasoline supplies to water treatment, sewage disposal, and more. Utilities must exercise tight control over all of the devices in their system.

SECTION 14-4 CHECKUP

1. What is the smart grid?
2. What is a microgrid?
3. What are three communication strategies for the smart grid?
4. Why is cybersecurity so important to the grid?

14-5   Power Transmission

Power transmission is the moving of electrical energy from the source to the end user. It includes the transmission substation and the distribution substation. The installation of new structures and facilities always raises public concerns that need to be addressed. Moving power is one of the issues accompanying renewable energy systems.

Around the world, the mix of power systems varies, but nearly all power is connected to a transmission grid at the source. Exceptions are some stand-alone renewable systems, including some small wind farms and solar farms. In many cases, the grid has been extended to their locations so they can be connected. Government agencies are involved in decisions to extend the grid to new locations because such work often crosses private land. In the United States, the agency responsible for these decisions is the federal Energy Regulatory Commission. Extending the grid involves expensive investments; in some cases, the cost is slowing down the expansion of new renewable energy production. For example, a major wind resource is located in the upper Midwest of the United States, but the grid capacity is limited in this area. To exploit this resource, new transmission lines must be installed or sections of existing grid must be expanded to allow for greater power transmission.

Issues surrounding expansion of the grid include land usage and construction of substations and other infrastructure. Renewable sources typically produce electrical power at much lower voltages than is necessary for transmission. For example, solar PV systems typically have an output of 600 to 1,000 V dc, which needs to be converted to ac and transformed to a much higher voltage for transmission. All of the issues that accompany new sources and new transmission lines need to be addressed, for example, the environmental issues of the sources and the environmental effects of locating long transmission lines on wildlife because of clearing the land and building access roads.

Substations

Previously, a substation was defined as a network of switching equipment, control equipment, and transformers used to convert voltage to a different level and eventually route it to the user. At the production site, a transmission substation steps up voltage for transmission; at the area where it will be used, it is stepped down by a distribution substation. Transmission substations use tapped transformers that step up the voltage to a predetermined level. Other equipment at the transmission substation includes various types of switches and relays to control power, as well as a variety of protection equipment such as surge arresters in case of a lightning strike. Switches for safely disconnecting and grounding lines are used when personnel are working on the lines. Large, oil-filled or compressed-air circuit breakers are in the line to trip in case line conditions exceed preset limits. Most of this equipment

FIGURE 14-15 Small Substation for Residential Area Users (*Source:* Tom Kissell.)

FIGURE 14-16 Large Substation for an Industrial Complex (*Source:* Tom Kissell.)

can be controlled and reset from remote locations with telecommunication signals and do not require onsite personnel.

At the distribution substation, the same type of equipment is present, but transformers are step-down types. Distribution substations can be used for residential installations and for large industrial plants. The main difference between a residential substation and a large industrial substation is the voltage and power levels involved. In most parts of the world, the transformers in residential substations typically have three-phase voltage on their primary side, and they provide single-phase voltage on their secondary side. Residential transformers use only two of the phases to produce single-phase voltage, so different transformers are wired for different phases to balance power in the three phases. Transformers in an industrial substation supply all three phases (typically at 480 V or more). Figure 14-15 shows a smaller substation for a residential area; Figure 14-16 shows an example of a larger substation for an industrial user.

Some transformers in both the residential and industrial substation are connected in parallel so that they can share power. In this case, each of the transformers must be similar in size when it comes to voltage and current so that none of the transformers become overloaded. When transformers are similar in size, they tend to share the load evenly. Control systems monitor the individual load of each transformer and the total load for the substation, and switches are used to isolate transformers or substations when problems occur.

Controlling the Load over Grid Transmission Lines

The grid transmission system has another important job routing electrical power to areas where it is needed. Electrical power consumption changes constantly throughout the day and from day to day. In general, the lowest consumption occurs during overnight hours. Many industrial users do not use as much energy during the night, and some companies that operate only a day shift use very little power during the night. Because the grid can move power across several time zones, the control system can move power to an area where it is needed during peak consumption times. Power is often moved when very high temperatures in an area cause additional power consumption because of heavy air-conditioning use.

SECTION 14-5 CHECKUP

1. Name two safety components at a substation.

2. How are multiple transformers in a substation connected so that they share power?

3. What is the main difference between the transformers in transmission substations and distribution substations?

4. What is a brownout? What measures can be taken to avoid or eliminate brownouts?

14-6   Connecting to the Grid

When producing electrical power for the grid, a renewable energy system must meet several criteria, including voltage, frequency, and power quality. This section covers compatibility issues and requirements for connecting to the grid.

Generation sites include both renewable and nonrenewable sources. If a renewable energy system is designed to produce power for the grid, it must meet certain specifications, which are different depending on the location. When electricity is produced from a renewable energy system, it may be a dc voltage from a photovoltaic source or unsynchronized ac from a wind machine. The voltage must be matched to the ac level, frequency, and phase of the grid. The exact level depends on the point on the grid where the connection is made. Small residential systems may send single-phase 240 V ac to the grid. Large systems may need to be stepped up to very high voltages for transmission lines.

Before a privately owned renewable energy system is connected to the grid, a permit to connect to the grid must be filed with the utility company. For privately owned systems, the owner can be reimbursed for excess power that the renewable energy system produces and returns to the grid. Various government entities have laws that allow or require utilities to permit renewable energy systems and other sources to be connected to the grid, and to provide some type of compensation for the energy that is produced. Some government agencies may require the utility to pay only a minimal amount for this energy, thus making the renewable energy less valuable. Other agencies require utilities to pay the same or nearly the same value as they charge for their electricity. This requirement varies around the world.

One way of providing the connection between a privately owned renewable energy system and the grid is with a utility meter called a net meter. A **net meter** can measure the difference between the power coming from or going back to the grid. Some versions of the net meter allow the excess generated electricity to be banked, or credited to the customer's account, which basically means that the electricity generated by the renewable energy system is sold at the same rate as the electricity that is used. The net meter allows the electric meter to spin forward and backward, which simplifies the process of measuring how much electricity was produced for or consumed from the grid.

Net Meter

(*Source:* Tom Kissell.)

The net meter measures electical power as it flows to an end user. It also measures any power that the end user produces from a solar energy system, a wind turbine, or other renewable energy system. The net meter measures current as it flows in either direction.

Power Quality Issues

The term **power quality** is a measure of the purity of the ac power, including the presence of unwanted harmonics and noise on the power line and the frequency. A **harmonic** is a frequency that is a whole-number multiple of another basic frequency. Harmonics on the power line are multiples of the power line frequency (50 Hz or 60 Hz). Under certain conditions, the electrical equipment can be damaged by harmonics. Harmonics are created both at the source and in nonlinear loads, and because of hysteresis in transformers or motors.

In addition to harmonics, nonlinear loads at a user's site can inject noise on the line that is transmitted to other users. Nonlinear loads include motor variable frequency drives, welders, arc furnaces, electronic ballasts, and more. Harmonics can affect the operation of equipment, particularly sensitive controllers and computers, and can increase losses in transmission. Their magnitude diminishes with frequency, and harmonics higher than the thirteenth harmonic (780 Hz for a 60 Hz fundamental frequency) are not included in measuring harmonic content. One way to reduce harmonic content in power lines is to use a pulse width modulation (PWM) converter at the user's site or filters to remove frequencies other than the power line frequency. A PWM converter can generate a signal of any desired shape or frequency in high-power applications.

In renewable systems, noise from the electrical equipment at the source can put poor-quality ac on the grid. In the case of PV solar systems, the dc from the solar panels is converted to ac by an inverter. Inverters are notorious for creating harmonic and radio frequency noise because of the high-speed switching commonly used. It can be made worse if the unit is not properly grounded. Filtering can help alleviate noise from inverters.

Noise can cause voltage transients and is difficult to deal with due to its random nature. It includes lightning and power interruptions. In critical cases where a brownout, blackout, or noise can be a serious problem (a computer center or hospital, for example), the customer can install an **uninterruptible power supply (UPS)**, which automatically switches to an auxiliary generator or batteries if utility power is lost or has other problems.

The ideal power quality exists when the voltage and current produced by the generating system has a perfectly clean sinusoidal waveform with no harmonic content or noise. The frequency must match the grid frequency exactly. The voltage of the ideal power source stays constant regardless of changing conditions. Of course, the ideal source does not exist, but high-quality sources come close to this ideal. If parameters are too far from the ideal power, they need to be corrected before connection to the grid. The major issues affecting power quality are frequency and voltage control, low-voltage ride through, and flicker. These issues are discussed in the next subsections.

Frequency and Voltage Control

Frequency and voltage on the grid are controlled by the individual power producers who put large amounts of electrical power into the grid. The grid receives voltage from large energy producers such as nuclear-powered generators, coal-fired generation systems, and hydroelectric generators, as well as from renewable energy systems. Large energy systems can control voltage and frequency easily, but some renewable energy systems may have difficulty because the energy from renewable sources, particularly wind and to some extent solar, may be variable. Because most of the electrical energy on the grid has constant voltage and constant frequency from the larger energy producers, the grid itself can help synchronize the frequency from renewable energy systems (for example by using doubly fed induction generators).

Low-Voltage Ride Through

Low-voltage problems in the grid may occur for several ac voltage cycles, or it may last for several hours. It may also be caused in one phase of a three-phase system or in all three phases. Low-voltage is a condition that exists any time the voltage on the grid drops 10% or more below the rated voltage level.

When low-voltage occurs at the point where the renewable energy system is connected, the renewable energy system must have the capability to ride through this condition or disconnect from the grid. It does not matter if the renewable energy system is causing the low voltage or if the low voltage is caused elsewhere on the grid. The safety system for this problem is called **low-voltage ride through (LVRT)**. When the low-voltage fault occurs in the grid, instrumentation measures the severity and the duration of the problem and determines the best solution for correcting it.

If the renewable energy system is causing the low-voltage condition, it must be able to correct the condition quickly or disconnect from the grid. The choice is to disconnect the source from the grid and automatically reconnect after the low-voltage condition has cleared, or to remain connected and ride through the low-voltage condition. Another possibility is to stay connected and help correct the problem by adding power from the renewable energy source's capacitive banks, which can help the low-voltage condition. The method used by the renewable energy system may depend on the type of problem the grid is experiencing. Obviously, corrective action is useful only for large systems because smaller systems cannot affect the grid enough to change the condition.

Flicker

Flicker is defined as a short-lived variation in electrical power. It is sometimes seen as a momentary dropout, as when a light varies. This problem is more noticeable when a renewable energy system is connected directly to a residential or commercial application as the sole source of power. If the renewable energy system uses an inverter, it typically has the ability to control the increase or decrease in voltage so that the change in voltage is kept to a minimum in both duration and amount.

1. What is a net meter?

2. If a low-voltage condition exists on the utility grid, how should a renewable energy system respond?

3. What issues concerning power quality must renewable energy systems address when connecting to the grid?

4. What is a UPS? Where is its use important?

CHAPTER SUMMARY

- Three-phase voltage is three ac voltages with the same magnitude but separated by 120°.

- The advantage of three-phase power is that it is more efficient to produce and less expensive to transmit than single-phase power. Industrial motors that use three-phase are more efficient than equivalent single-phase motors.

- Single-phase power for residential and smaller commercial customers can be derived from three-phase power.

- If the load is purely resistive in an ac circuit, the voltage and current are in phase with each other, and the power can be calculated by multiplying the voltage times the current. This power is called true power.

- The turns ratio is the ratio of the number of turns in a secondary winding of a transformer to the number of turns in the primary winding (defined for electronic power transformers, but other definitions exist, so the user needs to check).

- Transformers are rated in volt-amperes (VA).

- Reactive power is measured in a unit called volt-amp-reactive (VAR).

- The combination of reactive power and true power in an ac circuit is called apparent power.

- Phase shift occurs when the voltage phase leads the current phase, or vice versa.

- Three-phase transformers can be an integral unit (wound on a single core), or they can be constructed from three single-phase transformers.

- Windings on a three-phase transformer are configured either as a delta (Δ) or a wye (Y); the letter or symbol suggests the schematic shape.

- When working on power transformers, it is important to wear personal protective equipment.

- The electrical grid is an interconnected network of power stations, transmission lines, and distribution stations that transmit power from suppliers to consumers.

- The grid transmission system moves electricity from power plants to distribution substations.

- The grid distribution system is a delivery system for moving electricity from distribution substations to consumers.

- The smart grid is an electrical distribution technology that uses computer-based remote control and automation to improve the delivery efficiency of electricity and provide information to the customer for optimal power use.

- Government agencies typically monitor the electrical power demand and the capacity resources (available electrical power to use) to determine the capacity margin as a percentage.

- The actual transmission lines, towers, transformers, and switch gear are owned by individual companies called transmission companies (transcos).

- Grid companies (gridcos) are sets of companies that manage the grid function, which is the interconnecting and routing of electricity through the hardware cables so that areas receive the electricity they need.

- The grounding system for electrical renewable energy systems generally consists of one or more copper grounding rods that have been inserted into the earth to a depth of 2.4 m (8 ft).

- A ground fault interrupt (GFI) circuit breaker is a special circuit breaker used in hazardous locations (such as near water) that can trip if the hot and neutral current differ by more than a few milliamps.

- A microgrid is a small electric utility that can be connected to the larger grid or can stand alone as an independent system.

- A substation is a network of switching equipment, control equipment, and transformers that are used to convert voltage to a different level. Transmission substations can switch in various power sources and step up voltage to very high voltages. Distribution substations reduce the voltage for the distribution network.

- A net meter can measure electricity that flows from the grid to a customer or from the customer to the grid.

- The term *power quality* refers to issues dealing with the frequency of voltage and current, and different types of unwanted signals or noise, including electrical noise, dc injection, and harmonics. A harmonic is a whole-number multiple of a frequency.

- In situations where power delivery is critical, a customer can install an uninterruptible power supply (UPS) that supplies clean power with backup.

- Flicker is a short-lived voltage variation in the electrical power that might cause a load such as an incandescent light to flicker.

KEY TERMS

arc flash   A high-current, high-energy arc between two points that has the potential to cause injury or death.

distribution system   The delivery system that moves electricity from distribution substations to consumers.

electrical grid   An interconnected network of power stations, substations, transmission lines, and distribution systems that transmit power from suppliers to consumers.

flicker   A short-lived variation in the electrical power.

ground fault interrupt (GFI) circuit breaker   A special circuit breaker that can trip if the hot and neutral current differ by more than a few milliamps; this condition can occur when a person is receiving an electrical shock.

harmonic   A frequency that is a whole-number multiple of another basic frequency.

low-voltage ride through (LVRT)   A safety system in place to address a low-voltage fault in the grid. Electrical instrumentation measures the severity and the duration of the problem, and determines the best solution to correct it.

microgrid   A subset of the smart grid that has the ability to operate independently of the main grid (the macrogrid). A microgrid can be a small electric utility that is connected to the larger grid, or it can stand alone as an independent system.

personal protective equipment (PPE)   Special safety equipment needed to work on power system with voltages above 200 V. Personal protective equipment may consist of safety glasses, safety shields, gloves, hard hats, or suits to protect against arc flash.

power quality   A measure of the purity of the ac power, including the presence of unwanted harmonics and noise on the power line and the frequency.

smart grid   An electrical distribution technology that uses computer-based remote control and automation to improve the delivery efficiency of electricity and to provide information to the customer for optimal power use.

substation   A network of switching equipment, control equipment, and transformers that are used to convert voltage to a different level.

transmission system   The system that moves bulk electricity from the transmission substations located at power plants and renewable energy systems to distribution substations.

uninterruptible power supply (UPS)   A source that automatically switches to an auxiliary generator or batteries if utility power is interrupted.

FORMULAS

Equation 14-1	$PF = \dfrac{P_{true}}{P_a}$	Power factor
Equation 14-2	$P_{total} = \sqrt{3}\,(PF)P_a$	Total power delivered to a three-phase motor
Equation 14-3	$P_{pri} = P_{sec}$	Power delivered by an ideal transformer
Equation 14-4	$I_{phase} = \dfrac{I_{line}}{\sqrt{3}}$	Phase current in a delta connected transformer with a balanced load
Equation 14-5	$V_{phase} = \dfrac{V_{line}}{\sqrt{3}}$	Phase voltage in a wye connected transformer with a balanced load

CHAPTER TRUE/FALSE QUIZ

Determine whether each statement is true or false. Answers are at the end of the chapter.

1. In three-phase, the sinusoidal waveforms are separated by 120°.

2. In general, single-phase motors are simpler to maintain and more efficient than three-phase motors.

3. Split-phase refers to two hot wires with the same polarity and neutral.

4. A phasor is a rotating vector.

5. True power is always larger than apparent power.

6. Normally, three-phase is generated by conversion of single-phase.

7. Three-phase requires the utility to send L1, L2, and neutral to the customer.

8. Apparent power is measured in volt-amps (VA).

9. Power factor can be expressed as the ratio of true power to apparent power.

10. In a purely resistive load, the power factor is 0.

11. Three-phase transformers can be constructed from three single-phase transformers.

12. An ideal transformer does not dissipate any internal heat.

13. A common configuration for three-phase transformers is the wye-wye connection.

14. Phase voltage is the same as line voltage.

15. The volt-ampere (VA) rating for a transformer is calculated for either the primary side or the secondary side.

16. A key component in a transmission substation is a step-down transformer.

17. An ohmmeter reading of infinite resistance (∞) across a transformer winding indicates that the winding is shorted.

18. The electrical grid includes power stations, transmission lines, and distribution stations.

19. A GFI breaker trips if line current and neutral current are not the same in a single-phase system.

20. The smart grid collects information on all appliances in use in a home.

21. A microgrid does not include the ability to supply power in event of a main grid failure.

22. *Brownout* is a term used to describe a low-voltage situation in the electrical supply.

23. The smart grid has no means of identifying end users.

24. Distribution substations increase the voltage for transmission lines.

25. The purpose of a net meter is to monitor the Internet.

26. A harmonic of a given frequency is a whole-number multiple of that frequency.

27. An issue with power quality includes noise induced on lines.

CHAPTER MULTIPLE-CHOICE QUIZ

Complete each statement by selecting the one correct answer. Answers are at the end of the chapter.

1. An advantage of single-phase power for residential use is its
   a. increased delivery efficiency compared to three-phase
   b. savings for delivery
   c. higher voltage operation compared to three-phase
   d. all of these

2. Apparent power is
   a. smaller than true power
   b. measured in watts
   c. a combination of resistive and reactive power
   d. all of these

3. Power factor is
   a. true power divided by apparent power
   b. apparent power divided by true power
   c. true power multiplied by apparent power
   d. true power added to the apparent power

4. Three-phase power has three waveforms separated (ideally) by
   a. 90°
   b. 120°
   c. 180°
   d. 240°

5. The efficiency of an *ideal* transformer is
   a. 85%
   b. 90%
   c. 95%
   d. 100%

6. If the voltage is stepped down in a nonideal transformer, the current is
   a. stepped up and more power is delivered to the load
   b. stepped down and more power is delivered to the load
   c. stepped up and less power is delivered to the load
   d. stepped down and less power is delivered to the load

7. Which of the following cannot be used to determine the turns ratio for a transformer?
   a. The number of turns in the primary and secondary windings
   b. The ratio of voltage in the primary and secondary windings
   c. The ratio of the current in the primary and secondary windings
   d. The ratio of primary voltage to primary current

8. In a wye connection on a three-phase transformer, the coils are connected to a central point called the
   a. wye point
   b. center point
   c. center-tap
   d. ground point

9. An advantage to using three individual single-phase transformers instead of one integrated three-phase transformer is that
   a. the total weight is less
   b. one can be replaced if a failure occurs
   c. they are more efficient
   d. all of these

10. A good choice for adjusting a voltage that is consistently high or low on a power transformer is a
    a. tapped transformer
    b. variable transformer
    c. variable load
    d. divider network

11. Before troubleshooting a large power transformer, you should
    a. be wearing personal protective equipment
    b. have already undergone special training
    c. both of these

12. A GFI system senses and compares current in
    a. two hot wires
    b. one hot wire and neutral
    c. ground and neutral
    d. none of these

13. The smart grid sends power
    a. as well as cable TV
    b. and includes two-way communication about its use and related information
    c. filters the voltage to reduce noise
    d. all of these

14. The smart grid will include
    a. distribution control
    b. integrated communication
    c. advanced components
    d. all of these

15. The main difference between a transmission substation and a distribution substation is the
    a. type of surge equipment necessary
    b. voltage and power levels involved
    c. switching equipment
    d. safety disconnect equipment

16. Low-voltage ride through (LVRT) refers to
    a. a circuit that holds constant voltage even when disconnected from the grid

b. a circuit in a renewable energy system that drops out if the voltage from the grid does not match the renewable energy voltage

c. the ability of an electrical generation system to withstand a low voltage on the grid

d. a high-power regulating circuit

17. Flicker is when voltage
    a. has random variations
    b. is consistently low
    c. is out of phase with current
    d. is all of these

CHAPTER QUESTIONS AND PROBLEMS

1. Compare three-phase, single-phase, and split-phase voltages.

2. Draw a phasor diagram for a split-phase ac signal with 60 Hz, 120 V on the hot line.

3. The phasors in question 2 rotate once each cycle. How long does it take to complete one complete revolution?

4. Explain how 208 V is derived from two different line voltages in a three-phase system.

5. Determine the power factor for a circuit if the true power of the circuit is 12 kW and the apparent power is measured as 14.5 kVA.

6. A three-phase motor is running from 230 V and has a line current of 11.6 A.

   a. Determine the apparent power.
   b. If the power factor is 0.92, determine the total power.

7. Explain why transformers are rated with units of VA rather than W.

8. Explain why the secondary windings on a step-up transformer can be made with smaller-diameter wire.

9. Compare the line voltage and phase voltage on a three-phase transformer. On what type of winding are they the same?

10. Compare the line current and phase current on a three-phase transformer. On what type of winding are they the same?

11. A delta connected secondary has 230 V to three 175 Ω balanced resistive loads. Calculate the line voltage and phase

voltage across the secondary, the line current and phase current in each load, and the total power delivered to the three loads.

12. Calculate the source power delivered to the load in problem 11.

13. Calculate the phase voltage of a wye connected secondary if the line voltage is 480 V.

14. Give four examples of personal protective equipment.

15. Show a drawing of the main parts and functions of the power grid, from the source to the end user.

16. What are two major functions of a grounding system?

17. What is flashover and what are its causes?

18. Compare delivering power on overhead wires with underground wiring.

19. What is the definition of a microgrid? Why do you think it is important for renewable energy systems?

20. Describe the six implementation issues confronting the implementation of the smart grid.

21. How is a technology such as Li ion batteries able to save energy on the smart grid?

22. Explain the function of the substation and state the difference between a transmission substation and a distribution substation.

23. Identify the major issues for power quality.

24. Calculate the first five harmonics of 60 Hz.

25. What is a net meter? How can it help homeowners?

FOR DISCUSSION

Assume a homeowner wants to connect a small renewable system to the grid. What power issues must the homeowner address?

ANSWERS TO CHECKUPS

Section 14-1 Checkup

1. An electrical delivery method whereby two "hot" wires of opposite polarity and a neutral are supplied to the user from the power utility

2. L1 (line 1), L2 (line 2), and N (neutral)

3. It is more efficient to produce and transmit than single-phase voltage.

4. It can create a rotating magnetic field without starting windings and it can use smaller wiring than comparable single-phase motors.

5. True power is always dissipated in the circuit; apparent power is product of the voltage times the current, expressed in volt-amps, and includes resistive and reactive components.

6. The ratio of true power to apparent power; it is dimensionless.

Section 14-2 Checkup

1. A single-phase transformer has one primary winding and one secondary winding. A three-phase transformer has three primary windings and three secondary windings.

2. A delta connection has the three coils configured such that they could be arranged in the shape of the Greek letter delta (Δ), and a wye connection has the three coils configured such that they could be arranged in the shape of the letter Y.

3. Phase voltage refers to the voltage across a component that is within the wye or delta configuration; line voltage refers to the voltage between the lines brought out of the transformer.

4. A delta connection

5. A wye connection

Section 14-3 Checkup

1. Power stations, transmission lines, and distribution stations

2. Substations form the basic interface to the transmission system.

3. The delivery system from the substations to consumers

4. Voltage is increased so current can be lower, which reduces both power loss and allows smaller wire.

5. Ground fault interrupt

Section 14-4 Checkup

1. An electrical distribution technology that uses computer-based remote control and automation to improve the delivery efficiency.

2. A microgrid is a subset of the main grid and can operate as an independent system.

3. Broadband over power line, fiber to the home, and hybrid fiber coax (HFC)

4. To protect it from being hacked or interrupted by malicious intruders

Section 14-5 Checkup

1. Switches for safely disconnecting and grounding lines and large circuit breakers in case line conditions exceed preset limits

2. Transformers are connected in parallel so that voltage is the same.

3. Transformers for transmission substations are step-up transformers, and transformers for distribution substations are step-down transformers.

4. A brownout is a low-voltage condition on the grid; grid operators can remove some larger loads and may be able to reroute power from another sector.

Section 14-6 Checkup

1. A utility meter that can measure the difference between the power coming from or going back to the grid

2. The renewable energy system must have the capability to ride through this condition or disconnect from the grid.

3. Power quality issues include unwanted signals such as harmonics and noise being injected in power lines. Renewable energy systems that are connected to the grid need a clean sine wave, so inverters need to be checked and any sources of noise eliminated.

4. Uninterruptible power supply; it is important in critical cases where a brownout, blackout, or noise can be a serious problem (a computer center or hospital. for example).

ANSWERS TO TRUE/FALSE QUIZ

1. T 2. F 3. F 4. T 5. F 6. F 7. F 8. T 9. T   18. T 19. T 20. F 21. F 22. T 23. F 24. F

10. F 11. T 12. T 13. F 14. F 15. T 16. F 17. F   25. F 26. T 27. T

ANSWERS TO MULTIPLE-CHOICE QUESTIONS

1. b 2. c 3. a 4. b 5. d 6. c 7. d 8. a 9. b   10. a 11. c 12. b 13. b 14. d 15. b 16. c 17. a

List of variables and their meaning

Variable	Meaning
A	area
Ah	ampere-hour
B	magnetic flux density
B_{dod}	battery's maximum depth of discharge
C	capacitance
C_{emf}	counter emf
COP	coefficient of performance
C_p	Betz's coefficient
C_{tur}	turbine coefficient
E	irradiance (light energy)
f	frequency
ϕ	magnetic flux
F	force
FF	fill factor
F_m	magnetimotive force
g	gravitational constant
γ	specifc weight of water
h	height
H	magnetic field intensity
h_{eq}	equivalent head
h_h	heat produced
η_{inv}	efficiency of inverter and cabling
η_{sys}	efficiency of the system
h_w	electric energy input
I	current
I_a	armature current
I_{CH}	current rating of charge controller
I_{MPP}	maximum power point current
I_{SC}	short-circuit current
l	length
m	mass
N	number of turns
n	turns ratio
N_p	number of poles
P	power
p	pressure
P_a	apparent power
PF	power factor
$P_{out(max)}$	maximum power out

Variable	Meaning
P_{pri}	power delivered to primary
P_r	reactive power
P_{sec}	power delivered to the load from the secondary
P_{source}	source power
P_{true}	true power
Q	charge
Q_v	volumetric flow rate
R	resistance
ρ	density
$\mathcal{R}$	reluctance
R_a	armature resistance
ρ_a	air density
rpm	revolutions per minute
R_T	total resistance
T	period
t	time
T_q	torque
t_{solar}	average hours of sunlight
t_{store}	backup time required
V	voltage
v	velocity
v_{ind}	induced voltage
V_{MPP}	maximum power point voltage
V_{OC}	open-circuit voltage
Vol	volume
V_{OUT}	dc output voltage
V_{pri}	primary voltage
V_S	source voltage
V_{sec}	secondary voltage
v_w	wind velocity
W	work or energy
ω	rotational speed
W_{day}	daily energy requirement
W_{KE}	kinetic energy
W_{PE}	potential energy
μ	permeability
μ_0	permeability of a vacuum
μ_r	relative permeability

GLOSSARY

absorptance A dimensionless number that is the ratio of absorbed radiation to incident radiation.

absorption stage The stage of battery charging after the battery voltage reaches the maximum and the current through the battery begins to decrease. The voltage is held at the maximum value while the current decreases.

ac generator A rotating electromagnetic machine that produces a sinusoidal voltage.

acre-foot A volume of water that would cover an acre to a depth of 1 foot. It is equal to 1,233 m^3.

active stall control A control that adjusts the blade pitch, with the goal of taking control of the blade and rotor and bringing them to a complete stop if necessary.

active tracker A type of solar tracker that uses external power to move a solar collector throughout the day from east to west.

active yaw control On wind turbines, a motor that is responsible for moving the nacelle into the wind.

air density (ρ_a) A measure of how much mass is contained in a given volume of air.

airfoil A part such as a wing, propeller blade, or rudder that is designed to optimize air flow and usually optimize the lift-to-drag ratio.

algae Any of various aquatic, photosynthetic organisms that range in size from single-cell organisms to giant kelp.

alkaline fuel cell (AFC) A very efficient fuel cell that requires pure hydrogen fuel and pure oxygen. It uses an aqueous (water-based) electrolyte solution of potassium hydroxide (KOH) in a porous, stabilized matrix.

alternating current (ac) Current that reverses direction on a periodic basis.

alternation The portion of a sine wave that is either positive or negative. Each cycle has two alternations.

alternator An alternating current generator.

altitude angle The angle formed between a horizontal plane and an imaginary line pointed to the sun.

American Wire Gauge (AWG) A standard in which the diameter of wires are arranged according to gauge numbers. Decreasing gauge numbers indicate increasing wire size.

ampacity The maximum current a wire can carry under certain specified conditions.

ampere The unit of current symbolized by I.

ampere-hour (Ah) A unit of charge that specifies the capacity of a battery.

ampere-turn (At) The SI unit of magnetomotive force (mmf).

amplitude A measure of a sine wave from the center to the peak.

anaerobic digestion A bacterial fermentation process in which microorganisms break down biodegradable material in the absence of oxygen.

analemma The figure 8 pattern the sun makes in its annual motion if its position is observed daily at the same location and time.

anemometer A wind instrument that measures wind speed. It is mounted on the top of the wind turbine nacelle, usually near the back.

angle of attack The angle at which the wind strikes turbine blades.

anode (1) The electrode in which oxidation takes place. In a fuel cell, it is the terminal that breaks down hydrogen into ions and electrons, which pass through the external circuit. (2) one of the two terminals on a diode.

anti-islanding A protective feature of a grid-tie inverter that detects a power outage and disconnects the renewable energy source from the grid.

apparent power (P_a) The product of root-mean-square (rms) voltage and rms current without regarding phase difference. It is expressed in volt-amps (VA). In a purely resistive circuit, it is the same as true power.

aquaculture Farming of aquatic plants and animals, including fish, shrimp, oysters, and seaweed.

arc flash A high-current, high-energy arc between two points that has the potential to cause injury or death.

armature (1) In electric machines, the winding in which an alternating voltage is generated by virtue of its relative motion with respect to a magnetic flux field. (2) The moving part of an electrical device such as a relay that responds to a magnetic field.

asynchronous generator A generator that produces an alternating voltage in which the waveform that is generated is not synchronized to the rotational speed.

atmosphere (atm) A unit of pressure equal to 1.013×10^5 Pa. It is the pressure exerted under specified conditions by the atmosphere at sea level.

attenuator (1) A device that reduces the amplitude or power in a signal. Electronic attenuators are normally calibrated in decibels. (2) A device that extracts energy from wave power by converting relative motion between large semisubmerged cylindrical sections to electricity.

autoranging A feature in which the digital multimeter (DMM) selects the optimum range automatically for displaying the reading.

azimuth angle A horizontal angular distance from a reference direction, usually north.

balance point The temperature at which a heat pump's efficiency drops to a point where the energy used to drive the compressor is equal to the output heat energy.

band gap The amount of energy required to free an electron from the valence band of a silicon atom.

barometric pressure The pressure exerted by the atmosphere on the earth. Barometric pressure at sea level is 760 mm Hg, 101 kPa, or 14.7 psi.

barrage A relatively low dam that traps the inflow water of the tides and releases it through turbines to generate power.

Betz's law A formula originated by Albert Betz that states that the highest possible efficiency that a wind turbine can achieve is approximately 59%.

binary-cycle plant A geothermal power plant that uses the brine water or steam from the geothermal reservoir to heat and vaporize a secondary fluid with a lower boiling point to drive the turbine and electrical generator. Also known as *organic Rankine cycle (ORC) plant*.

binding energy The energy that holds the nucleus of an atom together. It is different for each atom and depends on the number of nucleons.

biodiesel A biofuel that is made from vegetable oil, animal fat, or cooking grease and is combined with alcohol and other ingredients to form the final product.

biogas A direct product of anaerobic digestion.

biomass Organic (carbon-containing) materials that have energy stored as carbohydrates, which can be used as feedstock for the production of renewable energy.

bitumen A black tarlike hydrocarbon classified as pitch; it can occur naturally or after refining petroleum.

black body A body that absorbs all radiation falling on it and then reemits it in a wavelength that is related to its absolute temperature.

blade pitch The position or rotation of a turbine blade as it is attached to the rotor or hub with respect to the wind turbine.

blade pitch control A control system that is used to change the orientation of the blades to moderate the speed of the turbine.

boundaries In earth science, the point at which tectonic plates interact with each other.

boundary-layer wind Winds in the lowest part of the atmosphere where turbulence and friction play a role in the behavior of the atmosphere.

breeder reactor A nuclear reactor designed to produce plutonium, which could extend uranium supplies considerably because it can convert the otherwise unusable ^{238}U into a fissionable fuel.

brownout Reduced voltage from the grid when demand exceeds supply (the voltage drops from 10% to 20%).

brushes In motors or generators, the electrical conductors that provide a connective path for current from a stationary part to a moving part.

bulk stage The stage of battery charging where the battery voltage increases at a constant rate.

capacitance (C) The amount of charge that a capacitor can store per unit of voltage across its plates.

capacitor An electrical device that stores energy in the form of an electric field established by electrical charge.

carbon One of the most abundant elements in the earth's crust, and the fourth most abundant element in the universe by mass after hydrogen, helium, and oxygen. Carbon is present in all known forms of life and is the chemical basis for all life on earth.

carbon cycle The process by which carbon atoms are endlessly cycled around the biosphere.

carbon reservoir A natural feature that stores carbon.

catagenesis The cracking process that results in the conversion of kerogens into hydrocarbons, including natural gas and oil.

catalyst A chemical that speeds up a reaction but emerges from the reaction unchanged.

cathode (1) The electrode in which reduction takes place. In a fuel cell, it is the terminal that combines oxygen and returning hydrogen ions to form water and the electrolyte. (2) one of two terminals of a diode.

celestial equator The projection of the earth's equator in the sky. The celestial equator traces a line at an angle above the horizon that is $90° - L$, where L is the latitude.

chain reaction A self-sustaining reaction. In nuclear reactors, the fission reaction that occurs when one neutron on average triggers another fission event.

charge A property of matter resulting from the presence of an extra electron or the absence of an electron in the atom.

charge controller A device that regulates and limits charging current to prevent overcharging batteries.

chemical conversion In reference to biomass renewable energy, it is the process that converts biomass into various forms of fuel without combustion.

circuit breaker A resettable protective device that detects excess current and opens the circuit.

circular mil (CM) English unit for cross-sectional area of wires.

clamp meter A type of meter for measuring alternating and direct current. It does not require the circuit to be broken to insert the meter.

closed-loop (1) In feedback theory, a condition where a portion of the output is returned to the input. (2) In water heating systems, a condition where the potable water is never exposed to the outside environment because a separate loop is used with a fluid that is heated.

closed-loop vertical heat pump system Heat pump system that uses vertical tubes where space is limited or when the soil is too shallow for trenching.

coefficient of performance (COP) A measure of efficiency for a heat pump. It is the ratio of the heat produced to the energy consumed, and it varies with the outside temperature.

co-generation Using heat for electrical generation and the lower temperature waste heat for another purpose. Also known as *combined heat and power. See also* combined heat and power (CHP).

combined heat and power (CHP) A process characterized by the production of both electrical power and heat.

combiner box A double-insulated box that allows several strings from solar modules to be connected together in parallel. It also houses fuses for the strings and includes surge and overvoltage protection from potential lightning strikes.

combustion The burning of a substance. It is a rapid exothermic reaction that is the oxidation of a fuel, combining the fuel with an oxidant (usually oxygen).

commutation Provision of a unidirectional current from a generator.

commutator A rotary electrical switch that periodically changes the direction of current from the rotor to the external circuit.

compound generator A generator that has two field windings: one in series and one in parallel. Typically, the output voltage tends to be independent of the load.

concentrating pholtovoltaic (CPV) A technology that uses lenses or curved mirrors to intensify sunlight on solar PV cells.

conductors Materials that allow the free movement of charge. They can be composed of solids, liquids, or gases.

conduit Tubes through which insulated wires are run that protect the wires.

continental crust The outermost layer of the earth's crust that runs from the surface to approximately 400 km deep.

continuity test A test of an electrical path to verify the path.

conventional current A convention that assigns a positive to negative direction for current.

convergent boundaries A tectonic boundary where two plates move toward each other.

convergent plates Tectonic plates on the outer surface of the earth that collide together.

Coriolis force A fictitious force that is used to explain the fact that the rotation of the earth creates an accelerating reference frame for an observer on earth. The observer concludes that the resulting air motion is due to a force.

coulomb (C) The unit of charge. The total charge possessed by 6.25×10^{18} electrons.

critical angle of attack The angle at which an airfoil is positioned to obtain maximum lift.

crystallinity Property of a material that indicates how perfectly ordered the atoms are in the crystal structure.

current The flow of electrical charge past a specified point in a circuit.

Darrieus wind turbine A type of vertical-axis wind turbine that rotates using lift forces to drive it.

delayed neutrons Neutrons that are emitted many seconds after a fission event by certain fission fragments. They are critical to maintaining control of a nuclear reactor.

delta A three-phase wiring method named for the Greek letter delta and takes its triangular shape in diagrams.

density (ρ) A measure of how much mass is contained in a given unit volume (density = mass/volume).

depth of discharge (DOD) The ratio, expressed as a percentage, of the quantity of charge (usually in ampere-hours) removed from a battery to its rated capacity. Also appears as B_{dod} in equations.

deuterium An isotope of hydrogen containing one proton and one neutron in the nucleus.

diagenesis The process of converting constituents to a different product through application of heat and pressure.

dielectric The insulating material between the plates of a capacitor.

diffuse horizontal irradiance (DHI) The portion of global horizontal irradiance that comes in indirectly from the sun (scattered radiation).

digital multimeter (DMM) An instrument that can measure voltage, current, and resistance.

diode A type of semiconductor device that conducts current in only one direction.

direct current (dc) Current that is uniform in one direction.

direct-drive wind turbine A type of turbine that does not use any transmission gears. The generator turns at the same speed as the turbine blades rotate.

direct-methanol fuel cell (DMFC) A fuel cell that is a subcategory of the proton exchange membrane fuel cell (PEMFC). Hydrogen is separated from methanol fuel by a steam reformer and supplied, with water and air, to the fuel cell. The electrolyte is a polymer similar to the PEMFC.

direct normal irradiance (DNI) The portion of global horizontal irradiance that comes in a straight line from the sun.

distribution system The delivery system that moves electricity from distribution substations to consumers.

district heat Heat generated in a centralized location that is distributed to residential and commercial users to meet heating requirements.

divergent boundaries A tectonic boundary where two plates move away from each other.

divergent plates Tectonic plates on the outer surface of the earth that move apart or have spreading centers.

doping The process used to increase the conductivity of a semiconductor in a precise and controlled way.

double-flash steam plant A geothermal plant with two pressure-reducing stages to create high-pressure and low-pressure steam. The high- and low-pressure steam is routed to two different turbines, which turn a generator.

doubly fed induction generator (DFIG) An ac induction generator that has a wound rotor connected to a different source of ac than the stator. AC from the grid or from an inverter is supplied to the fields in the stator. Also known as *double-excited induction generator.*

downwind turbine Turbine designed so that the wind blows over the nacelle and then over the blades.

drag The force that opposes the motion of the airfoil as it moves through the air.

drainback system A solar water heating system in which the circulating fluid is circulated only when heat is available at the collector; otherwise, the collector and exposed plumbing is drained.

drive train The components of a wind turbine that consist of the turbine blades, rotor, low-speed shaft, gearbox, high-speed shaft, and the generator

drive train compliance The correct alignment of all the components in the drive train of a wind turbine.

dry milling A process used to produce ethanol.

dry steam Superheated steam that has no liquid water in suspension. Also known as *superheated steam.*

dry-steam plant A geothermal electrical plant that uses superheated dry steam from a geothermal reservoir and routes it directly to a steam turbine and generator to produce electricity.

dual flash steam plant *See* double-flash steam plant.

dynamic braking Process that sends energy produced by the generator to a resistive load that makes the generator increase its load and slow down. When the generator produces electricity, it can be used to charge batteries or power any electrical load.

ecliptic The sun's path in the sky with reference to an observer on earth.

efficiency (PV cell) The ratio of light energy falling on the cell to the light energy that is converted to electrical energy.

electrical grid An interconnected network of power stations, substations, transmission lines, and distribution systems that transmit power from suppliers to consumers.

electrolysis The breaking down of a substance that contains ions by applying an electric current between two electrodes and separating the substance into its components. The electrolysis of liquid water breaks it down into hydrogen and oxygen gases.

electrolyte A substance that dissociates into ions and conducts electricity when it is dissolved in a suitable medium.

electromagnetic field The magnetic field produced when there is current through a conductor.

electron The basic atomic particle that accounts for the flow of charge in solid conductors.

emittance The total flux emitted per unit area from a material. It is related to the ability of the material to give off radiant heat.

end bells The end plates of a motor or generator that mount to the frame and include the bearings.

energy The ability or capacity for doing work.

enhanced geothermal system (EGS) A system in which a geothermal site that is deficient in water or permeability is made productive by artificial means.

enrichment A process that involves separating ^{235}U from natural uranium, which is mostly ^{238}U.

enthalpy The amount of energy in a system capable of doing mechanical work. It is a function of temperature, pressure, and volume.

equalization stage The process in which the battery is overcharged at approximately 1 V above float in order to equalize the charge on all the cells in the battery or battery bank.

ethanol An alcohol made by fermentation of sugars from corn, wheat, rice, sugar beets, sugar cane, sorghum, potatoes, and other starchy food sources.

exciter A smaller auxiliary generator that supplies field current for a larger generator.

fail-safe brakes Brakes designed so that they go to the set condition by the action of strong springs during an emergency or loss of energy on the wind turbine.

farad (F) The unit of capacitance.

feedstock (biomass) Raw material used to create an energy product, such as electricity, heat, or fuel.

fill factor (FF) The ratio of a cell's actual maximum power output to its theoretical power output.

finite-element analysis (FEA) A type of computer program that uses large mathematical algorithms to test a complex design. For wind turbines, FEA is used for turbine blade design.

fission The process of breaking a heavy nucleus into smaller fragments and releasing energy.

flash When the pressure on hot water (or brine) is reduced by allowing it to flow through a throttle valve, which in turn causes it to turn quickly into steam.

flashing the fields A procedure whereby an external source is applied to the field windings to start a self-excited generator.

flash steam (1) A mixture of pressurized hot water and steam that converts to steam when pressure is released. (2) A combination of both steam and water: the most common type of geothermal reservoir.

flash-steam plant A geothermal plant that creates steam from high-pressure hot water (brine) using a special control valve or orifice plate to reduce the pressure and cause some of the liquid to boil (flash) into steam. The steam is used to drive a steam turbine and generator to produce electricity.

flicker A short-lived variation in electrical power.

float stage The final maintenance or trickle charge stage with the purpose of offsetting any self-discharging of the battery.

float voltage The relatively constant voltage that is applied continuously to a battery to maintain a fully charged condition.

fossil energy replacement (FER) ratio The ratio of the energy delivered to the consumer to the fossil energy used at the production site.

fossil fuels Fuels that formed from decaying plant and animal matter and were primarily formed over millions of years. Fossil fuels include coal, oil (petroleum), and natural gas.

four-wire delta A wiring configuration for three-phase transformer whereby each of the three phases from the corner of a delta, and a center-tapped neutral from either L_1 or L_2 is included.

fracking The process of injecting water into drill holes to cause fractures in dense rocks such as shale to enable natural gas and oil to flow to a well.

Francis turbine A reaction water turbine that directs water from the outer circumference toward the center of a runner. Water flows through a scroll case, which is a curved tube that diminishes in size similar to a snail shell.

free electron An electron not bonded to an atom; a conduction electron.

frequency The number of complete cycles of a periodic wave that occur in one second.

front The boundary between two air masses.

fuel cell A device that converts electrochemical energy into dc directly by using a constant flow of fuel (usually hydrogen) from an outside source.

fuel cell electric vehicle (FCEV) A hydrogen-powered vehicle that uses electricity from a fuel cell to drive an electric motor to power the vehicle. The hydrogen fuel is generally obtained from natural gas or other fossil fuel.

fuel cell stack A group of fuel cells that are connected and bound together to provide increased electrical power.

fuel processing unit A portion of a fuel cell system that converts the input fuel into a form usable by the fuel cell.

fuse A nonresettable protective device that opens the circuit when there is excessive current.

fusion The process of putting together (or *fusing*) light nuclei to release energy.

gas-to-liquid (GTL) A refinery method for converting natural gas to gasoline or diesel fuel.

Gauss (G) A unit of magnetic flux density.

geopressurized reservoir A geothermal resource that consists of high-pressure hot brine (salt) water in a deep reservoir that is completely saturated with natural gas.

geothermal heat pump A heat pump that uses a standard refrigeration cycle; however, its heat source is glycol, which is circulated through a large amount of piping buried in the ground.

geothermal steam Steam that is drawn from deep within the earth.

geyser A natural hot spring that reaches temperatures high enough to boil and intermittently eject a column of water and steam into the air.

global horizontal irradiance (GHI) The total amount of shortwave radiation received on a horizontal surface.

gradient wind A wind that blows at a constant speed and flows parallel to imaginary curved isobars just above the earth's surface, where friction from irregularities such as mountains, trees, and buildings cause changes in the flow.

greenhouse gas A gas that contributes to the greenhouse effect by absorbing short wavelength infrared energy and reradiating it at longer wavelengths.

grid *See* electrical grid.

grid companies (gridcos) Companies that manage the grid function, which is the interconnection and routing of electricity through the electrical system.

grid-tie system An electrical generating system that is tied to the utility grid.

ground fault Current in the ground conductor rather than neutral, which normally should have no current.

ground fault interrupt (GFI) circuit breaker A special circuit breaker that can trip if the hot and neutral current differ by more than a few milliamps. This difference can occur when a person is receiving a shock.

ground fault protection device (GFPD) A device that (1) detects a ground fault, (2) interrupts the current in the line, (3) indicates that a fault has occurred with a visible warning, and (4) disconnects the faulty module.

grounding system An earthing system used as a safety measure to ensure that certain parts of a system are at the same electrical potential as the earth's surface. It generally consists of one or more copper rods that have been inserted into the earth to a depth of 2.4 m (8 ft) and bonded or connected with heavy copper conductors to parts of the system.

half-life The time required for one-half of a radioactive substance to decay.

harmonics The frequencies contained in a composite waveform. They are integer multiples of the repetition frequency (fundamental).

head The height from which water falls. The gross head is the total height. The net head is the equivalent height after equivalent friction losses in piping are subtracted.

heating system performance factor (HSPF) A measure of a heat pump's heating efficiency over an entire heating season. It represents the total heating (in Btu) compared to the total electricity consumed (in W-h) during the same period.

heat pump A device that utilizes a refrigeration cycle to move heat from a cooler region to a warmer one. It can operate like an air conditioner in the summer and/or heater in the winter.

heat recovery system A part of the fuel cell that processes the excess heat for another use such as heating water or creating steam.

heliostat A device that uses a movable mirror mounted so that it reflects the sunlight to a fixed target (such as a tower collector).

henry (H) The unit of inductance.

hertz (Hz) The unit of frequency.

high-concentration photovoltaic (HCPV) A photovoltaic cell that has concentrations of more than 300 suns. An HCPV typically incorporates dish reflectors and Fresnel lenses, but cone lenses are also used.

high-speed shaft The output shaft of the gearbox that is connected to the generator on a wind turbine.

high-speed shaft brakes The brakes that are attached to a disc on the high-speed rotor shaft. These brakes make the high-speed shaft stop turning.

holding torque With reference to a stepper motor, the maximum torque that can be applied externally to a stopped motor without causing it to rotate to the next step position.

hole A vacancy created in an atomic bond when a valence electron becomes a free electron.

horizontal-axis wind turbine (HAWT) A wind turbine in which the main rotor shaft is pointed in the direction of the wind to extract power.

hot dry rock The second warmest geothermal resource. It has a temperature gradient of more than 40°C per kilometer of depth.

hot springs A naturally occurring water resource in which geothermally heated water rises to the surface.

hot water reservoir A shallow geothermal resource that has water in a reservoir that is hot enough to create steam. Also known as *natural steam reservoir*.

hydrocarbon A molecule containing only hydrogen and carbon atoms.

hydrocarbon fuels Fuels that are produced from hydrocarbon material such as methane, natural gas, gasoline, diesel fuel, and coal gas.

hydroprocessing A general chemical engineering term for various chemical processes that react feedstocks (oils) with hydrogen at a high temperature and pressure to alter the chemical properties of the feedstocks. In renewable energy systems, hydroprocessing is focused on the catalytic conversion of feedstocks such as algae to produce fuels.

hydrospallation drilling A drilling method in which a stream of superheated high-pressure water is directed at the rock surface.

hydrothermal convection system A naturally occurring underground system that brings heated water or steam up to the surface without pumping.

hydrothermal resource A geothermal resource characterized by fluid, heat and permeability.

hysteresis A characteristic of a magnetic material whereby a change in magnetization lags the application of the magnetic field intensity (magnetizing force).

impulse turbine A rotary engine that changes the direction of a high-velocity fluid, thus converting kinetic energy into mechanical rotating energy.

induced voltage A voltage produced as a result of a moving magnetic field.

inductance The property of a wire conductor to oppose a change in current.

induction generator An asynchronous electrical machine that can function as a motor or as a generator.

infrared The portion of the electromagnetic spectrum with wavelengths between 700 nm and 1 mm, which is longer than visible light.

injection well The well on a geothermal site where water is returned to the earth after it is used in a geothermal plant.

inner core The deepest level of the earth. It starts at a depth of approximately 6,300 km and is located at the center of the earth.

insolation From *in*cident *sol*ar radi*ation*. A measure of the energy received on a surface in a specific amount of time. It can be measured in units of W/m^2.

insulators Materials that prevent the free movement of charge. Also known as *nonconductors*.

International Building Code (IBC) A standards document that addresses all aspects of the design and installation of building systems, including electrical systems.

International System of Units (SI) SI stands for Système International, from French. These units are the basic units of the metric system.

inverse square law A physics law that states that the flux from an isotropic point source is reduced by the square of the distance from the source to the receiver.

inverter A device that converts dc to ac.

islanding The situation when a grid-tie renewable energy source continues to operate and provide power to a certain location and remains connected to the electrical grid after the grid no longer supplies power.

jet stream A high-altitude river of air that marks the boundary between cold polar air and warmer mid-latitude air. Each hemisphere had one jet stream.

joule (j) The SI unit of energy. The work done when 1 newton of mechanical force is applied over a distance of 1 meter.

Kaplan turbine A reaction water turbine that uses propellers with adjustable blades. The turbine is usually placed in a spiral casing called a volute.

kerogen A mixture of organic chemicals that are part of the organic matter in sedimentary rocks.

kilowatt-hour (kWh) A unit of energy. The energy used when 1,000 watts of power are expended in one hour.

kinetic energy The energy of motion. An example is the energy of flowing water.

laminar flow The smooth flow of a gas or liquid in a regular path.

land breeze A breeze that causes the wind to blow from the land toward open water.

landfill A structure in the ground or on top of it in which trash is isolated from the environment.

latent heat of vaporization The heat absorbed or released during a change of state from a liquid to a gas.

latitude The angle in degrees formed by a line that extends from the center of the earth to the equator and from the center of the earth to a given point on the globe.

leachate Water or other liquid that has entrapped environmentally harmful substances from materials through which it has passed.

leading edge The front of a blade or wing.

left-hand rule A method of determining the direction of flux lines around a current-carrying conductor by imagining that you are grasping the conductor with your left hand, with your thumb pointing in the direction of electron flow. Your fingers indicate the direction of the magnetic lines of force.

Lenz's law The law that states that, when the current through a coil changes, an induced voltage is created across the coil in a direction that always opposes the change in the current.

lift A component of an aerodynamic force exerted on a body that is perpendicular to a fluid (such as air) flowing past it.

lift-to-drag ratio The ratio of the value of lift force to the value of drag force.

light-emitting diode (LED) A type of diode that emits light when current passes through it.

liquid-dominated reservoir A geothermal reservoir that uses a combination of liquid and vapor that forms when large volumes of water come in contact with extremely hot rocks deep in the earth.

lithosphere The outer surface of the earth consisting of the crust and upper mantle. It is made of a number of tectonic plates that move or float over the mantle.

load A component that uses the power. Any type of end-use device or appliance that has resistance.

load factor The net amount of energy an energy conversion system can actually produce when compared to the amount of energy it could produce if it operated at its rated output 100% of the time.

lockout A device placed on the handle of the disconnect switch after the handle is placed in the off position. It allows a padlock to be placed around it so it cannot be removed until the work is completed.

longitude Reference circles that are perpendicular to the equator and converge at the poles. Longitude increases in the positive direction in the easterly direction.

lower mantle Layer of the earth that is between the outer mantle and the upper mantle. It runs approximately from 650 km to 2,800 km deep.

low-peak time Time of day when energy demand is less. Also known as *off-peak time*.

low-pressure turbine A turbine specifically designed to produce electrical power when the steam is at a lower pressure.

low-speed shaft The shaft between the rotor and the gearbox on a wind turbine.

low-voltage ride through (LVRT) A safety system in place to address a low-voltage fault in the grid. Electrical instrumentation measures the severity and the duration of the problem, and determines the best solution to correct it.

magnetic field intensity (H) The magnetomotive force per unit length of the magnetic material. The unit is the ampere-turns per meter. Also known as *magnetizing force*.

magnetic flux The group of imaginary force lines going from the north pole to the south pole of a magnet.

magnetic flux density The amount of flux per unit area (A) perpendicular to the magnetic field.

magnetomotive force (mmf) The cause of a magnetic field. It is measured in ampere-turns.

manual ranging A digital multimeter (DMM) feature that requires the user to select an appropriate range for the measurement.

maximum power point tracking (MPPT) The process for tracking the voltage and current from a solar module to determine when the maximum power occurs in order to extract the maximum power.

mechanical parking brake A brake used on a wind turbine to lock the blades in position when the blades must be secured for maintenance or to protect them from high winds.

membrane electrode assembly The components in a fuel cell that includes the anode and cathode electrodes, the electrolyte, and the catalyst.

meridian A line of longitude that passes through both poles and a point straight over the observer's head.

microgrid A subset of the smart grid that has the ability to operate independently of the main grid (the macrogrid). A microgrid can be a small electric utility that is connected to the larger grid, or it can stand alone as an independent system.

microinverter A dc to ac converter that is sized to operate with a single solar module so it can provide maximum power-point tracking for the module and provide greater efficiency.

mid-latitude cell An atmospheric circulation path in each hemisphere in which air flows toward the pole and eastward at low elevations and toward the equator and westward at higher elevations. Also known as *Farell cell*.

millibar A unit of atmospheric pressure equal to 100 Pa.

moderator A substance (usually graphite or water) used to slow down prompt neutrons in a reactor.

molten-carbonate fuel cell (MCFC) A fuel cell similar to the solid oxide fuel cell (SOFC) but with carbonate ions as the charge carrier in a high-temperature solution of liquid lithium, potassium, or sodium carbonate as an electrolyte.

molten magma resources The hottest form of a geothermal resource that is approximately 2,000°C. It is most commonly found in volcanoes.

multijunction A type of thin-film PV cell that is basically two or more individual single-junction cells arranged in descending order of band gap.

nacelle An enclosure or housing on a wind turbine for the generator, gearbox, and any other parts of the wind turbine that are on the top of the tower.

National Electric Code (NEC) A set of codes and standards designed to protect the safety of electrical workers and the public.

National Solar Radiation Database (NSRDB) A document available from the National Renewable Energy Laboratory (NREL) that describes the amount of solar energy available at any location. It includes data for each component of solar radiation.

natural steam reservoir *See* hot water reservoir.

neap tide A tide that occurs when the sun and moon are at right angles to the earth at first and third quarter causing tides in which the difference between high and low tide is less than average. Neap tides occur twice a month.

net meter A utility meter that measures the difference between the power coming from or going back to the grid.

neutral A conductor that carries current in normal operation and is connected to earth ground.

nonsalient poles Poles that do not project out from the surface of the rotor.

normal geothermal gradient The rate at which temperature increases with depth in the earth. It is typically a 25°C increase per km of depth.

n-type impurity A type of element that has five electrons in its outer shell.

nucleon An atomic particle found in the nucleus of an atom.

offshore wind farm A group of wind turbines that are located in the ocean or a large body of water.

ohm The unit of resistance.

Ohm's law A circuit law that specifies the relationship among voltage, current, and resistance as a mathematical formula.

open-loop (1) A type of system that does not use feedback to adjust its parameters. (2) A solar water heating system in which the potable water is circulated through the collectors.

open-loop well water geothermal heat pump system A type of geothermal heat pump system that uses a well or large body of water as the heat exchange fluid. The water in this system is circulated up from a well or from a body of water.

organic Rankine cycle (ORC) Similar to the ordinary Rankine cycle except that thermal oil is used as the working fluid instead of water.

oscillating water column A fixed device for producing electrical power from waves. It consists of a large tube that extends over a cliff and into the ocean. Wave action causes water to rise in the tube and displace air, which rotates a wind turbine.

outer core Layer of the earth that is between the inner core and the lower mantle. It runs from approximately 5,100 km to a depth of 6,300 km and consists of molten material.

oxidation A chemical process that involves the loss of an electron.

parallel circuit A type of circuit connection that has two or more loads connected across a common voltage source. Each load provides a separate path for current.

partial oxidation (POX) reactor A method used to produce hydrogen from a hydrocarbon. It can use a catalyst at high temperature to complete the reaction. Hydrogen and carbon monoxide are products of the reaction.

pascal (Pa) The SI unit of pressure defined as 1 N of force exerted over 1 m².

passive pitch control For wind turbines, a blade pitch control system in which the blade is designed to stall at higher wind speeds to reduce rotation speed.

passive stall control A design feature of wind turbine blades that causes them to create turbulence on the back of the blades in high winds, thus reducing the lift force that drives the rotor.

passive tracker A solar tracking device that does not require external power to turn the axis.

passive yaw control A method of orienting the nacelle of a small wind turbine by using the power in the wind itself to orient the nacelle (typically with a tail fin).

peak demand The period of greatest electrical energy consumption during a specific period of time.

peaking power plant A power plant that produces electrical power only when the demand for power is highest.

peak power For a wind turbine, the amount of electrical power the wind turbine can produce at the highest rated wind speed.

Pelamis wave energy system An attenuator wave energy system that consists of five semisubmerged cylindrical sections. Relative motion between the cylinders pumps hydraulic fluid to a hydraulic motor and generator to produce electricity.

Pelton turbine An impulse turbine in which water moves under it (impulse) rather than water falling over it. It is among the most efficient types of water turbines.

penstock A pipe or channel for conveying water to a turbine or waterwheel.

period (T) The time required for one cycle.

permeability (1) A measure of the ease with which a magnetic field can be established in a given material. (2) A measure of the ability of a material to pass a fluid.

personal protective equipment (PPE) Special safety equipment needed to work on power system with voltages above 200 V. Personal protective equipment may consist of safety glasses, safety shields, gloves, hard hats, and/or suits to protect against arc flash.

perturb and observe (P&O) algorithm A procedure in which a variable is changed (perturbed) and the effect of the change on another variable is monitored (observed).

petroleum replacement ratio (PRR) The ratio of the energy delivered to the consumer in the form of biofuel compared to the petroleum energy used in the process.

phase shift Shift that occurs when one waveform leads or lags another.

phasor A rotating vector, which is a quantity with magnitude and direction.

phosphoric acid fuel cell (PAFC) A fuel cell that is equivalent in structure to the proton exchange membrane fuel cell (PEMFC), but it has liquid phosphoric acid as the electrolyte. The electrolyte is contained in a Teflon-bonded, silicon carbide matrix.

photobioreactor (PBR) A closed system that uses light and optimum conditions to enable high productivity of algae.

photon Packets of energy in sunlight.

photosynthesis A process in plants, algae, and some species of bacteria that uses energy from the sun to convert carbon dioxide and water into carbohydrate, which is stored in the plant, algae, or bacteria for food.

photovoltaic (PV) cell A thin layer or wafer of silicon that has been doped to create a *pn* junction.

photovoltaic effect The basic physical process by which a photovoltaic (PV) cell converts sunlight into electricity.

photovoltaic (PV) wire A special type of stranded copper wire that is sunlight-resistant and dedicated for the interconnection of PV modules.

pitch The rotational angle of the blades on a wind turbine.

pitch control A control that is used to change the angle of the blades on a wind turbine to help determine the speed of the turbine.

plate tectonics A theory in geology that explains how the earth's crust and upper mantle move and interact with each other.

pn junction The boundary created between an *n*-type and *p*-type semiconductor.

point absorber A floating wave energy converter that is in a fixed position. It bobs up and down from wave motion. The motion with respect to a fixed reference is captured and the energy is converted to electricity.

polar cell An atmospheric circulation path in each hemisphere in which cold, dry air from the poles flows away from the poles at low elevation and warmer air flows above it to replace it.

Surface winds in the polar cell are primarily weak easterly winds (polar easterlies).

pondage The storage area behind a run-of-the-river (ROR) system (not present in all ROR systems).

potential energy Stored energy.

potentiometer (pot) A three-terminal variable resistor.

power (P) The rate at which energy is expended.

power curve A graph that shows the wind speed and the output power of a wind turbine over a range of wind speeds from zero to the maximum wind speed for which the wind turbine is designed.

power electronic frequency converter An electronic circuit that converts ac to dc and back to ac at the precise frequency of the power utility.

power factor (1) The cosine of the phase angle between current and voltage. (2) The ratio of true power to apparent power.

power quality A measure of the purity of the ac power, including the presence of unwanted harmonics and noise on the power line and the frequency.

pressure gradient The change or variation in atmospheric pressure per unit of horizontal distance in the direction in which the pressure changes most rapidly. It is expressed in units of pressure per unit length.

pressurized water reactor (PWR) A nuclear reactor that includes fuel rods, control rods, and a moderator in a vessel that is filled with water under high pressure to prevent it from boiling. The water in the vessel serves as both a moderator and a means to move hot water to a heat exchanger and eventually a steam-driven turbine.

primary winding The input winding of a transformer.

prime meridian A great circle of longitude that passes through both poles and Greenwich, England.

prime mover The initial agent or source of energy in a system.

production well The well on a geothermal site where hot steam or brine is brought up from underground.

prompt neutrons High-energy neutrons emitted during the fission process.

proton exchange membrane fuel cell (PEMFC) A type of fuel cell that uses hydrogen fuel and oxygen (obtained from air) to produce pure water and electricity. It uses porous carbon electrodes that contain a platinum catalyst.

p-type impurity A type of element that has three electrons in its outer shell.

pulse width modulation (PWM) A process in which a signal is converted to a series of pulses with widths that vary proportionally compared to the signal amplitude.

pumped storage system A system of two dams, each with a reservoir. One is located at a much higher elevation than the other. Water is released from the higher reservoir to produce electrical power. The water is captured in the lower reservoir and pumped back during off-peak hours to the higher reservoir, where it is used again.

raceway pond A large, shallow, open water configuration that resembles a racetrack for growing algae. The water, with its algae and nutrients, is moved around the track by a rotating paddle assembly.

Rankine cycle An ideal reversible heat-engine cycle approximated by the operating cycle of an actual steam engine.

reactance The opposition to current in an ac circuit that is caused by inductance or capacitance.

reaction turbine A rotary engine that develops torque by reacting to the pressure of a fluid moving through the turbine, thus primarily converting potential energy into mechanical rotating energy.

reactive power (P_r) The rate at which energy is stored and alternately returned to the source by a capacitor or an inductor. The unit is the volt-amp-reactive (VAR).

reduction A chemical process that involves the gain of electrons.

reformer A device that extracts hydrogen from another fuel such as methane.

reforming The thermal or catalytic process of breaking down a large molecule into smaller ones.

regenerator In a Stirling engine, a wire mesh located between the heat exchangers and serves as temporary heat storage while the gas cycles between the hot and cold sides.

relative permeability The ratio of a material's permeability (μ) to the permeability of a vacuum (μ_0). Because μ_r is a ratio, it has no units.

reluctance ($\mathcal{R}$) The opposition to the establishment of a magnetic field in a material.

reservoir In geology, a natural underground accumulation of oil and natural gas.

resistance The opposition to current.

resistor Component designed to have a certain amount of resistance between its leads or terminals.

respiration The process in which oxygen is used to break down organic compounds into carbon dioxide (CO_2) and water (H_2O).

rest energy The equivalent energy of matter because it has mass.

retentivity The ability of a material, once magnetized, to maintain a magnetized state without the presence of a magnetizing force.

reversing valve A valve that changes the flow of the refrigerant when the heat pump is switched from cooling mode to heating mode.

rheostat A type of variable resistor used to vary current.

root-mean-square (rms) current The value of alternating current that produces the same power dissipation as the same value of dc.

rotor The rotating center part of a motor or generator.

rotor brakes The brakes on a wind turbine rotor that stop the blade rotor from turning.

run-of-the-river (ROR) A hydroelectric system that uses river flow to generate electricity. The system may include a small dam with storage for water, but many do not.

saccharification The process of breaking a complex carbohydrate (such as starch or cellulose) into simple sugars.

salient Projecting beyond a surface, level, or line.

salient poles Magnetic poles on a rotor that project beyond the surface and used on lower-speed rotors.

saturation In magnetics, the point where an additional increase of current in the wire does not produce any more flux lines.

Savonius wind turbine A type of vertical-axis wind turbine consisting of two or three scoops that rotate and using a difference in forces on the scoops to drive it.

sea breeze An onshore breeze created whenever there is a difference between the water temperature and the land temperature.

seasonal energy efficiency ratio (SEER) A measure of an air conditioner's or a heat pump's cooling efficiency over an entire cooling season. It represents the total cooling (in Btu) compared to the total electricity consumed (in W-h) during the same period.

seating The process of contouring the brush to the commutator.

secondary winding The output winding of a transformer.

self-excited shunt generator A generator that supplies its voltage from its armature to create the field current. The field and armature windings are parallel to one another.

semiconductors A crystalline material that has four electrons in its valence shell and has properties between those of conductors (metals) and insulators (nonmetals).

separately-excited shunt generator A dc generator with an external supply for the field current.

series circuit A type of circuit connection that has a single complete path (forming a string) from the voltage source through the load (or loads) and back.

shunt A parallel connection.

shunt generator A generator that has its field winding and armature connected in parallel. The armature supplies both the load current and the field current.

single-phase voltage An alternating voltage that is supplied with only one continuously varying sinusoidal waveform.

sinusoidal wave The cyclic pattern of ac voltage or current. Also known as *sine wave*.

smart grid An electrical distribution technology that uses computer-based remote control and automation to improve the delivery efficiency of electricity and to provide information to the customer for optimal power use.

solar array A combination of solar modules.

solar cell A type of diode specifically designed to allow sunlight to penetrate the semiconductor regions and thus generate a voltage.

solar concentrator A type of solar collector that collects light over a certain area and focuses it onto a smaller area.

solar constant The power emitted by the sun that falls on 1 m^2. It is generally cited as 1,368 W/m^2.

solar module Combinations of multiple photovoltaic (PV) cells connected to produce a specified power, voltage, and current output.

solid oxide fuel cell (SOFC) A fuel cell named for the solid oxide electrolyte that it uses. The fuel is hydrogen, which is supplied as a gas along with oxygen from the air. The electrolyte is a hard, nonporous, solid oxide or ceramic compound made of yttria-stabilized zirconia.

spillway A structure built into a dam that allows water to be released to avoid overfilling the reservoir. The excess water does not go to the turbines, so the energy is lost.

split-phase A single-phase electrical delivery method whereby two "hot" wires of opposite polarity and a neutral are supplied to the user from the power utility.

spring tide A tide that occurs when the sun, moon, and earth align to create a very high tide and a very low tide. Spring tides are not related to the spring season.

squirrel-cage rotor A rotor for an ac motor or generator that is made of a wire cage with sections of laminated steel pressed on the frame.

stall A reduction in the lift force as the angle of attack increases beyond some point.

stator The stationary part of a motor or generator.

step-down transformer A transformer in which the secondary voltage is less than the primary voltage.

stepper motor A motor that moves a rotor in discrete steps in response to a pulse signal.

step-up transformer A transformer in which the secondary voltage is greater than the primary voltage.

Stirling engine A type of heat engine that cools and compresses a gas in one portion of the engine and expands it in a hotter portion to obtain mechanical work.

substation A network of switching equipment, control equipment, and transformers that are used to convert voltage to a different level.

supervisory control and data acquisition (SCADA) A data collection system that provides information such as wind velocity, wind direction, and the amount of electrical power that is being produced from a wind turbine.

surface-water system A type of closed-loop geothermal system that uses a lake or pond system where one or more loops of plastic tubing are laid directly into the body of water that is near the building. Also known as *lake loop system.*

sweep area The amount of area the wind turbine blades cover when they rotate one revolution. Also known as *swept area.*

synchronization The process of producing a fixed phase relationship between two or more waveforms.

synchronous generator An ac generator in which the output frequency is synchronized to the position of the rotor.

syngas A synthetic gas that contains hydrogen and carbon monoxide, as well as small amounts of other gases and water.

synodic day The time for the earth to make one revolution on its axis. It is almost exactly 24 hours.

tectonics *See* plate tectonics.

terminating device In ocean energy systems, a fixed device that produces electrical power from waves.

Tesla (T) A unit of magnetic flux density.

thermal conversion Any of several processes that use heat to extract energy from biomass without combustion by changing its chemical form with chemical reactions and interaction with oxygen.

thermal neutrons Low-energy neutrons that are efficient at creating fission in a nuclear reactor.

thin-film photovoltaic Type of photovoltaic (PV) that uses layers of semiconductor materials from less than a micrometer (micron) to a few micrometers thick.

three-phase Three ac voltages that have the same magnitude and frequency but are separated by 120°.

thyristor A type of semiconductor device for controlling the amount of power delivered to a load.

tidal barrage dam A large dam that stretches completely across an estuary, harbor, or river that connects to a part of the ocean that has a tide. It is designed to capture energy from the tides.

tidal barrage system A system designed to convert tidal power into electricity by trapping water behind a dam, called a tidal barrage dam, and generating power from the inflow and/or the release of water.

tidal stream generator (TSG) An electrical generating system that uses a water turbine to turn a generator and produce electrical power when a stream or river caused by tides flows past it.

tip speed The velocity at the outer edge of a rotor blade as it moves.

tip speed ratio (TSR) The ratio of the tip speed to the wind speed.

Tokamak A fusion reactor used by researchers to investigate properties of plasmas with the goal of finding a fusion energy reactor that can be used for electrical power generation.

torque The product of a force, F, and the perpendicular distance, x, from a fixed point that tends to produce rotation about the point. It is measured in newton-meters in the SI system and pound-feet in the English system.

trailing edge The back part of a turbine blade.

transesterification A chemical process in which an organic ester reacts with an organic alcohol. The reaction is often catalyzed with an acid or a base catalyst.

transfer switch A switch used for connecting or disconnecting a source from the grid.

transform boundaries A tectonic boundary where two plates slide along each other.

transformer An electrical device constructed of two or more coils (windings) that are electromagnetically coupled to each other to provide a transfer of power from one coil to the other.

transform plates Two tectonic plates that slide laterally past each other.

transistor A type of semiconductor device that can be used as a switch or for amplification.

transmission companies (transcos) Individual companies that manage the hardware and own the actual transmission lines, towers, transformers, and switchgear.

transmission system The system that moves bulk electricity from the transmission substations located at power plants and renewable energy systems to distribution substations.

tritium An isotope of hydrogen containing one proton and two neutrons in the nucleus.

tropical cell An atmospheric circulation cell with hot air from the equator rising to higher elevations and moving toward the poles, with cooler air coming in from lower elevations to replace the hot air. Also known as *Hadley cell.*

true power (P_{true}) The power dissipated in a circuit, usually in the form of heat. True power is measured in watts.

turbine A rotary machine that takes energy from a moving fluid (water, steam, or gas) and turns a shaft. Usually the shaft is connected to an electrical generator.

turbulent flow An erratic flow of a fluid.

Turgo turbine A compact impulse turbine that uses half cups around the runner instead of full cups, like the Pelton turbine. The half cup on the Turgo turbine is better suited for medium and higher flow rates.

turns ratio For an electronic power transformer, the number of turns in a given secondary divided by the number of turns in the primary.

ultracapacitor An electrical component capable of storing hundreds of times more charge than a standard capacitor; they are useful for emergency backup applications.

Uniform Solar Energy Code (USEC) A set of standards for plumbing and mechanical requirements for solar energy systems.

uninterruptible power supply (UPS) A source that automatically switches to an auxiliary generator or batteries if utility power is interrupted.

upper mantle Layer of the earth that is between the lower mantle and the continental crust. It runs approximately from 400 km to 650 km deep and is partially molten.

upwind turbine A wind turbine designed so that the wind blows over the blades and then over the nacelle.

valence electrons Electrons in the outer shell of an atom.

vertical-axis wind turbine (VAWT) A wind turbine that has its main rotational axis oriented in the vertical direction.

volt The unit of voltage or electromotive force.

voltage (V) Energy per unit charge.

voltage regulator Electrical circuit that automatically maintains a constant output over variations in load.

volt-ampere (VA) rating A rating of the maximum allowed apparent power dissipated for a device found by multiplying the voltage (volts) by the current (amperes).

volt-amps-reactive (VAR) The unit of reactive power.

watt The unit of power.

Watt's law A circuit law that expresses the relationship among voltage, current, resistance, and power as a formula.

wavebob A floating point absorber device that captures wave energy from the relative motion between a floating torus and a submerged tank. The relative motion moves a fluid that drives a hydraulically driven motor and generator. It is currently no longer produced.

wave height The vertical distance on a water wave between the trough and the crest.

wavelength The horizontal distance between two consecutive points on a wave. In water, it is typically from a crest of one wave to the crest of the following wave.

wet milling The process in which a feedstock is soaked in water and usually dilute sulfurous acid to soften it before further processing.

wind direction indicator A vane blade that is moved by the wind until the vane is pointed in the same direction from which the wind is blowing. Also known as *wind vane*.

wind farm A group of wind turbines located in one area.

wind turbine peak performance A condition that occurs when the output of the wind turbine generator is at or above its rated output.

wind turbulence A condition in which the wind blows erratically.

wind vane *See* wind direction indicator.

wound rotor A rotor core assembly that has a winding made of individually insulated wires.

wye A three-phase wiring method that is named for the letter Y because its diagram is shaped like the letter Y.

yaw Direction the wind turbine blades and nacelle are facing.

yaw brakes Braking system that slows the yaw ring to a stop and inserts a series of interlocking pins into the yaw mechanism to secure it so it cannot move during maintenance or other safety condition. When the yaw motion mechanism needs to move again, the brakes and pins are released to allow the nacelle to rotate.

yaw control The control that manipulates the rotation of a horizontal-axis wind turbine around its tower or vertical axis.

yaw control assembly An assembly consisting of the yaw drive, which is a large gear, and the yaw motor, which has a gear mounted to its shaft that engages the larger yaw gear.

yaw drive An electric motor used to rotate the yaw ring that changes the position of the nacelle yaw on a wind turbine.

zenith A point directly overhead an observer.